D0843106

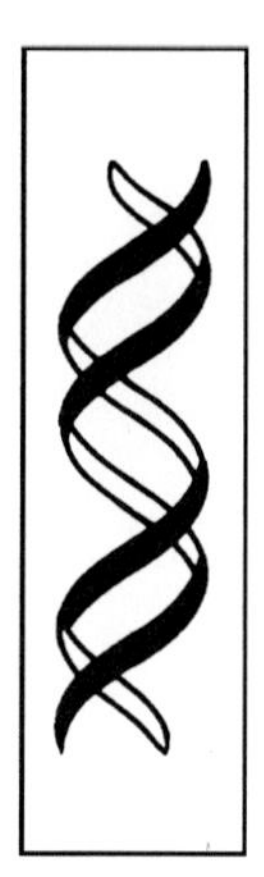

Perspectives on Properties of the Human Genome Project

Advances in Genetics, Volume 50

Serial Editors

Jeffery C. Hall
Waltham, Massachusetts
Jay C. Dunlap
Hanover, New Hampshire
Theodore Friedmann
La Jolla, California

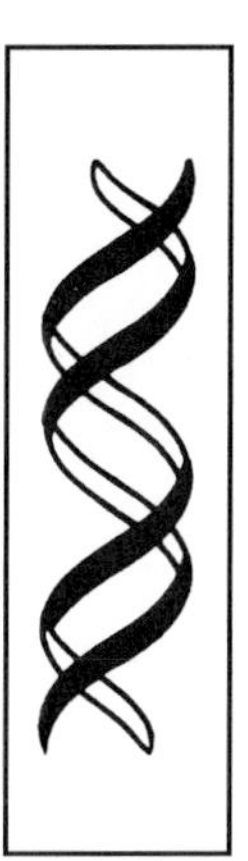

Perspectives on Properties of the Human Genome Project

Edited by
F. Scott Kieff
Associate Professor of Law
Washington University School of Law
St. Louis, Missouri

John M. Olin Senior Research Fellow in Law,
Economics, and Business
Harvard Law School
Cambridge, Massachusetts

ELSEVIER
ACADEMIC
PRESS

AMSTERDAM • BOSTON • HEIDELBERG • LONDON
NEW YORK • OXFORD • PARIS • SAN DIEGO
SAN FRANCISCO • SINGAPORE • SYDNEY • TOKYO
Academic Press is an imprint of Elsevier

This book is printed on acid-free paper. ♾

Elsevier Academic Press
525 B Street, Suite 1900, San Diego, California 92101-4495, USA
84 Theobald's Road, London WC1X 8RR, UK
http://www.academicpress.com

International Standard Book Number: 0-12-017650-5

PRINTED IN THE UNITED STATES OF AMERICA
03 04 05 06 07 08 9 8 7 6 5 4 3 2 1

Contents

Contributors

Numbers in parenthesis indicate the pages on which the authors' contribution begins.

Michael Abramowicz (231) George Mason University, Arlington, Virginia 22201

James Boyle (97) Duke University School of Law, Durham, North Carolina 27708

Dan L. Burk (305) University of Minnesota Law School, Minneapolis, Minnesota 55455

A. M. Chakrabarty (3) Department of Microbiology & Immunology, University of Illinois College of Medicine, Chicago, Illinois 60612

Iain M. Cockburn (385) Boston University School of Management, Boston, Massachusetts 02215

James H. Davis (427) Human Genome Sciences, Inc., Rockville, Maryland 20850-3338

Rochelle Cooper Dreyfuss (195) New York University School of Law, New York, New York 10012

Richard A. Epstein (153) The University of Chicago Law School, Chicago, Illinois 60637

Rebecca S. Eisenberg (209) University of Michigan Law School, Ann Arbor Michigan 48109

Justin Hughes (263) Benjamin N. Cardozo School of Law, Yeshiva University, New York, New York 10003

David A. Hyman (275) University of Maryland School of Law, Baltimore, Maryland 21201

Robin Jacob (449) High Court of England and Wales, Strand, London WC2A 2LL, United Kingdom

Horace Freeland Judson (483) Center for History of Recent Science, The George Washington University, Washington, DC 20052

F. Scott Kieff (125, 471) Washington University School of Law, St. Louis, Missouri 63130; Harvard Law School, Cambridge, Massachusetts 02138

Edmund W. Kitch (271) University of Virginia, Charlottesville, Virginia 22903

Mark A. Lemley (305) Boalt Hall School of Law, University of California at Berkeley, Berkeley, California 94720-7200

Edward T. Lentz (441) New Lisbon, New York 13415

Roderick R. McKelvie (459) Fish & Neave, Washington, DC 20006

Charles R. McManis (79) Washington University School of Law, St. Louis, Missouri 63130

Hon. Gerald J. Mossinghoff (13) Oblon, Spivak, McClelland, Maier & Neustadt, Alexandria, Virginia 22314

Pauline Newman (355) United States Court of Appeals for the Federal Circuit, Washington, DC 20439

Michael J. Meurer (399) Boston University School of Law, Boston, Massachusetts 02215

J. H. Reichman (289) Duke University School of Law, Durham, North Carolina 27708

Herbert F. Schwartz (361) Fish & Neave, New York, New York 10020

Gerald Sobel, Esq. (23) Kaye Scholer LLP, New York, New York 10020-3598

Joseph Straus (65) Max Planck Institute for Intellectual Property, Competition, and Tax Law, Munich, Germany D-80539

R. Polk Wagner (367) University of Pennsylvania Law School, Philadelphia, Pennsylvania 19118

Michele M. Wales (427) Human Genome Sciences, Inc., Rockville, Maryland 20850-3338

Acknowledgment

Greatest thanks for this volume is due to the "Conference on Intellectual Property and The Human Genome Project" held at Washington University in St. Louis on April 12–13, 2002, which was essential in shaping the final drafts and ideas that formed this book. The conference was supported by a grant from the Whittaker Foundation of St. Louis and sponsored by the Center for Interdisciplinary Studies at Washington University School of Law in conjunction with the Washington University School of Medicine. Special thanks are due John Drobak, the director of the School of Law's Center for Interdisciplinary Studies and Leslie Kerr, the center's associate director, for generously hosting the event, and Linda McClain, the center's office manager, for running such an elegant and well organized conference. I also thank Chuck McManis, the director of the school's Intellectual Property Program, Dan Ellis, the chair of the school's Intellectual Property Advisory Board, and Joel Seligman, the school's dean, for their continued support, vision, and guidance. I thank the many contributors for allowing their drafts to be made available to conference participants in advance of the conference so the event itself could be best used to bring these works towards their final state. I also thank Philip Berwick, the school's associate dean for Library Services, Aris Woodham, the Webmaster, and Brian Ingram, the network administrator, for setting up and maintaining the protected Web pages needed to facilitate the exchange of drafts that was so essential for this effort.

In addition to those whose written work appears inside, I thank the other conference participants for their many helpful comments on the papers and their general contributions to the dialogue. Special thanks are also due my administrative assistant Jan Houf, for making everything work, and my research assistant Anna Martina Tyreus for her remarkably effective help in every aspect of shaping the final manuscripts for delivery to the publisher, from last-minute editing and researching to careful version-control and formatting. And perhaps most importantly, I thank Hilary Rowe, Erin LaBonte-McKay, and their fantastic team at Academic Press for forging this text into a wonderfully printed, bound, and distributed work.

A listing of all invited conference participants follows. Although too many to be acknowledged individually here, such contributions were important and can be found in the full audio and video recordings of the conference, which are available for free on the Web at www.wulaw.wustl.edu/centeris/apr13agenda.html. I thank Darryl Barker, the school's director of Multimedia Technology, and Robyn Achelpohl and Bill Mathews of the Multimedia Department for their help making and posting these recordings.

Conference Participants

Keynote Speaker

- The Hon. Mr. Justice Robin Jacob, High Court Judge of England and Wales and Judge of the Patents Court of England and Wales

Paper Presenters

- Michael Abramowicz, Assistant Professor of Law, George Mason University School of Law
- James Boyle, Professor of Law, Duke University Law School
- Dan L. Burk, Julius E. Davis Professor of Law, University of Minnesota Law School
- Ananda Chakrabarty, Distinguished University Professor, Department of Microbiology and Immunology, University of Illinois, and inventor in United States Supreme Court case of Diamond v. Chakrabarty
- Iain Cockburn, Professor of Finance and Economics, Boston University School of Management
- Rebecca S. Eisenberg, Robert and Barbara Luciano Professor of Law, The University of Michigan Law School
- Richard A. Epstein, James Parker Hall Distinguished Service Professor of Law, University of Chicago Law School
- Horace Freeland Judson, Research Professor of History, Director, Center for History of Recent Science, The George Washington University
- F. Scott Kieff, Associate Professor of Law, Washington University School of Law, and John M. Olin Senior Research Fellow in Law, Economics, and Business, Harvard Law School
- Charles R. McManis, Professor of Law and Director, Program in Intellectual Property, Washington University School of Law
- Michael Meurer, Associate Professor of Law, Boston University School of Law
- Mark A. Lemley, Professor of Law and Director, Berkeley Center for Law & Technology, University of California, Berkeley, School of Law
- Jerome Reichman, Bunyan S. Womble Professor of Law, Duke University Law School
- Joseph Straus, Professor of Law and Director, Max-Planck-Institute for Foreign & International Patent, Copyright & Competition Law, Munich, Germany

Essay Presenters

- James H. Davis, Senior Vice President and General Counsel, Human Genome Sciences
- Rochelle C. Dreyfuss, Pauline Newman Professor of Law, New York University School of Law
- Wendy J. Gordon, Paul J. Liacos Scholar-in-Law Professor of Law, Boston University School of Law
- Justin Hughes, Visiting Professor of Law, UCLA School of Law, and former attorney-advisor, United States Patent and Trademark Office
- Edmund W. Kitch, Joseph M. Hartfield Professor of Law, University of Virginia School of Law
- F. Scott Kieff, Associate Professor of Law, Washington University School of Law, and John M. Olin Senior Research Fellow in Law, Economics, and Business, Harvard Law School
- Edward T. Lentz, Counsel, Morgan, Lewis & Bockius
- The Hon. Roderick R. McKelvie, U.S. District Judge, U.S. District Court of the District of Delaware
- The Hon. Gerald J. Mossinghoff, Former Commissioner of Patents and Counsel, Oblon, Spivak, McClelland, Maier & Neustadt
- The Hon. Pauline Newman, Circuit Judge, U.S. Court of Appeals for the Federal Circuit
- Stephen S. Rabinowitz, Partner, Pennie & Edmonds
- Herbert F. Schwartz, Partner, Fish & Neave, and Adjunct Professor of Law, University of Pennsylvania Law School
- Gerald Sobel, Partner, Kaye Scholer
- R. Polk Wagner, Assistant Professor of Law, University of Pennsylvania Law School

Discussants

- The Hon. James G. Glazebrook, U.S. Magistrate Judge, U.S. District Court for the Middle District of Florida
- David C. Hilliard, Partner, Pattishall, McAuliffe, Newbury, Hilliard & Geraldson
- David A. Hyman, Professor of Law, University of Maryland School of Law
- Elliott D. Kieff, Harriet Ryan Albee Professor of Medicine and Professor of Microbiology and Molecular Genetics, Harvard Medical School

While those attending the conference have been my teachers for so long, two teachers who are no longer with us deserve special mention. The first is Judge Giles S. Rich, who was the supreme teacher in that those around him learned without it seeming as though he were teaching. The century's preeminent lawyer, jurist, and scholar in the patent field, he was the kindest of gentlemen, and possessed enough curiosity for a thousand cats. I owe Judge Rich a particular thanks for supporting my shift into academia and allowing me to prepare my first book in his chambers on evenings and weekends, "moonlighting," as he called it, during my service as his law clerk. On his death in 1999, the country lost a national treasure and so many lost a dear friend. The second is S. Leslie Misrock, who was the senior partner of Pennie & Edmonds, the firm that transformed me into a lawyer and patent lawyer. Leslie was also a great teacher. He was a master of delegation and promotion—asking those more junior "to take the laboring oar" on increasingly important projects, hoping they would realize their fullest potential. While a student at MIT, I learned of Leslie as that alum who had most successfully made the shift from science to law. Comforted to learn the path he took, I followed his lead. I was grateful for the chance to work directly with him on so many projects and for his support in my shifts from private practice, first to work for Judge Rich, and then to become an academic. If we lost "Mr. Patent" on the death of Judge Rich, then on Leslie's death in 2001 we lost "Mr. Biotech." Possessing such diverse temperaments—Judge Rich quiet, pensive, and almost detached; Leslie aggressive, fast-acting, and almost overwhelming—they each overcame adversity to succeed in all respects while maintaining an uncommon kindness and potency that galvanized such productive networks, including contributors to this volume. To GSR and SLM, I owe thanks for their teaching, their friendship, this volume, and more.

FSK

Introduction

Although the multinational and multi-million-dollar effort to sequence the human genome that is aptly known as the Human Genome Project has successfully yielded a draft of the entire oeuvre, significantly more work must be done before the project's full range of promised scientific knowledge and medical applications is realized. Such downstream development and application are inherently cumulative endeavors that require countless exchanges among members of the basic biological science community. As a result, members of this community and those who study it must concern themselves with the applicable legal rules and norms that might operate to facilitate or frustrate such exchanges. It is generally agreed that this landscape of law and norms dramatically shifted over the last 20 years, as the assertion of patent rights in the field of basic biological research has become increasingly common since the landmark *Diamond v. Chakrabarty* decision by the U.S. Supreme Court in 1980, which held that there is no *per se* exclusion from patentability for subject matter that merely happens to be living or related to life, such as genes, gene products, and even live animals. It is hotly debated what the net impact of this shift has been. Simply put, the question at issue is whether patents facilitate or frustrate the promised developments and applications from the Human Genome Project.

This volume is devoted to exploring from diverse perspectives the many issues that arise when intellectual property rights—especially patents—are used to protect products and processes relating to the Human Genome Project. Building upon the leading research in the field, contributors to this volume offer both a look back at what ground has been covered and a look forward at where future efforts should be focused. They do so from their positions in the private, government, and academic sectors, and in the disciplines of law, medicine, science, history of science, economics, and business.

To begin framing the present debate requires at least an overview of the present use and recent history of intellectual property rights in the field of basic biological research. This is the topic of Part 1, which offers some basic primers on the law and technology in this emotionally and scientifically complex area. Ananda Chakrabarty reviews the past, present, and future of patents in the field, which were first asserted to cover manufactured machines and mass-produced chemical reagents, but which after the *Chakrabarty* decision were used much more expansively to cover many of the animals, cell lines, and nucleic acid and protein sequences that were the very focus of scientific and medical study. To be sure, the views expressed by the U.S. government about such patent rights have evolved over time, as discussed by Gerry Mossinghoff. The early government position can best be evidenced by the Patent Office decision to refuse patent protection in the *Chakrabarty* case, which was reversed by the appellate court in a decision that was affirmed by the Supreme Court in 1980. Also in 1980, Congress passed the Bayh-Dole Act, which encouraged individual recipients of federal grants to seek patent protection on the fruits of the very research that was funded by those grants. This shift in the federal government's views on biotechnology patents must be seen in the larger context of the parallel shift in the federal government's views on the general interface between patent rights, which confer a type of monopoly, and antitrust laws, which are designed to prevent monopolies, as elucidated by Gerry Sobel. In contrast, the Europen Union and its member states, as well as much of the developing world, have each taken a markedly less rosy view towards biotechnology patents, as explored in the comparative analyses by Joseph Straus and Chuck McManis—though, as discussed by McManis, the World Trade Organization's (WTO) Trade-Related Aspects of Intellectual Property Rights (TRIPS) agreement may be quite influential in this international debate. The theoretical underpinnings for these different approaches are to some extent influenced by larger debates about the limits of capitalism. For example, James Boyle argues that all that a well functioning patent system will deliver is whatever society values most, under a regime in which "value" is determined by ability and willingness to pay; and so if we measure value along some other metric then we might not be getting a good deal with patents. While different governments of the world have adopted quite different approaches towards biotechnology patents, what is certain is that at least the United States, for the 20 years since the *Chakrabarty* decision, there has been permission, if not encouragement, for a substantial expansion of patenting in this area. Although this reflects a consensus view about what actually has happened, in the descriptive sense, whether it is good or bad is a topic of the great normative debate that is discussed in the remainder of the volume.

The case in favor of a legal system that gives formal property right treatment to the products and processes of the Human Genome Project begins with a set of economic arguments about how markets over such property rights might operate. These arguments, pro and con, are the topic of Part 2. Scott Kieff presents the case in favor of property rights in this area as operating in accord with both the community norms of basic science and the general theory of property. In response, although Richard Epstein supports the general case for enforcing property rights using strong property rules—backed by a right to exclude—rather than using weaker liability rules—backed only by a take-

and-pay requirement—because of what he terms his general preference for "all or nothing solutions," he suggests that at least for some of the basic biological research tools, such as sequence fragments, the better result is to avoid property rights altogether. Furthermore, as Rochelle Dreyfuss points out, the economic arguments by both Kieff and Epstein are based to some extent upon assumptions that might not be valid, including for example about rational behavior and about the very goals we should have for our patent system. More specifically, Rebecca Eisenberg argues that a property owner's effort to reach through to downstream users in this highly cumulative and collaborative field may make negotiations too difficult. In response to such concerns about transaction costs, Michael Abramowicz argues that we should use nonproperty systems, like government grants, as a way to provide the incentives that patent proponents say must be provided while at the same time avoiding the costs of negotiating for patent licenses, which are the focus of many patent critics. Those like Justin Hughes, who see indeterminacy in each of these competing theories, suggest various empirical approaches to better understand the real-world problems facing progress in basic biological research. When we look to such real-world evidence, we must remember, as admonished by Ed Kitch, to consider the important roles played by the *ex ante* incentives caused by legal and social institutions. Moreover, when we gather data and study such data through models, we must remember, as admonished by David Hyman, to keep track of what exactly in the putatively real world can be fairly said to correspond to the data we elect to use and the results our models generate.

One approach for bringing real world input to this otherwise theoretical debate might be through comparisons with other technologies and other legal regimes, which are the topic of Part 3. Jerry Reichman proposes to cast the many problems with patents into a broader perspective by evaluating what he sees as the similar but deeper systemic crisis that has overtaken developed intellectual property regimes in the past 50 years. Focusing instead on differences rather than similarities, Dan Burk and Mark Lemley suggest that important differences exist in the way patent law has been applied to biotechnology as compared with other technologies; and they argue that this is because many of the core principles of patent law simply are not adaptable for use in the field of biotechnology. This proposition is not without its detractors. As Pauline Newman and Herb Schwartz suggest, the principles of patent law may not have been applied as Burk and Lemley argue. Furthermore, according to Polk Wagner, an application of the positive law view as offered by Burk and Lemley would not make normative sense.

An alternative approach for bringing real world input to this debate might be through comparisons between the actual transactions that we would expect to take place under different regimes of laws and norms, such as one in which property rights are allowed and one in which they are not. Part 4 focuses on the details of exchanges over the products and processes of the Human Genome Project and how they actually might take place in both the academic and in business settings under these different regimes. One comparison that might shed some light on the debate shows how patents can distort the scientific research side of the biological community by shifting the boundary between noncommercial and for-profit research in biomedical science, and contributing to the appearance at this interface of a venture funded, for-profit "tool sector" in the biotechnology industry. This is the topic explored by Iain Cockburn. Another comparison that might shed light on the debate shows how patents can distort the medical application side of the biological community by contributing to the explosion of the pharmacogenetics and genetic test sectors of the healthcare market. This is the topic explored by Michael Meurer. Yet, as James Davis and Michele Wales point out, despite potential distortions from the theoretical optimal, patents have made vast contributions to both sides of the biological community. Moreover, as shown by Ed Lentz, the lawyers who have to structure the actual legal arrangements made when patents are in play can deploy strategies that would facilitate, not frustrate, such exchanges for both the academic and the private sectors in ways that can offer real and positive achievements for the agendas of both the private, self-regarding licensor and a skeptical social ethicist.

While comparisons about transactions can shed some light on the debate, a full analysis also requires comparisons about dispute resolution under the same alternative regimes. Part 5 explores the details of the disputes that inevitably occur over the products and processes of the Human Genome Project and how they actually can and might take place in both the academic and business settings under different regimes. One such comparison is between a regime in which disputes are resolved by someone with technological expertise and one in which disputes in this technologically complex area are resolved the same way typical commercial disputes are resolved. This is the topic explored by Robin Jacob, who questions the merits of a one-size-fits-all approach, as is used in many respects by U.S. District Courts in civil cases. Roderick McKelvie responds with an analysis of some empirical evidence about how technologically complex issues are actually decided in U.S. District Courts. Scott Kieff then offers a theoretical explanation for why the ordinary judges and juries of U.S. District Courts might be best suited for deciding technologically complex questions in patent cases, even for those seemingly complex technological issues that arise when determining patentability over the prior art. In comparison, Horace Judson unpacks some of the complex problems the academic science community itself has found when it has been forced to have its own technologically trained members conduct legal-type decisionmaking over its own internal disputes.

This collection of diverse perspectives about the role of property rights in the Human Genome Project is presented with a writing and annotation style designed to accommodate diverse readership—from academics and students to the general public. A unique integrated collection of authors, views, and topics, it is essential reading for anyone interested in the timeless problems at the interface between biotechnology and property specifically, and the history, sociology, and organization of science generally.

Section 1

WHERE WE ARE AND HOW WE GOT HERE

1

Patenting Life Forms: Yesterday, Today, and Tomorrow

A. M. Chakrabarty
Distinguished University Professor
Department of Microbiology & Immunology
University of Illinois College of Medicine
Chicago, Illinois 60612

I. INTRODUCTION

I recently attended two interesting meetings involving biotechnology and ethical/legal considerations. The first was organized by the University of Pennsylvania in Philadelphia, entitled "Who owns life?" The second was organized by the University of Milan in Italy, entitled "Do judges make law?" Both are rhetorical questions for which simple answers do not exist. For the first, the definition of what life is or when it begins has eluded a satisfactory answer since the days of the abortion debate or even earlier, when philosophers and scientists tried to distinguish between living and nonliving objects. To complicate matters, the title had an additional intrigue: What is meant by ownership? In a sense, we all own life, if ownership means having the ability to breed new lives (as we do with cattle, chickens, fish, or plants) and to terminate such lives at will. The organizers, when they posed the question, however, meant human life. The question of ownership was not absolute, either.

0065-2660/03 $35.00

Ownership here meant having a patent on a life form that prevents others from using the life form for commercial purposes for a period of time. Simply put, the question was if we could, or should, obtain patents on whole humans or human parts, such as human cells, genes, organs, limbs, or any combination thereof, including an animal with some well-recognized human features.[1]

The second rhetorical question posed by organizers at the University of Milan appeared to be more straightforward: Do judges make law? Judges, of course, are not supposed to make law. They are supposed to interpret laws that are framed by legislative bodies, such as the U.S. Congress or Italian Parliament, based on the perceived intention of such a body. When scientific advances are rapid and border on areas that arouse strong public interest, religious feelings, or transcend accepted moral codes of conduct, they lead to conflicts that usually end up in courts of law. However, there may not yet be any law, governing such scientific advances. In the United States, the Supreme Court has authorized trial judges to be the gatekeepers of acceptable scientific evidence, so judges often face or adjudicate cases of complex science in the absence of a legal structure or a well-defined law. Since precedents are important in legal cases, such rulings often tend to be viewed as laws of the land. Should a judge or a body of judges adjudicate such cases or leave them for the legislative bodies for resolution? Should judges make law?[2]

II. PATENTING LIFE FORMS

My involvement with the two issues, *viz.*, patenting life forms or judicial law-making, goes back to the 1970s when I was a research scientist at General Electric R&D center in Schenectady, N.Y. There I developed a genetically improved microorganism that was designed to break down crude oil rapidly and was therefore deemed suitable for release into oil spills for their cleanup. Because such an environmental release would make them available to anyone who wanted to have them, GE filed for a patent for both the process of constructing the genetically manipulated microorganism and the microorganism itself. The Patent and Trademark Office (PTO) granted the process patent but rejected the claim on the microorganism based on the fact that it was a product of nature. GE appealed to the PTO Board of Appeals, pointing out that the genetically engineered microorganism was genetically very different from its natural counterparts. The Board conceded that it was not a product of nature

[1] Chakrabarty, A. M. (2002). *Patenting of Life Forms: From a Concept to Reality*, *in* WHO OWNS LIFE? (D. Magnus, A. Caplan, and G. McGee, eds.), Prometheus Books, Amherst, NY.

[2] Chakrabarty, A. M. (2002). *Genetics research and the judicial decision*, NOTIZIE DI POLITEIÀ **18,** 99–102.

but still rejected the claim on the patentability of the microorganism because of the fact that the microorganism was alive.

Convinced that such a rejection had no legal basis, GE appealed to the U.S. Court of Custom and Patent Appeals (CCPA). The CCPA ruled three to two in GE's favor. Judge Giles S. Rich, speaking on behalf of the majority, contended that microorganisms were "much more akin to inanimate chemical compositions such as reactants, reagents, and catalysts than they are to horses and honeybees or raspberries and roses." Another judge, Howard T. Markey, concurred, saying, "No congressional intent to limit patents to dead inventions lurks in the lacuna of the statute, and there is no grave or compelling circumstance requiring us to find it there." The PTO then appealed to the Supreme Court. Initially, the Supreme Court sent it back to the CCPA for reconsideration in light of another patent case, but the CCPA again ruled in favor of GE, this time emphasizing that the court could see "no legally significant difference between active chemicals which are classified as dead and organisms used for their chemical reactions which take place because they are live."

The CCPA decision prompted the Solicitor General to petition the Supreme Court for a ruling, which was granted. Several *amicus* briefs were filed by many individuals and organizations to support or oppose the case, which came to be known as *Diamond v. Chakrabarty* because Sidney Diamond was the new commissioner of patents. One *amicus* brief filed on behalf of the government came from People's Business Commission (PBC), which argued that patenting a life form was not in the public interest and that granting patents on a microorganism would inevitably lead to the patenting of higher life forms, including mammals and perhaps humans. The "essence of the matter," PBC argued, was that the "issuance of a patent on a life form was to imply that life has no vital or sacred property and it was simply an arrangement of chemicals or composition of matter."

In 1980, 8 years after the initial filing, the Supreme Court, by a vote of 5 to 4, agreed with Judge Giles Rich of the CCPA that the microorganisms in question were a new composition of matter, the product of human ingenuity and not of nature's handiwork, and thus a patentable subject matter. Justice William Brennan, arguing on behalf of the minority, pointed out that the case involved a subject that "uniquely implicates matters of public concern" and therefore belonged to congressional review and a new law. Ownership of life forms became a reality with interesting consequences!

III. WHAT HAPPENED NEXT?

The decision of the Supreme Court was narrowly focused on what was described as a "soulless, mindless, lowly form of life," even though the court emphasized that anything under the sun made by man could be patented as long as it met the criteria. The PTO took the decision as a broad interpretation of the patenting of life forms and has issued hundreds of U.S. patents covering microorganisms, plants, animals, fish, bird and human genes, mutations, and cells. Aside from the disputed oil-eating microorganism, the earliest patent for a living organism was awarded to Chakrabarty and Kellogg for "Bacteria capable of dissimilation of environmentally persistent chemical compounds," on August 13, 1985. The earliest genetically modified plant patent went to Kenneth Hibbard, Paul Anderson, and Mellanie Barker, for "Tryptophan overproducer mutants of cereal crops," issued on April 15, 1986. The first patented animal was the "Harvard Mouse," a mouse that was genetically altered to make it more susceptible to developing breast cancer, issued to Philip Leder and Timothy Stewart as "Transgenic non-human mammals" on April 12, 1988. Human cells, expressed sequence tags (ESTs), single nucleotide polymorphisms (SNPs), and cultivation and isolation of stem cells have also been patented. For example, a patent on a cell line derived from the spleen of a leukemia patient was issued to David Golde and Shirley Quan as "Unique T-lymphocyte line and products derived therefrom" on March 20, 1984. Another example is a patent issued to Ann Tsukamoto, Charles Baum, Ykoh Aihara, and Irving Weissman on October 29, 1991, "Human hematopoetic stem cell," which covered isolated human bone marrow stem cells.

The issuance of patents on animals and human body cells conferred a degree of ownership that was an uncharted territory with unprecedented legal ramifications. Many disputes involving patent infringement cases emerged because of questions related to obviousness, enablement, or the priority of invention that had to be decided by the courts. More difficult were the questions about ownership rights and privileges. For example, in the patent "Unique T-lymphocyte line and products derived therefrom," the inventors used the spleen of a patient, Mr. John Moore, who suffered from hairy cell leukemia and came for treatment to Dr. Golde at the University of California at Los Angeles (UCLA). As part of the treatment, Mr. Moore's spleen was removed and Dr. Golde developed a cell line with enriched T-lymphocytes that produced large amounts of lymphokines useful for cancer or AIDS treatment. Without Mr. Moore's initial knowledge or consent, but requiring Mr. Moore's repeated visits to the hospital, Dr. Golde and UCLA applied for a patent on the cell line derived from Mr. Moore's spleen, which was granted in 1984. Mr. Moore subsequently sued Dr. Golde and UCLA, claiming theft of his body part. The trial court ruled against Mr. Moore, but the ruling was reversed by the Court of Appeals, and the case finally ended up at the California Supreme Court. Both the Appeals Court and the Supreme Court recognized the novelty of Mr. Moore's claim, which was without any precedent in law. Nevertheless, the California Supreme Court ruled against Mr. Moore on the issue of conversion (unauthorized use of his body part) but recognized his right to be informed of what the physician was doing involving his health and well being.

Who should own intellectual property when a given invention not only requires human ingenuity but also human tissues? On October 30, 2000, a lawsuit was filed in Chicago federal court by the Canavan Foundation of New York City, the National Tay-Sachs and Allied Diseases Association, and a group of individuals. These individuals and organizations raised money, set up a registry of families, and recruited tissue donors to help develop a genetic test for the Canavan disease, a central nervous system disorder caused by a mutation on chromosome 17 in a gene encoding the enzyme aspartoacylase. The disease predominantly affects Ashkenazi Jewish children and a genetic test was deemed to be useful in screening potential parents harboring such mutations. The foundation provided diseased tissues and money to Dr. Reuben Matalon, who developed a genetic test while employed at the Miami Children's Hospital (MCH). Once the test was developed, a patent on it was obtained by MCH. However, MCH was alleged to charge a high fee for the test, restrict access to the test, and put a limit on the number of tests that a licensee can perform. This is contrary to the intention of the individuals and foundations, which was to help develop a cheap test that could be made widely available to prospective parents to prevent Canavan disease. The question of how important the contribution of the tissue donors is, without which tests cannot be developed and patents cannot be obtained, is going to resonate as more and more such cases end up in courts of law. Another important consideration is the cost of the tests. Diseases are often caused by mutations, deletions, or genetic rearrangements in human genes. Therefore, genetic screening for some diseases may involve several genetic tests to ensure detection of all possible genetic alterations. For a disease such as cystic fibrosis, where 70% of the patients harbor a single mutation (a trinucleotide codon deletion) in the *cftr* (cystic fibrosis transmembrane conductance regulator) gene, a single genetic test may not be enough and multiple mutations may have to be screened, thus greatly increasing the cost of genetic tests. Because detection of genetic alterations and development of such tests are time consuming and costly, the developers seek patents on such genetic tests and try to recoup the expense by assessing license fees for conducting the tests. Who should pay for these costs? Who decides the cost of such genetic tests? Should market forces determine who should live disease-free and who should suffer? The recent incidence in South Africa where drug companies allowed low-cost generic production of their patented AIDS drugs to be made available to HIV-infected patients is an interesting example of the societal responsibility of large drug manufacturers.

Then there is the question of patentability of genetic tests that do not detect all the mutations or genetic variations in a gene. Myriad Genetics of Salt Lake City obtained a European patent in 2000 on genetic testing of BRCA1 and BRCA2, called BRAC Analysis Technique, which used automated sequencing to scan for BRCA mutations and deletions. Mutations in BRCA1 and BRCA2 account for almost 10% of all breast cancers, and their early detection is important to treat breast cancer. However, certain deletions and genetic rearrangements in BRCA1, involving about 11.6 kb of DNA, cannot be detected by the Myriad tests and can only be detected by a patented process called combed DNA color bar coding. Such genetic deletions may comprise about 36% of all BRCA1 mutations. The claims of the Myriad Genetics European patent allegedly make it difficult for European clinicians to use the combed DNA color bar coding technique patented by Institut Pasteur, raising difficult legal questions on who should control or establish ownership of human genetic mutations and their detection and on human genetic makeup in general.

There are also privacy and civil rights issues in patented genetic tests. For example, the U.S. Equal Employment Opportunity Commission (EEOC) sued Burlington Northem Santa Fe Railroad for requiring genetic tests of its employees who filed claims for certain work-related hand injuries (carpal tunnel syndrome). The railroad wanted to determine which employees may be predisposed to this syndrome, which was believed to be due to a specific genetic deletion on chromosome 17. The railroad was alleged by the EEOC to have threatened to fire employees who refused the test, thus violating the employees' civil rights. This case has since been resolved to the satisfaction of the EEOC.

IV. WHAT ABOUT PATENTING HUMAN ORGANS?

As mentioned previously, human cells, including hematopoetic stem cells, have been patented. The characterization and culturing of human embryonic stem cells also have been patented by a group at the University of Wisconsin in 1998. The ability to purify and cultivate the pluripotent human embryonic stem cells from human fetal tissues in the laboratory presents exciting possibilities to allow these cells to differentiate into specific organs such as the liver, kidney, or heart. The human fertilized egg, within hours after fertilization, remains a totipotent cell because it has the potential to give rise to a human individual. Within a few days, however, the totipotent cells multiply to a blastocyst, which contains the pluripotent self-replicating stem cells that, under the right cellular cues, can differentiate into any of the 260 or so different tissues in the human body. These pluripotent embryonic stem cells, capable of indefinite replication, differ from adult stem cells, which do not replicate as easily and are often multipotent rather than pluripotent, meaning that they can give rise to some but not all cell types that constitute specific organs. Indeed, mouse embryonic stem cells have been shown to differentiate into insulin-secreting structures similar to pancreatic islets of Langerhans, as well as cell types constituting blood, the heart, muscles, and the brain. Similar differentiation of human embryonic stem cells to blood-forming cells has been reported in September of 2001 by researchers from the University of Wisconsin.

The ability of the patented human pluripotent embryonic stem cells to differentiate into blood-forming or potentially many other human tissue cells provides exciting possibilities to generate inexhaustible, virus-free blood supply or other human organs for transplantation or tissue regeneration purposes. Because such cells can be grown in culture and allowed to differentiate into specific cell types for tissue regeneration or organ transplantation in the presence of specific signals, it is likely that certain improvements can be made in the quality of such tissues and organs that might require patenting.

Much effort has gone into figuring out how to grow and develop an intact organ, such as a liver or a heart, from the differentiated cells. The current technology involves building a scaffold shaped like the patient's heart or liver. The scaffold will be made of a biopolymer, usually derived from seaweeds, that degrades slowly in water. The seaweeds have intricate capillaries that supply nutrients to all parts of the weed. The scaffold will be immersed in a bath of heart or liver cells in a culture obtained by inducing differentiation of human embryonic pluripotent stem cells. As the cells grow within the porous surfaces of the scaffold, they will take the shape of the scaffold. Using computer microchip engineering, capillaries will be built within the scaffold to ensure that all the cells have access to the nutrients. With time, the biopolymer will disintegrate, leaving behind an intact heart or liver composed of the patient's own cells. Various improvements can be made in the functionality of the organ, which can then be used for transplantation in case of need. Patents can then be obtained for the process of making such organs, as well as for the organs themselves. Specific mutations can be introduced into the cells to avoid transplant rejection, so that such organs can be mass produced and marketed worldwide.

The possibility of mass manufacture and worldwide trade in laboratory-made human organs raises important ethical and legal questions. Similar to blood banks or frozen embryo stores where human tissues or potential human lives can be stored and used, can laboratory-produced human organs and limbs be products of commerce? If these organs and limbs are to be patented, should the ownership be extended to multiple organs that are linked and inter-related? In case of organ failure or defective manufacture, should the liability be calculated on the status of the recipient or the lifestyle of the customer? If patents are granted for human organs, can they be extended to hybrids of humans and animals that may harbor little to large amounts of human cells, genes, and organs?

V. DEFINING HUMAN?

In 1998, a patent application was filed to the United States PTO for a human-animal hybrid. The hybrid was not actually made, but the concept was based on the fact that hybrids can be made out of sheep and goats (called geeps). It was pointed out that the DNA sequence identity between human and chimpanzees is of the same order as that between sheep and goats. Consequently, the applicants argued that it should be possible to make hybrids between humans and chimpanzees, and such hybrids could be useful for organ harvesting or other medical purposes. In reality, these applicants did not want the patents but were simply raising the issue to prevent future patents on engineering of human genes and human reproduction. The patent office rejected the application based on the 13th Amendment of the U.S. Constitution, which is an antislavery amendment that rejects the ownership of human beings.

Although the patent application has been rejected, it has raised some interesting questions. If the hybrid human-animal is a human and therefore cannot be owned, how much human genetic material or human phenotypic trait must be present in an organism or an animal to confer on it the characteristics of a human? Several human genes have been inserted in nonhuman organisms without raising problems of patentability. Is there an upper or lower limit of the presence of such human genes or traits to deter patentability based on the 13th Amendment? A Massachusetts company has claimed to have cloned a human embryo even though the embryo did not undergo enough cell division to give rise to a blastocyst. Given the large number of animals that have been cloned, it is likely that human cloning techniques can and will be perfected to generate human embryos. There are many uncertainties regarding the health and well-being of cloned animals, and certainly there would be enormous resistance to the cloning of human beings. Given scientists' curiosity and relative ease of cloning, it is likely that somebody, somewhere, would conduct nuclear transfer to enucleated human oocytes. What would happen if somebody transfers a chimpanzee nucleus into an enucleated human oocyte and implants it into a human uterus? Alternatively, one could take a chimpanzee egg and replace the nucleus with that of a human and then implant it in the womb of a chimpanzee. Even with the contribution of the cytoplasmic material, which controls gene expression, or the mitochondrial DNA of the egg, it is likely that the transferred nucleus will determine the primary genotype of the embryo. Is a baby chimpanzee delivered of a human mother a chimpanzee or a human? Alternatively, is a mostly human baby delivered of a chimpanzee mother a human or a chimpanzee? Can such babies be patented if they are not products of nature?

Of course it is not just the controversial products that might require patent protection. Processes could be as contentious. For example, considerable attempts are being made to develop a process during which an egg, upon stimulation by chemical or physical methods, can undergo cell division as if it has been fertilized by a sperm. The fertilized form of a human egg can, of course, be implanted in the womb of a woman, either the source of the egg or a surrogate mother, giving rise to a baby that is a product of the nucleus present in the egg (no other nucleus is involved and so no genetic shuffling can take place). Even though such a process is scientifically and technically farfetched, if or when such a process is perfected for human eggs and a patent applied for, is it within the domain of PTO to legitimize such a drastic process of human reproduction by issuing a patent on the process?

VI. EPILOGUE

This is both an exciting and a difficult time for a biologist. The technology of animal and human reproduction, as well as the techniques of genetic manipulation, are progressing so rapidly it creates situations that transcend our legal structure and directly affect our social and moral fabrics. It is high time that the United States Congress takes a serious look at where the science is going, where it needs to go to make a positive contribution, and perhaps define the boundaries of our venture into the unknown biological mysteries of nature. Of course, no Congressional mandate will ever cover all future scientific directions or orient human ingenuity, and the judiciary will play increasing roles in resolving conflicts involving human genetic reproduction, genetically manipulated plants and foods, and environmental restoration. There is thus a great need to continue dialogues between the judiciary, the legal community, the legislature, the interested public, and the scientific community to provide guidance in scientific developments that may have major impacts on society.

Finally, a word about patenting. Very few pharmaceutical companies tried to patent products or cultures before GE applied for the patent on the oil-eating microorganisms because these companies relied mostly on trade secrets. Patenting involves a detailed description of the invention, which would have required biological companies to divulge their secrets. GE is a corporation that relies heavily on new information and its dissemination. Thus the company was and remains today a major generator of patents. Patenting greatly facilitates information exchange and provides detailed descriptions of inventions that can be quickly accessed by both the academic and the industrial sectors. Indeed, the new generation of biotechnology companies relies heavily on their patents for raising capital and gaining scientific recognition. The *Diamond v. Chakrabarty* decision has immensely contributed to the growth of the biotechnology industry both by allowing patenting of life forms, as well as facilitating dissemination of scientific ideas, technology, and concepts.

Acknowledgments

I am indebted to Honorable Judge Dr. Pauline Newman, Dr. Franklin Zweig, and Professors John Witherspoon and Eliora Ron for their kind, constructive, and helpful comments and criticisms. The work in my laboratory is supported by two grants from NIH (ES 04050 and A1 45541).

2

The Evolution of Gene Patents Viewed from the United States Patent Office

Hon. Gerald J. Mossinghoff
Senior Counsel
Oblon, Spivak, McClelland, Maier & Neustadt
Alexandria, Virginia 22314

Former Assistant Secretary of Commerce and Commissioner of Patents and Trademarks

Professorial Lecturer
The George Washington University Law School
Washington, DC 2001

I. INTRODUCTION

I am honored to be asked to participate in this important conference on the intellectual property implications of the Human Genome Project. Honored—and thoroughly overwhelmed. As I told Scott Kieff when he extended his thoughtful invitation to me, I managed to get a high school education—from both the Jesuits and the Brothers of Mary here in St. Louis—a degree in electrical engineering from the Jesuits—at that other university in St. Louis—and a law degree, all without taking even an elementary course in biology. My advantage today, of course, is that I bring to this conference an uneducated layman's view of the subject.

0065-2660/03 $35.00

When I went to St. Louis University, every graduate—even those in engineering—in effect had to get a minor in philosophy, which in those days was fairly limited to studying the works of St. Thomas Aquinas. As I was preparing for this conference, I initially thought that a Thomastic approach to the patenting of parts of the human genome might prove to be helpful. At least in the popular press, it was directly related to the patenting of God's work, slavery, abortion, stem-cell research, and animal and human cloning. I would add parenthetically, one article on the subject advised those involved in the debate to "Keep Their Genes On!"

But Mrs. Mossinghoff did not raise any courageous sons here in St. Louis, so I decided not to delve into these areas, leaving it to others more qualified than I to address them. Instead I will address the key patent law elements in the debate regarding gene-related inventions.

II. ARE GENE-RELATED INVENTIONS PATENTABLE SUBJECT MATTER?

The answer is yes, without question. This answer would have been different a little more than two decades ago, when the United States Patent and Trademark Office (USPTO) under the Carter Administration opposed the grant of a patent on Dr. Ananda M. Chakrabarty's petroleum-eating enzyme. People can say what they will about the landmark *Diamond v. Chakrabarty* case,[1] but no one can seriously doubt that it led directly to the preeminence of the United States in biotechnology.

An interesting footnote to that case was the role the Upjohn Pharmaceutical Company played in its outcome. Two cases involving patent applications for living organisms were making their way though the USPTO and the courts in the late seventies: *Chakrabarty* and *In re Bergy*. The Bergy application was owned by Upjohn and concerned a purified version of an antibiotic-producing strain of streptomycin, a substance that appeared in nature but in an unpurified state. The purified version was invented by Dr. Malcolm E. Bergy. The USPTO Board of Appeals denied both applications based on the fact that the inventions involved living things. The Court of Customs and Patent Appeals (CCPA), predecessor of the present Court of Appeals for the Federal Circuit, ruled three to two that the patents should be granted because nothing in the patent laws barred the issuance of a patent to an otherwise patentable invention simply because it was alive.[2] Judge Giles E. Rich wrote the opinion for the majority of the CCPA, an opinion that ultimately had far-reaching implications.

[1] 447 U.S. 303 (1980).
[2] *In re Bergy*, 596 F.2d 952 (CCPA 1979).

The favorable ruling by the CCPA prompted the USPTO to appeal both *Bergy* and *Chakrabarty* to the Supreme Court. Both cases were sent back to the CCPA for reconsideration in light of the Supreme Court's opinion on mathematical algorithms in *Parker v. Flook.*[3] In early 1979, Judge Rich, again speaking on behalf of the majority of the CCPA, in *Bergy* and *Chakrabarty*, emphasized that biological conversions mediated by microorganisms are basically an application of basic chemistry and nothing in *Parker v. Flook* changed that opinion.

That second opinion of the CCPA was again appealed to the Supreme Court by the Carter Administration, with then-Patent Commissioner Sidney Diamond being named as Appellant. Meanwhile, Upjohn was persuaded to cancel its product claims covering the streptomycin producing the antibiotic lyncomycin. The thought prevalent among the patent bar was that it was far more likely that the Supreme Court would uphold the patenting of a modified living microorganism that did not occur in nature rather than one that did appear in nature but in a less purified form. In my view, that action by Upjohn, while perhaps not determinative, gave the five-to-four majority in *Chakrabarty* an uncluttered landscape on which to base its opinion.

Quite properly, the USPTO did interpret and continues to interpret *Chakrabarty* to require the patenting of living things whether they are genetically modified—like Dr. Chakrabarty's enzyme—or isolated and purified by the "hand of man"—like Dr. Bergy's streptomycin. This broad and proper reading of *Chakrabarty*, of course, plays directly into the decisions to grant more than 1000 patents on gene-related inventions and the hundreds of utility patents on genetically modified plants and animals.

In December 2001, the Supreme Court upheld utility patents on genetically enhanced plants, relying upon its decision in *Chakrabarty*—this time in a six-to-two opinion. In *J.E.M. Ag Supply v. Pioneer Hi-Bred International*,[4] Justice Clarence Thomas stated:

> As this Court recognized over 20 years ago in *Chakrabarty*, 447 U.S. at 308, the language of [35 U.S.C.] § 101 is extremely broad. "In choosing such expansive terms as 'manufacture' and 'composition of matter,' modified by the comprehensive 'any,' Congress plainly contemplated that the patent laws would be given wide scope." Ibid. This Court thus concluded in *Chakrabarty* that living things were patentable under § 101, and held that a man-made microorganism fell within the scope of

[3] 437 U.S. 584 (1978). *Parker v. Flook* is viewed by many scholars as a curious decision that was overruled by *Diamond v. Diehr*, 450 U.S. 175 (1981), which provided the legal basis for the thousands of U.S. patents on computer software.

[4] 534 U.S. 124, 122 S.Ct. 593 (2001).

> the statute. As Congress recognized, "the relevant distinction was not between living and inanimate things, but between products of nature, whether living or not, and human-made inventions." *Id. at 313.*

I would predict that that same line of reasoning would lead at least the *J.E.M.* majority to uphold patents on gene-related inventions when that issue comes before the Supreme Court—as I predict it will.

III. UTILITY OF GENE-RELATED INVENTIONS

For a person to be entitled to a patent, he or she must invent or discover something that is "new and *useful*." That single word "useful" has led to a whole body of case law.

Specific Guidelines for Determining Utility of Gene-Related Inventions were issued on January 5, 2001.[5] In brief outline, the document changed the "old" two-pronged test of *specific* and *credible* utility to a three-pronged test that adds *substantial* utility to the other two.

Examiners are instructed to:

> Review the claims and the supporting written description to determine if the applicant has asserted for the claimed invention any specific and substantial utility that is credible:
>
> (a) If the applicant has asserted that the claimed invention is useful for any particular practical purpose (i.e., it has a "specific and substantial utility") and the assertion would be considered credible by a person of ordinary skill in the art, do not impose a rejection based on lack of utility.
>
> (1) A claimed invention must have a specific and substantial utility. This requirement excludes "throw-away," "substantial," or "nonspecific" utilities, such as the use of a complex invention as landfill, as a way of satisfying the utility requirement of 35 U.S.C. § 101.
>
> (2) Credibility is assessed from the perspective of one of ordinary skill in the art in view of the disclosure and any other evidence of record (e.g., test data, affidavits or declarations from experts in the art, patents or printed publications) that is probative of the applicant's assertions. An applicant need only provide one credible assertion of specific and substantial utility for each claimed invention to satisfy the utility requirement.

[5] 66 Fed. Reg. 1092 (2001).

(b) If no assertion of specific and substantial utility for the claimed invention made by the applicant is credible, and the claimed invention does not have a readily apparent well-established utility, reject the claim(s) under § 101 on the grounds that the invention as claimed lacks utility. Also reject the claims under § 112, first paragraph, on the basis that the disclosure fails to teach how to use the invention as claimed.[6]

These guidelines, in my view, are consistent with the holding in *Brenner v. Manson* that ". . . a patent is not a hunting license. It is not a reward for the search, but compensation for its successful conclusion."[7]

IV. DO GENE-RELATED INVENTIONS SATISFY THE ENABLE AND WRITTEN DESCRIPTION REQUIREMENTS?

These two separate and distinct requirements are key to the basic constitutional *quid pro quo* of the patent system. 35 U.S.C. § 112, first paragraph states that:

> The specification shall contain a written description of the invention . . . as to enable any person skilled in the art to which it pertains, or with which it is most nearly connected, to make and use the same . . .

With respect to the *enablement requirement*, there are a number of factors that go into a determination. These include:

Breadth of claims

Nature of the invention

State of the prior art

Level of skill in the art

Level of predictability

Amount of direction/guidance

Presence/absence of working examples

Quantity of experimentation

[6] 66 Fed. Reg. 1098.
[7] 383 U.S. 519, 534 (1966).

With respect to the *written description requirement*, the basic question here, as in all other fields of technology, is: Can one skilled in the art reasonably conclude that the inventor was in possession of the claimed invention at the time the application was filed?

In general, there is, as there should be, a strong presumption that an adequate written description is present in the application as filed. Thus, the initial burden is on the examiner to establish a prima facie case of unpatentability; however, applicants should be able to show support for new and amended claims.

Factual considerations regarding the written description requirement of gene-related inventions, in view of the level of skill and knowledge in the art, include:

Complete or partial structure

Physical and/or chemical properties

Functional characteristics

Correlation between structure/function

Method of making

Combinations of the above

Since the level of skill and knowledge in any technology increases over time, it follows that the less mature the technology, the more evidence is required to show possession.

V. NOVELTY AND NONOBVIOUSNESS

An assertion is sometimes heard that in special cases—with e-commerce and gene-related patents usually being singled out—one should "raise the bar" to prevent "overbroad" patents. This leads to an idea, usually vaguely defined, to change the nonobviousness standard selectively.

Nonobviousness is the most important patentability requirement and perhaps the most difficult to apply. Section 103(a) provides:

> A patent may not be obtained though the invention is not identically disclosed or described as set forth in section 102 of this title, if the differences between the subject matter sought to be patented and the prior art are such that the subject matter as a whole would have been obvious at the time the invention was made to a person having ordinary skill in the art to which said subject matter pertains.

The enactment of § 103 in 1952 was a reaction to a line of Supreme Court cases in which U.S. patents were held to be invalid because they lacked "invention." In one celebrated case, Justice William O. Douglas went so far as to state that for a device to be patentable, it "must reveal the flash of creative genius."[8] The Supreme Court's anti-patent bias during the period leading up to 1952 was so pronounced that Justice Robert H. Jackson in a dissent complained "that the only patent that is valid is one which this Court has not been able to get its hands on."[9]

In his *Commentary on the New Patent Act*, Mr. P. J. Federico, a senior official of the U.S. Patent Office and one of the principal authors of the 1952 Act, stated as follows:

> There has been some discussion as to whether section 103 modifies the so-called standard of invention (which itself is an unmeasurable quantity having different meanings for different persons) in the courts and it may be correct to state that the printed record does not show an explicit positive command to the courts. While it is not believed that Congress intended any radical change in the level of invention or patentable novelty, nevertheless, it is believed that some modification was intended in the direction of moderating the extreme degrees of strictness exhibited by a number of judicial opinions over the past dozen or more years; that is, that some change of attitude more favorable to patents was hoped for. This is indicated by the language used in section 103 as well as by the general tenor of remarks of the Committees in the reports and particular comments.[10]

The Supreme Court did not reach the issue of the proper interpretation of § 103 until 1966, when the Court decided three patent cases, often referred to as the *Graham Trilogy*.[11] As stated by one leading patent law scholar:

> In *Graham*, the Court pointedly confirmed that Section 103 *codified* the judicially developed nonobviousness requirement. Congress did focus inquiry on objective obviousness and, in effect, directed abandonment of "invention," which the courts had previously used to encapsulate the

[8] *Cuno Engineering Corp. v. Automatic Devices Corp.*, 314 U.S. 84, 91 (1941). That case specifically prompted Congress to add a sentence to §103 that "Patentability shall not be negatived by the manner in which the invention was made."

[9] *Jungersen v. Ostby & Barton Co.*, 335 U.S. 560, 571 (1949) (dissenting opinion).

[10] P.J. Federico, *Commentary on the New Patent Act*, *reprinted in* 75 JOURNAL OF THE PATENT AND TRADEMARK OFFICE SOCIETY 161, 183 (1993).

[11] *Graham v. John Deere Co.* and *Calmar, Inc. v. Cook Chemical Co.*, 383 U.S. 1 (1966); *Adams v. United States*, 383 U.S. 39 (1966).

obviousness standard. "Invention" had led to conceptual confusion. But, according to the Court, Section 103 did not, and constitutionally could not, "lower" or fundamentally alter the patentability standard. On the merits, the Court held two of three patents invalid; it held a third patent valid, emphasizing that the invention, a battery that provided strong current with the addition of a water electrolyte, was met with initial skepticism by experts but later was used extensively by the United States government.[12]

In *Graham*—still the leading case interpreting § 103—the Supreme Court directed the lower courts and the USPTO to apply the following test:

> Under § 103, the scope and content of the prior art are to be determined; differences between the prior art and the claims at issue are to be ascertained; and the level of ordinary skill in the pertinent art resolved. Against this background, the obviousness or nonobviousness of the subject matter is determined. Such secondary considerations as commercial success, long felt but unsolved needs, failure of others, etc., might be utilized to give light to the circumstances surrounding the origin of the subject matter sought to be patented. As indicia of obviousness or nonobviousness, these inquiries may have relevancy.
>
> This is not to say, however, that there will not be difficulties in applying the nonobviousness test. What is obvious is not a question upon which there is likely to be uniformity of thought in every given factual context. The difficulties, however, are comparable to those encountered daily by the courts in such frames of reference as negligence and scienter and should be amenable to a case-by-case development. We believe that strict observance of the requirements laid down here will result in that uniformity and definiteness which Congress called for in the 1952 Act.[13]

Notwithstanding this guidance, the regional Circuit Courts of Appeals were all over the lot in interpreting the requirements of § 103. One of the issues was whether "synergism" in some form or another was required to satisfy the § 103 requirement. As noted by one patent law scholar, "prior to Federal Circuit analysis of the issue, confusion reigned among lower federal courts as to the proper role of synergism in evaluating nonobviousness."[14]

[12] Chisum *et al.* UNDERSTANDING INTELLECTUAL PROPERTY LAW 2-57 (Matthew Bender 1997 Reprint).

[13] 383 U.S. at 17.

[14] Peter D. Rosenberg, PATENT LAW BASICS 9–48 (Clark Boardman Callaghan 1995).

One of the principal areas of concern that led to the creation of the Federal Circuit was § 103 and the differences in its interpretation throughout the regional circuits. Although there are clear differences among the several judges serving on the Court of Appeals for the Federal Circuit in other areas of patent law, there are no major differences in the interpretation of § 103. In one celebrated case, the Federal Circuit relied upon § 103 when it vacated the Seattle District Court's Preliminary Injunction against Barnesandnoble.com on February 14, 2001.[15]

Thus, with respect to § 103 regarding nonobviousness, three factors have resulted in a workable standard of patentability both in the Patent and Trademark Office and in the courts: (1) the enactment of the section in 1952, (2) the authoritative interpretation of the section in the *Graham Trilogy* of cases, and (3) the creation of the Federal Circuit, which is doing an excellent job of interpreting § 103 on a case-by-case basis. There are now more than 700 Federal Circuit cases interpreting § 103 in dozens of technological contexts. If patent claims are said to be "overbroad," I assume that means that they would not be valid under 35 U.S.C. § 103 or § 112, as those sections are now written. Otherwise, I would have no idea what "overbroad" means.

To attempt now to amend § 103 to "raise the bar"—whatever that may mean in a given case—would, at the very least, result in a generation or two of uncertainty and confusion. Such an attempt would, in my view, be met with appropriate, vigorous, and successful opposition by high technology industry, inventor groups, and the organized patent bar.

VI. SUMMARY

In my remarks here, I have necessarily limited myself to general principles of patent law that are applicable to a consideration of gene-related inventions. Were I still Commissioner of Patents and Trademarks, I would not let electrical engineers anywhere near the examination of applications for patents for gene-related inventions, just as I would not let the USPTO's many Ph.D.s in microbiology or genetics near the examination of complex applications on computer architecture or programming. The same should hold true for attorneys who prosecute patent applications. That is why the U.S. patent system works so well. There can be no discrimination in the system by the field of technology—a principle enshrined in Article 27 of the World Trade Organization's Trade-Related Aspects of Intellectual Property. But we quite properly depend upon specialists to apply across-the-board general principles to very specialized technological fields of endeavor. This works very well in a system in which in the U.S., "everything under the sun made by humans" is patentable.

[15] *Amazon.com, Inc. v. Barnesandnoble.com, Inc.*, 239 F.3d 1343 (Fed. Cir. 2001).

3

Competition Policy in Patent Cases and Antitrust

Gerald Sobel, Esq.
Partner
Kaye Scholer LLP
New York, New York 10022-3598

0065-2660/03 $35.00

ABSTRACT

The article that follows examines the competition policy reflected in the decisions of the Court of Appeals for the Federal Circuit in its patent cases.[1] The court's views on this subject have been manifested most plainly in decisions that have transformed the law concerning infringement under the doctrine of equivalents and claim construction. In both categories, the court narrowed patent scope by reason of its desire to protect competitors. The article argues that the court's premise in prescribing narrower claim scope reflected an incomplete view of competition policy. The court's analysis overlooked the benefits to competition provided by patents, which stimulate inventions and their development. The article traces the development of antitrust jurisprudence and demonstrates how respect for the contribution of patents to competition and skepticism of free-riding has evolved, particularly beginning in the 1970s. The article draws a parallel between the Court's reasoning about competition policy, on the one hand, and the rejected views of Justices Hugo Black and William O. Douglas and abandoned patent-antitrust jurisprudence, on the other.

The Federal Circuit's decision in *Festo Corp. v. Shoketsu Kinzoku Kogyo Kabushiki Co., Ltd.*, 234 F.3d 558 (Fed. Cir. 2000), on the subject of equivalents is considered in the paper. In that decision, the majority adopted a new rule that completely barred infringement under the doctrine of equivalents of any claim limitation where, in prosecution, there had been a narrowing amendment relating to patentability. In the past, prosecution estoppel foreclosing equivalents had been subject to a "flexible bar," which, in some circumstances, allowed for equivalence notwithstanding such an amendment. The article points out that because almost all patents are amended during prosecution, the effect would be to allow widespread copying of patented inventions by trivial modifications of any narrowed claim limitation. The incentive to innovate in the future would be correspondingly diminished and the expectations of past patentees would be correspondingly altered.

[1] Published in 7 VA. J.L. TECH. 3 (Spring, 2002) and at www.vjolt.org. under copyright to the Journal.

Following the writing of the article, the Supreme Court decided *Festo* on *certiorari*, 122 S. Ct. 2519 (2002), and rejected the Federal Circuit's complete bar approach. The Court continued the flexible bar approach to equivalents, allowing for infringement despite a narrowing amendment. The doctrine of equivalents was explained in terms of "the nature of language," which "makes it impossible to capture the essence of a thing in a patent application." *Id.* at 1837. The Court announced a new rule, starting with a presumption that "the territory between the original claim and the amended claim" was disclaimed as to "readily known equivalents," *Id.* at 1842. The presumption could be overcome where:

> [1] "The equivalent may have been unforeseeable at the time of the application;" [2] "the rationale underlying the amendment may bear no more than a tangential relation to the equivalent in question;" or [3] "there may be some other reason suggesting that the patentee could not reasonably be expected to have described the insubstantial substitute in question." *Id.*

The Court recognized the effect of uncertainty from the doctrine of equivalents on competitors as a result of the flexible bar, but rejected that as a reason for a complete bar. *Id.* For example, "if competitors cannot be certain about a patent's extent, they may be deterred from engaging in legitimate manufactures outside its limits, or they may invest by mistake in competing products that the patent secures." *Id.* at 1837. The Court also quoted Justice Black's concern for competitors in his dissent objecting to the doctrine of equivalents in *Graver Tank & Mfg. Co. v. Linde Air Products Co.*, 339 U.S. 605 (1950). *Festo*, 122 S. Ct. at 1838. Despite a "delicate balance" between the interest of the patentees and those of the public, the Court squarely disagreed with the preference of the Federal Circuit, Justice Black, and competitors of the patentee. Each time the Court had addressed the subject over a 100-year period, it "affirmed the doctrine over dissents that urged a more certain rule." *Id.*

The Court's holding preserved placement of the burden of the uncertainty implicit in the doctrine of equivalents on competitors in favor of protecting patentees and the patent incentive to innovation. *Id.* At the same time, the effect of the Court's new rule was to impinge on the scope available to patentees. Requiring the inventor to write all knowable variations of his invention into the claims is a way of reducing the scope of equivalency. It is out of keeping with *Graver Tank*, which applied equivalents to an unclaimed variation described in the specification, as well as the thoughtful analyses of the Antitrust Division about the importance of a full patent reward. By the same token, applying equivalents to later-developed technology extends the patent to apply to the more creative competitor but inconsistently leaves the straightforward free-rider with a minor knowable variant outside the patent.

I. INTRODUCTION

The legal arena for competition extends to patents and, in particular, patent infringement litigation. Here competitors are often litigating over the right of one to compete with the other. Direct competitors square off, the patentee/innovator against the follow-on firms, which produce an improvement, a slightly different design that does the same thing, or a copy of the invention. Patent law determines the extent to which a second comer can emulate a patented product.

Motivated by a view of competition policy that reflects an overriding concern for the impact of patents on competitors of the patentee, the Federal Circuit has transformed patent law by restricting patent scope. It is the thesis of this paper that the court is applying a view of competition policy that is incomplete and has been abandoned in antitrust, where competition is the governing value. The Federal Circuit majority overlooks the role of patents in stimulating innovation and the competition engendered by it. This oversight is somewhat reminiscent of the views of Justices Black and Douglas in decisions of the Supreme Court in the 1940s and 1950s concerning both patents and antitrust. The Black and Douglas view was that patents were a special exception to a general scheme of competition and, as such, their scope was to be limited in favor of giving competitors a broad range. In antitrust, those views have been bypassed in favor of a policy which recognizes and emphasizes the role of patents as a stimulus for competition because of the incentive they create for inventors and investors.

We turn first to several fundamental changes made by the Federal Circuit in the law governing patent breadth driven by its view of competition policy, then to the parallel views of Justices Black and Douglas in the 1940s and 1950s, and finally to how competition policy has since been altered in the realm of antitrust in ways which undercut the Federal Circuit's premise.

II. THE FEDERAL CIRCUIT'S TRANSFORMATION OF THE LAW

A. Claim Construction

1. The Governing Law Before *Markman*

Prior to the Federal Circuit's 1995 decision in *Markman v. Westview Instruments, Inc.*,[2] the meaning of claims was routinely submitted to juries as questions of fact in light of expert testimony. The patent specification is required by statute "to enable any person skilled in the art . . . to make and use [it]."[3] The claims must "point out and distinctly claim the subject matter."[4] The meaning of a claim was typically a question of fact for the jury and a subject for expert testimony from those skilled in the art to which the patent was related.[5] The district courts risked reversible error by not considering expert testimony.[6] Jury verdicts concerning claim interpretation were reviewed by the Federal Circuit under the deferential standard[7] rather than *de novo*[8] review without deference to the decision below.

2. The Federal Circuit's *Markman* Decision

The Federal Circuit changed the law in two major respects in *Markman*. First, the court reclassified claim construction as a matter of law. This, according to the court's view, increases predictability by assigning the issue to the trial judge for decision rather than the jury[9] and allows the Federal Circuit to review the trial

---

[2] *Markman v. Westview Instruments, Inc.*, 52 F.3d 967 (Fed. Cir. 1995), *aff'd*, 517 U.S. 370 (1996).

[3] 35 U.S.C. § 112 (2001).

[4] *Id.*

[5] *See, e.g., Moeller v. Ionetics, Inc.*, 794 F.2d 653, 657 (Fed. Cir. 1986), *overruled by Markman*, 52 F.3d 967 (holding that the district court abused its discretion in excluding expert testimony on the meaning of the claims).

[6] In *Moeller*, for example, the Federal Circuit held that the district court abused its discretion in refusing to allow expert testimony as to the meaning of the claims directed to an electrode system for measuring cation concentration where it viewed "the interpretation of these claims as so 'simple' that the question could be resolved without expert testimony":

> In a patent case involving complex scientific principles, it is particularly helpful to see how those skilled in the art would interpret the claim. Indeed, the test of claim interpretation is directed to one skilled in the art, and it makes sense therefore to elicit testimony from such individuals. Though we do not establish that as a requirement in all cases, and leave it to the general discretion of the trial judge, we conclude that in this case the trial judge's failure to allow such testimony was an abuse of discretion.

Id. at 657 (citations omitted). Similarly, in *P. M. Palumbo v. Don-Joy Co.*, 762 F.2d 969, 975–76 (Fed. Cir. 1985), *overruled by Markman*, 52 F.3d 967, 975–76, the court, noting that expert testimony would be "helpful in ascertaining what one of ordinary skill in the art would consider," reversed the district court's grant of summary judgment and held that statements by one of skill in the art as to claim coverage could raise a genuine issue of material fact calling for a trial.

[7] *Markman*, 52 F.3d at 1018 ("no reasonable juror could have interpreted the claim in the fashion that supports the infringement finding. . . .").

[8] *See, e.g., Tol-O-Matic, Inc. v. Proma Produkt-Und Marketing Gesellschaft m.b.H*, 945 F.2d 1546, 1549–50 (Fed. Cir. 1991).

[9] *Markman*, 517 U.S. 370, 388–89 ("The construction of written instruments is one of those things that judges often do and are likely to do better than jurors unburdened by training in

court's determination *de novo* rather than by the deferential standard of review, which is applied to a jury's (or court's, when it acts as trier of fact) findings.

Second, in deciding the scope of patent claims, the court has made "intrinsic" evidence the "Bible" for construction. Though the test remained "what one of ordinary skill in the art at the time of the invention would have understood the term to mean,"[10] the publicly available claims and specification of the patent and its prosecution history were to govern against "extrinsic" evidence by experts on that subject.[11]

The court explained that it was motivated by its desire to protect competitors:

> [I]t is only fair (and statutorily required) that *competitors* be able to ascertain to a reasonable degree the scope of the patentee's right to exclude.... *They* may understand what is the scope of the patent owner's rights by obtaining the patent and prosecution history—"the undisputed public record,"—and applying established rules of construction to the language of the patent claim in the context of the patent. Moreover, *competitors* should be able to rest assured, if infringement litigation occurs, that a judge, trained in the law, will similarly analyze the text of the patent and its associated public record and apply the established rules of construction, and in that way arrive at the true and consistent scope of the patent owner's rights to be given legal effect.[12]

---

exegesis. Patent construction in particular 'is a special occupation, requiring, like all others, special training and practice. The judge, from his training and discipline, is more likely to give a proper interpretation to such instruments than a jury; and he is, therefore, more likely to be right....'") (citation omitted).

[10] *Markman*, 52 F.3d at 986.

[11] *Id.* at 979–81. Recently, a Federal Circuit case opened the door to "trustworthy extrinsic evidence" as a basis for foreclosing a meaning "inconsistent with clearly expressed, plainly apposite, and widely held understandings in the pertinent technical field." *Pitney Bowes v. Hewlett-Packard*, 182 F.3d 1298, 1308–09 (Fed. Cir. 1999). In any event, extrinsic evidence may always be examined to help [the district court] understand the underlying technology. *Id.* at 1308.

[12] *Markman*, 52 F.3d at 978–79 (citation omitted) (emphasis added). The facts of *Markman* were as follows: Markman invented a system to track articles of clothing and associated business transactions for use in the dry cleaning industry. *Id.* at 971–73. Markman's patent claims were drawn to an "inventory and control reporting system." *Id.* Westview produced and sold an electronic system capable of tracking cash and invoice totals, but not articles of clothing. *Id.* at 972–73. Markman sued Westview for infringement alleging that Westview's system, by virtue of tracking cash and invoices, fulfilled Markman's claim requirement of tracking "inventory." *Id.* Westview countered that cash and invoices do not constitute "inventory." *Id.* at 979–83. Based on its analysis of Markman's claim language, specification, and prosecution history, the Federal Circuit agreed with Westview's interpretation of the term "inventory." *Id.*

The Supreme Court affirmed, observing that "[i]t has long been understood that a patent must...'apprise the public of what is still open to them.'"[13]

Similarly, as the Federal Circuit later said on the same subject in *Vitronics*:

> *[C]ompetitors* are entitled to review the public record, apply the established rules of claim construction, ascertain the scope of the patentee's claimed invention and, thus, *design around the claimed invention.*[14]

The Federal Circuit later extended its procompetitor policy in this area by holding that when a claim remains ambiguous after consideration of the relevant evidence, and there is an equal choice between a broad and a narrower meaning of the claim, the narrow meaning governed.[15] Thus, in *Athletic Alternatives*, the Federal Circuit explained that the meaning of the claim term at issue had two conflicting possibilities, since the specification was silent and the prosecution history had "[t]wo strong and contradictory interpretive strands."[16] Confronted with "two equally plausible meanings of Claim 1," the court adopted the narrower meaning in view of the "fair notice function" of claims.[17]

B. Infringement by Equivalents and Prosecution History Estoppel

The doctrine of equivalents presents the most striking test of attitudes toward competitors versus the patentee. It assigns liability for utilizing a version of the patented invention even though the patent claims have been literally avoided. The push and pull of patentees' scope of protection and incentive to invent

[13] *Markman*, 517 U.S. at 373 (regarding notice to the public) (citing 141 U.S. 419 (1891)).

[14] *Vitronics Corp. v. Conceptronic, Inc.*, 90 F.3d 1576, 1583 (Fed. Cir. 1996) ("In those cases where the public record unambiguously describes the scope of the patented invention, reliance on any extrinsic evidence is improper. The claims, specification, and file history, rather than the extrinsic evidence, constitute the public record of the patentee's claim, a record on which the public is entitled to rely.") (citations omitted) (emphasis added).

[15] *Athletic Alternatives, Inc. v. Prince Mfg., Inc.*, 73 F.3d 1573, 1580–81 (Fed. Cir. 1996); *see also Ethicon Endo-Surgery, Inc. v. United States Surgical Corp.*, 93 F.3d 1572, 1581–82 (Fed. Cir. 1996) ("[t]o the extent that the claim is ambiguous, a narrow reading which excludes the ambiguously covered subject matter must be adopted....[W]e conclude that the term 'pusher assembly' unambiguously describes only the pusher bars and the cam bar retainer, but does not unambiguously cover any other portions of the '519 patent's firing means. In view of the above, Ethicon's claim 24 would read on a lockout which engaged the cam bar retainer, such as U.S. Surgical's open linear cutters, while it would not read on a lockout which engaged any other portion of the firing means, such as U.S. Surgical's endoscopic lockouts.").

[16] *Athletic Alternatives*, 73 F.3d at 1580.

[17] *Id.* at 1581.

versus competitors' freedom to operate play out in two areas: first, the test for equivalency, and second, prosecution history estoppel, the extent of surrender of scope of equivalency by reason of events in prosecution.

1. The Governing Law Before *Festo* as to Equivalents

The Supreme Court's first discussion of the doctrine of equivalents was in the nineteenth century in *Winans v. Denmead.*[18] The Court recognized the ease with which minor changes in a patent could avoid literal infringement and gave broader protection: "Where form and substance are inseparable, it is enough to look at the form only. Where they are separable; where the whole substance of the invention may be copied in a different form, it is the duty of courts and juries to look through the form for the substance of the invention...."[19] The governing decision of the Supreme Court between 1950 and 1997 was *Graver Tank.*[20] The Supreme Court held there, with Justices Black and Douglas dissenting, that infringement occurred even though the claim was not literally satisfied under the doctrine of equivalents.[21] As the Supreme Court wrote, this scope was allowed in order to hinder the "unscrupulous copyist" and to assure the patentee of the benefit of his invention.[22] Thus, a person could not imitate a patented invention by making some "unimportant and insubstantial" changes to the claimed invention.[23] The rationale for not limiting patents to their literal scope was described in practical terms: It was necessary to avert making patents into "a hollow and useless thing."[24] To prohibit only literal duplication "would place the inventor at the mercy of verbalism and would be subordinating substance to form."[25] The Court expressed concern for loss to the patentee of the "benefit of his invention" and for "foster[ing] concealment rather than disclosure of inventions, which is one of the primary purposes of the patent system."[26]

[18] *Winans v. Denmead,* 56 U.S. 330 (1853). The patent claimed a circular railroad car with a conical design. The accused device was octagonal but otherwise similar to a conical design and achieved the same benefits. *Id.* at 340.

[19] *Id.* at 343.

[20] *Graver Tank & Mfg. Co., Inc. v. Linde Air Prods. Co.,* 339 U.S. 605 (1950). The patent claims at issue involved an electrical welding composition employing a combination of an alkaline earth metal silicate and any other silicate. The Court held that the use of manganese (a non-alkaline earth metal) instead of magnesium (an alkaline earth metal) was a sufficiently insubstantial change which justified application of the doctrine of equivalents to find infringement.

[21] *Id.* at 612.

[22] *Id.* at 607.

[23] *Id.* at 607; *see also Lear Siegler, Inc. v. Sealey Mattress Co.,* 873 F.2d 1422, 1425 (Fed. Cir. 1989) (citing *Graver Tank,* 339 U.S. at 607).

[24] *Graver Tank,* 339 U.S. at 607.

[25] *Id.*

[26] *Id.*

Infringement by equivalency was held to apply "if two devices do the same work in substantially the same way, and accomplish substantially the same result, [because] they are the same, even though they differ in name, form, or shape."[27]

In 1995, in *Hilton Davis Chemical Co. v. Warner-Jenkinson Co.*,[28] the Federal Circuit *en banc* reconsidered the doctrine. The court affirmed, by seven to five, a jury's finding of infringement by equivalents and ruled that the doctrine of equivalents, which "requires proof of insubstantial differences" between the claimed and accused products, was an issue of fact for the jury[29] and is not a discretionary remedy.[30] The court remanded for consideration of possible application of prosecution estoppel.[31] As for the equivalency test, the court observed that the function-way-result test was not " 'the' test" but "often suffices."[32] The test for a finding of infringement under the doctrine of equivalents was declared to be a finding of "insubstantial differences" between the claimed and accused products or processes. The measure is objective, not subjective.[33]

On review, in 1997[34] the Supreme Court declined Warner-Jenkinson's invitation to "speak the death" of the doctrine of equivalents.[35] However, the Court found that "[t]here can be no denying that the doctrine of equivalents, when applied broadly, conflicts with the definitional and public-notice functions of the statutory claiming requirement."[36] The Court's response was to hold

[27] *Id.*

[28] *Hilton Davis Chemical Co. v. Warner-Jenkinson Co.*, 62 F.3d 1512 (Fed. Cir. 1995), *rev'd on other grounds*, 520 U.S. 17 (1997). The Court reaffirmed the doctrine of equivalents but reversed the Federal Circuit for a determination of whether prosecution history estoppel limited the scope of equivalents and thereby precluded a finding of infringement.

[29] *Hilton Davis*, 62 F.3d at 1522. The Federal Circuit "reviews a jury verdict on the fact question of infringement under the doctrine of equivalents for prejudicial error in the jury instructions, and lack of substantial evidence supporting the verdict." *Id.* at 1522. (citations omitted). If the Federal Circuit reviews a question of infringement from a bench trial, as in *Graver Tank*, it reviews the trial court's infringement finding for "clear error." *Id.* at 1521.

[30] *Id.* at 1521–22. *See also Lifescan Inc. v. Home Diagnostics Inc.*, 76 F.3d 358 (Fed. Cir. 1996) (where there is substantial evidence to support a factual conclusion, summary judgment is improper and the issue should be weighed by the trier of fact).

[31] The claim at issue was to a dye purification process utilizing a pressure of 200 to 400 p.s.i.g. and a pH from approximately 6.0 to 9.0. The jury found infringement by a pH of 5.0 under the doctrine of equivalents. The patentee's claims at issue recited a pH "from approximately 6.0 to 9.0." Warner-Jenkinson at times, according to the court, used a lower pH of 5.0 in its process. The claims at issue also recited a pressure of "approximately 200 to 400 p.s.i.g." Warner-Jenkinson used a pressure somewhere in a range of 200 to nearly 500 p.s.i.g. *Hilton Davis*, 62 F.3d at 1524.

[32] *Id.* at 1518.

[33] *Id.* at 1518–19.

[34] *Warner-Jenkinson Co. v. Hilton Davis Chem. Co.*, 520 U.S. 17 (1997).

[35] *Id.* at 21.

[36] *Id.* at 28.

(as the Federal Circuit had for many years)[37] that the determination of infringement under the doctrine is to be made on a claim-element-by-claim-element basis and not with respect to the invention as a whole: "[e]ach element contained in a patent claim is deemed material to defining the scope of the patented invention, and thus the doctrine of equivalents must be applied to individual elements of the claim, not to the invention as a whole."[38] With that requirement, the Supreme Court reasoned that "the doctrine will not vitiate the central functions of the patent claims themselves."[39] The Court stated:

> An analysis of the role played by each element in the context of the specific patent claim will thus inform the inquiry as to whether a substitute element matches the function, way, and result of the claimed element, or whether the substitute element plays a role substantially different from the claimed element.[40]

Further development in the particular wording of the test was left by the Supreme Court to the Federal Circuit.[41]

Before *Warner-Jenkinson*, the Federal Circuit had recognized the tension between the doctrine of equivalents and encouraging competitors to design around patents, but similarly opted to preserve the doctrine of equivalents.[42]

[37] In *Intellicall, Inc. v. Phonometrics, Inc.*, 952 F.2d 1384 (Fed. Cir. 1992), the court explained: "Phonometrics argues that the accused devices are equivalent overall to the claimed invention. That view of the doctrine of equivalents was rejected in *Pennwalt Corp. v. Durand-Wayland, Inc.*... As this court has *repeatedly* stated, infringement requires that *every limitation* of a claim be met literally or by a substantial equivalent." *Id.* at 1389 (emphasis in original) (citations omitted). The "all elements" rule had been adopted by the Federal Circuit in *Pennwalt Corp. v. Durand-Wayland, Inc.*, 833 F.2d 931, 938 (Fed. Cir. 1987), *cert. denied*, 485 U.S. 961, 1009 (1988), where the court held that a product does not infringe if it is missing an equivalent of a single limitation in a claim. The court decision stated that: "a court may not, under the guise of applying the doctrine of equivalents, erase a plethora of meaningful structural and functional limitations of the claim on which the public is entitled to rely in avoiding infringement..." *Id.* at 935. The dissent stated: "On the one hand, there is the historic right of affording the public fair notice of what the patentee regards as his claimed invention in order to allow competitors to avoid actions which infringe the patent and to permit "designing around" the patent. On the other hand, equally important to the statutory purpose of encouraging progress in the useful arts, is the policy of affording the patent owner complete and fair protection of what was invented."

Id. at 945. (Bennett, J., dissenting) (citations omitted).

[38] *Warner-Jenkinson*, 520 U.S. at 29. Before *Pennwalt*, the doctrine of equivalents was applied to the "claimed invention as a whole." *Hughes Aircraft Co. v. United States*, 717 F.2d 1351, 1364 (Fed. Cir. 1983) ("*Hughes I*").

[39] *Warner-Jenkinson*, 520 U.S. at 30.

[40] *Id.* at 40.

[41] *Id.*

[42] *See also Haynes Int'l, Inc. v. Jessop Steel Co.*, 8 F.3d 1573, 1581 (Fed. Cir. 1994) (Newman, J., concurring) ("Patent protection, if easily circumvented, does not enhance the incentive for industrial innovation." Richard C. Levin *et al.*, in *Appropriating the Returns from*

The Federal Circuit observed that the doctrine of equivalents represents an exception to "the requirement that the claims define the metes and bounds of the patent protection," but "we hearken to the wisdom of the Court in *Graver Tank*, that the purpose of the rule is 'to temper unsparing logic' and thus to serve the greater interest of justice."[43]

2. The Governing Law Before *Festo* as to Prosecution Estoppel

The Supreme Court in *Warner-Jenkinson* also addressed the companion doctrine of prosecution history estoppel.[44] Prosecution estoppel is designed to prevent the patentee from recapturing by equivalence what it surrendered in prosecution to obtain the patent. The Supreme Court held that the doctrine of prosecution history estoppel limits the reach of the doctrine of equivalents where claimed subject matter was surrendered "to avoid the prior art, or otherwise to address a specific concern—such as obviousness—that arguably would have rendered the claimed subject matter unpatentable."[45]

Industrial Research and Development, 3 BROOKINGS PAPERS ECONOMIC ACTIVITY 783 (1987), thus criticizes the effectiveness of the patent system as an innovation incentive. These policy issues are of particular concern to this court, which is charged with the body of law whose purpose is to support creativity and innovation. If it is desired to enlarge the restrictions on a patentee's recourse to the doctrine of equivalents, by stretching the grounds of estoppel, this should be explored by the technology community, not legislated by this court."); *State Indus., Inc. v. A. O. Smith Corp.*, 751 F.2d 1226, 1235–36 (Fed. Cir. 1985) ("Conduct such as Smith's, involving keeping track of a competitor's products and designing new and possibly better or cheaper functional equivalents is the stuff of which competition is made and is supposed to benefit the consumer. One of the benefits of a patent system is its so-called 'negative incentive' to 'design around' a competitor's products, even when they are patented, thus bringing a steady flow of innovations to the marketplace. It should not be discouraged by punitive damage awards except in cases where conduct is so obnoxious as clearly to call for them. The world of competition is full of 'fair fights,' of which this suit seems to be one."); *London v. Carson Pirie Scott & Co.*, 946 F.2d 1534, 1538 (Fed. Cir. 1991) ("Although designing or inventing around patents to make new inventions is encouraged, piracy is not. Thus, where an infringer, instead of inventing around a patent by making a substantial change, merely makes an insubstantial change, essentially misappropriating or even 'stealing' the patented invention, infringement may lie under the doctrine of equivalents.") (citations omitted).

[43] *Texas Instruments Inc. v. United States Int'l Trade Comm.*, 805 F.2d 1558, 1572 (Fed. Cir. 1986).

[44] The patent before the Court in *Warner-Jenkinson* disclosed a process for purifying dyes. *Warner-Jenkinson*, 520 U.S. at 21. During prosecution, the patentee amended the claims to recite that the process is carried out "at a pH from approximately 6.0 to 9.0." *Id.* at 22. The accused process was carried out at a pH of 5.0. *Id.* at 23. In light of these facts, the Supreme Court embarked on an "endeavor to clarify the proper scope of the doctrine" of equivalents. *Id.* at 21. The court noted that, although the parties did not dispute that the upper pH limit of 9.0 was added to avoid the prior art, "the reason for adding the lower limit of 6.0 is unclear." *Id.* at 32.

[45] *Id.* at 30. The Court implied that prosecution history estoppel may not apply to claim amendments made for other reasons, citing a number of its earlier decisions which involved claim amendments made to overcome prior art. *See, e.g., Exhibit Supply Co. v. Ace Patents Corp.*, 315 U.S.

The Court in *Warner-Jenkinson* set forth a new procedure for determining whether there is a basis to apply prosecution history estoppel. This procedure created a new burden on the patent-holder to show that an amendment was *not* made for a reason that would give rise to an estoppel. Where the file history does not provide the reason for the amendment, there is a rebuttable presumption against the patentee that the amendment causes an estoppel.[46]

The rule typically followed by the Federal Circuit was that the doctrine of equivalents was not foreclosed from applying to an amended claim, including claims amended to avoid prior art. The Federal Circuit had expressly recognized that it was necessary to scrutinize the prosecution to determine what had been surrendered. The standard for determining what was surrendered in applying prosecution history estoppel had been held by the Federal Circuit to be "an objective one, measured from the vantage point of what a competitor was reasonably entitled to conclude, from the prosecution history, that the applicant gave

---

126, 136 (1942); *Keystone Driller Co. v. Northwest Eng'g Corp.*, 294 U.S. 42, 48 (1935); *Smith v. Magic City Kennel Club, Inc.*, 282 U.S. 784, 788 (1931); *Computing Scale Co. of America v. Automatic Scale Co.*, 204 U.S. 609, 618-20 (1907); *Hubbell v. United States*, 179 U.S. 77, 83 (1900); *Sutter v. Robinson*, 119 U.S. 530, 541 (1886).

The Court stated that reasons for an amendment which would avoid prosecution history estoppel were provided in the amicus brief filed by the United States. *Warner-Jenkinson*, 520 U.S. at 31–32 ("As the United States informs us, there are a variety of other reasons why the PTO may request a change in claim language.") (citing Brief for United States as amicus curiae at 22–23). In its amicus brief, 1996 WL 172221 (Apr. 11, 1996), the United States explained that amendments to reflect the proper scope of enablement or to add precision to the claims—as opposed to amendments to avoid the prior art—do not necessarily, but may, raise an estoppel. *Warner-Jenkinson*, 520 U.S. at 22–23.

[46] The Supreme Court explained

> [W]e think the better rule is to place the burden on the patent holder to establish the reason for an amendment required during patent prosecution. The court then would decide whether that reason is sufficient to overcome prosecution history estoppel as a bar to application of the doctrine of equivalents to the element added by that amendment. Where no explanation is established, however, the court should presume that the patent applicant had a substantial reason related to patentability for including the limiting element added by [an] amendment. In those circumstances, prosecution history estoppel would bar the application of the doctrine of equivalents as to that element. The presumption we have described, [that] one [is] subject to rebuttal if an appropriate reason for a required amendment is established, gives proper deference to the role of claims in defining an invention and providing public notice, and to the primacy of the PTO in ensuring that the claims allowed cover only subject matter that is properly patentable in a proffered patent application.

Warner-Jenkinson, 520 U.S. at 33–34. The Court remanded for the Federal Circuit to determine whether a reason exists for applicants' amending the claims to set the lower limit of the pH range at 6.0 and whether that reason would give rise to prosecution history estoppel precluding infringement by a pH of 5.0. *Id.* at 34.

up to procure issuance of the patent."[47] Under the doctrine of equivalents, an applicant was estopped from recovering subject matter surrendered during patent prosecution to secure issuance of the claim.[48] The court said in 1998 in *Litton*: "[t]he common practice of amending a claim during prosecution, even amending to overcome prior art, does not necessarily surrender all subject matter beyond the literal scope of the amended claim limitation."[49] The scope of subject matter surrendered by an amendment extended to "that which was deemed unpatentable in view of the prior art" and " 'trivial' variations of such prior art features."[50]

3. The *Festo* Decision

The Federal Circuit majority, in an *en banc* decision in *Festo Corp. v. Shokestu Kinzoku Kogyo Kabushiki Co.*,[51] took another major step toward limiting patent scope based on its view of what is desirable for competitors. The court abolished the doctrine of equivalents for the large number of patent claim limitations for which there had been a narrowing amendment[52] for any reason relating to any statutory criterion for obtaining a patent.[53] As for these claim limitations,

[47] *Hoganas AB v. Dresser Indus., Inc.*, 9 F.3d 948, 952 (1993).

[48] *See, e.g., Hughes Aircraft Co. v. United States*, 717 F.2d 1351 (Fed. Cir. 1983), *overruled by Festo Corp. v. Shoketsu Kinzoku Kogyo Kabushiki Co.*, 234 F.3d 558 (Fed. Cir. 2000).

[49] *Litton Sys. Inc. v. Honeywell Inc.*, 140 F.3d 1449, 1455 (Fed. Cir. 1998).

[50] *Id.* at 1462.

[51] *Festo Corp. v. Shoketsu Kinzoku Kogyo Kabushiki Co.*, 234 F.3d 558, 619 (Fed. Cir. 2000), *cert. granted*, 121 S. Ct. 2519 (2001).

[52] *Id.* at 586. The facts in *Festo* were as follows: Festo Corporation sued Shoketsu Kinzoku Kogyo Kabushiki Co. (SMC) alleging infringement of its two patents on magnetically coupled cylinders. *Id.* at 578. The Festo patents claimed devices with a pair of sealing rings. The Stoll patent, in addition, claimed a sleeve made of a magnetized material. *Id.* at 580. SMC's devices contained a single seal and had an outer portion of the sleeves made of an aluminum alloy, not a magnetizable material. *Id.* at 582. Stoll had amended claim 1 during prosecution for § 112 to recite "sealing rings" and "a cylindrical sleeve made of a magnetized material." *Id.* at 583. During reexamination of the Carroll patent, Festo canceled claim 1, which did not recite "a sealing ring," and added claim 9, which did recite such element. *Id.* at 590. The district court found defendants to have infringed both patents under the doctrine of equivalents. *Id.* at 585. The decision was affirmed on appeal. *Festo Corp. v. Shoketsu Kinzoku Kogyo Kabushiki Co.*, 72 F.3d 857 (Fed. Cir. 1995). SMC had petitioned the Supreme Court for a writ of certiorari, which the court granted, vacating the judgment and remanding the case for consideration in light of *Warner-Jenkinson*. *See Shoketsu Kinzoku Kogyo v. Festo Corp.* 520 U.S. 1111 (1997). On remand, the Federal Circuit, hearing the case *en banc*, reversed the judgment and barred the application of the doctrine of equivalents on the ground of prosecution history estoppel. The court found that the amendments in the Stoll patent that added a "magnetizable sleeve" or a "sealing ring" narrowed the literal scope of the claims. *Festo*, 234 F.3d at 588–89. Similarly, in the Carroll patent, the addition of the "sealing ring" narrowed the scope of the claims. The Federal Circuit held that the amended claim elements were not entitled to a range of equivalents and could not be infringed. *Id.* at 590–91.

[53] The statutory requirements of patentability extend beyond distinguishing over prior art and encompass definiteness of claim language, disclosure of the best mode, enablement, and written description—virtually all amendments that occur in prosecution. *Id.* at 566.

everything but their literal terms was automatically surrendered. Since patent applications are routinely amended during the examination process,[54] equivalence protection under *Festo* is available only in the relatively small number of instances where there has been no amendment of the claims during patent prosecution. The court acknowledged that this was "a different conclusion" than its earlier rulings.[55]

The majority explained its concern for competitors:

> Thus, under the complete bar approach, technological advances that would have lain in the unknown, undefined zone around the literal terms of a narrowed claim under the flexible bar approach will not go wasted and undeveloped due to fear of litigation. The public will be free to improve on the patented technology and design around it without being inhibited by the threat of a lawsuit because the changes could possibly fall within the scope of equivalents left after a claim element has been narrowed by amendment for a reason related to patentability. This certainty will stimulate investment in improvements and design-arounds because the risk of infringement will be easier to determine.[56]

The freedom the new rule gave to competitors to copy with minor variation from the amended patent claims was the subject of extensive scrutiny by both the majority and minority opinions. The majority acknowledged the impact on patentees, but weighed the benefit from competitors more highly.[57] The rule announced by the *Festo* majority, as Judge Michel explained in dissent, facilitates copying of an invention because an infringer can review a patent prosecution history, identify an amended limitation, and then make a trivial change to bring the product outside of the literal meaning of the amended claim

[54] As the Federal Circuit itself has observed, "[a]mendment of claims is a common practice in prosecution of patent applications," and "comparatively few claims [are] allowed exactly as originally filed...." *Hughes 1*, 717 F.2d at 1363; *accord Festo*, 234 F.3d at 638 n.21 ("[F]or simple inventions, at most 10–15% of patents are granted without claim amendment.... For complex inventions the percentage of unamended applications is vanishingly small.") (Newman, J., concurring in part and dissenting in part).

[55] *Festo*, 234 F.3d at 574.

[56] *Id.* at 577.

[57] "Although a flexible bar affords the patentee more protection under the doctrine of equivalents, we do not believe that the benefit outweights the costs of uncertainty. The Supreme Court noted in *Warner-Jenkinson* that the doctrine of equivalents has 'taken on a life of its own, unbounded by the patent claims.' *Warner-Jenkinson*, 520 U.S. at 28–29. A complete bar reins in the doctrine of equivalents, making claim scope more discernible and preserving the notice function of claims." *Festo*, 234 F.3d at 578.

limitation.[58] The concurring opinion by Judge Lourie responded by acknowledging that "[I]n the future, a competitor may more closely approach the limits of the claims in a patent in which a narrowing amendment has been made without fear of liability."[59] Judge Lourie favored the benefit from competitors over "occasional injustices" to patentees: "I believe that such occasional injustices will be greatly outnumbered by competitors who will be able to introduce innovative products outside the scope of claims without fear of unjustified, protracted, and expensive litigation."[60]

The dissenters had a negative view of the ease with which competitors could rely upon minor variations and "free ride" on patents. Judge Rader's dissent identified troubling consequences in light of "the primary role of the doctrine,"[61] notably in rapidly developing technology fields for "after-arising technology":[62]

> Rather, the majority's new bright line rule, by constraining limitations amended for a statutory purpose to their literal terms, is likely to encourage insubstantial changes to an established product rather than investment in breakthrough technological advancements. Such a rule, therefore, promotes free riding and undercuts the return on a patentee's investment.[63]

[58] *Festo*, 234 F.3d at 616–17. "Under the majority's approach, anyone who wants to steal a patentee's technology need only review the prosecution history to identify patentability-related amendments, and then make a trivial modification to that part of its product corresponding to an amended claim limitation. All the other limitations may be copied precisely. The competitor will then be free to make, use, or sell an insubstantial variant of the patentee's invention." *Id.* at 600–01.

[59] *Festo*, 234 F.3d at 597 (Lourie, J., concurring).

[60] *Id.*

[61] *Festo*, 234 F.3d at 619 (Rader, J., concurring in part and dissenting in part).

[62] *Id.* In *Hughes Aircraft v. United States*, the Federal Circuit applied the doctrine of equivalents to reach technological alternatives not developed until after the patent application. The court held in *Hughes I* that a claim to a satellite navigation system in which the ground crew used data transmitted from the satellite to calculate existing and desired orientation was infringed under the doctrine of equivalents by a satellite which used an on-board computer to calculate the orientations. *Hughes I*, 140 F.3d at 1475. The court decided that, at the time of infringement, this was an insubstantial change in the claimed way. The court observed: "[T]he change in the S/E devices was the result of a technological advance not available until after the patent issued." *Id.* "This is a case in which a 'subsequent change in the state of the art, such as later-developed technology, obfuscated the significance of the limitation at the time of its incorporation into the claim.'" *Id.*

[63] *Festo*, 234 F.3d at 627 (Linn, J., concurring in part and dissenting in part, joined by Rader, J.).

Judge Michel explained the impact in biotechnology: "A protein molecule can only be claimed as the complete and specific sequence of amino acids comprising the protein. *See* 37 C.F.R. § 1.821 (2000). The particular amino acids that comprise a protein chain are frequently interchangeable with other amino acids without changing the protein or its functions. As our court noted with respect to a patent claiming the protein erythropietin, "over 3,600 different protein analogs can be

Judge Newman, in dissent, observed the same conflict between "the creation and commercialization of new technologies" and the "appropriation of the creative product":[64]

> The inventor and the imitator are affected by quite different economic considerations. The innovator takes the risk of commercial success or failure of new things in new markets—the risk of unfulfilled expectations, obsolescence, regulation, technologic failure. The imitator bears none of these risks; he is interested only in the successful products, not in the failures; he is interested only in the profitable products, not the marginal ones; he moves in only after the invention has been made and tested and the market developed, and can operate at lower margins. The patent system provides weight on the side of the innovator, aided by the doctrine of equivalents and its inhibition of close copying, establishing an incentive whose value has been tested by time.[65]

Judge Newman singled out the majority's "error of the 'simplified model,' which assumes a continuing supply of new products, and ignores the prior steps of invention and commercialization."[66]

4. Other Federal Circuit Efforts to Confine or Eliminate the Doctrine of Equivalents

In a process begun before *Warner-Jenkison*, the Federal Circuit had grafted several limiting doctrines onto the determination of equivalency.

---

made by substituting interchangeable acids at only a single amino acid position, and over a million different analogs can be made by substituting three amino acids." *Amgen, Inc. v. Chugai Pharm. Co.*, 927 F.2d 1200, 1213, 18 U.S.P.Q.2D (BNA) 1016, 1026 (Fed. Cir. 1991). Many such analogs are functionally identical to the claimed protein. Thus, a competitor seeking to make, use, or sell a protein that is protected by a patent containing an amended claim limitation will only have to substitute at a particular location in the chain an interchangeable amino acid for the particular amino acid recited in the patent claim as occupying that location." *Festo*, 234 F.3d at 617 (Michel, J., concurring in part and dissenting in part).

Judge Lourie responded: "As for the biotechnology example hypothesized by one of the dissenters, I believe the concern is largely theoretical. The first inventors in a field are only entitled to claim what they can describe and enable, and I am confident that competent patent attorneys can readily craft their claims to cover that subject matter so that estoppel can be avoided. Moreover, subsequent inventors will be better able to find and develop improved products without fear of lawsuits. Predictability will be enhanced." *Festo*, 234 F.3d at 597–98 (Lourie, J., concurring).

[64] *Festo*, 234 F.3d at 640 (Newman, J., concurring in part and dissenting in part).

[65] *Id.*

[66] *Id.* at 641.

a. Specific Exclusion

The Federal Circuit's decisions preceding *Warner-Jenkinson* include cases that held that the doctrine of equivalents cannot be used to cover structure in an accused device that the patent claim language, as the court characterized it, "specifically excludes" from a claim element. The doctrine was hard to distinguish from the exclusion of everything not literally claimed. One such case is *Dolly. Inc. v. Spalding & Evenflo Cos.*[67] The claim in *Dolly* covered a portable, adjustable chair for children that required a back panel, a seat panel, two side panels, and "a stable rigid frame which is formed in part from said side panels and which along with said seat panel and said back panel provides a body supporting feature. . . ."[68] The Federal Circuit held that the claim language specifically required that the "stable rigid frame [be] formed of components *other than* the seat and back panels."[69] Because the claim language specifically excluded the back panel and the seat panel from being part of the stable rigid frame ("along with said seat panel and said back panel") the patentee could not assert that the back and seat panel in an alleged infringer's chair without a stable rigid frame was an equivalent to the stable-rigid-frame claim element.[70]

[67] *Dolly, Inc. v. Spalding Evenflo Cos.*, 16 F.3d 394 (Fed. Cir. 1994).

[68] *Id.* at 396.

[69] *Id.* at 399 (emphasis added).

[70] *Id.* Another case in this line is *Athletic Alternatives*. The claim related to the pattern of strings on a tennis racket, called a "splay pattern." *Athletic Alternatives, Inc. v. Prince Mfg., Inc.*, 73 F.3d 1573, 1577–78 (Fed. Cir. 1996). The Federal Circuit interpreted the claim to require a "splay pattern" where the distance between strings must have at least three difference values: "a minimum, a maximum, and at least one intermediate value." *Id.* at 1581. In contrast, the tennis racket accused of infringement had a splay pattern with only two distances between adjacent strings. Under this claim interpretation, the Federal Circuit held as a matter of law, that "the [accused tennis racket's] two-distance splayed string system was 'specifically excluded from the scope of the claims.'" *Id.* at 1582 (citation omitted). The intermediate value string structure specified by the claim was absent. The court stated

> Were we to allow [plaintiff] successfully to assert the broader of the two senses of 'between' against [defendant], we would undermine the fair notice function of the requirement that the patentee distinctly claim the subject matter disclosed in the patent from which he can exclude others temporarily. Where there is an equal choice between a broader and a narrower meaning of a claim, and there is an enabling disclosure that indicates that the applicant is at least entitled to a claim having the narrower meaning, we consider the notice function of the claim to be best served by adopting the narrower meaning.

Athletic Alternatives, 73 F.3d at 1581; *see also Wiener v. NEC Elecs., Inc.*, 102 F.3d 534, 541–42 (Fed. Cir. 1996), *overruled by Cybor Corp. v. FAS Techs.*, 138 F.3d 1448 (Fed. Cir. 1998) (Noting that words of claim "specifically excluded" the accused memory. "Even if the accused device performs a substantially similar function and reaches a substantially similar result, this court cannot overlook the clear language of this limitation of the claims.") (citation omitted). In *Eastman Kodak*, the Federal Circuit, citing *Dolly*, held that a claim which required that a process be performed in an "inert gas atmosphere," where the specification referred to an "airtight seal," specifically excluded the process with heated air as an equivalent. The court said that the specification "suggests. . .that the. . .[reaction] should take

If each claim were to "specifically exclude" all alternatives not literally within it, the doctrine of equivalents would disappear. Accordingly, in *Ethicon Endo-Surgery*, the Federal Circuit retreated from its holdings that the words of the claims "specifically excluded" the device accused of equivalence.[71] "[A]ny analysis of infringement under the doctrine of equivalents necessarily deals with subject matter that is 'beyond,' 'ignored' by, and not included in the literal scope of a claim."[72] The court shifted its explanation of the summary judgments in *Dolly* to a factual determination that no reasonable jury could have found infringement by equivalence.

b. Foreseeability

The Federal Circuit invoked a test of foreseeability in another effort to limit the scope of equivalency. In *Sage Prods., Inc. v. Devon Indus., Inc.*, the Federal Circuit, in dictum, suggested that foreseeable alternatives not claimed should be excluded from equivalence.[73] "It is the patentee [not the public] who must bear the cost of its failure to seek protection for this foreseeable alteration of its claimed structure."[74] This dictum in *Sage* is at odds with the Supreme Court's decision in *Graver Tank*, where the Court held that the accused element was equivalent to the element found in the specification and expressly relied on two prior art patents, which taught the use of the accused element as a substitute for the claimed element.[75] Under the rationale suggested in *Sage*, because the prior art taught that the accused element was a substitute for the claimed element, such a substitution necessarily would have been foreseeable and therefore would have precluded application of the doctrine of equivalents. The Federal Circuit itself said the opposite in the earlier *Pall* case: "It is not controlling whether the inventor foresaw and described [the] potential equivalent at the time the patent application was filed."[76]

---

place without air." *Eastman Kodak Co. v. Goodyear Tire & Rubber Co.*, 114 F.3d 1547, 1560 (Fed. Cir. 1997), *overruled by Cybor Corp. v. FAS Techs.*, 138 F.3d 1448 (Fed. Cir. 1998).

[71] *Ethicon Endo-Surgery, Inc. v. United States Surgical Corp.*, 149 F.3d 1309 (Fed. Cir. 1998).

[72] *Id.* at 1317.

[73] *Sage Prods., Inc. v. Devon Indus., Inc.* 126 F.3d 1420 (Fed. Cir. 1997). The patent "discloses a disposal container that allows a user to deposit hazardous medical waste without touching waste already in the container." *Id.* at 1422. The court rejected the patentee's argument "that having two constrictions below the top of the container is the same, for purposes of infringement, as having one constriction above and one constriction below" because the patentee was seeking "to remove entirely" clear structural limitations, by reading out the top of the container and over said slot claim limitations. *Id.* at 1424. In addition to relying on the literal words of the claim, the court cited the lack of evidence of an insubstantial difference. *Id.*

[74] *Id.* at 1425 (citations omitted).

[75] *Graver Tank*, 339 U.S. at 611–12.

[76] *Pall Corp. v. Micron Separations*, 66 F.3d 1211, 1220 (Fed. Cir. 1995) (citation omitted).

c. The Patent Specification Precludes Equivalence[77]

In *Maxwell v. Baker, Inc.*,[78] the Federal Circuit held that a patentee may not obtain coverage under the doctrine of equivalents of subject matter described in the specification. The court was concerned about encouraging a patent applicant "to present a broad disclosure in the specification of the application and file narrow claims, avoiding examination of broader claims that the applicant could have filed. . . ."[79]

The Supreme Court in *Warner-Jenkinson*, however, suggested the opposite when it rejected the argument that "the doctrine of equivalents should be limited to equivalents that are disclosed within the patent. . . ."[80] The Court implied that disclosure of an alternative does not preclude being reached by equivalents. The Federal Circuit has since retreated from *Maxwell* in a

[77] The court's sympathetic view of the latitude available to competitors of patentees extends to other areas as well. An example is the court's recent decision holding that a first inventor's demonstration that a method of treatment was effective for cancer, rather than ineffective, was in the public domain. In *Bristol-Myers Squibb Co. v. Ben Venue Labs., Inc.*, 246 F.3d 1368 (Fed. Cir. 2001), the Federal Circuit recently applied the inherency doctrine to deny patentability to discoveries of a new biological activity and result. The claim to a cancer treatment method was held to be inherently anticipated by a prior art reference disclosing the use of taxol within the claimed dosage range to treat cancer, but reported that no antitumor response was observed in the failed experiment. *Id.* at 1372. The court rejected Bristol's argument that "its inventors achieved success, where Kris had assertedly failed. . . ." *Id.* at 1380. The court affirmed summary judgment invalidating claims to a method of treating cancer with an efficacious amount, within a stated dosage range, of taxol (an anticancer drug derived from the Pacific Yew tree). *Id.*; *see also Eli Lilly v. Barr Laboratories*, 251 F.3d 955, 970 (Fed. Cir. 2001) (patent claim invalid for double patenting, since administration of a compound in the earlier claim inherently caused the mechanism described in the later claim.) Dissenting in *Lilly*, Judge Newman wrote: "[E]very biological property is a natural and inherent result of the chemical structure from which it arises, whether or not it has been discovered. To negate the patentability of a discovery of biological activity because it is 'the natural result' of the chemical compound can have powerful consequences for the patentability of biological inventions."

Id. at 976.

[78] *Maxwell v. Baker, Inc.*, 86 F.3d 1098 (Fed. Cir. 1996), *cert. denied*, 520 U.S. 1115 (1997). In *Maxwell*, the plaintiff devised a method to keep shoe pairs together in a store without having to punch a hole through the shoes. *Id.* at 1101. The plaintiff's system involved inserting "tabs along the inside of each shoe [that] connected the shoes with a filament threaded through . . . each tab." *Id.* In her patent, plaintiff claimed "a fastening tab . . . and means for securing said tab between said inner and outer soles." *Id.* at 1102. Plaintiff also "disclosed in the specification, without claiming them, alternatives in which the fastening tabs could be 'stitched into the lining seam of the shoes.'" *Id.* at 1108. The court stated that "[b]y failing to claim these alternatives, the Patent and Trademark Office was deprived of the opportunity to consider whether these alternatives were patentable." *Id.* The court continued by stating that "[a] person of ordinary skill in the shoe industry, reading the specification and prosecution history, and interpreting the claims, would conclude that [the plaintiff], by failing to claim the alternative shoe attachment systems in which the tabs were attached to the inside shoe lining, dedicated the use of such systems to the public." *Id.* Therefore, the court concluded that the defendant could not infringe the plaintiff's patent by using one alternate shoe attachment system that plaintiff had given to the public. *Id.*

[79] *Id.* at 1107.

[80] *Warner-Jenkinson*, 520 U.S. at 37.

post-*Warner-Jenkinson* decision. In *YBM Magnex*,[81] the Federal Circuit expressly limited *Maxwell* to its "particular facts."[82] The court reversed the ITC's decision applying *Maxwell* to exclude from equivalence an alternative described in the specification. The court pointed out that *Maxwell*, as applied by the Commission, "would place *Maxwell* into conflict with Supreme Court and Federal Circuit precedent."[83] In *Graver Tank*, the Supreme Court upheld infringement by equivalence for an alternative disclosed in the specification.[84] The Federal Circuit has since agreed to *en banc* review of a case presenting the same issue.[85]

Nevertheless, when an important function is identified for a claim element in the patent specification, the Federal Circuit has denied infringement by equivalence where the accused device lacked that "one key function."[86] The majority opinion expressed concern that "other players in the marketplace are entitled to rely on the record made in the Patent Office in determining the meaning and scope of the patent."[87]

III. COMPETITION POLICY IN THE TREATMENT OF PATENTS

A. The Black/Douglas View of Patents in the Context of Antitrust and Competition Policy in the 1940s and 1950s

Competition policy powerfully influenced the treatment of patents by the Supreme Court in both patent and antitrust decisions, particularly in the decades of the 1940s and 1950s. The Supreme Court frequently—Justices Black

[81] *YBM Magnex, Inc. v. Int'l Trade Comm'n*, 145 F.3d 1317 (Fed. Cir. 1998).

[82] *Id.* at 1321. The *YBM Magnex* court described *Maxwell* as an unusual case where the patentee "disclosed two distinct, alternative ways in which pairs of shoes are attached for sale, and only claimed one of them." *Id.* at 1320. It then explained that "[i]n view of the distinctness of the two embodiments, both of which were fully described in the specification, [the *Maxwell* court] denied [patentee] the opportunity to enforce the unclaimed embodiment as an equivalent of the one that was claimed." *Id.*

[83] *Id.* at 1320.

[84] *Graver Tank Mfg. Co. v. Linde Air Prods. Co.*, 339 U.S. 605 (1950).

[85] *Johnson & Johnston Assoc. Inc. v. R.E. Service Co.*, Nos. 99–1076, 99–1179, 99–1180, 2001 U.S. App. LEXIS 4038 at 1347 (Fed. Cir. Jan. 24, 2001). The *en banc* questions posed by the court are: "(1) Whether and under what circumstances a patentee can rely upon the doctrine of equivalents with respect to unclaimed subject matter disclosed in the specification. (2) Whether in this case the jury's finding of infringement should be reversed because the patentee was foreclosed from asserting the doctrine of equivalents with respect to unclaimed subject matter disclosed in the specification."

Id.

[86] *Vehicular Tech. Corp. v. Titan Wheel Int'l, Inc.*, 141 F.3d 1084, 1093 (Fed. Cir. 1998) (Newman, J. dissenting in view of "new rule of law" contrary to precedent barring equivalency if the equivalent "does not possess the unclaimed advantages or functions described in the specification.")

[87] *Id.* at 1091.

and Douglas always—reflected a high regard for the interest of competitors in narrow patents. As explained previously, the Federal Circuit's view has been similar, assigning a higher value to the interests of competitors than those of patentees with respect to patent scope.

1. Patent Law

The Supreme Court initially limited patent rights in this period by applying a high standard of invention under the patent laws. In holding that patentable inventions must reveal a "flash of creative genius," Justice Douglas, writing for the Supreme Court majority in *Cuno Engineering Corp. v. Automatic Devices Corp.* expressed concern "lest. . .the heavy hand of tribute be laid on each slight technological advance in an art."[88] The alternative of a lower standard of patentability, he wrote, "creates a class of speculative schemers who make it their business to watch the advancing wave of improvement, and gather its foam in the form of patented monopolies, which enable them to lay a heavy tax upon the industry of the country, without contributing anything to the real advancement of the arts."[89] The result of such patenting affected competitors, since it "embarrasses the honest pursuit of business with fears and apprehensions of concealed liens and unknown liabilities to lawsuits and vexatious accountings for profits made in good faith."[90]

In 1950, Justices Douglas and Black, in *Automatic Radio*, summarized the Supreme Court's recent history of giving priority to the impact of patents on competitors rather than on the patentee:

> The Court in its long history has at times been more alive to that policy [constitutional patent clause re: "Progress of Science and Useful Arts"] than at other times. During the last three decades it has been as devoted to it (if not more so) than at any time in its history. I think that was due in large measure to the influence of Mr. Justice Brandeis

[88] *Cuno Eng. Corp. v. Automatic Devices Corp.*, 314 U.S. 84, 92 (1941). The Supreme Court in *Cuno Engineering* reversed the judgment of the lower court, finding that the respondent had a valid patent that was infringed. "[T]he new device, however useful it may be, must reveal the flash of creative genius, not merely the skill of the calling. If it fails, it has not established its right to a private grant on the public domain." *Id.* at 91; *see also Sears, Roebuck & Co. v. Stiffel Co.*, 376 U.S. 225, 230 (1964) ("a genuine 'invention' or 'discovery' must be demonstrated 'lest in the constant demand for new appliances the heavy hand of tribute be laid on each slight technological advance in an art.'" *Cuno Eng. Corp. v. Automatic Devices Corp.*, 314 U.S. 84, 92 (1941); *see Great Atl. & Pac. Tea Co. v. Supermarket Equip. Corp.*, 340 U.S. 147, 152–153 (1950); *Atl. Works v. Brady*, 107 U.S. 192, 199–200 [1883].).

[89] *Cuno Eng.*, 314 U.S. at 92.

[90] *Id.*

> and Chief Justice Stone. They were alert to the danger that business—growing bigger and bigger each decade—would fasten its hold more tightly on the economy through the cheap spawning of patents and would use one monopoly to beget another through the leverage of key patents. They followed in the early tradition of those who read the Constitution to mean that the public interest in patents comes first, reward to the inventor second.[91]

Justices Black and Douglas, again in dissent, cited the burden on competitors for their strong disagreement with the majority's view upholding the "doctrine of equivalents in *Graver Tank*." Justice Black wrote:

> The Court's ruling today sets the stage for more patent "fraud" and "piracy" against business than could be expected from faithful observance of the congressionally enacted plan to protect business against judicial expansion of precise patent claims. Hereafter a manufacturer cannot rely on what the language of a patent claims. He must be able, at the peril of heavy infringement damages, to forecast how far a court relatively unversed in a particular technological field will expand the claim's language after considering the testimony of technical experts in that field. To burden business enterprise on the assumption that men possess such a prescience bodes ill for the kind of competitive economy that is our professed goal.[92]

[91] *Automatic Radio Mfg. Co., Inc. v. Hazeltine Research, Inc.*, 339 U.S. 827, 837 (1950), (Douglas, J., dissenting). Similarly, Justice Black, dissenting in another case, expressed concern for competitors if the patentability standard was not high enough: "Those who strive to produce and distribute goods in a system of free competitive enterprise should not be handicapped by patents based on a 'shadow of a shade of an idea.'" *Goodyear Tire & Rubber Co., Inc. v. Ray-O-Vac Co.*, 321 U.S. 275, 279 (1944) (*citing Atl. Works v. Brady*, 107 U.S. 192, 200). The Supreme Court in *Goodyear* affirmed the judgment for respondent, holding that its patent for a flashlight battery cell was valid and infringed "since the petitioners' substitutions of structure and material were no more than the choice of mechanical alternatives and did not avoid the practice of the principle disclosed by the patent." *Goodyear*, 321 U.S. at 723–24.

[92] *Graver Tank & Mfg. Co. v. Linde Air Prods. Co.*, 339 U.S. 605, 617 (1950) (Black, J., dissenting). *See also Winans v. Denmead*, 56 U.S. 330, 347 (1853) ("The patentee is obliged, by law, to ... particularly 'specify and point' out what he claims as his invention. Fullness [sic] clearness, exactness, preciseness, and particularity, in the description of the invention, its principle, and of the matter claimed to be invented, will alone fulfill [sic] the demands of Congress or the wants of the country. Nothing, in the administration of this law, will be more mischievous, more productive of oppressive and costly litigation, of exorbitant and unjust pretensions and vexatious[sic] demands, [and] more injurious to labor than a relaxation of these wise and salutary requisitions of the act of Congress." (Campbell, J., dissenting)). *See also Festo*, 234 F.3d at 628, n.17–18 (citing *Winans* and *Graver Tank*) (Linn, J., concurring in part, dissenting in part).

While the Supreme Court ultimately rejected a "flash of creative genius" as the standard for patentability,[93] the Court, in a concurring opinion by Justice Douglas, in *Great Atlantic & Pacific Tea Co.* complained about the Patent Office and the lower courts' liberality in upholding patents: "The fact that a patent as flimsy and as spurious as this one has to be brought all the way to this Court to be declared invalid dramatically illustrates how far our patent system frequently departs from the constitutional standards which are supposed to govern."[94]

The lower courts responded. By 1973, the Second Circuit observed that more than 80% of patents reviewed on appeal resulted in a decision that the patents were invalid.[95]

2. Antitrust Law

In the area of antitrust policy, the exclusive rights given to patentees were treated by the Supreme Court of Justices Black and Douglas as a "special privilege"[96] to be "narrowly construed"[97] in a more general, more fundamental scheme of competition.[98] In the *Morton Salt* case of 1942, the Supreme Court, in an opinion by Justice Stone, created a "patent misuse" doctrine which denied legal relief to a patentee against infringers if it attempted to extend the scope of its patent monopoly.[99] The right to enforce the patent would be withdrawn,[100]

[93] *Graham v. John Deere Co.*, 383 U.S. 1, 15 ("It also seems apparent that Congress intended by the last sentence of § 103 to abolish the test it believed this Court announced in the controversial phrase 'flash of creative genius,' used in *Cuno Corp. v. Automatic Devices Corp.*, 314 U.S. 84 (1941).").

[94] *Great Atl. & Pac. Tea Co. v. Supermarket Equip. Corp.*, 340 U.S. at 158.

[95] *Carter-Wallace, Inc. v. Davis-Edwards Pharmacal Corp.*, 443 F.2d 867, 872 (2d Cir. 1971), *cert. denied*, 412 U.S. 929 (1973).

[96] *Morton Salt Co. v. G.S. Suppiger Co.*, 314 U.S. 488, 492 (1942) (patent owner may not claim the protection of a patent to secure an exclusive right outside the scope of the patent by a tie-in) (Black & Douglas, J.J., on the opinion); *see also United States v. Line Material Co.*, 333 U.S. 287, 310 (1948) ("The monopoly granted by the patent laws is a statutory exception to [the] freedom for competition [of the Sherman Act] and consistently has been construed as limited to the patent grant.") (Black & Douglas, J.J., on the opinion).

[97] *Mercoid Corp. v. Mid-Continent Inv. Co.*, 320 U.S. 661, 665 (1944) ("[T]he limits of the patent are narrowly and strictly confined to the precise terms of the grant.") (Black & Douglas, JJ., on the opinion).

[98] However, the Supreme Court recognized as lawful under § 2 of the Sherman Act the achievement of monopoly power by "a new discovery or an original entry into a new field," *Am. Tobacco Co. v. United States*, 328 U.S. 781, 786 (1946), or "development ... of a superior product, [or] business acumen," *United States v. Grinnell Corp.*, 384 U.S. 563, 571 (1966).

[99] *See Morton Salt*, 314 U.S. at 492–93.

[100] For example, the Court of Appeals for the Fourth Circuit in *Lasercomb American, Inc. v. Reynolds*, 911 F.2d 970, 978 (1990) applied patent misuse principles to copyright misuse in favor of

regardless of whether the patentee had violated the antitrust laws[101] or caused injury to anyone,[102] and even though there was a separate antitrust remedy against it, which the government or an injured party could pursue if there had been a violation of the law.[103] A similar attitude was reflected in Justice Douglas's opinion for the Supreme Court describing the role of the Declaratory Judgment Act in allowing an early challenge to a patent as lifting "the heavy hand of that tribute from the business."[104] Applying the same policy in 1969, the Supreme Court in *Lear v. Adkins* set aside "the technical requirements of contract doctrine" in favor of permitting licensees to challenge patents licensed to them regardless of an agreement to the contrary.[105]

This approach also led to a series of Supreme Court decisions which held or suggested the illegality of certain patent license restrictions where patent leverage was arguably used to gain an advantage outside the patent. For example, in *Northern Pacific Railway v. United States*,[106] tie-ins were held to be *per se* illegal.[107] This principle was applied to patent tie-in cases such as *International Salt Co. v. United States*,[108] in which a licensing agreement that required the licensees to purchase their unpatented salt from the defendant as a condition to leasing his patented salt-making machinery was held to be illegal.[109] The Supreme Court ruled that the "seller must have 'sufficient economic power with respect to the tying product to appreciably restrain free competition in the market for the tied product,'" and that the economic power necessary to impose a tie-in was "presumed when the tying product is patented."[110]

an infringer who was not affected by the offending license clause: "The question is not whether the copyright is being used in a manner violative of the antitrust law . . . but whether the copyright is being used in a manner violative of . . . public policy. . . ."

[101] The question is "not necessarily whether [the patentee] has violated the Clayton Act, but whether a court of equity will lend its aid to protect the patent monopoly." *Id.* at 490.

[102] A misuse defense can be asserted even when actual damages did not result from the misuse and even in the absence of a showing of impact upon the party asserting the defense. *Id.* at 493–94.

[103] Clayton Act § 4, 5, 15 U.S.C. § 4, 5 (1997).

[104] *Altvater et al.* v. *Freeman et al.*, 319 U.S. 359, 365 (1943).

[105] *Lear, Inc. v. Adkins*, 395 U.S. 653, 670–71 (1969).

[106] *Northern Pacific Railway v. United States*, 356 U.S. 1 (1958). The Court found that to show a tie-in, one need not show "more than sufficient economic power to impose an appreciable restraint on free competition in the tied product." *Id.* at 11. (Black, J.).

[107] A "tie-in" is an arrangement in which a seller conditions his sale of a product upon a buyer's purchase of a separate product from the seller or a designated third party. The anti competitive vice is that competitors are denied access to the market for the tied product. *See id.* at 5–6.

[108] *Int'l Salt Co. v. United States*, 332 U.S. 392 (1947) (Black & Douglas, J.J., on the opinion).

[109] *Id.* at 396 (holding that it is unreasonable, *per se*, to foreclose competitors from any substantial market).

[110] *United States v. Loew's, Inc.*, 371 U.S. 38, 45–47 (1962) (quoting *N. Pac. Ry. Co.*, 356 U.S. at 6).

The limitation on the patent right that it was exhausted by sale of the patented article was a long-established principle.[111] The frequently cited *Adams v. Burke* case held that an assignment of the right under a patent to make, use, and sell coffin lids but limit to within 10 miles of Boston was not enforceable against the purchaser of the coffin without condition.[112] This principle was confirmed in *United States v. Univis Lens Co.*[113]

With respect to price restrictions on articles manufactured under patent licenses, the Supreme Court had difficulty reaching a consensus. In *United States v. General Electric Co.*, the Court found that the imposition of price limitations on a single manufacturing licensee's sales "is reasonably within the reward which the patentee by the grant of the patent is entitled to secure."[114] Subsequent decisions held price limitations unlawful in a variety of other circumstances, however. In *United States v. United States Gypsum Co.*, it was held that a patentee cannot, "acting in concert with all members of an industry, . . . issue substantially identical licenses to all members of the industry under the terms of which the industry is completely regimented. . . ."[115] In *United States v. Line Material Co.*, the Supreme Court held that two or more patent owners cannot combine their patents and fix prices,[116] and in *Newburgh Moire Co. v. Superior Moire Co.*, the Third Circuit held that a patentee may not issue a plurality of licenses fixing licensees' selling price.[117] However, the *General Electric* doctrine survived a challenge by the Department of Justice by an equally divided Supreme Court in 1948.[118]

In the area of enforcement, the Department of Justice's Antitrust Division created a list of "Nine No-Nos," reflecting the view of patents prevalent in these cases. These were license and sale restrictions associated with patents that, in the department's view, were automatically illegal. These included, for example, the following:

[111] *See, e.g., Adams v. Burke*, 84 U.S. (17 Wall.) 453 (1873).

[112] *See id.*

[113] *United States v. Univis Lens Co.*, 316 U.S. 241, 249 (1942) ("An incident to the purchase of any article, whether patented or unpatented, is the right to use and sell it, and upon familiar principles the authorized sale of an article which is capable of use only in practicing the patent is a relinquishment of the patent monopoly with respect to the article sold.").

[114] *United States v. General Electric Co.*, 272 U.S. 476, 489 (1926).

[115] *United States v. United States Gypsum Co.*, 333 U.S. 364, 400 (1948) ("Patents grant no privilege to their owners of organizing the use of those patents to monopolize an industry through price control, through royalties for the patents drawn from patent-free industry products, and through regulation of distribution.").

[116] *United States v. Line Material Co.*, 333 U.S. 287, 308–10 (1948).

[117] *Newburgh Moire Co. v. Superior Moire Co.*, 237 F.2d 283, 293–94 (3rd Cir. 1956).

[118] *See United States v. Line Material Co.*, 333 U.S. 287 (1948); *see also United States v. Huck Mfg. Co.*, 382 U.S. 197 (1965) (where the outcome was the same).

> "No-No" Number 1: "It is clear that it is unlawful to require a licensee to purchase unpatented materials from the licensor"..."No-No" Number 3: "The Department believes it is unlawful to attempt to restrict a purchaser of a patented product in the resale of that product"..."No-No" Number 9: "[T]he Department of Justice considers it unlawful for a patentee to require a licensee to adhere to any specified or minimum price with respect to the licensee's sale of the licensed products."[119]

B. The Renunciation of the Black/Douglas View

A great change occurred in antitrust doctrine beginning in the 1970s, which led to an abandonment of the Black/Douglas view.

1. Economic Beginnings

In 1950, Joseph Schumpeter identified innovation as an important source of competition and called technological change a "gale of creative destruction."[120] Robert Solow was awarded the Nobel Prize for a work published in 1957 that demonstrated the positive impact of technological change and innovation on

[119] Abbott B. Lipsky, Jr., *Current Antitrust Division Views on Patent Licensing Practices, Remarks Before the American Bar Association Antitrust Section*, 50 ANTITRUST L. J. 515, 518, 520, 523 (1981) [hereinafter Lipsky, *Views*] (citing Bruce B. Wilson, "Law on Licensing Practices: Myth or Reality?" Department of Justice Luncheon Speech (Jan. 21, 1975)). The other "No-Nos" were: "No-No" Number 2: "The Department views it as unlawful for a patentee to require a licensee to assign to the patentee any patent which may be issued to the licensee after the licensing arrangement is executed." "No-No" Number 4: "A patentee may not restrict his licensee's freedom to deal in the products or services not within the scope of the patent." "No-No" Number 5: "The Department believes it to be unlawful for a patentee to agree with his licensee that he will not, without the licensee's consent, grant further licenses to any other person." "No-No" Number 6: "The Department believes that mandatory package licensing is an unlawful extension of the patent grant." "No-No" Number 7: "The Department believes that it is unlawful for a patentee to insist, as a *condition* of the license, that his licensee pay royalties in an amount not reasonably related to the licensee's sales of products covered by the patent—for example, royalties on the total sales of products of the general type covered by the licensed patent." "No-No" Number 8: "It is pretty clearly unlawful for the owner of a process patent to attempt to place restrictions in his licensee's sales of products made by the use of the patented process." *Id.* at 519–22.

[120] Joseph Schumpeter, CAPITALISM, SOCIALISM, AND DEMOCRACY, 84 (3rd ed. 1950); *see also id.* at 83–110; 1 Joseph A. Schumpeter, BUSINESS CYCLES. 84–192 (1939). Schumpeter's theory did not focus on either patents or technologic innovations, but his analysis and conclusions are readily transposed to these situations. Schumpeter considered the term *innovation* to include the development of new consumer goods, forms of industrial organization, markets, and methods of production. *Id.* at 82–84. He further defined innovation as putting new technologies into practice, as well as any combination of resources that establishes a "new production function." *Id.* at 87.

growth in productivity in the U.S. economy.[121] Edward Denison later reported that increases in productivity were primarily attributable to advances in scientific and technological knowledge.[122]

2. The Cases

It was only after the Black and Douglas era that a trend toward respecting patents began. In 1974, the Supreme Court in *Kewanee* observed that the patent system stimulates invention and commercialization by providing a 17-year exclusive right to the inventor.[123] The Court held that trade-secret law was not preempted by the patent law, and that they could coexist. Taking a different view than that of the Black and Douglas Supreme Court, the Court recognized that patent exclusivity was "an incentive to inventors to risk the often enormous costs in terms of time, research, and development," and had a "positive effect on society through the introduction of new products and processes of manufacture into the economy, and the emanations by way of increased employment and better lives for our citizens."[124]

A significant change occurred in antitrust law in connection with vertical restrictions in the *GTE-Sylvania* decision in 1977, when the Supreme Court recognized the negative effects of "free riding" and held that a restriction on competition was justified because it had a *pro*-competitive effect.[125] The Supreme Court overruled its earlier absolutist *Schwinn* decision, which had held that a vertical restriction on a product that had been sold to a wholesaler or retailer was *per se* unlawful as a "restraint on alienation."[126] The Court upheld a television set supplier's restriction on the geographic area in which its distributor could resell the sets.[127] The restriction encouraged promotion of competition among different brands by encouraging the distributor to promote the supplier's brand—something he might not do if he feared that a neighboring distributor

[121] Robert M. Solow, *Technical Change and the Aggregate Production Function*, 39 REV. OF ECON. & STAT. 312 (1957). Solow determined that increases in capital intensity accounted for only 12.5 percent of the measured increase in output per labor hour while the other 87.5 percent increase in growth was attributed to technologic changes such as improvements in production practices and equipment. *Id.*; *see also* F. M. Scherer & David Ross, INDUSTRIAL MARKET STRUCTURE AND ECONOMIC PERFORMANCE, 613–14 (3rd ed. 1990) (detailing Solow's study and resulting conclusions).

[122] Scherer & Ross, *supra* note 121, at 613 (citing Edward F. Denison, TRENDS IN AMERICAN ECONOMIC GROWTH, 1929–1982 30 (1985)) (68%).

[123] *Kewanee Oil Co. v. Bicron Corp.*, 416 U.S. 470, 480–81 (1974).

[124] *Id.* at 480.

[125] *Continental T. V., Inc. v. GTE Sylvania Inc.*, 433 U.S. 36, 54–55 (1977).

[126] *United States v. Arnold, Schwinn & Co.*, 388 U.S. 365 (1967) (holding that vertical restrictions on areas in which, and persons to whom, a product could be sold after the manufacturer had parted with ownership were *per se* violations of the Sherman Act).

[127] *GTE Sylvania*, 433 U.S. at 59.

would "free ride" on his promotional effort and use it to take sales away from him.[128] The recognition that a restriction on competition benefits competition by deterring free-riding and promoting more vigorous selling is analogous to the idea that a patent benefits competition by deterring free-riding by others on a patented invention and promoting the risk-taking involved in research.

In 1978, Judge Newman decided *SCM Corp. v. Xerox Corp.*[129] after a year-long antitrust trial concerning plaintiff's contentions that Xerox had acquired and maintained a monopoly in plain-paper office copiers. The claims against Xerox were predicated on its acquisition of thousands of patents by internal invention and external acquisition. It was alleged that the patents were used to exclude competition in the market for plain-paper office copiers; Xerox ultimately achieved a 100% share. The trial explored in great detail the origin and development of xerography and the Xerox Company, as well as the role of research, development, risk, patents, and investment in creating and excluding competition in copiers.

Judge Newman's decision is noteworthy for three reasons. First, his assessment of the proper interplay of patents and antitrust laws was based on an exhaustive evidentiary record which told how a new and major industry was created. Second, the possibility of conflict between patents and antitrust laws was framed in stark terms since Xerox ultimately was determined to have achieved a monopoly (100% of the relevant market) by reason of the exclusionary power of acquired patents. Third, straightforward application of a Supreme Court precedent concerning § 7 of the Clayton Act would have resulted in a violation of law. The decision recognized that "[t]his case presents important issues concerning the relationship between the patent laws and the antitrust laws."[130] Rather

[128] *Id.* at 51–52 n. 19, 55. "Vertical restrictions promote interbrand competition by allowing the manufacturer to achieve certain efficiencies in the distribution of his products. These "redeeming virtues" are implicit in every decision sustaining vertical restrictions under the rule of reason. Economists have identified a number of ways in which manufacturers can use such restrictions to compete more effectively against other manufacturers. . . . For example, new manufacturers and manufacturers entering new markets can use the restrictions in order to induce competent and aggressive retailers to make the kind of investment of capital and labor that is often required in the distribution of products unknown to the consumer. Established manufacturers can use them to induce retailers to engage in promotional activities or to provide service and repair facilities necessary to the efficient marketing of their products. Service and repair are vital for many products, such as automobiles and major household appliances. The availability and quality of such services affect a manufacturer's goodwill and the competitiveness of his product. Because of market imperfections such as the so-called "*free rider*" *effect*, these services might not be provided by retailers in a purely competitive situation, despite the fact that each retailer's benefit would be greater if all provided the services than if none did." *Id.* at 54–55 (citations omitted) (emphasis added).

[129] *SCM Corp. v. Xerox Corp.*, 645 F.2d 1195 (2d Cir. 1981), *cert. denied*, 455 U.S. 1016 (1982), *aff'g* 463 F. Supp. 983 (C.D. Conn. 1978).

[130] *Id.* at 985.

than the patent-competition conflict hypothesized by the Black/Douglas view, the court observed that "[e]conomic arguments can be made that these statutes have a common goal of maximizing wealth by facilitating the production of what consumers want at the lowest cost."[131] The court observed that the patent laws grant broad exclusionary power including, unlike other property, "the exclusive power to prevent anyone else from using the patented invention for economic gain, even a person who discovered or created the invention entirely independent of the patent owner."[132] Achieving the purpose of the patent law "to 'promote' the Progress of Science and useful Arts.' U.S. Const., Art. 1 § 8, Cl. 8. . . .is not limited to internally developed inventions," but extends to the "traditional approach. . .[of] assignment by the inventor of an exclusive license or the patent itself to a company willing to risk the investment needed for commercial success."[133] The court's principal holdings concerned the acquisition of key improvement patents by Xerox from the Battelle Institute, a contract research organization, at an early stage in the development of xerography. The patents were acquired in 1956, 4 years prior to the production of Xerox's first plain-paper copier and at least 8 years prior to the appearance of the relevant market.[134]

The court upheld Xerox's refusal to license the acquired patents to competitors. Not only is it "a premise of the patent laws that a company employing inventors must have substantial incentive to spend money for research that may lead to patentable inventions," but " 'the Progress of Science and Useful arts' is also aided by enabling a company, prior to the time it has developed a marketable product and thereby acquired any market power, to acquire patents from others, especially from non-competitor research entities."[135] This followed because "[s]uch internal and external accumulations of patents may well lead to the development of inventions and products using inventions."[136] The court cited the evidence at trial in support of its reasoning: "The history of plain-paper copiers, as evidenced by the record in this case, is an extraordinary example of such a development."[137] The court rejected SCM's § 7 of the Clayton Act[138] and § 2 of the Sherman Act[139] claims based on the company's acquisitions in 1956 on similar grounds.

The court also rejected the contention that a § 7 violation occurred in 1969 when the exclusionary power of the acquired patents resulted in Xerox's

[131] *Id.* at 996.
[132] *Id.* at 997.
[133] *Id.* at 1001.
[134] *Id.* at 993, 995.
[135] *SCM Corp. v. Xerox Corp.*, 463 F. Supp. at 1013.
[136] *Id.*
[137] *Id.*
[138] *Id.* at 1003–04.
[139] *Id.* at 1014–15.

100% share of the relevant market. The court was confronted by the Supreme Court decision in *United States v. E.I. DuPont de Nemours & Co.*,[140] where DuPont had purchased shares in General Motors at a time when it imposed no restraint. Then, after holding the shares for 13 years, GM grew and DuPont's ownership of the shares was held to violate § 7. The court, however, distinguished *DuPont* because, even if the holding theory of § 7 could create liability for the holding of some assets, the court concluded that the "proper reconciliation of the patent and antitrust laws precludes the application of such a doctrine to patents."[141]

The Court of Appeals for the Second Circuit affirmed the district court.[142] It recognized that "the public benefits from ... the increased competition the patented product creates in the marketplace."[143] Not only were inventors important in the innovation process, but investors played a "key role" in "both the funding of research that leads to inventions and the promotion that necessarily must follow to achieve successful commercialization...."[144] Agreeing with the trial court, the court held that the acquisition of patents covering inventions which were later successfully developed, leading to monopoly power, was lawful.[145] The court declined to follow *United States v. E.I. DuPont de Nemours & Co.*,[146] since illegality of Xerox's continued "holding" of the acquired patents for 13 years after the acquisition "would unduly trespass upon the policies that underlie the patent law system. The restraint placed upon competition ... must, in deference to the patent system, be tolerated throughout the duration of the patent grants."[147] In short, "to impose antitrust liability upon Xerox would severely trample upon the incentives provided by our patent laws and thus undermine the entire patent system."[148]

The courts also moved the law away from its prior strict views on patent misuse, which did not examine injury or anticompetitive effect. The Federal Circuit has ruled that there should be a showing of anticompetitive effects associated with challenged conduct which extends the patentee's statutory rights, but is not a *per se* misuse.[149] The Federal Circuit summed up its

[140] *United States v. E.I. DuPont de Nemours & Co.*, 353 U.S. 586 (1957).

[141] *SCM Corp.*, 463 F. Supp. at 1003.

[142] *SCM Corp.*, 645 F.2d at 1195.

[143] *Id.* at 1203.

[144] *Id.* at 1206 n.9.

[145] *Id.* at 1204.

[146] *DuPont*, 353 U.S. at 586.

[147] *SCM Corp.*, 645 F.2d at 1212.

[148] *Id.* at 1209.

[149] *Virginia Panel*, 133 F.3d 860, 869 (Fed. Cir. 1997) (citing *Mallinckrodt, Inc. v. Medipart, Inc.*, 976 F.2d 700, 708 (Fed. Cir. 1992)). The Federal Circuit pointed out that practices which are *per se* patent misuse include "tying" arrangements in which a patentee conditions a license under the patent on the purchase of a separable, staple good, *see, e.g.*, *Morton Salt Co. v. G.S.*

analysis in *Virginia Panel*.[150] A court must determine if such a practice is "reasonably within the patent grant, *i.e.*, that it relates to subject matter within the scope of the patent claims."[151] If the practice is within the patent grant, it does not have the effect of broadening the scope of the patent claims and cannot constitute patent misuse.[152] "If, on the other hand, the practice has the effect of extending the patentee's statutory rights and does so with an anti-competitive effect, that practice must then be analyzed in accordance with the 'rule of reason.' "[153]

With respect to tie-ins, a showing of market power in the relevant market for the tying product was added as a requirement. The Supreme Court had presumed that ownership of a patent conferred sufficient market power.[154] In *Jefferson Parish Hospital v. Hyde*,[155] tying was said to be *per se* illegal if "the seller's exploitation . . . [forces] the buyer into the purchase of a tied product that the buyer either did not want at all, or might have preferred to purchase elsewhere on different terms."[156] The five-member majority (in dictum) agreed with the earlier rule that "if the government has granted the seller a patent or similar monopoly over a product, it is fair to presume that the inability to buy the product elsewhere gives the seller market power."[157] That point, however, was sharply disputed by four Justices in a concurring opinion written by Justice Sandra Day O'Connor: "[A] high market share indicates market power only if the product is properly defined to include all reasonable substitutes for the

---

Suppiger Co., 314 U.S. 488, 491 (1942) and arrangements, in which a patentee effectively extends the term of his or her patent by requiring postexpiration royalties, *see, e.g.*, *Brulotte v. Thys Co.*, 379 U.S 29, 33 (1964). *Virginia Panel*, 133 F.3d at 869 (citations omitted).

[150] *Virginia Panel*, 133 F.3d at 869; *see also Mallinckrodt*, 976 F.2d at 708 (In the context of restriction on use of patented medical device after sale, "[t]he appropriate criterion is whether [the patentee's] restriction is reasonably within the patent grant, or whether the patentee has ventured beyond the patent grant and into behavior having an anticompetitive effect not justifiable under the rule of reason. . . . should such inquiry lead to the conclusion that there are anticompetitive effects extending beyond the patentee's statutory right to exclude, these effects do not automatically impeach the restriction."); *Windsurfing Int'l Inc. v. AMF, Inc.*, 782 F.2d 995, 1001–02 (Fed. Cir. 1986) (In context of rejecting misuse based on restriction on use of patent licensor's trademarks, "[t]o sustain a misuse defense involving a licensing arrangement not held to have been *per se* anti-competitive by the Supreme Court, a factual determination must reveal that the overall effect of the license tends to restrain competition unlawfully in an appropriately defined relevant market.").

[151] *Virginia Panel*, 133 F.3d at 869 (citing *Mallinckrodt*, 976 F.2d at 708).

[152] *Id.*

[153] *Id.*

[154] *See Zenith Radio Corp. v. Hazeltine Research*, 395 U.S. 100, 137 (1969); *United States v. Loew's*, 371 U.S. 38, 45 (1962).

[155] *Jefferson Parish Hosp. v. Hyde*, 466 U.S. 2 (1984).

[156] *Id.* at 12 (holding that a 30% market share for tying product is insufficient to prove market power).

[157] *Id.* at 16 (citing *Loew's*, 371 U.S. at 45–47).

product" and therefore "a patent holder has no market power in any relevant sense if there are close substitutes for the patented product."[158] After *Hyde*, more lower federal court cases have ruled that the ownership of a patent creates no presumption of market power.[159]

The patent misuse defense based on a tie-in was altered by a statute in 1988 that required that "the patent owner has market power in the relevant market for the patent or patented product . . ."[160] Senator Patrick Leahy told the Senate the legislation was intended "to support enhancement of the intellectual property rights" and to limit the application of the misuse doctrine.[161] In the area of postsale restrictions, the axiom that a sale of a patented product

[158] *Id.* at 37 n.7.

[159] *Compare In re Indep. Serv. Org. Antitrust Litig.*, 203 F.3d 1322, 1325 (Fed. Cir. 2000) ("A patent alone does not demonstrate market power."), and *In Re Pabst Licensing*, 2000 WL 1145725 at *6 (E.D. La. Aug. 11, 2000) (finding no presumption of market power solely by ownership of a patent), and *Schafly v. Caro-Kann Corp.*, No. 98–1005, 1998 U.S. App. LEXIS 8250, at *19 (Fed. Cir. April 29, 1998) ("Mere possession of a patent ... does not establish a presumption of antitrust market power"); *Little Caesar Enters. v. Smith*, 34 F. Supp.2d 459, 466 n.10 (E. D. Mich. 1998) (finding that ownership of a patent is only one factor towards a showing of market power), and *Northlake Mktg. & Supply Inc. v. Glaverbel*, 861 F. Supp. 653, 662 (N. D. Ill. 1994) ("mere possession of a 'patent does not establish presumption of market power . . .' ") *with Grid Sys. Corp. v. Texas Instruments, Inc.*, 771 F. Supp. 1033, 1037 n.2 (N.D. Cal. 1991) ("the presumption of economic power, when the tying product is a patent or copyright, survives"), and *Klo-Zik Co. v. General Motors Corp.*, 677 F. Supp. 499, 505–06 (E.D. Tex. 1987) ("The fact that a product is patented is usually enough to create a presumption of market power."), and *Digidyne Corp. v. Data General Corp.*, 734 F.2d 1336, 1340 n.4 (9th Cir. 1984) ("when the tying product is patented or copyrighted . . . sufficiency of economic power is presumed"), *cert. denied*, 473 U.S. 908 (1985).

The Supreme Court in *Eastman Kodak Co. v. Image Tech. Serv., Inc.*, 504 U.S. 451 (1992), confirmed the necessity for true market power in a relevant market as to the tying product. It stated that an arrangement violates § 1 of the Sherman Act if the seller has "appreciable economic power in the tying [product] market," as in "the power 'to force a purchaser to do something that he would not do in a competitive market.' It has been defined as 'the ability of a single seller to raise price and restrict output.' The existence of such power ordinarily is inferred from the seller's possession of a predominant share of the market." *Id.* at 464 (citations omitted) (citing *Hyde*, discussed *infra*, and other cases).

[160] 35 U.S.C. § 271(d)(5) (2001). "No patent owner . . . shall be . . . guilty of misuse . . . by reason of . . . (5) condition[ing] the license of any rights to the patent or the sale of the patented product on the acquisition of a license to rights in another patent or purchase of a separate product, unless in view of the circumstances, the patent owner has market power in the relevant market for the patent or patented product on which the license or sale is conditioned." *Id.* Congress rejected the automatic inference of market power where a patent covers the tying product. It required consideration of a patent owner's actual power in the relevant market to determine whether a tie-in constitutes patent misuse: "If the alleged infringer cannot prove that the patent owner has market power in the relevant market for the patent or patented product, the tying product, then there can be no patent misuse by virtue of the tie-in, and that is the end of the inquiry."

[161] 134 Cong. Rec., S17147 (daily ed. Oct. 21, 1988) (statement of Sen. Leahy).

necessarily exhausted the patent monopoly was abandoned. In *Mallinckrodt*,[162] the Federal Circuit held that a "single use only" restriction that was imposed on the sale of a medical device was enforceable as a matter of patent law in a suit for infringement against a contractually bound purchaser-user in disregard of the limitation. The court construed cases previously viewed as foreclosing postsale restrictions on patented goods by exhaustion as limited to unconditional sales.[163] The court distinguished earlier tying or price-fixing cases since the case before it was neither tying nor price-fixing, and was not *per se* illegal.[164] The court concluded that "should the restriction be found to be reasonably within the patent grant, *i.e.*, that it relates to subject matter within the scope of the patent claims, the inquiry is over."[165]

Concerning price restrictions on articles manufactured under patent licenses, the Department of Justice was, again, unable to persuade a majority of the Supreme Court that a price limitation on a single licensee ("No-No Number 9") was outside the patent grant and illegal. In *Huck Mfg. Co.*,[166] as it had been in *Line Material*[167] 17 years earlier, the Supreme Court was divided equally on whether to overrule *General Electric*.[168]

In sum, the antitrust cases began to recognize the harm free-riding does to competition and the contribution patents made to competition and courts

[162] *Mallinckrodt*, 976 F. 2d 700 (Fed. Cir. 1992). In *Jazz Photo Corp. v. ITC*, 264 F.3d 1094 (Fed. Cir. 2001), the Court held that words on a camera box did not amount to a "single use" contract imposed on the sale of a camera. The Federal Circuit overturned an ITC ruling that importing refurbished cameras infringed patents of Fuji under § 337 of the Tariff Act. Statements on the box instructed the purchaser that the camera will not be returned to the purchaser after processing. The Federal Circuit held that "these package instructions [were] not in the form of a contractual agreement by the purchaser to limit reuse of the cameras. There was no showing of a 'meeting of the minds.'" *Id.* at 1108. Accordingly, there was not a single-use license which had been exceeded by the purchasers. *Id.* at 1107.

[163] *Mallinckrodt*, 976 F.2d at 706; *e.g.*, *Adams v. Burke*, 84 U.S. (17 Wall.) 453 (1873) (assignment of right to make, use, and sell coffin lids within 10 miles of Boston was not enforced against purchaser of coffin without condition). The Federal Circuit also relied upon *General Talking Pictures*, *supra*, which gave effect to a restriction on manufacture and sale of a patented amplifier for home radio use against a purchaser from the licensee with notice of the restriction. *See Mallinckrodt*, 976 F.2d at 705.

[164] *Mallinckrodt*, 976 F.2d at 708.

[165] *Id.* The Court continued: "However should such inquiry lead to the conclusion that there are anticompetitive effects extending beyond the patentee's statutory right to exclude, these effects do not automatically impeach the restriction. Anticompetitive effects that are not *per se* violations of law are reviewed in accordance with the rule of reason. Patent owners should not be in a worse position, by virtue of the patent right to exclude, than owners of other property used in trade." *Id.*

[166] *Huck Mfg. Co. v. NLRB*, 693 F.2d 1176 (5th Cir. 1982).

[167] *United States v. Line Material Co.*, 333 U.S. 287 (1948).

[168] *United States v. General Electric Co.*, 272 U.S. 476 (1926).

began to give more latitude to patentees to enjoy the exclusive rights provided by patents and engage in restrictive practices.

3. The Antitrust Division's View of the Role of Innovation in Competition

About two years after the district court's decision in favor of defendant in *Xerox*, the Antitrust Division of the Department of Justice undertook a reexamination of the role of patents in competition. The result was a sharp departure from the past. The agency adopted policies that were emphatically propatent based on the positive role of patents in stimulating innovation, competition, and economic benefits. The department's speeches articulated the interaction of these factors in a depth that went beyond many of the cases and well beyond what the Federal Circuit majorities have attempted.

First, in 1981, the Department of Justice announced its reexamination of the prior department's negative view of patent license restrictions[169] with a speech by Deputy Assistant Attorney General Lipsky. He discredited the "Nine No-Nos."[170] He said that the " 'Nine No-Nos,' as statements of rational economic policy, contain more error than accuracy."[171] The speech contrasted a proper antitrust treatment of patents with antitrust in other areas, recognizing that the market power conferred by the patent was not "of independent competitive concern."[172]

By 1985, Deputy Assistant Attorney General of the Antitrust Division, Charles Rule, characterized the changed point of view as "a second revolution."[173] He recognized the link between economic welfare and technological progress and cited the data:

[169] *See* Lipsky, *Views*, *supra* note 119, at 515. Lipsky introduced his remarks by stating: "For the better part of the last decade, Division enforcement policy toward patent licensing has been advertised using a list of forbidden practices commonly known as the 'Nine No-No's.' Each of these practices is thought to be especially deserving of antitrust condemnation by virtue of some inherently anticompetitive feature." *Id.* (footnote omitted).

[170] *Id.*

[171] *Id.*

[172] Lipsky stated that: "While the antitrust analyst is at liberty outside the patent field to dabble in such issues as whether the market power of the seller was lawfully or unlawfully acquired, or whether the extent of that market power is or ought to be of independent competitive concern, the patentee comes to judgment with those questions settled according to constitutional and congressional instructions.... Thus, the independent decisions of the patentee regarding the means by which an invention is to be combined with other productive inputs ought to be regarded as having no inherent anticompetitive import." *Id.* at 519.

[173] Deputy Assistant Attorney General Charles F. Rule, "Technology Licensing and the Second American Revolution: Storming the Ramparts of Antitrust and Misuse." Statement Before the John Marshall Law School (Feb. 22, 1985) [hereinafter, Rule, Technology Licensing].

> It is estimated that during the past 80 years, technological progress has accounted for almost one-half of the growth in per capita real income. Companies that invest heavily in the research and development of new technologies have about three times the growth rate, twice the productivity rate, nine times the employment growth, and only one-sixth the price increases as companies with relatively low investments in R&D.[174]

Rule commented on the "basic failure" of past courts and the Department of Justice "to recognize some fundamental facts about the nature of intellectual property and the beneficial role that technology licensing plays in a healthy, competitive economy." Andewelt observed that:

> The creation and development of sophisticated new technology is an enormously expensive and risky undertaking.[175] There is never any guarantee that years of hard work and bushels of money poured into R&D will result in usable technology or, even if it does result in usable technology, that commercial success will follow.[176]

He recognized that "we must provide an adequate incentive" for this effort and that is the role of the patent laws.[177] Rule then explained that the patent system prevented "free-riding" competitors from "appropriat[ing] to themselves much of [a patent's] value, denying to the inventor the full fruits of his creation ... Unless the 'free rider' problem is somehow addressed, those who might otherwise undertake risky and expensive R&D will not do so. Fewer technologies will be developed and consumers will face higher prices and fewer choices."[178]

The Division's Chief of the Intellectual Property Section, Roger Andewelt, pointed out that rather than "the patent grant [being] inconsistent with our free enterprise system[179] ... the patent grant often can have a

[174] *Id.*

[175] *See generally* Arrow, *Economic Welfare and the Allocation of Resources for Invention*, IN THE RATE AND DIRECTION OF INVENTIVE ACTIVITY, 610–14 (1962).

[176] Rule, *supra* note 173.

[177] *Id.*

[178] *Id.*

[179] In 1986, Assistant Attorney General Rule explained that the antitrust laws and the procompetitive policy they represent and the patent laws do not conflict: "That said, I would like to turn to the relationship between patent law and antitrust law. Contrary to popular opinion, the two do not conflict [contrary to some statements, including mine]. Nevertheless, the perception that they do appears to be predominant in the antitrust bar. For example, only a few years ago the Antitrust Section of the ABA held a National Institute on patent licensing, at which one of the authors began with the bald assertion that "[the] antitrust and patent laws conflict at their interface." In addition, in *International Wood Processors v. Power Dry, Inc.*, 792 F.2d 416 (4th Cir. 1986), the

significant positive effect on competition."[180] Patents benefit competition by encouraging inventions to reach the market[181] and by encouraging the disclosure of inventions that otherwise would have been kept secret, making the disclosed information available to competitors.[182]

Rule, in a 1986 lecture, elaborated on two ways in which the patentee's right to exclude fosters competition. First, by preventing free-riding, "the patent creates a market-based set of incentives for creating new technologies and for developing and exploiting those technologies once created." The patentee can obtain a *quid pro quo* from others that want to use it "and the promise of that return creates the incentive for investments in research and development." The results are those of increased competition—"they bring down costs that enable society to use its limited resources more efficiently and effectively to meet consumers' demands." Second, by converting information into a legally defined piece of property that can be transferred according to articulated legal rules, intellectual property substantially reduces the cost of market transactions that move information to its most highly valued uses. The availability of patent property rights accounts for technology "in its current abundance . . . and without [them] the market in technology would be unable to function."[183]

As for whether patents as an incentive for invention are a negative force for antitrust and competition, or a positive one, Rule concluded, "[t]here

---

Fourth Circuit cited that assertion and rendered a decision that overlooked the potential economic benefits from the patent-related conduct at issue in the case." Deputy Assistant Attorney General Charles F. Rule, "The Antitrust Implications of International Licensing: After The Nine No-Nos," Remarks Before the Legal Conference Sponsored By The World Trade Association And The Cincinnati Patent Law Association (Oct. 21, 1986) [hereinafter Rule, Antitrust Implications].

[180] Roger B. Andewelt, "Basic Principles to Apply at the Patent–Antitrust Interface," Remarks to the Houston Patent Law Association (Dec. 3, 1981).

[181] "The availability of exclusive patent rights increases the possible reward for R&D. It thereby results in the development of some inventions that otherwise would not have been discovered and developed at all or, at least, not nearly as early as they were. For such inventions it is illogical to talk in terms of the patent grant conflicting with a competitive economic system. If there were no patent grant these inventions would not have reached the marketplace; therefore, the availability of a patent served only to benefit competition—to make additional or less expensive choices available to consumers." *Id.*

[182] "The patent grant also can yield significant procompetitive benefits when it results in disclosure of inventions that otherwise could and would have been kept secret. If the inventions were kept secret, competitors could not have copied the invention and competed with the original inventor. The grant of a patent on such an invention, therefore, would not halt imitative competition that otherwise would occur. The grant of a patent, however, will result in disclosure of the invention, which is pro-competitive. Disclosure may permit competitors to "invent around" or invent improvements of the patented invention. Disclosure also adds to the body of available technology and scientific knowledge and, by so doing, may precipitate other inventions not anticipated or even dreamed of by the patentees." *Id.*

[183] Rule, Antitrust Implications, *supra* note 174.

are few aspects of our legal and economic system that are more consistent with the objectives of antitrust than patent protection."[184]

In 1995 the Antitrust Guidelines for the Licensing of Intellectual Property, issued by the U.S. Department of Justice[185] and the Federal Trade Commission, underscored the point that the "incentives for innovation and its dissemination and commercialization" in the intellectual property laws and the antitrust laws share the common purpose of promoting innovation and enhancing consumer welfare.[186] By the same token, the guidelines recognize the negative effect of "rapid imitation" because it "would reduce the commercial value of innovation and erode incentives to invest, ultimately to the detriment of consumers."[187]

[184] *Id.*

[185] ANTITRUST GUIDELINES FOR THE LICENSING OF INTELLECTUAL PROPERTY, U.S. Dep't of Justice and the Fed. Trade Comm'n (April 6, 1995). These Guidelines supersede section 3.6 in Part I, Intellectual Property Licensing Arrangements, and cases 6, 10, 11, and 12 in Part II of the U.S. Department of Justice 1988 Antitrust Enforcement Guidelines for International Operations.

[186] "[T]he aims and objectives of patent and antitrust laws may seem, at first glance, wholly at odds. However, the two bodies of law are actually complementary, as both are aimed at encouraging innovation, industry, and competition." *Atari Games Corp. v. Nintendo of America, Inc.*, 897 F.2d 1572, 1576 (Fed. Cir. 1990).

[187] ANTITRUST GUIDELINES FOR THE LICENSING OF INTELLECTUAL PROPERTY, U.S. Dep't of Justice and the Fed. Trade Comm'n (April 6, 1995). Amicus curiae briefs of the Department of Justice in *Warner-Jenkinson* and *Festo* favored retaining equivalency, even for narrowed claims. In *Warner-Jenkinson* the government argued for the doctrine of equivalents set out in *Graver Tank*. Amicus Brief for U.S. at *13, *Warner-Jenkinson Co. v. Hilton Davis Chemical Co.*, 520 U.S. 17 (1997) (No. 95–728, 1996 WL 172221) ("As the court observed in *Graver Tank*, '[o]utright and forthright, duplication is a dull and very rare type of infringement.' Patentees should be protected against imitations that involve only colorable or trivial deviation from the literal terms of the patent claims, and not placed 'at the mercy of verbalism.'") (citations omitted). The division also supported a prosecution history estoppel which depended on a close examination of what had been surrendered. Amicus Brief for U.S., *Warner-Jenkinson*, at *20–21. ("[A]s the court correctly noted in *Insta-Foam Products, Inc. v. Universal Foam Systems, Inc.*, 906 F.2d 698, 703 (Fed. Cir. 1990), '[w]henever prosecution history estoppel is invoked as a limitation to infringement under the doctrine of equivalents, 'a close examination must be made as to, not only what was surrendered, but also the reason for such a surrender.'") In *Festo*, the United States, in 2001, also disagreed with the majority, and would not abolish prosecution history estoppel for amended claim limitations. Rather, the government suggests a rebuttable presumption, Amicus Brief for U.S. at 10, that can be overcome, for example, by a showing that the accused device is using a later-developed technology or could not reasonably have been included in the claims as drafted. *Id.* at 25–26. The department's brief offers no supporting competition analysis. While inconsistent with the Federal Circuit majority, the department's position would significantly limit the doctrine. Placing a heavy burden on the inventor to write all knowable variations of his invention into the claims is a way of reducing the scope of equivalency, out of keeping with *Graver Tank* (which applied equivalence to an unclaimed variation described in the specification) and the earlier thoughtful analyses of the Antitrust Division about the importance of a full patent reward. By the same token, applying equivalence only to later-developed technology extends the patent to apply to the more creative competitor, but leaves the straightforward free-rider outside the patent reward.

In short, the view of the Antitrust Division and the Federal Trade Commission the antitrust enforcement agencies expected to be most aggressive in their concern for competition,[188] favored the competition created by the patentees' innovations. The free-rider, on the other hand, was recognized as a negative force.

IV. PRECISE NOTICE TO COMPETITORS AND RULE OF REASON ANALYSIS UNDER THE ANTITRUST LAWS

Given the Federal Circuit's emphasis on precise notice of claim scope to competitors, it is informative to compare the precision the Federal Circuit is seeking with the precision of notice available to businesses concerning which practices violate the antitrust laws. The Federal Circuit's concern is to charge patentees and favor competitors with the imprecision in claim language by forcing narrow definitions of claim terms by the court and by eliminating a penumbra of equivalency. If competition policy is comfortable with vagueness in antitrust, one asks whether perhaps some imprecision (as in equivalency, for example) is tolerable from the perspective of competition policy in the patent law.

The analysis under the antitrust laws necessary for determining whether particular restrictive practices unreasonably restrict competition under § 1 of the Sherman Act is, with a few exceptions,[189] the rule of reason. The long-accepted test under the rule of reason has, at times, been recognized to be vague.

The general contours of rule of reason analysis were articulated by Justice Brandeis in 1918 in *Chicago Board of Trade v. United States*: "The true test of legality is whether the restraint imposed is such as merely regulates

[188] "[T]he mission of the Antitrust Division [of the Department of Justice] has been to promote and protect the competitive process—and the American economy—through the enforcement of the antitrust laws." Overview of the Antitrust Division of the Department of Justice, *available at* http://www.usdoj.gov/atr/overview.html The mission of the FTC Federal Trade Commission is to enforce "a variety of federal antitrust and consumer protection laws. The Commission seeks to ensure that the nation's markets function competitively, and are vigorous, efficient, and free of undue restrictions. The Commission also works to enhance the smooth operation of the marketplace by eliminating acts or practices that are unfair or deceptive." Federal Trade Commission's Visions, Missions and Goals, *available at* http://www.ftc.gov/mission.htm

[189] *See White Motor Co. v. United States*, 372 U.S. 253, 265 (1963) (Brennan, J., concurring) (noting that "the *per se* rule of prohibition has been applied to price-fixing agreements, group boycotts, tying arrangements, and horizontal divisions of markets."); *Continental Airlines, Inc. v. United Airlines, Inc.*, 126 F. Supp. 2d 962, 971 (E.D. Va. 2001) ("*Per se* analysis applies to . . . price-fixing, market-allocation agreements among competitiors, tying arrangements, and group boycotts") (footnotes omitted).

and perhaps thereby promotes competition or whether it is such as may suppress or even destroy competition."[190] The test invited exploration of "the facts peculiar to the business to which the restraint is applied; its condition before and after the restraint was imposed; the nature of the restraint and its effect, actual or probable. The history of the restraint, the evil believed to exist, the reason for adopting the particular remedy, the purpose or end sought to be attained. . . ."[191]

This remained the principal general statement of the rule of reason until the Supreme Court's decisions in the late 1970s in *Continental T.V., Inc. v. GTE Sylvania Inc.*[192] and *National Society of Professional Engineers v. United States.*[193] In *GTE Sylvania,* the Court explained that under the rule of reason "the fact-finder weighs all of the circumstances of a case in deciding whether a restrictive practice should be prohibited as imposing an unreasonable restraint on competition."[194] And in *NCAA v. Board of Regents,*[195] "the essential inquiry [is] ... whether or not the challenged restraint enhances competition." The courts have observed that a restraint's effect on competition depends on the restraint's effect on output,[196] price,[197] resource allocation,[198] quality, and

[190] *Chicago Board of Trade v. United States,* 246 U.S. 231 (1918). The rule of reason, which can be traced back to common law, was first applied in the Sherman Act context in *Standard Oil Co. v. United States,* 221 U.S. 1, 58 (1911).

[191] *Chicago Board of Trade,* 246 U.S. at 238.

[192] *GTE Sylvania,* 433 U.S. at 36 (1977).

[193] *National Soc'y of Prof'l Eng'r v. United States,* 435 U.S. 679 (1978).

[194] *GTE Sylvania,* 433 U.S. at 49; *accord Business Elecs. Corp. v. Sharp Elecs. Corp.,* 485 U.S. 717, 723 (1988); *Metro Indus. v. Sammi Corp.,* 82 F.3d 839, 843 (9th Cir. 1996), *cert. denied,* 117 S. Ct. 181 (1996); *Wisconsin Music Network v. Muzak Ltd. Partnership,* 5 F.3d 218, 222 (7th Cir. 1993); *Capital Imaging Assocs. v. Mohawk Valley Med. Assocs.,* 996 F.2d 537, 543 (2d Cir.), *cert. denied,* 510 U.S. 947 (1993); *United States v. All Star Indus.,* 962 F.2d 465, 468 (5th Cir. 1992), *cert. denied,* 506 U.S. 940 (1992); *United States v. Suntar Roofing, Inc.,* 897 F.2d 469, 472 (10th Cir. 1990); *Murrow Furniture Galleries v. Thomasville Furniture Indus.,* 889 F.2d 524, 527 (4th Cir. 1989); *DeLong Equip. Co. v. Washington Mills Abrasive Co.,* 887 F.2d 1499, 1507 (11th Cir. 1989), *cert. denied,* 494 U.S. 1081 (1990); *Thurman Indus. v. Pay 'N Pack Stores,* 875 F.2d 1369, 1373 (9th Cir. 1989).

[195] *NCAA v. Board of Regents,* 468 U.S. 85, 104 (1984).

[196] *Id.* at 99, 114, 120; *Broadcast Music, Inc. v. CBS,* 441 U.S. 1, 19–20 (1979) (appearing to equate anticompetitive effect with a restriction on output and increase in price); *see also Chicago Prof'l Sports Ltd. P'ship v. NBA,* 961 F.2d 667, 673 (7th Cir. 1992) (noting in dictum that "a 'restraint' that in the end expands output serves the interests of consumers and should be applauded rather than condemned"), *cert. denied,* 506 U.S. 954 (1992); *Schachar v. American Academy of Ophthalmology, Inc.,* 870 F.2d 397, 399 (7th Cir. 1989) ("Antitrust law is about consumers' welfare and the efficient organization of production. It condemns reductions in output that drive up prices."); *Martin v. American Kennel Club,* 697 F. Supp. 997, 1002–03 (N.D. Ill. 1988). *But see Rothery,* 792 F.2d at 231 ("I think it premature to construct an antitrust test that ignores all other potential concerns. . .except for restriction of output and price raising.") (Wald, J., concurring).

[197] NCAA, 468 U.S. at 109 n.38; *Wilk v. American Med. Ass'n,* 895 F.2d 352, 359 (7th Cir. 1990), *cert. denied,* 496 U.S. 927 (1990).

consumer choice.[199] As a result, commentators have observed that the standard "is always something of a sliding scale in appraising reasonableness, but the sliding scale formula deceptively suggests greater precision that we can hope for.... The quality of proof required should vary with the circumstances."[200] From the businessmen's perspective, "without further elaboration, reasonableness is too vague to guide the businessman's actions or the judge's discretion. Such openness is a mixed blessing. Unbounded by technical limitations, it reaches every evil. But unless disciplined by the purposes of the antitrust laws, it is a vagrant standard."[201]

In short, our competition policy has accepted some vagueness in a fundamental antitrust principle. It follows that, from the same perspective, some imprecision in claims should be tolerable as well.

V. CONCLUSION

The Federal Circuit majority has dramatically changed the law of claim construction and the doctrine of equivalents based on its guiding principle of giving competitors of patentees wide breadth and confining the patent scope available to the patentee narrowly. In claim construction, the court left in place the rule that claims are to be read as those of persons of ordinary skill in the particular art at the time of the invention would understand them. Nevertheless, the court declared that expert testimony as to what the claims meant would be irrelevant or secondary and the jury's role in resolving the parties' differences concerning meaning would be dropped. In order to maximize freedom for competitors of the patentee to avoid its patented invention, the written prosecution record would govern claim scope and judges would construe that record.

Similarly, rather than continuing to require a determination of the scope of what was surrendered in a narrowing claim amendment during prosecution, the court erased the doctrine of equivalents altogether for the many

[198] *Les Shockley Racing, Inc. v. National Hot Rod Ass'n*, 884 F.2d 504, 508 (9th Cir. 1989) ("only when the restraining force of an agreement. . .affecting trade becomes unreasonably disruptive of market functions such as price setting, resource allocation, market entry or output designation is a violation of the Sherman Act threatened").

[199] *E.g.*, *Brown Univ.*, 5 F.3d at 668 (identifying "reduction in output,. . .increase in price [and] deterioration in quality" as anticompetitive effects in rule of reason analysis); *Tunis Bros Co. v. Ford Motor Co.*, 952 F.2d 715, 728 (3d Cir. 1991) ("An antitrust plaintiff must prove that challenged conduct affected the prices, quantity or quality 'of goods or services.'"), *cert. denied*, 505 U.S. 1221 (1992).

[200] Phillip E. Areeda, ANTITRUST LAW: AN ANALYSIS OF ANTITRUST PRINCIPLES AND THEIR APPLICATION ¶ 1507, at 402 (1996).

[201] *Id.* at 362.

patent claims which had been narrowed in prosecution. In order to maximize the freedom of competitors to avoid the patented invention, any minor change by the competitor in a claim element that was the subject of a narrowing amendment in prosecution would avoid the patent.

In both instances, increased quality in the notice provided to competitors of the patent comes at the expense of diminishing the patentee's reward for its innovation and its ability to stop infringers. This, however, reflects but one of the procompetition policies underlying the patent law. The other is to promote progress and competition by providing the patentee with a robust reward as an incentive to innovate.

The Federal Circuit has embraced an approach that counts only competitors of the patentee and not the patentee's contribution to competition and the reward necessary to elicit it. This approximates the view that patents are an exception to a general scheme of competition by emulation and are to be narrowly confined. This principle is somewhat reminiscent of views about the relationship between patents and competition of Justices Black and Douglas in the 1940s and 1950s.

The Black/Douglas restrictive view of patents developed in light of the competition policy reflected in the antitrust laws. However, the Black/Douglas approach has been left behind in that realm. A sharply different view took hold beginning with *Kewanee* and manifested particularly in *SCM v. Xerox*. The *SCM v. Xerox* decision stressed the contribution of patentees to competition from the unique position of having a full record of the role of patents in one of the major innovations of the twentieth century.[202]

The perspective introduced in *SCM v. Xerox* was developed at length by the Antitrust Division. The new approach, termed a "revolution" by the Antitrust Division, recognized the need for solid patent protection to encourage innovation by patentees because innovation is important to competition and economic welfare. This approach led to the reversal of the division's previous restrictive view of patent licensing practices and caused the division to criticize the courts' misunderstanding of the role of patents in promoting competition by the patentee.

Beginning with the *GTE* case, antitrust policy identified free-riding as a negative. The idea that free-riding detracts from the incentive of businesses to promote their products led to reversal of the *Schwinn* decision which had condemned all vertical restrictions on resale. By the same token, rapid imitation by competitors of a patentee was viewed as a negative by the Antitrust Division in the patent context.

[202] The year-long trial was probably the most in-depth microeconomic industry study ever done. In contrast to the landmark industry studies done by *Mansfield*, the court had the benefit of exhaustive access to the parties' and competitors' internal records and full testimony in an adversary proceeding.

Later antitrust cases and the 1988 statute concerning patent misuse revised the Supreme Court's earlier antitrust doctrines, giving patentees more leeway to impose restrictions. This affected patent misuse, tie-ins, and postsale restrictions.

The Federal Circuit majority's dedication to narrowing the scope of patent protection and allowing competitors to design around patents with minor changes of the claimed invention promotes a form of free-riding. The Federal Circuit majority reached that decision by using an incomplete calculus that measures benefits to competition by how readily competitors are permitted to closely imitate the patent. That standard, however, omits weighing the contribution of patentees through promoting competition by their innovations. This omits a procompetitive force powerful enough to have changed the law and its enforcement in antitrust where promoting competition is the governing principle. It also fails to recognize the dramatic changes in antitrust law that the contribution of patentees to competition has caused.

4 Product Patents on Human DNA Sequences: An Obstacle for Implementing the EU Biotech Directive?

Joseph Straus
Managing Director
Max Planck Institute for Intellectual Property, Competition, and Tax Law
D-80539 Munich, Germany

Professor of Law
University of Ljubljana
SI-1000 Ljubljana, Slovenia

Professor of Law
University of Munich
D-80539 Munich, Germany

Marshall B. Coyne Visiting Professor of International and Comparative Law
The George Washington University Law School
Washington, DC 20052

I. INTRODUCTION

After 10 years of tense and controversial debates, the European Union, on July 6, 1998, adopted the Directive 98/44/EC of the European Parliament and the Council on the Legal Protection of Biotechnological Inventions.[1] Aimed at

[1] OJ EC No. L 213/13 of 30.7.1998.

0065-2660/03 $35.00

securing effective and harmonised protection of biotechnological inventions throughout the EU-member states in order to maintain and encourage investment in the field of biotechnology (Recital 3), without creating a separate body of law (Recital 8), the Directive determines the subject matter eligible for, as well as explicitly excluded from patent protection (Recitals 20 24, 38 42, 45, Articles 3-6), provides for some specific rules on the scope of protection (Recital 25, Articles 8–11), on compulsory cross-licensing (Article 12) and, finally, on deposit, access and re-deposit of biological material (Article 13). The deadline for implementing the Directive into national laws of the member states lapsed on July 31, 2000. However, as yet only Denmark, Finland, Greece, Ireland, Spain, and the United Kingdom have implemented the Directive into their national laws. The decision of the European Court of Justice (ECJ) of October 9, 2001,[2] which explicitly confirmed the *légalite* of the Directive, seemingly had no impact on national lawmakers of the member states. On the contrary, in January of 2002, the Lower House of the French Parliament passed a bill, which, if adopted also by the Senate, would ban from patent protection all inventions related to or involving biological material of human origin.[3] If one assumed that the Directive remained without impact on the patent practice in Europe, however, that would be equally incorrect. Although not bound by any legal mechanisms to the EU legal system, the Administrative Council of the European Patent Organization on June 16, 1999[4] decided to adapt the Implementing Regulations to the European Patent Convention (EPC) to the rules set forth in the Directive to the extent they relate to patentability issues. As a result, new Rules 23b-23e and 28 (6) were introduced into the Implementing Regulations. Under Article 164 (1) EPC, the Implementing Regulations are an integral part of the EPC and therefore are equally binding on the EPO's Boards of Appeal (Article 23 (2) EPC) and on national courts. Hence, for practical application of the EPC, only the interpretation of its provisions laid down in the Implementing Regulations is binding. A different interpretation of the EPC would be possible only if it is specifically demonstrated that a particular rule of interpretation is inconsistent with the EPC itself.[5] Since the

[2] Case C-377/98—Kingdom of the Netherlands, supported by Italian Republic and by Kingdom of Norway vs. European Parliament and Council of the European Union supported by Commission of the European Communities. The ECJ rejected all the pleas of the applicants.

[3] Projet de Loi adopté par l'Assemblée Nationale en première lecture relative à la bioéthique, texte adopté no. 763 of January 22, 2002 (http://www.assemblee-nationale.fr/ta/ta0763.asp). Article 12bis (new) of the bill reads as follows: Un élément isolé du corps humain ou autrement produit par un procédé technique, y compris la séquence ou la séquence partielle d'un gène, ne peut constituer une invention brevetable." This wording literally contradicts Article 5 (2) of the Directive.

[4] OJ EPO 1999, 425.

[5] Notice of the European Patent Office of 1 July 1999 concerning the amendment of the Implementing Regulations to the European Patent Convention, OJ EPO 1999, 545 ss., under No. 11.

overwhelming majority of biotech patent applications in Europe is filed with the European Patent Office (EPO), this eventually means that they are examined and patents issued according to the principles set forth in the EU Directive.

II. PRESUMABLE OBSTACLES FOR IMPLEMENTATION

The reasons why the great majority of EU member states resisted implementing Directive 98/44/EC into its national laws may vary from country to country. Reasons may stem from differences in the understanding of ethics, as demonstrated by the action of the Lower House of the French Parliament, as well as some specific economic interests.[6] The Organization for Economic Cooperation and Development (OECD) in a workshop held in June of 2001 identified the following concerns, which might be responsible for hesitations of the EU member states: dependency resulting from DNA patents in general and from undue broad claims specifically; reluctance of researchers to enter fields with already patented genes; genetic testing; monopolistic genetic testing practices; royalty stacking; and explosion of legal disputes.

Despite the fact that, by 2001, the EPO has issued some 400 patents on DNA sequences of human origin,[7] an empirical study of the Max Planck Institute for Foreign and International Patent, Copyright, and Competition Law commissioned by the German Federal Ministry for Education and Research[8] could not confirm the concerns identified by the OECD. Interviews carried out in 25 institutions—large pharmaceutical companies, biotech start-ups, universities, and other publicly funded research institutes and clinical institutions involved in genetic testing—have revealed that all involved can cope with the system as is satisfactorily. The only sign of a negative impact of product patents on DNA sequences: Researchers, in most cases, refrain from researching further uses of a gene once they realize that it has already been patented for a third party. The perspective of being dependant on a third party's dominant patent once a

[6] In its action, the Kingdom of Netherlands referred to the "express request of the Netherlands Parliament, in the light of the opposition expressed to genetic manipulation involving animals and plants and to the issuing of patents for the products of biotechnological procedures liable to promote such manipulations." (*cf*. No. 4 of the ECJ Judgement).

[7] Some 7,700 patent applications for biological material (*i.e.*, "any material containing genetic information and capable of reproducing itself or being reproduced in a biological system"—Article 2 (1) a) of the EU Directive) were pending in the EPO in 2001. By 1999, the EPO has granted some 3,000 patents for biotechnological inventions which fall within the scope of the Directive. At the same time some 1,500 applications related to transgenic plants and some 600 to transgenic animals and around 2,000 DNA sequences were pending in the EPO (*cf*. Notice of the EPO, OJ 1999, 425 ss. No. 7).

[8] Straus/Holzapfel/Lindenmeir, Empirical Survey on "Genetic Inventions and Patent Law," Munich, January 2002.

commercially usable product has been developed is clearly disliked.[9] In a broader context, this impact is well reflected in concerns expressed by leading scientists in Germany. Professor Hubert Markl, President of the Max Planck Society for the Advancement of Science, stated in a recently published article[10]:

> I want to make it perfectly clear that I am not against the patenting of genetic information, even from humans, as long as these basic conditions are fulfilled[[11]] We should, however, be carefully aware that it is not the chemical formula of a gene but the functional information which is the truly patentable content and only by providing access to such a functional process—and not from the genetic substance itself, as a natural biochemical compound, such as an antibiotic—can the marketable benefit be derived. If this defined and useful function is clearly proven, I can see no legitimate legal or moral argument against issuing patents for genetic sequences derived from the human genome, which may—as already mentioned—anyway be identical to sequences from other organisms.[12] . . . I also wish to emphasize strongly that it is definitely desirable to limit the patent rights given to a well-defined and clearly proven useful function, since we know now that each single gene—of which we may have only 25,000 to 40,000 altogether in one human genome—may be involved in the production of a ten- or twentyfold number of functional proteins, and many such proteins may be enmeshed in a number of different functions of an organism. Assigning broadly defined patent rights to a specific gene plus its protein, for which only one function has been described, in such a way that all additional functions described in the future are also covered—even though common practice when granting traditional patents on chemical or pharmaceutical substances—could be ruinous to an economic landscape of biotech start-ups, because the actual inventor of a completely new marketable use would immediately be subjected to serfdom under license to someone who did not contribute his own creative intellectual or practical effort to a new development in that field. Sweeping genetic technology patents could thus all too easily brush a promising new industry down the drain.[13]

[9] Straus/Holzapfel/Lindenmeir, op. cit., footnote 8, No. 6c.

[10] *Who Owns the Human Genome? What can Ownership Mean with Respect to Genes?* 33 IIC 1 ss. (2002).

[11] *i.e.*, Strict patentability requirements [J.S.].

[12] Op. cit., p. 3.

[13] Op. cit., p. 4.

This line of arguments not only reflects thoughts expressed by prominent scientists, such as professor Ernst-Ludwig Winnacker, President of the German Research Organization (DFG),[14] and the Nobel Laureate Hartmut Michel[15] but also early warnings of some highly respected U.S. experts such as Thomas Kiley,[16] cofounder of Genentech, and Professor Thomas Caskey, the former president of the Human Genome Organization (HUGO).[17]

On October 18, 2000, when the German federal government decided to implement the EU Directive into national law, it seemingly foresaw the Parliament's resistance to amending the Patent Act. It therefore, at the same time, announced that it would undertake the necessary steps at the European level to initiate changes and improvements in the European patent law, which, in view of the enormous progress in the sequencing of the human genome and in biomedical research, might prove necessary. As areas of special interest the German government identified: the reach (extent) of product patents; patentability requirements for genes, gene sequences, and parts thereof; ethical boundaries for patenting of parts of the human body; and adequate rules for the relationship between patent rights and plant variety protection.[18]

The first consultations of the bill in the German Parliament confirmed the expected resistance which nearly exclusively focused on the issue of product patents on DNA sequences of human origin. In a nearly never-experienced unanimity, representatives of all parties rejected that idea as unacceptable and unjustifiable.[19] Symptomatic is the statement of Margot von Renessee, chairwoman of the Enquête Commission on "Law and Ethics in Modern Medicine" who, in principle, strongly supports the efforts of the government aimed at implementing the Directive, but nonetheless objected in this respect as follows:

> [W]e would like to adapt our national patent law within the EU Patent Directive as much as possible ... in a way by solving ... the problem which, in my view, marks the system boundary, namely the risk of overrewarding the person who obtains a product patent, particularly in the field of natural products. This is particularly evident in the case of

[14] *Cf.* 406 NATURE 111 (2000).

[15] *Cf.* his Letter to the Editors, 407 NATURE 285 (2000).

[16] *Patents on Random Complementary DNA Fragments?* 257 SCIENCE 215 ss. (1992).

[17] *The Great Gene Patent Race*, CHEMISTRY & INDUSTRY 825 (16 October 1995).

[18] The Federal Ministry of Justice on August 16, 2000, published a bill for an "Act Implementing the Directive on the Legal Protection of Biotechnological Inventions" (Ref. No.: 3620/6. Reasons for the bill followed on September 14, 2000. The bill submitted to the Parliament was published as Parliament Print No. 14/5642).

[19] Minutes of the session are available on the Internet under: www.bundestag.de. *See for details also* Straus, *Produktpatente auf DNA-Sequenzen—Eine aktuelle Herausforderung des Patentrechts*, GRUR 2001, 101 ss.

genes, because they are multi-functional. Granting a product patent for one function alone ultimately carries the risk of over-rewarding, which is why we should move so far as possible in the direction I have described. I would also like the patent lawyers to help by telling us how we can limit the product patent to the function discovered, whilst not sacrificing anything by so doing.[20]

III. THE CASE OF PRODUCT PATENTS ON DNA SEQUENCES

It appears necessary that the arguments raised against product patents on DNA sequences by such a diverse group of opponents be first examined against the background of the rules set forth in the Directive and the case law related to patenting in the area of pharmaceuticals and chemicals.

First, it must be observed that it was generally welcomed that the Directive clarified that the human body, at various stages of its formation and development, and the simple discovery of one of its elements, including the sequence or partial sequence of a gene, cannot constitute patentable inventions (Recital 16, Art. 5 (2)). This merely confirms the prevailing interpretation of the patent law.[21] This was not the case with another rule of the Directive, which equally only confirmed the former case law: Namely, that elements isolated from the human body or otherwise produced by means of a technical process, including the sequence or partial sequence of a gene, may constitute a patentable invention, even if the structure of that element is identical to that of a natural element (Recital 20, Art. 5 (2)). To opponents, it seems already disturbing that DNA sequences of human origin are treated along the same criteria that the German Federal Patent Court in 1977 in the *Antamanide* case[22] established for distinguishing between a nonpatentable discovery and a patentable invention with regard to naturally occurring substances. According to *Antamanide*, a patentable invention not only requires the product to be identifiably disclosed for the first time, *i.e.*, that the public had no previous knowledge of it, but that the technical teaching on the process for producing such product and its possible

[20] For the views of Renessee cf. also M. von Renessee/K. Tanner/D. von Renessee, DAS BIOPATENT—EINE HERAUSFORDERUNG AN DIE RECHTSETHISCHE REFLEXION, MITTEILUNG DER DEUTSCHEN PATENTANWÄLTE (Mitt.) 2001, 1 ss. (at 3).

[21] The German Reich's Patent Office already in the 1930s explicitly adhered to the principle that man may not be made the object of technology (Patent Office of 15.12.1933, Mitt. 1934, 19 (20). Cf. Beier/Straus, GENETIC ENGINEERING AND INDUSTRIAL PROPERTY RIGHTS, Int. Prop. 1986, 447 ss.

[22] Decision of 28.7.1977, GRUR 1978, 238 - *Naturstoffe (Antamanide)*—with comments by Uttermann.

uses must also be disclosed.[23] However, the position has become altogether negative when it was also considered that according to the case law of the German Federal Supreme Court (Bundesgerichtshof, or BGH), this "product protection" is in principle "absolute," *i.e.*, resulting in the right of the patent-holder to prohibit any commercial use of the patented product—chemical compound—whether he has recognized and disclosed such use or not.[24] And this applies even if such further uses identified by third parties are inventive and can also be patented.

In the somewhat emotional public debate on the patenting of DNA sequences, not all the aspects of the *Imidazoline* decision of the German Federal Supreme Court, with which this ultimately result-oriented case law on absolute product protection was reasoned, appear to have been clarified in the light of the new rules adopted in the biotech Directive. The views expressed so far in legal doctrine[25] have obviously not reached the lawmaker. In any case, they were not reflected in the debates of the German Parliament.

[23] *Ibidem*. Because of their significance in the context of interest here, the holdings of the Court relating to novelty and inventive step should be recalled: "If we assume the identity of the antanamide claimed with the substance contained in Death Cap (Amanita phalloides) this does not affect the novelty of the present invention. . . . But on the date of application, no expert was able to use the cyclic decapeptide antanamide with the formula given in Claim 1. They did not know the substance as such, let alone its valuable property, namely the effect of neutralizing an absolutely lethal dose of phalloidine. . . . The substance claimed also reveals the advance and the inventive step necessary for the granting of a patent. The effect justifying the patent lies in the fact that the substance claimed has the effect. . . . This effect could not be expected because the cyclic hexapeptide Phallin A, partly made up of the same aminoacids, proved to be ineffective against the toxin of Death Cap. Further, a period of 30 years has passed between the first publication of the constituents of Death Cap in 1936 and the present application, during which time experts have overlooked the valuable substance of antamanide, although this area has been very intensively researched" (*cf.* 31st Report in Liebig's Annals of Chemistry 700 (1966)) (*Ibidem* p. 239). *Cf. also* Federal Patent Court of 24.7.1978, GRUR 1979, 702—*Menthothiols*.

[24] BGH of 14.3.1972, GRUR 1972, 541 (544)—*Imidazoline*.

[25] *Cf.*, *e.g.*, Meyer-Dulheuer, *Der Schutzbereich von auf Nucleotid-oder Amonisäure Sequenzen gerichteten biotechnologischen Patenten*, GRUR 2000, 179 ss. Dörries, *Patentansprüche auf DNA Sequenzen: Ein Hindernis für die Forschung? Anmerkungen zum Regierungsentwurf für ein Gesetz zu Umsetzung der Richtlinie 98/44/EG*, Mitt. 2001, 15 ss.; Nieder, *Die gewerbliche Anwendbarkeit der Sequenz oder Teilsequenz eines Gens—Teil der Beschreibung oder notwendiges Anspruchsmerkmal von EST-Patenten?* Mitt. 2001, 97 ss.; *Nieder, Gensequenz und Funktion—Bemerkungen zur Begründung des Regierungsentwurfs für ein Gesetz zur Umsetzung der Richtlinie 98/44/EG*, Mitt. 2001, 238 ss.; Feuerlein, *Patentrechtliche Probleme der Biotechnologie*, GRUR 2001, 561 ss.; Oser, *Patentierung von (Teil-) Gensequenzen unter besonderer Berücksichtigung der EST-Problematik*, GRUR Int. 1998, 648 ss.; von Pechmann, *Wieder aktuell: Ist die besondere technische, therapeutische oder biologische Wirkung Offenbarungserfordernis bei der Anmeldung chemischer Stofferfindungen?* GRUR Int. 1996, 366 ss.; Straus, *Abhängigkeit bei Patenten auf genetische Information—Ein Sonderfall?* GRUR 1998, 314 ss.

First, it has to be recalled that in the *Imidazoline* decision, the BGH did indeed state that "product protection ('*Stoffschutz*,' *i.e.*, protection of substance, *i.e.*, protection of chemical compound) is absolute in principle" and that this does not prevent the applicant from claiming purpose-bound *Stoffschutz*. But the BGH also added that it did not have to decide whether there are exceptional cases where the claimed *Stoffschutz* could be granted in a way to be limited to a specific purpose (use), because the case at hand in no way gave reasons for such considerations. The BGH based its position in this respect on the fact that in the case of chemical compound inventions, the technical problem (object) of the invention is to be seen in providing a new chemical compound with a constitution defined in more detail. According to this case law, the indication about the technical or therapeutic effect does not form part of the invention itself (*Stofferfindung*) and therefore does not need to be included already in the documents submitted with the application, but could be submitted later, in the course of the examination proceedings.[26] It seems of particular interest in this respect that since then there has been an almost unanimous consensus that, in the case of chemical compound inventions, the basic patenting requirement of the inventive step is usually met with surprising properties and effects that the new product has, compared with comparable known chemicals and which a person skilled in the art could not anticipate. This also applies if the product itself is obvious and the process even known.[27]

A look at the EU Biotech Directive shows that there are two reasons why BGH case law related to product patents on chemical compounds can no longer apply in regards to the patenting of DNA sequences of human origin. On the one hand, the disclosure of a mere DNA sequence without indication of a function does not contain any technical information and is therefore not a patentable invention (Recital 23), even if the method of manufacture is indicated. And on the other hand, the industrial applicability of the DNA sequence—in other words, its function—cannot be submitted in the course of the examination process, but has to be *specifically* disclosed already in the patent application as filed (Recital 22, last sentence, Art. 5 (3)). Where the use of a sequence or partial sequence of a gene for making a protein or part of a protein is claimed, the protein or part protein and its function must be specified (Recital 24). If therapeutic or diagnostic uses are claimed, the disorder to be diagnosed or treated must be specifically indicated. Thus, the European legislator has made the function of a claimed DNA sequence an integral part of the notion of an

[26] GRUR 1972, 544.

[27] Reference is merely made to Busse (Keukenschrijver), PATENTGESETZ, 5th edition, Berlin/New York, 1999, section 4, note 91; Benkard (Bruchhausen), PATENTGESETZ, GEBRAUCHSMUSTERGESETZ, 9th edition, Munich 1993, section 4, note 39.

invention (inventive concept) of a chemical compound invention, at least in this area, and has thereby deprived the previous arguments by the BGH of their basis in this respect.[28]

However, at the same time, the EU Biotech Directive makes it clear that it is not creating any special patent law for biotechnological inventions; it is only making adaptations and amendments that the national legislator must also implement in order to take adequate account of the developments that use biological material but at the same time also fulfill the patentability requirements (Recital 8).[29] This is achieved, in addition to the clarifying amendment to the notion of an invention in case of DNA sequences, by the explicit clarification that the natural preexistence of biological material alone does not constitute a patentability obstacle (Art. 3 (2)). The latter principle thus has only confirmed the former case law of the German Federal Patent Court. Two other provisions of the Directive relating to the scope of protection and its effects need to be mentioned. First, the provision that the protection of a product which consists of or contains genetic information, *e.g.*, a DNA sequence, extends to any product—*except man*—in which this product is incorporated and in which the genetic information is contained and performs its function (Art. 9). In order to counteract the problem of far-reaching dependencies arising from the scope of product protection, the interpretation rule is then set forth in one Recital, according to which patented DNA sequences, which only overlap in parts that are not essential to the invention, have to be considered as independent in terms of patent law (Recital 25).[30]

[28] In his final analysis, Meyer-Dulheuer, GRUR 2000, 181 too, refers to the indication of function in claims referring to DNA sequences as "the technical feature constituting the invention." Cf. also Nieder, Mitt. 2001, 99; same Mitt. 2001, 238; Feuerlein, GRUR 2001, 563 s. Whereas Meyer-Dulheuer and Nieder take the view that the indication of function "must have place in the patent claims as the feature that justifies patentability and scope of protection" (Meyer-Dulheuer, op.cit.), Feuerlein believes that according to the Directive "the applicability of a sequence or partial sequence of a gene" does not have to be included in the patent claims. Evidently, Feuerlein is referring to the indication of the function because, in the next sentence of his article, he emphasizes that for all other inventions in the area of chemistry, the indication of the function of the claimed compound continued not to form a mandatory part of the description (GRUR 2001, 564). For more on the issue of whether the indication of the function belongs in the description, see *infra*.

[29] ECJ in its judgment (No. 25) also explicitly emphasized that the Directive does not introduce a new right. Attorney General Jacob, in his brief, made it also clear that no changes in classical patentability requirements of novelty, inventive activity, and industrial applicability have been introduced by the Directive (No. 39). (The brief of Attorney General Jacobs of June 14, 2001, is accessible under www.curia.eu.int/engl/index.htm).

[30] *Cf.* for more Straus, *Biotechnology, and Patents*, 54 CHIMIA 293 ss., at 297 (2000); Nieder, Mitt. 2001, 238.

IV. THE EU BIOTECH DIRECTIVE IN THE LIGHT OF THE LATEST SCIENTIFIC DEVELOPMENTS

The European legislature adopted the Directive in July of 1998. It can be assumed that the European Parliament, the Commission, and the Council have not only endeavored to take account of current developments in science and technology but that they have also actually taken them into account. One should not fall prey to illusion and assume that any legislature could regulate in advance developments of such dynamic progress and speed that have taken and continue to take place in the field of genomics.[31] It is not by chance that, as yet, the U.S. government by and large[32] has resisted all demands to include special rules relating to biological inventions into the Patents Act (35 U.S.C.).[33] Instead, it trusts the guidelines of the U.S. Patent and Trademark Office (US PTO)[34] and in the art of interpretation by the Court of Appeal for the Federal Circuit (CAFC) and, of course, the U.S. Supreme Court.[35]

If Europe had a central patent jurisdiction of the U.S. type, the Directive could be regarded dispensable. A court of this type could be responsible both for a Europewide harmonized interpretation of patent law and for adapting it to the changing conditions in science and technology. For example, both in the *Red dove*[36] decision and the *Rabies virus* decision,[37] the BGH has

[31] Even the decision of the German federal government of October 18, 2000 could create this impression. It states, *inter alia* "that in view of the pioneering advances in the decoding of . . . the European patent law has not found final answers to the challenges of this new area of technology on all points."

[32] Amendments to section 271 (e) and 287 (c) of 35 U.S.C. can be disregarded in the context of interest.

[33] However, mention should be made of two bills submitted to the 107th Congress 2nd Session on January 31 by Representative Lynn Rivers, namely the "Genomic Research and Diagnostics Accessibility Act of 2002" and the "Genomic Science and Technology Innovation Act of 2002." If adopted, the first amendment would provide for an exemption from infringement for the use of "any patent for or patented use of genetic sequence information for purposes of research."

[34] *See Revised Interim Written Description Examination Guidelines of 21.12.1999*, 64 FED. GOV. 71434 and *Revised Utility Examination Guidelines of 12.12.1999*, 64 FED. GOV. 71440, which now require a "specific, substantial and credible" usefulness. *Cf.* Cantor, *Using the Written Description and Enablement Requirements to Limit Biotechnology Patents*, 14 HARVARD JOURNAL OF LAW & TECHNOLOGY 267 ss. (Fall 2000); Meigs, *Biotechnology Patent Prosecution in View of PTO's Utility Examination Guidelines*, 83 y PTOS 451 ss. (July 2001); Worrall, *The 2001 PTO Utility Examination Guidelines and DNA Patents*, 16 BERKELEY TECH. L.J. 123 ss. (2001).

[35] The most recent example of its interpretation of the statute relating to biotechnological inventions being Case No. 99-1996, *J.E.M.AG Supply, Inc., DBA Farm Advantage Inc. et al v. Pioneer Hi-Bred International, Inc., of December 10, 2001, 63 PTCJ 144 (12-14-01)*, confirming the patentability of plants and plant varieties under 35 U.S.C. § 101.

[36] GRUR 1969, 672.

[37] GRUR 1987, 231.

provided impressive proof that it is both able and willing to interpret patent law dynamically dependent on the actual state of science and technology, taking into account the purpose of patent protection. Nothing illustrates this better than its statement: "Bearing in mind the purpose of patent law, which is to promote technical progress and encourage the inventive spirit of industry in a beneficial way, when interpreting and applying patent law, more account must be taken of practical needs than of theoretical considerations."[38]

In any case, it is evident that in the case of the United States, the interpretation and application of the 35 U.S.C. has not worked to its disadvantage. The United States remains the undisputed worldwide leader in this field. All the leading genomics companies are United States companies; 76% of genomics start-ups publicly traded and 71% of start-ups in private hands are United States companies, and a strong concentration of the ownership of patents and of other intellectual property rights in the United States is widely accepted. Europe and Asia carry far less weight in this respect.[39] It can also be assumed that the European legislator was well aware of this situation when it decided to adopt the Biotech Directive in order to achieve a harmonized interpretation of the patent law throughout Europe and to introduce protection standards in line with those of United States and Japanese law.

It should be added here that at present DNA is primarily used as the source for the production—expression in various hosts—of therapeutically effective proteins. The determination of the complex three-dimensional structure of proteins and their function has long been an extremely time-consuming and cumbersome undertaking. It took Nobel Laureate Max Perutz, for instance, 22 years to clarify the structure of hemoglobin. Based on Xray crystalography and Nuclear Magnetic Resonance (NMR) technologies, researchers more recently have achieved enormous progress. The protein data bank contains already 13,000 different structures into which proteins can fold. Using three-dimensional models, scientists are experimenting towards optimizing structures found in nature.[40] Identifying proteins that are involved in human disease as well as determining their three-dimensional structure is of enormous importance.[41] This importance is best demonstrated by comparing genomics and proteomics: "The genome tells you what could theoretically happen inside the cell.

[38] GRUR 1987, 233.

[39] *Cf.* Cook-Deegan/Chan/Johnson, World Survey of Funding for Genomic Research, Final Report to the Global Forum for Health Organisation, Washington DC 2000, http://www.stanford.edu/~bobcd/, p.3.

[40] *Cf.* Service, *Structural Genomics Often High-Speed Look at Proteins*, 287 SCIENCE 1954 ss. (2000); Abbot, *Studies by Numbers*, 408 NATURE 130 ss. (2000).

[41] See for the respective efforts, *e.g.*, Stevens/Yokoyama/Wilson, *Global Efforts in Structural Genomics*, 294 SCIENCE 89 ss. (2001); Baker/Sali, *Protein Structure Prediction and Structural Genomics*, 294 SCIENCE 93 ss. (2001).

Messenger RNA tells you what might happen, and the protein tells you what is happening."[42]

Also, it should be known that whereas the human genome does not vary much between individuals (99.9% homology), the expression of proteins demonstrates remarkable variations. Protein expression varies from tissue dramatically. This is true even for the expression of proteins in the tissue of a specific person in the course of the process of aging. Even at times when the existence of some 100,000 genes was still assumed, reference was made to the fact that 10–20 million chemically distinguishable proteins form the contents of a single cell.[43]

V. LESSONS TO BE DRAWN FROM THESE NEW DEVELOPMENTS

Taking into account the purpose of the patent law as defined by the BGH and the mandate of the German Federal Constitutional Court to the patent legislator—not only to protect individual interests but also "to impose the necessary limits on individual powers and rights in the interests of the common good and to create the correct balance between the rights of the individual and the rights of the general public"[44]—these findings cannot remain without consequences for the patenting of DNA sequences.

It must be assumed that with the state of technology that has existed for some time, as a rule—but by no means always—no activity is required for the clarification of the structure of a gene, *i.e.*, the exact analysis of its DNA sequence, which would represent an inventive step. This will also apply increasingly to the clarification of its functions. In this case, as a rule, there will be no qualitative difference as regards the inventive step needed for clarification and discovery of the first function, *e.g.*, medical use and of the further functions of a gene.

Contrary to the *Imidazoline* decision of the BGH, the high level of interactivity and interdependence of the processes at the genome level does not allow the inventive step, in relation to the first making available of the DNA molecule, to be acknowledged in the event of an obvious clarification of the DNA structure based solely on the discovery of the first function of the structure. In view of the consensus concerning the present state of sequencing technology, the technical problem (object) solved by such an invention cannot be seen in isolation or chemical synthesis of a new chemical product of a more closely defined structure, but only in conjunction with the discovery of the one or several functions of a product that exists in nature, whose discovery and

[42] Service, *Can Celera Do it Again?* 287 SCIENCE 2136 ss. (2000).
[43] *Cf.* Service, 287 SCIENCE 2137 (2000).
[44] Decision of 10.5.2000, GRUR 2001, 43, 44—Clinical trials.

structure clarification are not based on any inventive activity. In this respect, the *Imidazoline* case and other cases decided by the BGH are not comparable with the new situation. Thus, limitation (purpose bound) of the protection to the function(s) disclosed is clearly necessary.[45] In this situation, it is only logical that the function(s) must be specified in the patent claims, which will automatically affect the protective scope of the patent.[46] In this connection, *Meyer-Dulheuer* rightly concludes from the known *Antivirus agent* judgment of the BGH:

> If the critical creative contribution is to be seen in finding the function of a sequence, and the invention can only be applied with prior knowledge of this function, a "final" element is just as inherent in the biotechnological sequence claim as in the specific chemical compound claim [*Stoffanspruch*]. If this function is not used or not achieved, encroaching on its scope of protection is also ruled out.[47]

However, such a limitation is neither appropriate nor necessary if the finding and structure clarification of the DNA sequence as *such* are based on an inventive step. In this case, the technical problem (object) of the invention actually has to be seen in the making available of the DNA molecule for the first time, with the aid of which it will be possible to discover for the first time other functions in addition to the function(s) identified by the inventor, which—again, according to the decision of the legislator—represents the necessary integral component of the invention and which must be specifically indicated in the application.[48] In my opinion, the question of whether the function(s) of a DNA sequence that is structurally identical to that occurring in nature have to be included into the patent claims cannot be answered generally and in advance;

[45] *Also Meyer-Dulheuer*, GRUR 2000, 181: "Therefore, the object and solution of a biotechnological sequence patent are not already exhausted in the provision of a new sequence." *Also Nieder*, REPORT 2001, 238. However, both link these conclusions with the relevant invention concept according to the Directive without limitation.

[46] *Also Meyer-Duhlheuer*, GRUR 2000, 181; Nieder, REPORT 2001, 238 A. A. Feuerlein, GRUR 2001, 563, who, however, overlooks the fact that according to the construction of the Directive, the indication of the function is not only intended as an essential criterion of industrial applicability, but that it has also been raised to an integral component of the invention concept itself.

[47] *Meyer-Dulheuer*, GRUR 2000, 181. According to local opinion, the fact that the "antivirus agent" case concerned the use of a known active ingredient in no way changes the relevance of this statement.

[48] Di Cataldo puts forward similar considerations for inventions of chemical compounds in general, *Cf*. Some Considerations on the Inventive Step and Scope of Patents for Chemical Inventions in: Straus (ed.), Aktuelle Herausforderungen des geistigen Eigentums, Festschrift for Friedrich-Karl Beier on his 70th birthday, Cologne, etc. 1996, p. 11 ff (16).

it depends on where the critical contribution to the invention, *i.e.*, that based on the inventive activity, is seen. If the provision of the DNA sequence—*e.g.*, of a certain combination of the exons of a gene responsible for the expression of a protein is already based on an inventive activity,[49] it would not be justified to require that the function is indicated in the patent claims. On the other hand, if the examination of the relevant state of technology reveals that the inventive merit "merely" lies in the clarification of the function, the applicant may be required in the course of the examination, opposition, and even the appeal proceedings to include the details of the function of the claimed DNA sequence into the patent claims accordingly, *i.e.*, restrict them to it and thereby limit the scope of the patent accordingly. Anyone who uses this DNA section for other purposes is no longer encroaching upon this patent. But if the critical contribution lies in the provision of the DNA sequence itself, the traditional, absolute product—chemical compound—protection, with the resultant dependencies, is justified and complies with the system. In view of the present state, this will apply only in exceptional cases. However, the dependence rule (Recital 25 of the Directive) relating to partly overlapping sequences also gets a say here. If the phrase "essential to the invention" is interpreted within the meaning of the actual disclosure in the patent application and not objectively, as proposed by HUGO,[50] this could result in a further considerable reduction in the dependencies.[51] The result, which purely and simply depends on the correct application of the patentability requirements by patent offices and courts, may well contribute to a balanced application of patent law and also comply with the EU Biotech Directive and the Agreement on Trade Related Aspects of Intellectual Property Rights.

[49] Which, according to the general understanding, will rarely and increasingly seldom be the case.

[50] HUGO Statement on Patenting of DNA Sequences, 6 GENOME DIGEST No. 1, 10 f. (October 2000).

[51] *Cf.* Straus, 54 CHIMIA 297 (May 2000).

5

Patenting Genetic Products and Processes: A TRIPS Perspective

Charles R. McManis
Thomas & Karole Green Professor of Law
Washington University School of Law
St. Louis, Missouri 63130

At the heart of the debate over patenting genetic products and processes is the question of whether granting patents on the results of upstream genetic research will promote or undermine the norms of the biological research community and/or promote or retard the development of downstream commercial products and processes.[1] One way to develop an empirical basis for resolving this question, of course, would be to identify national patent regimes that do and do not provide

[1] *Compare* Michael A. Heller & Rebecca S. Eisenberg, *Can Patents Deter Innovation? The Anticommons in Biomedical Research*, 280 SCIENCE 698 (1998); Arti Kaur Rai, *Regulating Scientific Research: Intellectual Property and the Norms of Science*, 94 NW. U. L. REV. 77 (1999) *with* F. Scott Kieff, *Facilitating Scientific Research: Intellectual Property Rights and the Norms of Science—A Response to Rai and Eisenberg*, 95 NW. U. L. REV. 691 (2001).

0065-2660/03 $35.00

patent protection for the results of upstream genetic research and determine what effect these patent systems are having on the norms of the biological research community and the development of downstream products and processes in those particular countries.

A complicating factor in any such study, however, is that the national patent regimes of many countries are currently undergoing a profound transformation due to the requirements imposed by the Agreement on Trade Related Aspects of Intellectual Property Rights (generally referred to as the TRIPS Agreement), one in a bundle of agreements hammered out during the Uruguay Round of Multilateral Trade Negotiations among members of the General Agreement on Tariffs and Trade (GATT) and now administered by the World Trade Organization (WTO).[2] The TRIPS Agreement is unprecedented in the field of international patent law because it has, for the first time, established international minimum standards for the granting of patent protection, and most members of the WTO are obligated to have implemented these standards by January 1, 2005.[3] Thus, any attempt to address the ongoing debate regarding the patenting of genetic products and processes must take into account the rapidly changing face of national and international patent law.

Paradoxically, under the new TRIPS regime, it may be easier (and thus more essential) to obtain patent protection for the results of upstream genetic research than for the resulting downstream commercial products and processes. This is so because Article 27 of the TRIPS Agreement is drafted in such a way that it will arguably require the members of the World Trade Organization to provide patent protection for the results of genetic research at the microbiological and submicrobiological level but it will leave members free to grant or exclude patent protection for commercially valuable "macrobiological" products (*i.e.*, plants and animals), as well as for any associated biological processes (other than microbiological processes) for producing the same.[4] Thus, in a post-TRIPS world, the only reliable avenue for securing patent protection with respect to downstream macrobiological products and processes may turn out to be through the prompt filing of upstream patents at the microbiological or submicrobiological level, as

[2] *See* Final Act Embodying the Results of the Uruguay Round of Multilateral Trade Negotiations, done at Marrakesh, Morocco, 15 April 1994, reprinted in *The Results of the Uruguay Round of Multilateral Trade Negotiations—The Legal Texts* (GATT Secretariat (ed.) 1994) 2–3 [hereinafter *Results of Uruguay Round*]; Marrakesh Agreement Establishing the World Trade Organization, Annex 1C: Agreement on Trade-Related Aspects of Intellectual Property Rights, April 15, 1994 [hereinafter TRIPS Agreement], reprinted in Results of Uruguay Round, *supra*, at 16–19, 365–403.

[3] *See* TRIPS Agreement, *supra* note 2, Articles 27–34 (Patents) and 65–66 (transitional arrangements).

[4] *Id.*, Article 27. For a discussion of the provisions of Article 27, *see infra* notes 18–61 and accompanying text.

innovations at this level will be the heart of the downstream products and processes in any event. By way of explaining this puzzling TRIPS perspective on the patenting of genetic products and processes, this article will examine 1) the multilateral trade negotiation process that produced Article 27 of TRIPS; 2) the specific provisions of Article 27 that dictate this curious result with respect to genetic innovation; 3) the interpretive questions and political fallout that the language of Article 27 will likely generate as the transitional period for compliance with TRIPS draws to a close; and 4) the implications of all this for human genomic research and patent protection.

I. THE URUGUAY ROUND OF GATT NEGOTIATIONS

To put the TRIPS Agreement—and Article 27 in particular—into its larger political context, one must first understand that the multilateral trade negotiations leading up to the TRIPS Agreement were from beginning to end a bare-knuckled "North-South" confrontation.[5] The industrialized world demanded higher levels of intellectual property protection in the developing world in return for continued, and to some extent improved, developing country access to markets (particularly for textiles and agricultural products) in the industrialized world.[6] The industrialized world also insisted on an improved multilateral dispute settlement process that would guarantee developing country compliance with the new international minimum standards for intellectual property protection, while at the same time providing the developing world a measure of relief from unilateral threats of trade sanctions by industrialized countries such as the United States.[7] The reason that intellectual property protection was injected

[5] For previous discussions of the negotiations that led to the TRIPS Agreement, *see* Charles R. McManis, *Taking TRIPS On the Information Superhighway: International Intellectual Property Protection and Emerging Computer Technology*, 41 VILLANOVA L. REV. 207, 211–214 (1996) [hereinafter McManis, *Taking TRIPS*]; Charles R. McManis, *Intellectual Property and International Mergers and Acquisitions*, 66 U. CINCINNATI L. REV. 1283, 1286–1289 (1998); and Charles R. McManis, *The Interface Between International Intellectual Property and Environmental Protection: Biodiversity and Biotechnology*, 76 WASH. U. L.Q. 255, 257–259 (1998) [hereinafter McManis, *Biodiversity and Biotechnology*]. *See also* Frederick M. Abbott, *The WTO TRIPS Agreement and Global Economic Development* [hereinafter Abbott], in Frederick M. Abbott & David J. Gerber (eds.) Public Policy and Global Technological Integration 39–57(1997); J. H. Reichman, *From Free Riders to Fair Followers: Global Competition Under the TRIPS Agreement*, 29 N.Y.U.J. INT'L L. & POL. 11 (1996–1997) [hereinafter Reichman, *From Free Riders to Fair Followers*]. *See generally* Terence P. Stewart (ed.), The GATT Uruguay Round: A Negotiating History (1986–1994) [hereinafter Stewart], Vol. IV: The End Game (Part I) (1999).

[6] *See* Abbott, *supra* note 5, at 41.

[7] *Id.* at 42. The dispute settlement provisions are contained in Article 64 of TRIPS, which incorporates by reference the disputed settlement provisions contained in Articles XXII and XXIII of the 1994 update of GATT. *See* TRIPS Agreement, *supra* note 2, Article 64. For a discussion

into the Uruguay Round of multilateral trade negotiations in the first place was because of the previous failure of the World Intellectual Property Organization (WIPO), the international organization historically responsible for international intellectual property policy, to achieve any consensus on revising existing intellectual property treaties. The resulting deadlock was brought about, in large measure, by the rapidly increasing numbers of newly independent WIPO-member countries hailing from the developing world.[8]

An equally important point about the TRIPS component of the Uruguay Round of multilateral trade negotiations, however, is that, as the negotiations progressed and industrialized countries sought to delineate the precise minimum intellectual property standards that they wished to impose on the developing world, particularly with respect to the patenting of biotechnology products and processes, the debate increasingly became a "North-North" debate, exposing some fundamental differences among the intellectual property regimes of industrialized countries that would have to be reconciled if any agreed-upon international minimum standards were to be achieved.[9] The specific provisions of Article 27 of TRIPS are as much a product of this "North-North" debate as they are of the larger confrontation between the North and South.

For example, the negotiating proposals of the United States, Japan, the Nordic countries, and Switzerland initially advocated requiring broad patent coverage without exclusion for plants or animals.[10] However, most members, industrialized as well as developing, had opposed the patentability of plants and living organisms.[11] Indeed, the European Patent Convention (EPC)[12] is the obvious model for the exceptions to patentability eventually included in Article 27.2 and 27.3. Article 52(4) of the EPC specifies that methods of treatment of the human or animal body by surgery or therapy and diagnostic methods practiced on the human or animal body shall not be regarded as inventions, while Article 53 spells out an *ordre public* exception and goes on to specify that

---

of two relevant interpretive controversies likely to be resolved through that process, *see infra* notes 36–61 and accompanying text.

[8] *See* McManis, *Taking TRIPS*, *supra* note 5, at 214–215; Stewart, *supra* note 5, Vol. II: Commentary, 2249–2255 (1993); Nuno Pires de Carvalho, The TRIPS Regime of Patent Rights 50 (2003) [hereinafter Carvalho], noting that "the TRIPS Agreement, unlike the WIPO Treaties, should not be seen as an isolated agreement: it should rather be considered as a part of a package, under which developing countries have made concessions in the field of intellectual property in exchange for concessions in the area of goods, particularly agriculture."

[9] *See* Stewart, *supra* note 5, Vol. II: Commentary, 2294–2295 (1993).

[10] *Id.* at 2294.

[11] *Id.*

[12] European Patent Convention, Oct. 5, 1973, as amended Dec. 21, 1978, *reprinted in* Marshall A. Leaffer (ed.) INTERNATIONAL TREATIES ON INTELLECTUAL PROPERTY LAW 675–734 (2d Ed. 1997) [hereinafter Leaffer].

patents shall not be granted for plant or animal varieties or essentially biological processes for the production of plants or animals, other than microbiological processes or the products thereof.[13] The United States itself acknowledged during the course of the negotiations that there was no international consensus as to whether genetically engineered life forms should be patentable.[14]

At the same time, the European Union joined the United States and Switzerland in expressing concern over widespread patent exclusions for food, chemical, and pharmaceutical products, and also expressed concern over the limited recognition of *sui generis* protection of plant breeders' rights—the preferred European method for protecting plant innovation.[15] Developing countries, on the other hand, expressed concern that they might be prevented from securing access to modern technology due to overprotection of intellectual property rights.[16] The eventual upshot of this multifaceted debate was that, in return for increased developing country access to the agricultural and textile markets of the industrialized world, developing countries reluctantly agreed to Article 27 of TRIPS, as tempered by the transitional patent provisions contained in Article 65.4, which were in turn qualified by the "mailbox" and "exclusive marketing rights" provisions of Article 70.[17]

II. ARTICLE 27 OF THE TRIPS AGREEMENT

Article 27.1 of TRIPS establishes a very broad international minimum standard for the subject matter of patent protection, requiring that patent protection be made available for any inventions, whether products or processes, in all fields of

[13] *Compare* Arts. 52(4)–53, Leaffer, *supra* note 12, at 690–691, *with* TRIPS Agreement, *supra* note 2, Article 27.2 and 27.3. The text of EPC Articles 52(4) and 53 are as follows: "Article 52: Patentable Inventions (4) Methods for treatment of the human or animal body by surgery or therapy and diagnostic methods practiced on the human or animal body shall not be regarded as inventions which are susceptible of industrial application within the meaning of paragraph 1. This provision shall not apply to products, in particular substances or compositions, for use in any of these methods. Article 53: Exceptions to Patentability European patents shall not be granted in respect of: (a) inventions the publication or exploitation of which would be contrary to "ordre public" or morality, provided that the exploitation shall not be deemed to be so contrary merely because it is prohibited by law or regulation in some or all of the Contracting States; (b) plant or animal varieties or essentially biological processes for the production of plants or animals; this provision does not apply to microbiological processes or the products thereof."

For the text and a discussion of the comparable provisions of the TRIPS Agreement, *see infra* notes 20–21 and accompanying text.

[14] *See* Stewart, *supra* note 5, Vol. IV: The Endgame, at 473.

[15] *Id.*

[16] *Id.* Vol. II: Commentary, at 2267.

[17] *See* Abbott, *supra* note 5, 41–42. For a detailed explanation of the transitional provisions and the mailbox and exclusive marketing rights provisions, *see infra* text accompanying notes 31–33.

technology, provided that they are new, involve an inventive step, and are capable of industrial application.[18] By way of emphasizing the significance of the requirement that patents be made available for "any" inventions, the second sentence of Article 27.1 goes on to state that patents shall be available and patent rights enjoyable without discrimination as to (among other things) the field of technology involved.[19]

At the same time, Articles 27.2 and 27.3 of TRIPS create important subject-matter exceptions to the broad rule of Article 27.1. Article 27.2 permits member countries to exclude from patentability any "inventions, the prevention within their territory of the commercial exploitation of which is necessary to protect *ordre public* or morality, including the protection of human, animal, or plant life or health or the avoidance of serious prejudice to the environment."[20] Article 27.3 further permits WTO member countries to exclude two specific classes of subject matter from patentability: 1) diagnostic, therapeutic, and surgical methods for the treatment of humans or animals; and 2) plants and animals other than microorganisms, and essentially biological processes for the production of plants or animals other than nonbiological and microbiological processes.[21]

[18] *See* TRIPS Agreement, *supra* note 2, Part II: Standards Concerning the Availability, Scope, and Use of Intellectual Property Rights, Article 27.1, the first sentence of which states that: "Subject to the provisions of paragraphs 2 and 3, patent shall be available for any inventions, whether products or processes, in all fields of technology, provided that they are new, involve an inventive step and are capable of industrial application."

A footnote to this sentence explains that for purposes of Article 27, the terms "inventive step" and "capable of industrial application" (the European version of the standards for patentability) may be deemed by a member to be synonymous with the terms "nonobvious" and "useful" respectively (the U.S. standards).

[19] *Id.* The text of this sentence states that, subject to the transitional provisions contained in Article 65.4, the Article 70.8 provisions (*i.e.*, the "mailbox" and "exclusive marketing" provisions) governing the protection of existing subject matter (discussed *infra* notes 31–33 and accompanying text), and Article 27.3 (discussed *infra* notes 25–31 and accompanying text), "patents shall be available and patent rights enjoyable without discrimination as to the place of invention, the field of technology, and whether products are imported or locally produced."

[20] *Id.* The entire text of Article 27.2 states that:

> Members may exclude from patentability inventions, the prevention within their territory of the commercial exploitation of which is necessary to protect *ordre public* or morality, including to protect human, animal, or plant life or health or to avoid serious prejudice to the environment, provided that such exclusion is not made merely because the exploitation is prohibited by their law.

[21] *Id.* The first sentence of Article 27.3 states that:

> Members may also exclude from patentability:
> (a) diagnostic, therapeutic and surgical methods for the treatment of humans or animals;

As open-ended as the Article 27.2 exception initially sounds, it is in fact subject to two important limiting conditions: First, the exception applies only if a prohibition against the *commercial exploitation* of the invention is necessary to protect *ordre public* or morality.[22] Second, (as is made clear by a concluding proviso) the exception applies only if the exclusion from *patentability* will likewise contribute to the protection of *ordre public* or morality.[23] By contrast, the two exclusions from patentability permitted by Article 27.3 are subject to no such limiting conditions.[24] For that reason, the two exclusions found in Article 27.3 will be the principal focus of this article.

It is important to note that neither of the subject matter exclusions permitted under Article 27.3 extends to *product* patents at the genetic or microbiological level of research.[25] Article 27.3(a) merely permits the exclusion of all commercially valuable diagnostic, therapeutic, and surgical *methods* of medical and veterinary treatment, while Article 27.3(b) permits the exclusion of "macrobiological" *products* and any associated *biological processes* (other than microbiological processes) for producing the same—namely any commercially valuable plants and animals (other than microorganisms), and any commercially valuable but essentially biological processes (other than microbiological processes) for the production of plants and animals. Thus, whereas Article 27.3(a) permits the exclusion of all genetic *methods* of medical and veterinary treatment, Article 27.3(b) is the provision that will apparently have a more significant adverse impact on the patenting of downstream genetic *products*, at least at the macrobiological level, though arguably not at the microbiological or submicrobiological level.[26]

Scientifically unsound though it may be,[27] this fundamental distinction in TRIPS between the micro- and macrobiological realms seems to echo the

---

(b) plants and animals other than micro-organisms, and essentially biological processes for the production of plants or animals other than non-biological or microbiological processes.

For the concluding provisions of Article 27.3(b), *see infra* notes 29–30 and accompanying text.

[22] *See* Carvalho, *supra* note 8, at 170–173; Oarlos M. Correa, INTELLECTUAL PROPERTY RIGHTS, THE WTO AND DEVELOPING COUNTRIES: THE TRIPS AGREEMENT AND POLICY OPTIONS, 63 (2000) [hereinafter Correa]. Correa notes that "it is debatable whether the exception can be applied while at the same time permitting the distribution or sale of the invention, or whether there is a need for an actual ban on commercialization," and cites to conflicting authorities on this point.

[23] *See* Carvalho, *supra* note 8, at 172; Correa, *supra* note 22, at 63.

[24] For the text of the first sentence of Article 27.3, *see supra* note 21. For the concluding provisions of Article 27.3, *see infra* notes 28–30 and accompanying text.

[25] For a discussion of the possible meanings of the terms "microorganisms" and "microbiological," *see infra* note 60 and accompanying and following text.

[26] *Cf. infra* notes 51–74 and accompanying text.

[27] *See* J. H. Reichman, *Universal Minimum Standards of Intellectual Property Protection under the TRIPS Component of the WTO Agreement* [hereinafter Reichman, *Universal Minimum IP*

contrast that Professor Rebecca Eisenberg notes, at the outset of Chapter 10, between the absence of public controversy in the United States during the 1980s over patenting DNA sequences and the later heated debates over the patenting of living subject matter generally, and higher life forms in particular. This same contrast is captured in James Boyle's image of the current "double-decker" policy debate in Chapter 6, comprised of "a lower deck of tumultuous popular and nonlegal arguments about everything from the environment to the limits of the market, and a calm upper-deck which is [preoccupied with] fine-tuning the input-output table of the innovation process." While Article 27.3 of TRIPS clearly attempts to balance opposing viewpoints in the raucous debate occurring on the lower deck of the bus over the patenting of higher life forms, neither the broad language of Article 27.1 nor its apparent applicability to the patenting of genetic sequences seems to have aroused any particular attention during the TRIPS negotiations, thus enabling the latter issue to become comfortably ensconced (for the moment, at least) on the upper deck, unaffected by the fray going on below.

Exemplifying the delicate balance that the TRIPS Agreement attempts to achieve with respect to the patenting of higher life forms, a proviso contained in the penultimate sentence of Article 27.3 specifies that, although WTO members are not obliged to provide patent protection to higher life forms, they are to provide either patent or *sui generis* protection or some combination of the two for plant varieties.[28] The final sentence of Article 27.3 goes on to state that both of the exceptions permitted by Article 27.3(b) are subject to a mandatory review by the TRIPS Council 4 years after entry into force of the Agreement Establishing the World Trade Organization (a review that is currently under discussion).[29] This concluding "built-in agenda" item is obviously designed as a sop to the United States, as Article 27.3 adopts an essentially European approach to the protection of plant innovation.[30]

Another delicate balance that the TRIPS Agreement is attempting to achieve, of course, is between the interests of industrialized and developing

Standards], in Carlos Correa & Abdulqawi Yussuf (eds.), INTELLECTUAL PROPERTY AND INTERNATIONAL TRADE: THE TRIPS AGREEMENT, 21, 39 (1998).

[28] The text of this sentence provides that "Members shall provide for the protection of plant varieties either by patents or by an effective *sui generis* system or by any combination thereof." TRIPS Agreement. *supra* note 2, Article 27.3(b) (second sentence).

[29] *Id.* (third sentence). The responsibilities of the TRIPS Council are specified in TRIPS Article 68.

[30] *See supra* text accompanying notes 12–13. *See also* text accompanying notes 64–65. By the time this built-in review was due to begin, however, it was the developing world that was clamoring for the review, while the United States and the rest of the industrialized world were having second thoughts about revisiting the issue. *See generally infra* notes 64–67 and accompanying text.

country members with respect to the implementation of the patent provisions of TRIPS. The relevant transitional provision is to be found in Article 65.4, which permits developing countries to delay until January 1, 2005 the granting of product patent protection for any area of technology not previously protected in that country.[31] That exception, in turn, is itself subject to the mailbox and exclusive marketing rights provisions of Article 70.8 and 70.9, which require developing countries to accept applications for pharmaceutical and agricultural chemical patents and grant 5 years of *de facto* patent protection in the form of exclusive marketing rights for any pharmaceutical and agricultural chemical products that have been granted patent protection by other WTO members.[32] This extended transitional period for granting product patent protection not previously granted by a WTO developing country member largely explains why the economic and political fallout from the patent provisions in TRIPS, particularly as these provisions relate to biotechnology, is only beginning to capture the international headlines.[33]

As a practical matter, the subject matter exclusions permitted by Article 27.3(b) of TRIPS are likely to have a far more profound (or at least a more immediate) impact on the development of downstream agricultural and plant products and processes than on the development of downstream medical products, as the latter impact will likely be limited to allowing the exclusion of such inventions as the Harvard Oncomouse.[34] On the other hand, the future of patent protection in the medical and agricultural fields of biotechnology is sufficiently interdependent to warrant a more detailed examination of the probable impact of the TRIPS Agreement on both fields of biotechnology.

III. INTERPRETIVE CONTROVERSIES AND POLITICAL FALLOUT LIKELY TO ARISE OUT OF ARTICLE 27

As the transitional period for developing countries to provide across-the-board patent protection comes to a close, the WTO dispute settlement process (and the quasi-judicial WTO Appellate Body in particular) will face a number of interpretive controversies generated by the language of Article 27. Likewise, in

[31] *See* TRIPS Agreement, *supra* note 2, Article 65.4. *See also supra* notes 3 & 17 and accompanying text.

[32] *Id.*, Article 70.8 and 70.9.

[33] *See infra* notes 35, 66–74 and accompanying text.

[34] For the varied fortunes of the Harvard Oncomouse in the United States, Europe, Japan, and most recently Canada, *see* Jim Brown, "No patent for Harvard mouse, court rules," TORONTO STAR (Dec. 5, 2002), *available at* www.TheStar.com (reporting that, although Harvard has long since obtained patents in the U.S., Japan, and much of Europe, the Supreme Court of Canada refused to award a patent for the genetically modified mouse).

the mandated review built into Article 27.3(b), the quasi-executive TRIPS Council will have to confront the international political fallout that the patent provisions of TRIPS has already produced—as evidenced, for example, by the recent Doha Ministerial Declaration and the separate Declaration on the TRIPS Agreement and Public Health, as well as the events which led up to the declarations.[35]

A. Interpretive Controversies Concerning Article 27

As for the interpretive questions likely to arise over the meaning of Article 27, one need only to consult any of a number of commentaries that set forth policy options for developing countries in implementing the TRIPS Agreement, some of which serve as virtual guides to what Jerome Reichman characterizes as the gray areas existing in the TRIPS text that can be exploited by developing countries as they implement the metamorphosis from "free riders to fair followers."[36] Among the best of these are a book-length study by Professor Carlos Correa, entitled *Intellectual Property Rights, the WTO and Developing Countries: The TRIPS Agreement and Policy Options* (2000),[37] a more recent report of the U.K. Commission on Intellectual Property Rights (on which Professor Correa served), entitled *Integrating Intellectual Property Rights and Development Policy* (2002),[38] and Professor Reichman's article, *From Free Riders to Fair Followers: Global Competition Under the TRIPS Agreement.*[39]

If Professor Correa's book serves as an accurate guide, one of the most likely TRIPS controversies confronting the WTO Appellate Body will be over what constitutes an "invention," within the meaning of Article 27.1, particularly with respect to biological materials that preexist in nature. According to Correa, there is no common international understanding of the term "invention."[40] One of the areas where the lack of a uniform concept of invention is

[35] *See* Doha WTO Ministerial 2001: Ministerial Declaration, WT/MIN(01)/DEC/1, Nov. 20, 2001, adopted Nov. 14, 2001, para. 17 and 19 [hereinafter Doha Declaration]; Doha WTO Ministerial 2001: Declaration on the TRIPS Agreement and Public Health, WT/MIN(01)/2 Nov. 20, 2001, adopted Nov. 14, 2001) [hereinafter Doha TRIPS & Public Health Declaration], *available at* http://www.wto.org (last visited April 12, 2003). For a discussion of the events leading up to these declarations and their specific provisions, *see infra* notes 66–73 and accompanying text.

[36] *See* J. H. Reichman, *From Free Riders to Fair Followers, supra* note 5, at 16.

[37] *See* Correa, *supra* note 22.

[38] *See* http://www.iprcommission.org/graphic/documents/finalreport.htm (last visited April 12, 2003).

[39] J. H. Reichman, *From Free Riders to Fair Followers, supra* note 5. *See also* Reichman, *Universal Minimum IP Standards, supra* note 27.

[40] Correa, *supra* note 22, at 51. *See also* "Elements of a *Sui Generis* System for the Protection of Traditional Knowledge," WIPO Intergovernmental Committee on Intellectual Property and Genetic Resources, Traditional Knowledge and Folklore, WIPO/GRTKF/IC/3/8, March 29,

most relevant, he says, relates to the distinction between patentable "inventions" and unpatentable "discoveries."[41] He defines a discovery as the mere recognition of what already exists; namely finding the causal relationships, properties, or phenomena that objectively exist in nature, while an invention is said to entail developing a solution to a problem by the application of technical means.[42]

As Professor Correa points out, in the United States, an isolated or purified form of a natural product is patentable according to principles developed for chemical patents, so long as the invention meets the standards of novelty, utility, and nonobviousness.[43] Moreover, novelty is not defined to mean "not preexisting in nature" but rather novel in a prior art sense, meaning that an unknown but natural substance existing only in an unpurified form will not exclude an isolated or purified form of the product as patentable subject matter.[44] Similarly, under the European Patent Convention, a patent will be granted when a substance found in nature may be characterized by its structure, the method of obtaining it, or by other criteria, if it is new in the sense that it was not previously available to the public.[45] A similar approach has been followed in Japan.[46]

However, Professor Correa also points out that a more restrictive approach has been applied in some developing countries. According to the Brazilian Patent Law, for example, no invention can be claimed with respect to "the whole or part of natural living beings and biological materials found in nature, or isolated therefrom, including the genome or germplasm of any natural living being, and any natural biological processes."[47] Some developing countries

2002, noting at paragraph 11 that one reason a precise definition of "traditional knowledge" is not necessary for identifying the legal elements of a mechanisms for its protection is because most patent laws "do not define the concept of 'invention;' equally, international harmonization and standard-setting in patent law have proceeded without specific or authoritative international definitions of this fundamental concept. . . ."

[41] *Id.*

[42] *Id.* at 52. *Compare* Correa's use of these terms *with* Article I, section 8, clause 8 of the United States Constitution, which empowers Congress "To promote the Progress of . . . useful Arts, by securing . . . to . . . Inventors the exclusive Right to their respective . . . Discoveries," which likewise suggests that only those "discoveries" capable of being "invented" are constitutionally within the subject matter of federal patent protection. *But cf.* section 100 of the U.S. Patent Statute, 35 U.S.C. 100, which defines the statutory term "invention" to mean "invention or discovery," suggesting that the statutory terms "invention" and "discovery" are merely being used to distinguish between a product (an invention) and a process (a discovery).

[43] Correa, *supra* note 22, at 53.

[44] *Id.*

[45] *Id.*

[46] *Id.*

[47] *Id.* at 54, citing Brazilian Patent Law Article 10.1X, Law 9,279, (1996).

take this position, Professor Correa says, because they feel that, though rich in biological materials, they are at a disadvantage in relation to industrialized countries that possess the technology and resources to unveil and extract the value embodied in such materials.[48] (But query: How else will these countries create local incentives to innovate in a sustainable way with their own wealth of biological materials if not through patenting processes for extracting the value resident in such materials and any resulting useful products?)

While according to Professor Correa, either of the foregoing approaches to the definition of an "invention" would be consistent with the TRIPS Agreement,[49] it could nevertheless be argued that the Brazilian definition of invention, both in design and effect, discriminates against biotechnology as a field of technology, thus violating the nondiscrimination provision of Article 27.1 of TRIPS.[50] To be sure, even in the United States, it has long been held that a "product of nature" cannot be patented because it does not constitute a machine, composition of matter, or article of manufacture, even if the claimant is the first person to discover or identify the product.[51] But the product of nature doctrine has a long and checkered history in United States patent law and, notwithstanding its occasional misapplication to what in fact are nonnovelty, claim overbreadth, inadequate written description, or nonenabling disclosure problems, it has been refined to the point that it actually bars the patenting of very little other than a few marginal examples of "true" products of nature,[52] while at the same time permitting patents on isolated or purified forms of a product existing in nature to be patented as articles of manufacture or compositions, so long as they are sufficiently new (in a prior art sense), useful, and nonobvious.[53]

[48] *Id.*

[49] *Id. But cf.* Reichman, *From Free Riders to Fair Followers, supra* note 5, at 36–39; and Reichman, *Universal Minimum IP Standards, supra* note 27, at 38–41, who does not echo Correa's position on this point.

[50] *See supra* note 19 and accompanying text.

[51] 1 Chisum on Patents § 1.02[7](2003) [hereinafter Chisum].

[52] *See id.* § 1.02 [7][a], citing as examples of "truly natural products" such subject matter as the "cellular tissues of the Pinus australis [tree] eliminated in full lengths from the silicious, resinous, and pulpy parts of the pine needles and subdivided into long, pliant filaments adapted to be spun and woven" and a patent claim based on new strain of microorganism "from a soil sample taken from the Val-de-Marne department of France" and used to produce a known antibiotic through a known cultivation procedure. *See also id.* § 1.02[7][b], which points out that the Supreme Court's decision in *Funk Brothers Seed Co. v. Kalo Inoculant Co.*, 333 U.S. 127 (1948), invalidating a patent for a mixture of inoculants of the bacteria genus (*Rhizobium*) as a "work of nature," 333U.S. 177, 130, was said by concurring Justice Felix Frankfurter to be better understood as an invalidation of the claim for overbreadth, as the claim was for any mixture of compatible strains of the bacteria but failed to identify the particular strains that the patentee had discovered.

[53] *See generally* Chisum, *supra* note 51, § 1.02[7].

Even before 1980, when the U.S. Supreme Court ruled in *Diamond v. Chakrabarty*[54] that a living, genetically altered microorganism constituted patentable subject matter as either an article of manufacture or a composition, scientists had begun filing patent applications disclosing and claiming not only the processes but the products of biotechnology, and were granted patents on sequences of DNA that encode for the production of therapeutically useful proteins either in isolated form or as part of a vector or transformed cell, just as scientists before them were able to obtain a patent on a particular Vitamin B_{12}-active composition useful for treating pernicious anemia, even though it is found in minute quantities in the rumen of cattle.[55] Indeed, as the Supreme Court noted in *Chakrabarty*, the U.S. Patent Office, as far back as 1873, granted Louis Pasteur a patent on "yeast, free from organic germs or disease, as an article of manufacture."[56]

If the product of nature doctrine can be used to deny patent protection for all of the classes of subject matter listed in the Brazilian Patent Law,[57] and that statutory language is interpreted broadly to bar patents on isolated genetic sequences as such or their artificial insertion into other organisms for therapeutic purposes, it is difficult to resist the suspicion that one of the subject-matter exclusions that was supposedly banned by TRIPS Article 27.1—namely exclusions for pharmaceutical product patents[58]—has just been allowed to sneak back in by way of a slightly smaller back door. Arguably, this is precisely the sort of discrimination against a particular field of technology that Article 27.1 seeks to prevent.[59] If such a broad swath of pharmaceutical products, or microbiological products and processes generally, can be excluded as products of nature, it is difficult to understand why the same could not be said for many chemical products and processes as well. Nor can one easily explain why Article 27.3 of the TRIPS Agreement would go to such lengths to spell out more limited subject matter exceptions for diagnostic and therapeutic methods and product

[54] 447 U.S. 303 (1980).

[55] *See* Chisum, *supra* note 51, § 1.02[7][c], *citing* Merck & Co. v. Olin Mathieson Chemical Corp. 253 F.2d 156 (4th Cir. 1958).

[56] 447 U.S. 303, 314 n. 9.

[57] *See supra* text accompanying note 47.

[58] That TRIPS explicitly bans exclusions of pharmaceutical product patents can be extrapolated from Article 70.8, the mailbox provision of TRIPS, which specifies that "[w]here a member does not make available as of the date of entry into force of the WTO Agreement patent protection for pharmaceutical and agricultural chemical products commensurate with its obligations under Article 27," the member must, "notwithstanding the transitional provisions contained in Part VI, provide as from the date of entry into for the WTO Agreement a means by which applications for patents for such inventions can be filed..." TRIPS Agreement, *supra* note 2, Article 70.8. For a discussion of the mailbox provision and its relationship with the transitional provisions of TRIPS, *see supra* notes 31–32 and accompanying text.

[59] *See supra* note 19 and accompanying text.

patents for plants and animals (*i.e.*, macroorganisms). Why bother creating such narrow escape hatches from obligatory patent protection, after all, if the product of nature doctrine allows for the removal of an entire wall?

A second likely (but more limited) interpretive controversy (if Professor Correa is again a reliable guide) will be over the precise scope of the TRIPS obligation to provide patent protection for microorganisms and microbiological processes. Professor Correa notes that while "microorganism" has been extensively interpreted so as to embrace any cell and subcellular elements, a narrower definition according to the scientific concept could limit microorganisms to bacteria, fungi, algae, protozoa, or viruses.[60] However, even if these terms can be defined as narrowly as Professor Correa suggests, it is difficult to see how such definitions, either by themselves or in conjunction with an expansive product of nature doctrine, could be construed to permit a flat ban on the issuance of any product patent where a genetic sequence (which is, after all, nothing more than a particular chemical sequence, albeit one coding particular genetic information) is extracted from one organism and artificially inserted into another organism that contains no such sequence in nature. To exclude any product patent in such a case would seem to create a *de facto* subject-matter exclusion of precisely the sort explicitly prohibited by Article 27.1, and one that also renders largely superfluous the more limited subject-matter exclusions explicitly permitted by Article 27.3.

In any event, as these two interpretive controversies go, so goes the fate of international patent protection for much upstream microbiological research. Perhaps the most important point to be made about these two interpretive controversies and their probable outcome is that they will both ultimately be decided by the Appellate Body of the WTO Dispute Settlement Process, a decision-making body that is clearly situated on the upper deck of what Professor Boyle has described as the current double-decker policy debate over the patenting of genetic products, and one that typically rules in favor of the industrialized world in such interpretive disputes.[61] Given the track record of the Appellate Body up to this point, it is quite likely that the developed world will prevail on both of these issues.

What, then, of the fate of the downstream macrobiological products and processes that Article 27.3 currently allows to be excluded from patent protection? The answer to this question will largely depend on the outcome of TRIPS Council's mandated review of the provisions of Article 27.3(b) and the

[60] *See* Correa, *supra* note 22, at 68, *citing* J. Coombs (*Macmillan Dictionary of Biotechnology*) 198 (1886). *But cf.* (*Webster's Unabridged Dictionary of the English Language*) 1215 (Random House 2001), defining "microorganism" as "any organism too small to be viewed by the unaided eye, as bacteria, protozoa, and some fungi and algae."

[61] For a discussion of the establishment of the WTO Appellate Body and a summary of its patent decisions to date, *see* Carvalho, *supra* note 8, at 277–291.

rough-and-tumble politics that have already begun on the lower deck of the policy bus, beginning with the failure of the tear-gas beclouded WTO Ministerial Conference in Seattle in 1999, followed by the more recent Doha Ministerial Declaration and the separate Doha Declaration on the TRIPS Agreement and Public Health.

B. The Political Fallout Produced by Article 27

As Professor Correa points out, Article 27.3(b) is the only provision of the TRIPS Agreement subject to an early review.[62] So far, however, there has been no agreement in the TRIPS Council on the nature of this review.[63] As originally envisioned, the mandatory review provision was obviously designed as a sop to the United States, as Article 27.3(b) adopted an essentially European solution to the debate over the patenting of plants and animals.[64] Not surprisingly, the United States was quick to indicate that the TRIPS Council should "consider whether it is desirable to modify the TRIPS Agreement by eliminating the exclusion from patentability of plants and animals and incorporating key provisions of the UPOV agreement regarding plant variety protection."[65]

However, the international political climate began to change in the wake of the clouds of tear gas that engulfed the aborted 1999 WTO Ministerial Conference in Seattle,[66] and since then the developed world has been thrown on the defensive with respect to the mandatory review of Article 27.3(b), as well as with respect to the direction of the new Doha Round of multilateral trade negotiations more generally. Developed countries now increasingly hold that Article 27.3 merely calls for a "review of implementation,"[67] while developing countries argue that the review should include the possibility of revising the text, not so much to strengthen but rather to loosen Article 27's requirements with

[62] *See* Correa, *supra* note 22, at 69, 211.

[63] *See* Carvalho, *supra* note 8, at 50, n. 166, stating that the same broad trade-related spectrum of interests that led to the success of the TRIPS negotiations, even though the earlier WIPO negotiations had failed (*see supra* note 8 and accompanying text) also explains why the negotiations on the built-in agenda have thus far failed: "Being exclusively dedicated to intellectual property, the TRIPS Council does not facilitate across-the-board negotiations, as the General Council and the Ministerial Conference of the WTO do."

[64] *See supra* notes 13 & 30 and accompanying text.

[65] *See* Correa, *supra* note 22, at 211, n. 2. "UPOV" is the French acronym for the International Convention for the Protection of New Varieties of Plants, Dec. 2, 1961, as revised at Geneva on Nov. 10, 1972, on Oct. 23, 1978, and on Mar. 19, 1991, *reprinted in* Leaffer, *supra* note 12, 47–85.

[66] *See* Statement by WTO Director-General Mike Moore, "It is vital to maintain and consolidate what has already been achieved," Press Release, Press/160, Dec. 7, 1999 (commenting on the suspension of talks).

[67] Correa, *supra* note 22, at 211.

respect to patentable subject matter, and in any event to make this provision of TRIPS more sensitive to the needs and interests of the developing world.[68]

For example, several proposals have been made by developing countries and non-governmental organizations to revise Article 27.3(b) to ensure that naturally occurring materials, including genes, are not patentable, and to recognize some form of protection for the "traditional knowledge" of local and indigenous communities.[69] Moreover, developing countries generally wish to ensure that the exception for plants and animals is retained, and that they continue to have the flexibility to develop *sui generis* regimes on plant varieties, suited to the seed-supply systems of the countries concerned.[70] Finally, developing countries have pressed specific proposals, both in the TRIPS Council and before the WIPO, to establish international minimum patent standards requiring members not to grant or to cancel *ex officio* or upon request any patent or other intellectual property rights on any biological materials obtained 1) from collections held in international germplasm banks and other depositories where such materials are publicly available; or 2) without the prior consent of the country of origin and/or indigenous or local communities providing the materials.[71] More generally, developing countries are demanding compliance with the obligations contained in the Convention on Biological Diversity (which the United States has yet to ratify) to share the benefits with the country of origin of any patented biological material.[72] The latter concern was explicitly recognized at the Fourth Ministerial Conference of the WTO in Doha, as the Doha Ministerial Declaration instructed the TRIPS Council to examine, *inter alia*, "the relationship between the TRIPS Agreement and the Convention on Biological Diversity, the protection of traditional knowledge and folklore, and other relevant new developments raised by members pursuant to Article 71.1 [which authorizes the TRIPS Council to undertake reviews in the light of any new developments which might warrant modification or amendment of this agreement]."[73]

[68] *Id.*

[69] *Id. See generally* Charles R. McManis, *Intellectual Property, Genetic Resources and Traditional Knowledge Protection: Thinking Globally, Acting Locally*, 11 CARDOZO J. INT'L & COMP. LAW 2201 (2003) [hereinafter McManis, *Thinking Globally, Acting Locally*]; McManis, *Biodiversity and Biotechnology*, *supra* note 5.

[70] Correa, *supra* note 22, at 211.

[71] *Id.* at 212. *See generally* McManis, *Thinking Globally, Acting Locally supra* note 71, at 221–2219.

[72] Correa, *supra* note 22, at 212; McManis, *Thinking Globally, Acting Locally*, *supra* note 71, at 2210–2211.

[73] *See* Doha Declaration, *supra* note 35, para. 19. *See also id* para. 17 (stressing the importance of implementing and interpreting the TRIPS Agreement "in a manner supportive of public health, by promoting both access to existing medicines and research and development into new medicines," in connection with which the Doha Ministerial issued a separate, and more detailed declaration. *See* Doha TRIPS & Public Health Declaration, *supra* note 35.

In short, given the current international political climate, it is highly unlikely that exceptions from patentability permitted by Article 27.3(b) will soon be eliminated. It is far more likely that in the upcoming round of multilateral trade negotiations at least some of the foregoing demands of the developing world will be met.

IV. CONCLUSION: IMPLICATIONS FOR HUMAN GENOMIC RESEARCH AND PATENT PROTECTION

So, what are the implications of all this for the future of genomic research and patent protection? Certainly, the immediate implications are not quite as profound for human genomic research and patent protection as they are for plant genomic research and patent protection. But as has been suggested above, given the particular language of Article 27 of the TRIPS Agreement, the future of biotechnology patent law in both fields is essentially interdependent, at least as far as the TRIPS Agreement is concerned. Moreover, while it is true that not all upstream genetic research is necessarily microbiological, and not all downstream commercial products and processes are necessarily macrobiological, there is nevertheless a sufficient correspondence to suggest that, given the likely decisions by the WTO Appellate Body in the two interpretive controversies discussed above, and the probable course of the political process about to unfold in the TRIPS Council's built-in review of Article 27.3, as well as in the new Doha Round of multilateral trade negotiations more generally, the international prospects for patent protection on upstream, microbiological or submicrobiological genetic research look a good deal more promising in the post-TRIPS world than the prospects for patent protection on downstream macrobiological commercial products.

However, this may not be the end of the matter. Once the WTO Appellate Body rules on the two interpretive controversies discussed above, it will not take long for those representing the developing world in the rough-and-tumble policy debates occurring on the lower deck of the policy bus to absorb the practical implications of these rulings emanating from the upper deck of the policy bus, as the WTO Appellate Body rulings will essentially moot much of the debate occurring on the lower deck with respect to the patentability of higher life forms. If isolated genetic sequences as such or isolated genetic sequences as inserted in an organism in which they do not appear in nature are indeed patentable subject matter, it will make little practical difference whether the larger macroorganisms into which they are inserted are or are not themselves patentable, as their genetic sequences, in any event, cannot be made, used, offered for sale, sold, or imported without the consent of the patent owner.[74]

At this point, the tumultuous popular and nonlegal arguments occurring on the lower deck of the policy bus may indeed fuse with the upper-deck fine-tuning of the input-output table of the innovative process, or at least be joined by a common staircase, as Professor Boyle urges—though perhaps not in quite the manner that he envisions. What springs to mind is the image of policy makers from both decks suddenly jamming the staircase, each frantically attempting to commandeer the bus—or in any event prevent the other deck from succeeding in doing so. It has the makings of a very wild ride.

[74] *See* TRIPS Agreement, *supra* note 2, Article 28, specifying the exclusive rights that must be conferred on patent owners.

6

Enclosing the Genome: What Squabbles over Genetic Patents Could Teach Us

James Boyle
William Neal Reynolds Professor of Law
Duke University School of Law
Durham, North Carolina 27708

0065-2660/03 $35.00

I. INTRODUCTION

In other writings,[1] I have argued that we are in the middle of a second enclosure movement. The first enclosure movement involved the conversion of the "commons" of arable land into private property. The second enclosure movement involves an expansion of property rights over the intangible commons—the world of the public domain, the world of expression and invention. Quite frequently it has involved introducing property rights over subject matter—such as unoriginal compilations of facts, ideas about doing business, or gene sequences—that were previously said to be outside the property system and therefore uncommodifiable, "essentially common," or part of the common heritage of mankind.

The justifications given for the first enclosure movement were often, though not always, centered on the need for single-entity private-property rights over land to encourage development and investment, prevent over and under use, and, in general, avoid the phenomena which we refer to today as "the tragedy of the commons." Enclosure's defenders argue that to increase agricultural production and, in the long run, to generate an agricultural surplus sorely needed by a society whose population had been depleted by the mass deaths of the sixteenth century. Private property saved lives. Though "overuse" is rare in the intellectual commons, the rest of the arguments are exactly the same ones used to support the second enclosure movement. Intellectual property is needed to encourage development and investment. This argument is made when defending drug patents against compulsory licensing claims, in the debates over the creation of new intellectual property rights over data and business methods, and in the rhetoric of support for the Digital Millennium Copyright Act.[2]

But as this conference shows, and as a cursory study of the newspapers demonstrates, it is the idea that the genome has been turned over to private ownership that has fueled real public attention. Again, the supporters of enclosure have argued that the state was right to step in and extend the reach of property rights; that only this way can we guarantee the kind of investment of the time, ingenuity, and capital necessary to produce new drugs and gene therapies. To the question, "Should there be patents over human genes?" the supporters of enclosure would answer again, "Private property saves lives." Again, the opponents of enclosure have claimed that our common genome "belongs to everyone," that it is "the common heritage of humankind," that it should not and perhaps in some sense *cannot* be owned, and that the consequences of turning over the human genome to private property rights will be

[1] James Boyle, *Fencing Off Ideas*, Daedalus 13 (Spring 2002); *The Second Enclosure Movement and the Construction of the Public Domain*, 66 LAW AND CONTEMP. PROBS. 33 (2003). The Introduction to this chapter draws heavily on the latter article.

[2] For details see Boyle, *The Second Enclosure Movement*, *supra* note 1 at 33–44.

dreadful, as market logic invades areas which should be the farthest from the market. What damaged view of the self and what distorted relationship between human beings and the environment flows from a world in which our genetic code itself is fenced off, made both alienable and alien at the same moment?

The analogy is not perfect, of course; the commons of the mind has many different characteristics from the grassy commons of Old England. Some would say that we never had the same *traditional* claims over the genetic commons that the victims of the first enclosure movement had over theirs; this is more like a newly discovered frontier land or perhaps a privately drained marshland than it is like well known common land that all have traditionally used. In this case, the enclosers can claim (though their claims are hotly disputed) that they discovered or simply made usable the territory they seek to own. The opponents of gene patenting turn more frequently than the farmers of the seventeenth century to religious and ethical arguments about the sanctity of life and the incompatibility of property with living systems. Importantly, too, the genome—like most of the subjects of the second enclosure movement, such as software, data, digital music, and text—is nonrival; unlike the "commons" enclosed during the first enclosure movement, my use does not interfere with yours. I can work it, and you can work it, too. All of these differences might be relevant, or might not, depending on the underlying normative framework chosen to assess property claims. But for the moment, think of the critics and proponents of enclosure—locked in battle; hurling at each other incommensurable claims about progress, efficiency, traditional values, the boundaries of the market, the saving of lives, and the disruption of a *modus vivendi* with nature. We have done this before, and perhaps we can learn something from the process.[3]

In an earlier work,[4] I concentrated on the economic and utilitarian arguments for the second enclosure movement. In sum, I concluded that, for a variety of reasons, the embrace of the logic of enclosure to justify the expansion of intellectual property across so many fields and so many dimensions was probably a mistake. I argued that while the basic argument for intellectual property protection remains as strong as ever, we have adopted an asymmetric analytic framework which 1) overestimates the applicability of the general logic of enclosure to the special case of intellectual material; 2) undervalues the importance of the

[3] The analogy to the enclosure movement has been too succulent to resist. To my knowledge, Ben Kaplan, Pamela Samuleson, Yochai Benkler, David Lange, Christopher May, and Keith Aoki have all employed the trope, as I have myself on previous occasions. For a particularly thoughtful and careful development of the parallelism, see Hannibal Travis, "Pirates of the Information Infrastructure: Blackstonian Copyright and the First Amendment," 15 BERKELEY TECH. L.J. 777 (2000). It is also worth noting that Jeremy Rifkin, one of the most outspoken critics of gene patents, has recently turned to this analogy as well. *See* Jeremy Rifkin, THE BIOTECH CENTURY (1998).

[4] *See* note 1 *supra*.

public domain and the commons to intellectual production; and thus also 3) focuses only on the (very real) arguments in favor of private property while neglecting the role of the raw materials out of which future innovation is constructed. The resulting policies also 4) overestimate the potential threats and underestimate the potential benefits of the technologies of cheap copying to existing intellectual property rights and 5) fail to take seriously enough the important potential for various types of distributed production, which require a rather different intellectual property environment in order to flourish. Some of those conclusions, 1) and 2) in particular, seem strongly applicable to fundamental gene patents—particularly those which may operate to block or channel future research. Others, for example, 5), have a possible but more dubious relevance.

In this chapter, however, I turn my attention to the other arguments over enclosure, the ones that economic historians—and intellectual property scholars for that matter—tend to note briefly and somewhat dismissively before turning their attention to the real meat of the economic incentives set up by property systems. My question, then, is not, "Will licensing solve the potential monopolistic problems threatened by fundamental gene patents?" Or, "Will this particular bit of surgery on the Bayh-Dole Act cure the incentive system so as to encourage fundamental research to yield real products while discouraging 'blocking patents' across the arteries of research?" Those questions are vital ones on which important work has already been done.[5] But I want to ask a different question: one about the fundamental structure of intellectual property discourse. What can the debate over gene patents teach us about the structure of our discipline, about our pattern of inquiry? What does that debate reveal about both the selective focus and the selective blindness in the way that intellectual property scholars analyze the question? In short, this article treats the debate over gene patents as a rhetorical case study, as a place to reflect on the limits of the discipline. It does not attempt to enter that debate on one side or the other.

II. YOU CAN'T OWN A GENE

Like the debate over the first enclosure movement, the debate over the enclosure of the genome is not one debate, but many. The phrase "You can't own a gene" covers a remarkably broad range of arguments and beliefs. These arguments are not necessarily consistent, of course, and intellectual property

[5] John Walsh, Ashish Arora, & Wesley Cohen, *The Patenting of Research Tools and Biomedical Innovation*, draft (on file with author); Arti K. Rai & Rebecca S. Eisenberg, *Bayh-Dole Reform and the Progress of Biomedicine*, 66 LAW & CONTEMP. PROBS. 289, 289 (Winter/Spring 2003). For the seminal study of the early days of the genome project and the conflicts over gene patenting, which goes beyond economic considerations to consider some of the ethical objections made, see Robert Cook-Deegan, THE GENE WARS: SCIENCE, POLITICS AND THE HUMAN GENOME (1994).

lawyers would disclaim many as being beyond their purview. In addition, the arguments vary importantly depending on the subject matter of the patent; the debate is not merely, or even most interestingly, about patents on genes. Claims to patents on different types of genetic sequences raise very different issues. What's more, I mean here to catalogue types of objections rather than investigate them in any detail, which requires painting with a distressingly broad brush. Sweeping all of these qualifications aside for a moment, let me attempt a breathless summary of the most commonly heard objections to genetic patents. My hope is that, for the *cognoscente*, it may be useful to review territory so well known it is seldom examined. For the newcomer, this may serve as a rough guide to the territory to be deserted later in favor of more detailed maps.

A. The Sacred

"You can't own a gene because it would be against the tenets of my faith. God wrote those sequences of C, G, A, and T. It is heresy, or at least plagiarism, for you to claim to do so." In a more modest version, the accusation is not heresy or plagiarism, but simply ludicrous *hubris*: Who are *you* to claim a patent on a *human* genetic sequence, even a "purified" version of a genetic sequence with the introns spliced out?[6] (A version of this argument will reappear as an argument about whether genetic sequences fit the statutory and constitutional requirements of patent law.)

B. The Uncommodifiable

"You can't own a gene because it would be immoral; some things should be outside of the property system—babies, votes, kidneys; if they become commodified, then the market seeps into aspects of our lives that should be free

[6] "In the 19th century, another great debate ensued with abolitionists arguing that every human being has intrinsic value and "God-given rights" and cannot be made the personal commercial property of another human being. The abolitionists' argument ultimately prevailed, and legally sanctioned human slavery was abolished in every country in the world where it was still being practiced.... Now, still another grand battle is unfolding on the eve of the Biotech century. This time of the question of patenting life—a struggle whose outcome is likely to be as important to the next era in history as the debate over usury, slavery, and involuntary indenture have been to the era just passing. In May of 1995, a coalition of more than 200 religious leaders, including the titular heads of virtually every major Protestant denomination, more than 100 Catholic bishops, and Jewish, Muslim, Buddhist, and Hindu leaders, announced their opposition to the granting of patents on animal and human genes, organs, tissues, and organisms.... The coalition, the largest assemblage of U.S. religious leaders to come together on an issue of mutual interest in the 20th-century, said that the patenting of life marked the serious challenge to the notion of God's creation in history. How can life be defined as an invention to be profited from by scientists and corporations when it is freely given as a gift of God, asked the theologians? Either life is God's creation or a human invention, but it can't be both." Jeremy Rifkin, THE BIOTECH CENTURY, 64 (1998).

from market logic."[7] Sometimes the proponents of this argument rely on a teleological view of human flourishing, at other times on a theory of "spheres of justice" within a liberal state. At still other times they conjure up a dystopian future in which transgenic, subhuman creatures are traded as chattel. Our reverence for thinking beings will have been undermined by the twin expansions of genetic science and market logic.

C. The Environmental Ethic

"You can't own a gene; to do so is to embrace a system in which nature, even our own nature, is to be manipulated, traded, and commodified." While the environmentalist critics of gene patenting generally have concerns that stretch beyond the human genome alone, they see it as a particularly revealing example. Many argue that the frame of mind which would permit such property rights is one which displays a breathtaking willingness to tinker with delicate environmental systems.[8] Others claim that the availability of such juicy state-granted monopolies establishes a set of incentives which invite corruption of the legislative process and the undermining of rational environmental policy.

D. The Common Heritage of Mankind

"You (*alone*) can't own a gene because the genome belongs to all of us. Like the deep-sea bed, the history of the human species, or outer space, the genome is

[7] "Now, the most intimate Commons of all is being inclosed and reduced to private commercial property that can be bought and sold on the global market. The international effort to convert the genetic blueprints of millions of the years of evolution to privately held intellectual property represents both the completion of a half millennium of commercial history closing of the last remaining frontier of the natural world." Rifkin at 41. Not all agree. "The reductive objectification of life has enlarged understanding of vital processes, but it does not relegate living creatures to the category of objects. We understand that animals are physics and chemistry. But we can realize that they are sentient creatures and grant them rights that they do not inherently possess. We can realize that people perform physical and chemical functions, but we need not treat them as physical and chemical machines." Daniel Kevles, *Vital Essences and Human Wholeness: The Social Readings of Biological Information* 65 S. CAL L. REV 255, 277.

[8] "Biotechnology is not simply another mechanical or chemical procedure aimed at making the world better for us. With biotechnology, we are not reshaping matter, but harnessing life. We take a 3,500 million year old process that shaped our existence and the existence of every other organism on the planet and restructure it for our benefit. We need a more thoughtful conceptualization of this technology and more careful control over its development and use than is allowed by gung-ho biopatent policies. Biotechnology does offer promise and hope for bettering human life and perhaps other life as well. Opposing biopatents does not entail opposing biotechnology. Organism and gene patents should be resisted not because biotechnology should be resisted, but rather because these biopatents are a morally dangerous and inappropriate way of thinking about and encouraging biotechnology." Ned Hettinger, *Patenting Life: Biotechnology, Intellectual Property, and Environmental Ethics*, 22 B.C. ENVTL. AFF. L. REV. 267, 304 (1995).

"the common heritage of mankind," one of the quintessential resources that must be held in common."[9] Sometimes this view rests on some Teilhard de Chardin-esque teleology of the human species, while at other times it represents a straightforward appeal to international distributive justice: If the technologically advanced countries can secure property rights over resources that only advanced technology will reach, then patent rights over the genome are a form of second colonial expansion. The moon, the deep-sea bed, or the genome should not become the next place to be claimed for Queen Isabella of Spain.

E. The Rights of Sources

"You can't own *this* gene because I owned it first. My genetic information is my property. Your gene sequences came originally from a source and the source's claims should be recognized, either instead of or as well as the person seeking the patent." This objection is sometimes raised on behalf of those who unwittingly provided genetic material to a research project from which some desirable genetic sequence was derived.[10] At other times, the claim comes from the families of those with a particular genetic disorder. Here the genetic material was donated knowingly, but families are using property and contract claims to ensure that the development of tests and treatments for the disorder protects the interests of the patients involved.[11] Other variants include attempts by

[9] "The application of the Common Heritage concept to the genome would balance the interests of developed and developing countries, thus holding true to the traditional purpose. As stated previously, most developing countries are opposed to patenting genes, and consider it a form of neo-colonialism. Because the genome technically belongs to the citizens of LDCs (less developed countries) to the same extent as citizens of the developed countries, the Common Heritage Principle is necessary to balance the property rights of both LDCs and developed country citizens despite their different interests." Melissa L. Sturges, *Who Should Hold Property Rights to the Human Genome? An Application of the Common Heritage of Humankind*, 13 AM. U. INT'L L. REV. 219 251–2 (1997); *See also* J. M. Spectar, *The Fruit of the Human Genome Tree: Cautionary Tales About Technology, Investment, and the Heritage of Humankind*, 23 LOY. L. A. INT'L & COMP L. REV. 1. On both sides of these arguments, the background assumptions are often hard to parse. "From the perspective that genes are our common, universal possession symbolizing humankind's collective heritage, genes seem an inappropriate substance in which to grant individual intellectual property rights. Yet, in light of patent availability in most other scientific research fields, denying gene patents appears to be inequitable." Patricia A. Lacy, *Gene Patenting: Universal Heritage vs. Reward for Human Effort*, 77 OR. L. REV. 783.

[10] The classic case is that of John Moore. *See* James Boyle, SHAMANS SOFTWARE AND SPLEENS; LAW AND THE CONSTRUCTION OF THE INFORMATION SOCIETY, 97–107 (1996).

[11] "When the Terry family of Sharon, Massachusetts, realized that disease researchers didn't have the family's interests at heart, they took control of their bodies' intellectual property. The result? Do-it-yourself patenting was born. . . . Boyd (who hasn't been funded by PXE International) agreed with Terry's morally driven, but practical, view that the laws of this market left patients no choice but to seek control of the intellectual property. 'This was a way to ensure that the

particular groups or populations whose genetic information is somehow of interest to prevent what they see as "gene piracy."[12]

F. Patentable Subject Matter

"You can't own a gene because you can't patent it; it doesn't satisfy the basic requirements of the patent law and the constitution." The challenges differ depending on the type of genetic material involved. The main objections are as follows:

1. Novelty

Many genetic patents cover material that is not "novel" because they are patents on naturally occurring products; if patents are being given for laborious transcription of a naturally occurring substance, that hardly meets the novelty standard. Even if the sequences are "purified," for example by removing non-coding segments, critics argue that this should not meet the standard for novelty, and might also run into problems because of the "obviousness" of the process.

2. Nonobviousness

In general, the standard methods of genetic sequencing are such that each step in the process of cell line isolation, purification, and sequencing may be obvious from the prior step in the process. One of the larger concerns being expressed here, as in the discussion of novelty, is that intellectual property rights are being given in raw data, based on the labor required to produce it. Such "sweat-of-the-brow" claims are supposed to be anathema to the law of copyright and patent, while the subject matter here—genetic sequences—seem to be straightforward arrangements of factual data; a subject matter that the constitutional patent and copyright power does not reach.

test isn't going to cost an arm and a leg,' notes the scientist. . . . By laying legal claim to their bodies, the Terry's have advanced research enough that doctors genuinely may be able to arrest the progress of PXE before their children go blind, which often happens to patients in their late 30s. Ian Terry, now 11, already has initial signs of eye trouble. At the very least, the parents' organizing has inspired Elizabeth Terry, now 13, to say she'll become a patent lawyer—though her mother says she's also flirting with criminal law. No doubt, Elizabeth should pursue what she wants: By the time she graduates from law school, the patient advocacy movement should have plenty of patent attorneys on its side." Matt Fleischer, *Patent Thyself*, AMERICAN LAWYER June 2001.

[12] Kara H. Ching, *Indigenous Self-Determination in an Age of Genetic Patenting: Recognizing an Emerging Human Rights Norm*, 66 FORDHAM L. REV. 687.

3. Utility

Some of the genetic patents have been shotgun claims over large number of sequences, without clear knowledge as to what function these sequences actually have. Others cover sequences the main function of which is as probes in research so that their "utility" is as research intermediaries rather than useful end products.

As is often (and appropriately) the case for patent law, debates about the meaning of the statutory and constitutional requirements for patents frequently devolve into questions about the effects that a particular interpretation of the subject-matter requirements would have on innovation. Because the patent system is, in its constitutional and theoretical origins, guided explicitly by utilitarian concerns, almost all would concede that the ultimate benchmark for any interpretation of the subject-matter requirements should be its effect on the encouragement and promotion of the progress of the useful arts.[13] In the case of patents on genetic sequences, this means defining the requirements of patentable subject matter with the assumption that those requirements exist so as to encourage both abstract research and final practical deployment of some therapy or test. Thus, discussions of patentable subject matter tend to segue into the free-standing debate that already exists about the effects of genetic sequence patents on innovation.

G. Innovation Policy

"You can't own a gene because if you did so, it would actually hurt research and innovation, the very things you are trying to encourage." Here the arguments are particularly subtle and nuanced. The basic difference between the genetic commons and the earthy commons of old England, at least from an innovation policy point of view, is its nonrivalrous quality. The danger feared here is not

[13] Unfortunately, the Court of Appeals for the Federal Circuit, (CAFC) which handles patent appeals nationwide, does not appear to share this point of view. To this external observer it seems that the CAFC lurches from formalism to utilitarian analysis and back again, guided by some muse of its own. [For a related critique *see* Arti K. Rai *Specialized Trial Courts: Concentrating Expertise on Fact*, 17 BERKELEY TECH. L.J. 877 (2002)] The court also appears gratifyingly indifferent to academic opinion; many distinguished patent scholars appeared to think the State Street bank opinion [*State St. Bank & Trust Co. v. Signature Fin. Group*, 149 F.3d 1368, 1373 (D.C. Cir. 1998)] was the jewel in the crown of bad patent decisions. *See, e.g.*, Robert P. Merges, *As Many as Six Impossible Patents Before Breakfast: Property Rights for Business Concepts and Patent System Reform*, 14 BERKELEY TECH. L.J. (1999).] Yet, despite the storm of academic protest, one judge on the court recently remarked, with clear self-satisfaction, that everyone was clearly now comfortable with the court, so that perhaps its jurisdiction should now be expanded to cover copyright as well! Being ignored like this is clearly good for the otherwise nonexistent humility of legal academics, but since the court appears instead to take its lead from the patent bar, which has a something of a self-interest in maximalist protection, one wonders whether it is also a prescription for a good patent law jurisprudence.

unrestrained overuse but underinvestment in research and product development. Moreover, as is generally the case, intellectual property deals with a subject matter in which future products are generally assembled from the pot-shards of prior efforts. Each property right tacked onto your stream of intellectual outputs increases the cost of my stream of intellectual inputs. Even if, in principle, I could either "invent around" prior protected innovation,[14] or secure licenses on those chunks which proved impossible to bypass, the transaction costs involved in the process of identification, weighing the relative costs of inventing around, negotiating, and so on, might undercut a large number of innovations at the margin.[15] It is the work of intellectual property scholars to work at this point of delicate equipoise; the balance between, on the one hand, failing to guarantee sufficient protection (or government funding) in order to encourage both research and commercialization; on the other, slowing down the process of research and commercialization with a thicket of property rights. The best work on the subject refines this point of balance, pointing out counterintuitive processes at work within it,[16] expanding on institutional dynamics, and bringing in the insights of other disciplines and of empirical research on the innovation process.[17]

III. THE LIMITS OF INTELLECTUAL PROPERTY POLICY

As promised, this has been a somewhat breathless summary of the arguments about patenting human genetic sequences. It is far from complete. In particular, I have not expanded on the many thoughtful counterarguments that can be raised to each of these objections because I do not mean to enter the debate so much as to point out something about it. It is my hope that the summary I have given helps us to see that the academic intellectual property debate has tended to be much narrower than the popular or policy debate. First, patent scholars and patent lawyers have largely viewed all but the last two sets of criticisms as

[14] "Inventing around" is normally a safety valve in intellectual property policy, the possibility of which operates to reduce the drag on progress caused by prior rights and to create downward pressure on the price of licenses. One particular (though disputed) concern with the human *genetic* patents is the extent to which there may be no feasible way to innovate around a particular patent, precisely because of the fundamental quality of the subject matter.

[15] Michael A. Heller & Rebecca S. Eisenberg, *Can Patents Deter Innovation? The Anticommons in Biomedical Research*, SCIENCE, May 1, 1998, at 698.

[16] *See, e.g.*, Pamela Samuelson & Suzanne Scotchmer, *The Law and Economic of Reverse Engineering*, 111 YALE L. J. 1575, 1608–13 (2002); F. Scott Kieff, *Property Rights and Property Rules for Commercializing Inventions*, 85 MINN. L. REV. 697 (2001).

[17] *See, e.g.*, Wesley Cohen *et al.*, *Protecting Their Intellectual Assets: Appropriability Conditions and Why U.S. Manufacturing Firms Patent (or Not)*, NATIONAL BUREAU OF ECONOMIC RESEARCH, Working Paper No. 7552 (2000); Mark Lemley & John Allison, *Who's Patenting What? An Empirical Exploration of Patent Prosecution*, 53 VANDERBILT LAW REVIEW 2099 (2000).

outside their purview;[18] the legislature decides what is and is not patentable, they argue, so to address arguments about the immorality or environmentally hurtful quality of gene patents to the Patent and Trademark Office (PTO), to the patent bar, or even to patent scholars is to mistake the institutional function of those entities. The role of patent specialists is to make positive claims about what the law is and make restricted normative claims about how current or proposed patent rules will further the goals of the patent system, which in turn are understood in means-ends terms as a utilitarian calculation of the best way to foster technologically usable "progress." Thus, patent scholars do not eschew normative analysis altogether; most of them simply restrict it to the consensus utilitarian goals of the system.

Patent scholars will happily debate whether current PTO utility guidelines or the PTO's practices in reviewing patents on Expressed Sequence Tags are moving intellectual property rights so far "upstream" in the research cycle that they will actually discourage future research and innovation. But for most, claims about the effect of commodification on attitudes toward the environment, about distributive justice and the common heritage of mankind, or the moral boundaries of the market are simply outside the field. (The claims about the relationship of the genome to the realm of the sacred are even more so.) Ironically, some patent scholars, who in other areas of scholarship might criticize the utilitarian, means-ends metric as a partial or a "thin and unsatisfactory epistemology" through which to understand the world,[19] are, in the patent field, particularly wedded to it. A brief digression is necessary to explain why.

A. The Bipolar Disorders of Intellectual Property Policy

At its crudest, the basic division in the intellectual property field is between "maximalists" or high protectionists, on the one hand, and "minimalists," or those with a heightened concern about the public domain, on the other. While actual positions are considerably more nuanced, if one had to construct an "ideal type" of these two positions, it would run something like this: The maximalists favor expansive intellectual property rights. They tend to view exemptions and privileges on the part of users or future creators as a tax on rights holders and have sympathy for thinly disguised "sweat-of-the-brow" claims. They exhibit a kind of economic bipolar disorder: They are deeply pessimistic about market

[18] I would happily acknowledge that there are salutary exceptions to this claim. See e.g. Molly A. Holman & Stephen R. Munzer, *Intellectual Property Rights in Genes and Gene Fragments: A Registration Solution for Expressed Sequence Tags*, 85 IOWA L. REV. 735 (2000). I would argue, however, that the general claim holds true.

[19] This description of the philosophical shortcomings of economic analysis is, amusingly enough, Richard Posner's. Richard Posner, THE PROBLEMS OF JURISPRUDENCE XIV (1990) (citing the words of Paul Bator).

functioning around potential public goods problems in the *absence* of intellectual property rights, and yet strikingly, even manically, optimistic about our ability to avoid transaction costs and strategic behavior "anticommons effects" that might be caused by the *presence* of intellectual property rights.

The minimalists have exactly the opposite set of attitudes. They start from the presumption that the baseline of American law is that "the noblest productions of the human mind are upon voluntary disclosure to others, free as the air to common use." Thus intellectual property rights, understood as "monopolies," look like dangerous state-granted subsidies which should be confined, in amount and extent, to the minimum demonstrably necessary. Minimalists exhibit their own economic bipolarity; they are often deeply optimistic about the ability of creators and innovators to adapt business methods so as to gain returns on innovation without recourse to legal monopolies. (For example, by being "first to market," relying on the provision of tied services to generate revenue, or by adopting physical and virtual methods of exclusion in order to avoid the underlying public goods problem.) They are also relatively optimistic about the impact of technology on the innovation marketplace; they see new technologies that lower the cost of copying, such as the Internet or PCR, as providing benefits as well as costs to rights holders, so that every reduction in the cost of copying need not be met with a corresponding increase in the level of protection. On the other hand, the minimalists are deeply, almost tragically, pessimistic about the ability of rights holders to bargain around the potential inefficiencies and transaction costs that their rights introduce into the innovation process. Here, private action to avert potential market failure seems much harder, for some reason, than it did when we were discussing the adoption of possible business methods to gain a return on an innovation.

The maximalists' arguments are, in general, weaker, because the minimalist account is more faithful to thoughtful innovation economics and to the historical and constitutional traditions of American intellectual property law. More importantly, years of relentless expansion in intellectual property rights mean that the disputed frontier is now very far out indeed; maximalists must now defend rights that are, frankly, silly. However, what the maximalists lack in argumentative power they more than make up for in political power; their views are extremely hospitable to a variety of film, recording, software, and pharmaceutical companies who can supplement dubious economic *analysis* with indubitable economic *payments*—in many cases, payments direct to legislators. More interestingly, there are also some philosophical reasons why we tend to undervalue the public domain[20] and why strong intellectual property rights breed even stronger intellectual property rights.[21]

[20] *See, e.g.*, James Boyle, SHAMANS SOFTWARE AND SPLEENS; LAW AND THE CONSTRUCTION OF THE INFORMATION SOCIETY (1996). James Boyle, *A Politics of Intellectual Property: Envir-*

To be sure, the minimalist and maximalist I have just described are exaggerated ideal types, seldom met in their pure form. (And certainly no such caricatured scholars are present in this volume: indeed, a number of the scholars here have written important pieces that point out the dangers of both under- and overprotection.[22]) Still, the distinction catches something important. Now notice the irony. Minimalists might seem like the natural allies of those who are skeptical about gene patents for religious, environmental, moral, or distributive reasons. Yet inside the world of intellectual property policy, *the minimalists are often the ones who are put in the position of insisting on the need to confine analysis to the economic need for intellectual property rights as incentives to progress.* Normally this happens when a minimalist confronts some intellectual property claim built on Lockean labor principles, Hegelian ideas of personality, or even on the fecklessly unattractive behavior of some alleged "infringer" who is scarfing up a work created with the investment of much sweat and capital.

In the face of these appealing arguments, the minimalists insist sternly that the Constitution and the American tradition of intellectual property law forbid us such normative appeals. We must confine ourselves to incentives; no matter how hardworking or morally attractive the potential recipient of intellectual property rights, only the encouragement of the *next* creator is important. Thus when the issue turns to gene patents, the minimalists are happy to point out the dangers in moving intellectual property rights too far upstream, or the need to keep some teeth behind the utility, nonobviousness, and novelty requirements. Yet their relentlessly utilitarian framework makes it much harder for them to consider arguments that fuel much of the general popular, journalistic, and policy debate, *even when those arguments are not deployed to argue for the extension of intellectual property rights, but for their circumscription.* Having insisted on the necessity of a means-ends analysis focused on producing incentives to progress, it is much harder to engage the claims about the moral limits of the market in a liberal democracy, or the argument that the growth economics of a consumer society so permeates our assumptions as to make it impossible to consider the environmental consequences of our property systems. To be sure, there is excellent scholarly work that avoids this potential pitfall, but nevertheless the ambit of discussion in the intellectual property debate is a relatively narrow one, and I believe the means-end analysis is at least one of the reasons why.

---

onmentalism for the Net. James Boyle, *A Politics of Intellectual Property: Environmentalism for the Net?* 47 DUKE L. J. 87 (1997).

[21] James Boyle, *Cruel, Mean or Lavish? Economic Analysis, Price Discrimination and Digital Intellectual Property* 536 VANDERBILT LAW REVIEW 2007 (2000).

[22] Compare Rebecca Eisenberg, *Proprietary Rights and the Norms of Science in Biotechnology Research*, 97 YALE L. J. 177 (1987) & Heller & Eisenberg, *supra* note 14.

B. Reasons to be Narrow: Take 3

So far I have argued that the debate over gene patents allows us to reflect on a curious fact: A number of factors have come together to produce a remarkably narrow disciplinary self-conception in the world of intellectual property scholarship. I argued that our concentration on the clash between maximalist and minimalist visions of intellectual property has produced an unintended side effect: a curious methodological tunnel vision. The critical scholars most likely to question the ambit of new rights are, paradoxically, firmly wedded to the notion that the only legitimate rubric for intellectual property policy is the maximization of innovation. All other normative criteria are exiled beyond the pale of the discipline.

But is such narrowness a bad thing? First, are these debates *worth* participating in? For example, one does not have to be a logical positivist to believe that policy debates based on religious faith tend to collapse fairly quickly into mutually incompatible thunderous denunciations. There *might* be a principled deal to be brokered between the Darwinians and the theists on the status of the human genome, but the odds are against it. Does this caution hold true for most of the objections to genetic patents?

Second, even to the extent that there are fruitful intellectual debates to engage with here, do intellectual property scholars have anything to add to them? What do *we* know of environmental ethics, of the moral limits of market under liberal capitalism, or of the distributional or internationalist commitments raised earlier? Shouldn't the cobbler stick to his last?

Third, is this whole question not moot? We may believe that there are reasons, both inside and outside the legal system, why genetic patents should not have been granted, but the genie is now out of the bottle; at least in the United States. A multibillion dollar industry has been erected on the premise that quite fundamental genetic patents are available. In practice, neither the Supreme Court nor Congress nor the PTO is going to upset that premise, even if some have their doubts about whether it was ever correct either as a matter of law or as a matter of desirable innovation policy.

These are weighty objections, but I think that there is a response—or at least a partial response—to each of them.

First, let me concede that the religious objections present the hardest case for the benefits of disciplinary openness (and are thus frequently used as a proxy to suggest that none of the noneconomic arguments are worth considering, a conclusion that hardly follows.) Let me note, though, the fact that debates turn on matters of nonfalsifiable religious faith is hardly anything new to the world of legal scholarship. Indeed, it is often the case that such debates—abortion is the obvious example—attract an astounding amount of scholarly attention, some of it actually *premised* on the claim that the controversy is faith-based; some scholars have argued the fact that a policy is faith-based, or that the moral claims raised are incommensurable, has concrete legal or constitutional ramifications.[23] In other words, the religious components of fundamental policy debate have just as frequently in the past been reasons *for* scholarly intervention—wise or unwise though that intervention might be. One might also note that it is hard to draw a clean line between religious faith-based systems and secular moralities which developed out of them or away from them, and which also rest on *a priori* claims. And if it is hard to justify our conceptual apartheid even in the strongest case—straightforward claims to religious faith—then surely it must be very hard indeed to justify exclusion of the environmental, distributional, and other normative concerns, which certainly form an important part of more conventional property scholarship. Put differently, if one reverses the question and asks, "Is it appropriate for intellectual property scholarship to exile all forms of morality other than wealth-maximizing utilitarianism?" then boundary lines of our discipline look considerably less justifiable. But even if the frankly religious component of the debates drops out—a result which would not displease me, I confess—or the religious component is admitted only through some "purifying" secular filter of abstract moral philosophy, there is still much of substance remaining that fits within the ambit of the constitutional basis of American intellectual property law, yet lies beyond the current bounds of intellectual property scholarship, policy, and debate.

Think for a moment of more conventional property scholarship and policy debates. Imagine that we were debating whether or not to allow the commodification of nonreproducible organs, such as kidneys or corneas, or a debate about the sale of contracts of employment, or the question of ownership of Indian relics. Think of a discussion of the possible partition and sale of America's national parks. Or, for that matter, of the deep-sea bed, the surface of the moon, or of licenses to pollute. There are those, of course, who would pursue these questions solely through a narrow utilitarianism, in some framework of wealth maximization or theoretical Pareto superiority. But most would find such an approach a mark of a sadly crabbed legal and philosophical imagination. This is not to say that one would *neglect* the economic questions; in fact, they would often play a vital role. But they would surely do so in the context of a much wider analysis. Such analyses would, in all likelihood, be marked by explicit consideration of the effects of the property regime on human dignity and the environment, by consideration of the legitimacy of traditional property claims, and by discussion of the requirements of international distributive justice. Even within the utilitarian analysis, they would be marked by a skepticism about the normative force of extrapolations from the existing distribution

[23] *See, e.g.*, Ronald Dworkin, FREEDOM'S LAW: THE MORAL READING OF THE AMERICAN CONSTITUTION, 29 (1996).

of wealth, and would in all likelihood involve a frank choice among multiple possible measures of efficiency. *Is a discussion of patent rights over the genome so different that it warrants a completely different set of methodological assumptions?* Does being an intellectual property scholar offer a license to be parochial?

What of the second objection, that modesty should cause intellectual property experts to stay away from the environmental, moral, or distributive objections to genetic patents? Modesty is a great virtue to be sure, but one must note that it is altogether an uncharacteristic one for *legal* scholars. In fact, following their discipline's longstanding imperialistic urges, legal scholars have exercised eminent domain over a remarkable range of questions and specialties—from Rawlsian moral theory through game theoretic analysis, to literary theory, neoclassical economics, deconstruction, queer theory, and chaos theory. Why stop here? It may be that intellectual property is the only place where one can find a modest legal scholar, though that would not be the descriptive term that immediately leaps to mind to characterize the group. But we surely need some explanation of why the arguments for methodological modesty are so dominating here that they win *almost without discussion*, and yet so weak elsewhere.

More particularly, as I just pointed out, if one looks at parallels in the scholarship over real and personal property, one finds no such reticence. Debates over commodification and the limits of the market are commonplace,[24] as are discussions of the rights of indigenous peoples and the extent to which property has a fundamental status within our political tradition that demands greater constitutional protection.[25] If they can do it, we can, too. It is worth pointing out that the link between environmental policy and property law has been particularly fruitful; both in terms of practical explorations of the intersection, in the law of nuisance for example, and theoretical explorations of the juncture in the context of such issues as tradeable emissions rights, ecosystem services, or "the substitutes for a land ethic."[26] There is no reason to believe that the same number of insights, provocations, and engagements could not be found in the fight over the enclosure of the intangible genetic commons, the intersection between intellectual property policy and environmental policy, or the search for a Leopoldian "substitutes for a genome ethic."[27]

The third objection is that the entire question is moot because a multibillion dollar industry has been built on the premise of gene patents. Even if those were initially of questionable validity, and I think there is a respectable

[24] For an uncommon example, *see*, Margaret Jane Radin, CONTESTED COMMODITIES (1996).

[25] *See, e.g.*, Richard A. Epstein, TAKINGS: PRIVATE PROPERTY AND THE POWER OF EMINENT DOMAIN (1985).

[26] Aldo Leopold, A SAND COUNTY ALMANAC 210–211 (1949).

[27] *See infra* note 32.

argument in law and policy to say that this is the case, no court is now going to upset the financial arrangements that have already been built. This is an important and legitimate pragmatic consideration. Still, scholars are not supposed to give up their inquiries, even if there are strong practical reasons why the social arrangements they are challenging look particularly stable. This point is, or at least should be, axiomatic. Moreover, the insights that come from engaging the broad range of arguments for and against genetic enclosure are not limited to some up-or-down vote on gene patents. The same issues are presented at every level of the analysis, from the debate over the scope of stem cell patents, the PTO's utility guidelines, the exercise of governmental march-in rights, the next round of the Agreement on Trade Related Aspects of Intellectual Property (TRIPS) negotiations, the revision of Bayh-Dole, the debate over database protection, and so on. Those issues continue to arise and it will ill-serve us if we have a double-decker structure of policy debate about them; a lower deck of tumultuous popular and nonlegal arguments about everything from the environment to the limits of the market, and a calm upper-deck which is once again fine-tuning the input-output table of the innovation process. We need a fusion of the two, or at least a staircase between them.

IV. RECONSTRUCTING SCHOLARSHIP

Let me assume optimistically that a couple of the arguments I have made here will cause you at least a moment's thought before they are rejected. Perhaps you find it hard to explain the disparity in methodological breadth between property scholarship and intellectual property scholarship when it comes to questions of commodification. Alternatively, perhaps you are struck by the fact that even much of the most "critical" intellectual property scholarship tends to be relentlessly utilitarian, whether the question being debated is the expansion of rights or their circumscription. It could even be that you notice how seldom intellectual property scholarship deals explicitly with questions of distributive justice, or even with more expansive definitions of economic efficiency. Those questions, after all, are meat and drink to scholars of tort law, property law, taxation, and so on, even those operating in a largely economistic framework. Can they be irrelevant here? Let us say, in other words, that you are willing to entertain the *possibility* that the methodological boundaries of the discipline might be a little narrow and that the gene patenting controversy could illustrate some of that narrowness. I would count this chapter a success if it raised even this much of a doubt.

The next reasonable question is, "What would we be looking at if we agreed with you?" What kinds of potentially productive study are suggested if we take seriously some of the points developed here? It is a daunting task first to imagine and then defend an entire body of scholarship. Nevertheless, here are two lines of inquiry that strike me as particularly fruitful and, while they go far beyond the debate over gene patents, seem to be suggested by some of the arguments we find there.

A. Questioning and Refining the Ideal of Perpetual Innovation

Like most scholars of intellectual property, I use innovation as my touchstone. For both constitutional and economic reasons, the ideal of continued innovation in science, technology, and culture lies at the heart of the discipline. We worship at the church of innovation. We take it as an *a priori* good. But is the concept as unproblematic as our work would make it seem? In other contexts, environmentally inspired economics has raised powerful concerns about the different but related ideal of perpetual growth; is our ideal to build a market that makes more and more *stuff*, and in the process makes us (unequally) richer so we can afford to buy more *stuff*, fueling yet another cycle of productive expansion and relentless consumption? The environmentalists ask us whether such a conception of economics is environmentally sustainable or best suited to promote human flourishing? It would be hard to disagree that our conceptions of economics and growth policy are stronger as a result of their objections. Are there a similar set of questions about the ideal of perpetual innovation?

To put it more precisely, the environmental critique of perpetual growth depends on an empirical challenge to the assumption of infinite resources, an epistemological challenge to the notion that human welfare is reducible to consumer surplus, and a deontological or teleological challenge to the "form of life" implied by consumer materialism. But for most scholars, the response to these thoughtful challenges is not to reject growth as a goal. It is, rather, to talk of sustainable development, disaggregate the notion of growth, and talk about those forms of growth that are desirable. Can we, should we, do the same thing with innovation, seeking to disaggregate it and fine-tune our policies to achieve the type of innovation we wish, whether it be to preserve the environment, promote "human flourishing," honor the logic of the original position, or take account of the diminishing marginal utility of wealth? Is our discipline's goal of innovation really reducible to the attempt to "develop new stuff that sells?"

For those who practice the economics of the Chicago school, current revealed consumer preferences (based on the existing distribution of wealth and the pattern of ability and willingness to pay what that distribution yields) have an almost totemic power. Those preferences and only those preferences give rise to "neutral" or "scientific" descriptions of human welfare. Attempts to interfere with the pattern of innovation called for by those preferences (for example, by shifting research away from male pattern baldness and toward malaria and sleeping sickness) represent a "political choice" or a preference for equity over efficiency, and thus possibly a market distortion and resulting welfare loss. The decision blindly to follow the demand dictated by the current distribution of rights does not represent such a political choice. (To which one can only respond, "Humbug!") Thus, the Chicago school economist would effectively answer that innovation *is* reducible to the attempt to develop new stuff that sells. This is a neutral account of innovation and an appropriate guide to intellectual property policy. Any other notion, for example, seeking those forms of innovation that might be called for by a more equal distribution of wealth, or those forms which hold the best prospect of preserving the biosphere, represents a contentious political choice and is an inappropriate focus for scholarly inquiry. *Scholars should not be choosing among or between preferable* types *of innovation, but simply seeking to promote innovation—understood as a black box which takes its preference inputs from the existing distribution of wealth and its constraints from the existing pattern of rights and regulation.*

But the Chicago school account is, to use a precise term of art vouchsafed to me by my children, "bogus." There is no neutral account of innovation or efficiency. We must, and we already do, pick between different visions. For example, we choose between those visions of efficiency which do and those which do not rely on ability and willingness to pay as the sole measure of value. These would yield very different patterns of efficient innovation policy. Our analysis of innovation is influenced by market structures and ideals of progress that are partly caused by the very rules we are analyzing. We must choose whether to hold those structures constant, or put them, too, under the microscope of analysis. Consider, for example, the disparate treatment of investment in compiling factual data, on the one hand, and investment in invention or expression on the other. Are these differences to be taken as given, as exogenously determined rights which structure the market we study, or as a part of the field to be questioned? Our work is influenced by implicit selections *between* types of innovation already prefigured in the legal doctrine. Consider, for example, the treatment of pure research tools under the utility guidelines, or the different treatment of innovative parody and satire in copyright law. Some types of innovation are to be hailed, others not. My point is that, *implicitly*, scholars are already selecting among types of innovation.[28] We hail certain

[28] I take it that some economists are perfectly prepared to agree to certain versions of this argument. *See, e.g.*, Brett Frischmann, *Innovation and Institutions: Rethinking the Economics of U.S. Science and Technology Policy*, 24 VT. L. REV. 347 (2000). Yochai Benkler's work comes closest towards unpacking the meanings of innovation and selecting among and between them based on an analysis that is informed by economics but goes beyond the narrowly economic. Yochai Benkler, A *Political Economy of the Public Domain: Markets in Information Goods vs. The Marketplace of Ideas* in EXPANDING THE BOUNDS OF INTELLECTUAL PROPERTY: INNOVATION POLICY FOR THE KNOWLEDGE SOCIETY (R. Dreyfuss, D. Zimmerman, H. First eds.) (2000); *Free as the Air to Common Use: First Amendment Constraints on Enclosure of the Public Domain* 74 N.Y.U. LAW REVIEW 354 (1999).

market structures as more innovative when we really mean they promote more desirable forms of innovation when judged against some other metric—based on a particular and contentious conception of efficiency, distributional justice, or what have you.

If the gene controversy, where all these issues are forcibly thrust to the surface of the debate, forced us to disaggregate innovation and then articulate and defend the precise incarnation of it that we believe to be justified—whether on environmental, distributional, or deontological grounds—then it would have done us a favor. It is striking to see how uncritically appeals to big "P" progress and big "E" efficiency are accepted in our discipline, where in other areas of legal scholarship there is much more of an attempt to clarify which type of efficiency or which type of progress is being appealed to, an attempt which must reach outside the charmed circle of economic analysis in order to be meaningful. To be sure, early Chicago school law and economics scholarship talked as if there were one uncontestable vision of efficiency, and all other visions of the concept could be exiled to the touchy-feely world of "equity claims"—claims which were always to be pursued *somewhere else* than in the current analysis. But it would be hard to find many contemporary scholars who echo those claims, or who treat wealth maximization within the current distribution of wealth as an unproblematic definition of efficiency. Is that not, though, what we do all too often in intellectual property scholarship—at least when we worship at the church of innovation without clarifying what implicit choices about types of desirable innovation are being concealed by the very structure of our analysis? To say this is not to automatically nominate any particular successor notion of innovation. My own preference would be for considerably more scholarship to be done that compares our system of current innovation against one that, in questions of basic human needs such as access to medicines that cure fatal diseases, stipulates a certain minimum valuation to human life, even among the global poor. There are many possible solutions to the failures *and inefficiencies* this analysis would reveal in our current intellectual property system—ranging from supplementing the patent system with government bounties or prizes to offering dual zone patents to directly subsidizing research. Intellectual property scholars have an extremely valuable role to play in working out the bugs in such solutions, and they are less likely to play that role while they are in the thrall of the black-box model of innovation. This leads to my second suggestion.

B. From Public Goods to Public Choice

We look at intellectual property through a lens constructed around the focal point of rational choice theory. We care deeply about the structure of incentives that it sets up, the possible free rider effects, moral hazards, and so on. But we are curiously reticent in turning that lens back on the structure of incentives that intellectual property policy sets up *in the policy making process itself.*[29] We should study the market for intellectual property *legislation* with the same care and rigor that we display in studying the market for intellectual property. To put it differently, one environmentalist critique of gene patents is that they create a powerful incentive for rights-holders to drive state scientific policy toward the utilization of the technologies in which the right-holders have exclusive rights, and do so quickly, before the patent term expires, regardless of whether the environmental consequences of those technologies are well understood. This kind of phenomenon ought to be well within the intellectual property scholar's field of inquiry.

More generally, we need models of the market for intellectual property legislation that can help us understand how to build structural protections of the public interest into the policymaking process itself. Intellectual property is a legal device to avoid a "public goods" problem, meaning the problem of providing adequate incentives to produce goods which are nonrival and nonexcludable. But there are also a different variety of "public goods" in the market for intellectual property legislation. We know that legislation built around the self-interest of existing and aspirant monopolists will protect a variety of private goods, namely those of the firms and interests at the table. We know also that it will fail to protect certain kinds of interests—most notably those of large numbers of unorganized individuals with substantial collective but low individual stakes in the matters being discussed. It will fail to protect the interests of businesses that, as Jessica Litman puts it, are "outsiders who can't be at the negotiating table because their industries haven't been invented yet."[30] This will be even more true if those businesses are likely to challenge the business plans of the companies that *are* at the table. It will fail to protect the public domain, even as a source for future creativity, except in the limited cases where those around the table can clearly see that their *particular* form of production will work better with a rich commons-based system rather than a private property system tilted to give their particular form of production an advantage.

These are the dysfunctions that we know our market for intellectual property legislation will yield. Why do we not spend some of the time we currently invest into trying to remedy dysfunctions in information and innovation markets thinking of remedies for the dysfunctions in our legislative process? How about an intellectual property ombudsman to represent the interests of the public and the public domain? Or a required "information environment impact statement" by the General Accounting Office justifying each new

[29] For a general background, *see* Maxwell L. Stearns, PUBLIC CHOICE AND PUBLIC LAW; READINGS AND COMMENTARY (1997); Daniel A. Farber & Philip P. Frickey, *The Jurisprudence of Public Choice*, 65 TEX. L. REV. 873 (1987). For a similar suggestion about intellectual property policy, *see* Arti Rai, *Addressing the Patent Gold Rush: the Role of Deference to PTO Patent Denials* 2 WASH. U. J. L. & POL'Y 199, 215, note 51 (2000).

[30] Jessica Litman, Digital Copyright (2001).

grant of monopoly right in terms of its long-term economic effects? There are better ideas than these, but it is the focus on the policymaking process that I wish to defend. Again, attention to the gene patenting controversy proves instructive. It may be that the complaints of the people on the bottom of the bus have more to teach those on the top than we would like to admit.

V. CONCLUSION

I started by comparing the current progress of intellectual property expansion over the human genome to the English enclosure movement. As with the first enclosure movement, we are faced with a conflicting array of claims—claims about the essential commonness of the property in question or the moral and traditional reasons why it should not be shifted to individual ownership, claims about the effect of newfound property rights on environmental attitudes, claims about the limits of the marketplace, about economic effects, about encouragement of innovation, and so on. As with the first enclosure movement, some of these claims are essentialist, romantic, or just analytically sloppy. As with the first enclosure movement, we have available to us a cooler language of economic incentives that promises a calmer and smoother discussion about the encouragement of progress. (Though the economic arguments for the second enclosure movement strike me as considerably weaker than those for the first.) And, as with the first enclosure movement, to retreat from the full range of issues into those calmer, smoother waters would be a mistake, *even if we ended up supporting the furthest reaching extensions of property rights currently being proposed.*

In this article I have tried to summarize—without endorsing—a broad range of criticisms of gene patenting. My point was *not* to say that this broad range of criticisms was correct; indeed, I find many of them unpersuasive. Nor was it to reject the analytical structure of innovation economics that forms the backbone of intellectual property scholarship. Instead, I used the gene patenting debate as an example to make the larger point that intellectual property scholars are mistaken to write off most nonutilitarian criticisms as outside their purview; concentrating instead only on a positive exegesis of patent law and a restricted normative exegesis of which patent policy will best ensure continued research and development of commercially desirable products. Even if one is unconvinced by Rifkin or the environmentalists, in other words, when one focuses on the *pattern* of disciplinary exclusion rather than the particular substantive arguments, it becomes apparent how narrow and how contentious our disciplinary assumptions are. What's more, I argued, the major professional division among intellectual property scholars—between maximalists and minimalists—has the strong but probably unintended effect of helping to reinforce this narrowness over the methodological tenets of the discipline.

The minimalists are those with the strongest incentive to challenge the expansion of intellectual property rights over the genome. Yet because of the structure of the rhetoric in the discipline, they are also those most committed to the claim that only utilitarian arguments about the encouragement of future innovation are legitimate parts of the discourse. Minimalists are accustomed to fighting off covert "sweat-of-the-brow" claims, concealed appeals to natural right, and Hegelian notions of personality made manifest in expression—all deployed to argue that rights-holders should have their legally protected interests expanded yet again. Against these rhetorics, minimalists insist on both constitutional and economic grounds that the reason to extend intellectual property rights can only be the promotion of innovation. When we turn to the question of whether there are noneconomic reasons for *curtailing* the reach of patents over genetic material, minimalists' hard-wired reflex is to restrict their analysis to the same utilitarian domain. Beyond the debate between maximalists and minimalists, the gene patenting controversy throws other articles of faith into doubt, from the various possible presumptions behind the idea of perpetual innovation to the assumed operation and dysfunction of the political process. We have much to learn from the gene wars even though they may currently seem to yield more heat than light.

I think most intellectual property scholars will find this claim unconvincing. Told that they must supplement a discipline in which answers seem possible—even if disputed—with greater attention to debates in which they have much less familiarity and in which answers seem much more contested, they will understandably cavil. What's more, the intellectual biographies of many patent scholars mean that they find the intersection of a particular complex technology with a set of different innovation markets to be a much more personally hospitable terrain than the broader landscape I am suggesting.

I cannot prove that they are wrong, of course. For epistemological reasons, if nothing else, the proof of this pudding is in the eating. But I will offer two analogies to suggest that a broader debate about genetic patents within intellectual property scholarship would be of particular benefit.

First, consider the analogous effect of environmental thinking on both the debates within the analytical structure of tangible property scholarship.[31] The increase in the breadth of the analysis, the need for a multigenerational framework, the importance of externalities, and the complexity and interrelationship of the natural systems within which schemes of public and private law land regulation intersect all have been highlighted by the brush with environmental concerns. Disciplines do not develop in a vacuum; in this case the analytical toolbox itself has been changed by the encounter. Even the practitioner of a "just-the-facts-Ma'am" style of property scholarship, uncomfortable

[31] Carol Rose's work illustrates my point here particularly well.

with the fuzziness and enthusiasm of all these tree-huggers, has seen his discipline improved in the process. There is no reason to believe that the same thing would not be true of intellectual property scholarship as it confronts the very real questions posed by the consequences to environmental policy of patents on genetic material, including human genetic material. And this thought, in turn, suggests that in other areas the broader debate over genetic patents may have much to teach intellectual property scholars, in the same way that the debate over commodification changed property scholarship in the 1980s and 1990s. In particular, we intellectual property scholars need to grapple more thoroughly than we have with the limitations as well as the advantages of the utilitarian, progress-enhancing analytical framework we adopt. In other areas of scholarship, we are much more willing to point out the limits of utilitarian analysis, the weakness of a "willingness and ability to pay" model of social worth, the impact of wealth effects on the efficiency calculus, the intergenerational moral concerns hidden by the framework of growth economics, and the problems with always assuming endogenous preferences or exogenously determined rights. Just as we may have more to offer the debate over gene patents than a careful analysis of effects of exclusive rights at different points in the research stream, so, too, the broader debate over gene patents may have things to show us about the relative complacency of our methodological assumptions.[32] Ironically, as I suggested earlier, this point may be particularly important, and particularly difficult, for the minimalists.

For a second analogy, consider the justice claims that have recently caused "access to essential medicines" to become a fundamental part of drug patent policy both domestically and internationally. Again, these are a set of issues that fit poorly within conventional intellectual property scholarship; but the arguments are not mere exhortations to take drugs away from companies and hand them over to the poor and sick. The essential-medicine questions are not simple, either economically or institutionally, and—after some initial reticence—the academy now seems to be turning its eyes to the complicated points of treaty interpretation, regional institutional design, international price discrimination, and alternative patent regimes that this particular and real moment of human suffering forces us to think about. Can we really believe that our scholarly focus will be somehow *weaker* as a result of the forced encounter

[32] I have struggled with this point particularly in my scholarship on the public domain. In that scholarship, I rely heavily on the basic tools of innovation economics. But I am always troubled when I read these lines from Aldo Leopold, from a chapter of A SAND COUNTY ALMANAC entitled "Substitutes for a Land Ethic." "One basic weakness in a conservation system based wholly on economic motives is that most members of the land community have no economic value. . . . When one of these noneconomic categories is threatened, and if we happen to love it, we invent subterfuges to give it economic importance. . . It is painful to read those circumlocutions today." Aldo Leopold, A SAND COUNTY ALMANAC 210–211 (1949).

with claims of distributive justice and human rights? In fact, with any luck, the intensity of feeling about a particular controversy over AIDS drugs may actually force us to acknowledge the single greatest weakness behind a patent-driven drug development policy; a patent-driven system for drug development will, if working correctly, deliver drugs on which there is a high social valuation—measured in this case by ability and willingness to pay. To put it another way, to have a patent-driven drug policy is to *choose* to deliver drugs that deal with obesity, male pattern baldness, rheumatoid arthritis, and heart disease. It is to choose *not* to have a system that delivers drugs for tropical diseases or any disease suffered overwhelmingly by the poor.

To say this is not to condemn drug patents; it is rather to suggest precisely the two lines of inquiry that I argued for in this article. First, if our goal is truly to help to eliminate human suffering, then we should spend more time thinking about alternative and supplementary ways of encouraging pharmaceutical innovation beyond the drug patent system. Second, when we talk about innovation and progress in the intellectual property system, we quickly and easily substitute some universal imagined ideal of Progress for the actual specific version of "progress" toward which our current distribution of entitlements and rights will push us. Many policies that might seem justified by the promotion of large "P" progress might seem more questionable if they were instead pushing us towards the specific vision of progress that held latently within the pattern of demand established by our current distribution of rights and wealth. To quote Amartya Sen, "there are plenty of Pareto optimal societies which would be perfectly horrible places to live."[33]

Now, if these lessons can be taught to us in a concrete and unforgettable way by the debate over drug patents, is there any reason to believe that the larger debate over gene patents will offer us any less insight or provocation? It *could* be, of course, that the end result would be exactly the same; perhaps all of us would find our conclusions unchanged—even if we were a little more critical about worshiping at the church of innovation, even if we clarified our definitions of that concept and the notion of efficiency that underpins it, even if we broadened our scholarly focus to include the kind of institutional and environmentalist inquiries I suggest here and made our discussion of commodification a little more similar to that which occurs in conventional property scholarship. Perhaps this change in methodology would leave our substantive positions unchanged, though I doubt it. Perhaps its effects would only be found in other areas, such as the essential-medicines question, the question of the goals of basic

[33] The source was an interview with an uncomprehending MSNBC interviewer at the time of Sen's receipt of the Nobel Prize in Economics in 1998. The reader will have to trust the author's faulty memory for the precision of the quote.

science policy, or the question of the redesign of the institutional framework through which intellectual property policy is made. But even if all that were true, the gene patenting debate could still teach intellectual property scholars a set of lessons we sorely need. At least, that is, if we have the courage to enter it.

Acknowledgments

Thanks to Lauren Dame, Bob Cook-Deegan, and Alex Rosenberg for their comments and criticisms. Bob, in particular, should not be held responsible for the results, but his comments helped me greatly by pointing out how *alien* the idea of taking the rhetorical structure of a discipline seriously is to the world of pragmatic intellectual property policy. Thus, if nothing else, my focus should be clearer, even if it is ultimately judged to be no less mistaken.

Section 2

THE CASE FOR PROPERTY RIGHTS

7

Perusing Property Rights in DNA

F. Scott Kieff
Associate Professor of Law
Washington University School of Law
St. Louis, Missouri 63130

John M. Olin Senior Research Fellow in Law, Economics, and Business
Harvard Law School
Cambridge, Massachusetts 02138

ABSTRACT

Many practical questions have arisen since the use of patents relating to DNA has become progressively more common after the landmark case of *Diamond v. Chakrabarty* in 1980: What ownership rights do patents confer? What is owned?

0065-2660/03 $35.00

Who owns it? What does ownership mean to nonoweners? The central theme of the chapter is that much of the debate about patent rights relating to DNA is needlessly burdened by concerns that are inapposite, given an informed understanding of how things actually work in this area; and that as a result, insufficient attention is paid to exploring propositions that might elucidate or even alleviate pertinent problems in need of solutions. The approach of the chapter is more analytical than prescriptive. It endeavors to unpack how property rights in this area actually operate and concludes by making a few modest prescriptions about areas for future study, including efforts to explore the mechanisms that might be at play when actual transactions in basic biological materials and protocols break down. The intuition behind the chapter is that although many of the problems identified by patent critics do exist and, in theory, might be pernicious in creating some inefficiency in the form of dead-weight loss, in practice these problems turn out to be less pernicious than those likely to be caused by alternative regimes and are mitigated through various behaviors of market participants.

I. INTRODUCTION

The assertion of property rights in DNA has been a topic of increasing debate as patents relating to DNA have become progressively more common since the landmark case of *Diamond v. Chakrabarty* in 1980.[1] The *Chakrabarty* case held by a mere five-to-four majority that living things are patentable subject matter, which generated countless questions about the use of patents in this area. Some are broad and linked to policy: Is this good? Is this bad? These questions are beyond the reach of the chapter. Other questions are more limited and practical: What types of ownership are we talking about, or what rights does ownership confer? What is owned? Who owns it? What does ownership mean to nonowners? This is where the chapter hopes to offer some help, because an analysis of these more mundane questions may help elucidate approaches to their more lofty counterparts.

The central theme of the chapter is that much of the debate about patent rights relating to DNA is needlessly burdened by concerns that are

[1] For a collection of sources and a general overview of issues relating to patents on DNA, *see* Donald S. Chisum, Craig Allen Nard, Herbert F. Schwartz, Pauline Newman, & F. Scott Kieff, PRINCIPLES OF PATENT LAW (2nd Ed. 2001), pp. iii-vi (Judge Rich on patents), pp. 58–81 (economics of patents), pp. 290–295 (introduction to recombinant DNA technologies), pp. 295–319 (important cases on recombinant DNA patents claims), pp. 319–322 (lawyers' insight on DNA patent claims), pp. 714–727 (utility requirement and DNA patent claims), pp. 747–763 ((*Diamond v. Chakrabarty*) case and Sidebar, pp. 839–842 (strategy and risk in patent claims), pp. 1043–1049 (strategies for bringing and avoiding suit), pp. 1097–1099 (claim breadth), pp. 1211–1217 (some FDA issues), and pp. 1217–1222 (governmental immunities and impact on biotechnology)). *See also*, Dan L. Burk, *Introduction: A Biotechnology Primer*, 55 U. PITT. L. REV. 611 (1994).

inapposite, given an informed understanding of how things actually work in this area; and that as a result, insufficient attention is paid to exploring propositions that might elucidate or even alleviate pertinent problems in need of solution. In support of this theme, the chapter proceeds in four parts: Part II explores the types of ownership and the rights they each convey; Part III explores the object of the ownership interest; Part IV explores the owners of the ownership interest; Part V explores the implications of ownership for nonowners.

The approach of the chapter is more analytical than prescriptive. It endeavors to unpack how property rights in this area actually operate and concludes by making a few modest prescriptions about areas for future study, including efforts to explore the mechanisms that might be at play when actual transactions in basic biological materials and protocols break down. The intuition behind the chapter is that although many of the problems identified by patent critics do exist and, in theory, might be pernicious in creating some inefficiency in the form of dead-weight loss, in practice these problems turn out to be less pernicious than those likely to be caused by alternative regimes and are mitigated through various behaviors of market participants.

II. THE TYPES AND RIGHTS OF OWNERSHIP IN DNA

The operation of patent rights in DNA can be best understood in view of both some differences and some interactions between patents and other more common forms of property rights. When thinking about the differences, consider that patents are more limited than other property rights along a number of dimensions. When thinking about the interactions, consider that neither set of rights operates in a vacuum.

A. The Differences between These Types of Property

The differences between patents and other forms of property rights can be seen on the conceptual level by considering the bundle of rights conferred by these different forms of property. Ownership in a form of ordinary personal property, like a car, or real property, like some hypothetical Blackacre, confers on its owner a broad bundle of rights generally conceived of as a right to use. To ensure the owner's access for use, this bundle is understood to include the right to exclude others from using. But the right to exclude is not the entirety of the bundle. The bundle is generally understood to include at least some affirmative rights of use. For example, even government interference with a private land owner's right to use her own land may give rise to claims of taking without just compensation under the Fifth and Fourteenth

Amendments and their prohibitions on the federal and state governments, respectively.[2]

In contrast, a patent confers only the right to exclude use of whatever is covered by the patent claim.[3] Patents do not give patentees any right to use. The owner of a patent on a particular technology may or may not be able to actually practice that technology. Consider that other patents may also prevent use. For example, a patentee of some improved technology such as a DNA vector carrying a particular allele of a certain gene would not be able to use that technology without permission from any patentees having claims to more basic components, such as that vector. Similarly, consider that other legal regimes may also operate to regulate and in some cases prohibit the use of certain technologies, whether they happen to be patented or unpatented. For example, some regulation by the Environmental Protection Agency on the use of chemicals by the Food and Drug Administration on the use of drugs may prevent a patentee from ever administering a new drug it or perhaps even testing it.

The distinction between a right to use and a right to exclude is more than semantic. The distinction leads to a different impact when the ownership claim of the property regime comes into contact with some other regulatory restriction on use. For example, a patent right to exclude use will itself not directly be impacted while a full-blown property right might be so impaired that a takings claim might arise. Conversely, the patent right to exclude use does not interfere with other regulatory restrictions on use. There is no direct interference because the patent only creates an additional right to exclude the use that is otherwise already restricted by the regulation. There also is no indirect interference because the ordinary incentives to create and commercialize inventions that patents provide will be dissipated by other regulatory restrictions on use. Therefore, while there may be many legitimate debates about which uses of technologies should be restricted for various public policy reasons—from nuclear energy to nucleic acids—those debates need not concern themselves with the patent arena.

There are additional differences between patents and other forms of property rights that can be seen on a more palpable level by considering simple matters like the cost to initiate the rights. While personal property rights begin automatically upon capture of otherwise unowned subject matter, patent rights can only be awarded by the Patent Office;[4] and the cost of applying for a patent can be quite substantial—around

[2] Such takings claims can be raised when either a full taking through eminent domain has occurred or when even a substantial diminution in value through a regulatory action has occurred.

[3] United States patents give to the patentee only a right to exclude others from using whatever is claimed in the patent. *See* 35 U.S.C. § 154 (a) ("Every patent shall contain . . . a grant to the patentee . . . of the right to exclude others").

[4] *See* Chisum *et al.*, *supra* note 1, at 91–160 (discussing procedures for obtaining a patent including the processes of preparing an application and prosecuting the application before the Patent Office).

$25,000 for simple cases, and much more for cases involving complex technologies in fast-moving fields like recombinant DNA, gene therapy, etc.

Differences also are evident when considering the duration of the rights. While personal property rights generally last as long as someone is exerting dominion and control over the subject matter, patent rights last only for 17 years on average.[5] Moreover, although patents typically have a 17-year term, patents are not enforceable if certain maintenance fees are not paid at 4 statutorily set times within that term.[6]

These more tangible limitations on the patent right collectively exert at least one significant influence on who will end up having ownership of a patent: Patent owners will be only those who place—or who can convince investors to place—such a sufficient value on ownership that they will have sufficient willingness and ability to pay the costs of obtaining and maintaining ownership. Therefore, the only inventions that are likely to be subject to patent protection at any given time are those that are able to capture and hold the attention of a willing owner. All of this increases the chance that patent owners will be responsive to banal business concerns about money—perhaps as opposed to more esoteric individualized concerns, over which transactions might be more difficult. Therefore, the chance of breakdowns in transactions over patented technologies may be less likely than what others suggest, as explored later in the discussion of the implications of ownership for nonowners.[7]

The 17-year limit on term helps clear the register of patents in force at any given time. This facilitates bargaining between patentees and those seeking to use patented technologies by decreasing the number of patentees with whom each user will need to negotiate.[8]

[5] *See* Chisum *et al.*, *supra* note 1, at 831–836 (discussing patent term and explaining that after the GATT amendments that became effective in 1995, patents expire 20 years from filing date, which yields an effective patent term of 17 years because the examination process takes an average of 3 years to complete).

[6] Not surprisingly, many patents are allowed to lapse before full term for failure of the patentee to pay the fees. For a recent discussion of this issue, *see* Mark A. Lemley, *Rational Ignorance At The Patent Office*, 95 NW. U. L. REV. 1495, 1503 (2001) (footnotes omitted) from which the following text is an excerpt:

> A surprisingly large number of issued patents lapse for failure to pay required maintenance fees. Payment of these fees, which are relatively low, is a prerequisite to bringing a patent lawsuit: failure to pay maintenance fees is effectively an abandonment of the patent. The evidence presented in Table 3 suggests that nearly two-thirds of all issued patents lapse for failure to pay maintenance fees before the end of their term: nearly half of all patents are abandoned in this way before their term is half over. Most of these patents aren't litigated or licensed during the short time they are in force.

[7] *See generally infra* Section V. *See also infra* text accompany notes 67 and 82.

[8] *See* F. Scott Kieff, *Property Rights and Property Rules for Commercializing Inventions*, 85 MINN. L. REV. 734 (2001) (discussing impacts of longer or shorter terms).

Finally, patents and other forms of property differ in one important way in which the patent counterpart is more expansive than the ordinary personal property counterpart. There is no national registry of ownership in personal property. Even land ownership is only easily traceable through tax records in the local jurisdiction. In contrast, patent records, including date of issuance, date of expiration, inventor, and assignee, are all fully computer-searchable for free on the web page of the Patent Office. This feature of patents also facilitates bargaining between patentees and those seeking to use patented technologies.

B. The Interactions between These Types of Property

Turning now to the interactions between patents and other forms of property, it must be kept in mind that one person's ownership of one type often exists in the context of someone else's ownership of the other. For DNA, this may mean that one person owns an actual physical sample of a piece of DNA, while another person owns a patent that covers that DNA.

Consider, for example, the case of a patient suffering from a genetically based disease and a treating doctor who seeks a patent on a piece of DNA or a cell line that is derived from the patient during treatment. This is not an unreal hypothetical. Two highly publicized cases follow this general pattern. One is the lawsuit presently pending in the Northern District of Illinois between a patient, family members, and the Canavan Foundation as plaintiffs, and a treating doctor and his hospital, Miami Children's, as defendants. The patent covers a test for the defective gene that causes an illness called Canavan's disease, designed to help to educate potential parents about the level of risk to their potential children. The plaintiffs argue that they provided funding, tissue samples, and even the idea to try to create a genetic test. The other case is the recent decision in *Moore v. Regents of the University of California* from the California Supreme Court. That case involved a patient with hairy-cell leukemia whose doctor obtained a patent on a cell line derived from the cells taken from the patient's spleen during treatment, unbeknownst to the patient.

It is important to see who brings what property interest to bear in cases like these. The Court in the *Moore* case affirmed the patient's right to bring a cause of action for breach of fiduciary duty and lack of informed consent. After that case, because a patient must be informed about whether his or her doctor will be using biological materials removed during treatment to obtain patents, such a patient could elect to shop among doctors for one who will give the patient a "cut" of any profits made from such patents. In this sense, the individual can be viewed as having a form of limited personal property right to the parts of the individual's own body once they are removed, including his or her own cells, and DNA.[9]

In the Canavan's disease case, the plaintiffs appear to be going farther, basing their argument in part on the source of the money and the idea to use patient DNA to create a genetic test. The money argument would require a court to consider the law of contracts relating to inventions made for hire. Where an inventor has been employed to invent, or where there are express terms in the contract to so provide, an employer may indeed own the invention of his or her employee.[10] The idea argument would require a court to determine who enjoys the legal status of "inventor" in the first instance, which requires a determination of how much intellectual contribution was made to the subject matter of the patent claim.[11] The touchstone of inventorship under patent law is significant contribution to the conception or reduction to practice of the actual invention itself. Merely providing the cash and the broad idea may not be enough to enjoy the status of "inventor."[12]

Informed by the rights each party is likely to bring to such a relationship, individual actors and policy planners can better sculpt the relationships that may evolve in the future. More specifically, when informed by these baselines, it may then be worth exploring the default terms and optional variants that parties might want to—or might be able to—include in the contractual relationships between patients having genetic disorders and their treating doctors who might seek patents. To be sure, in individual cases, the exigencies of a particular situation might not be conductive to meaningful bargaining, but on a group level, legislative, regulatory, or collective bargaining efforts might be of some use. Regardless of which approach is adopted, a more explicit confrontation of these issues from the contract perspective might allow parties to best structure these transactions. Such a contract approach provides an option that might be an addition or alternative to approaches based in property law or patent law. Putting this another way, if we are upset with the results in these cases, we must consider what changes we should explore making

---

[9] But this right is also limited. For example, the 13th Amendment prevents parents from owning their biological children as property, even though they originate from one cell from each parent.

[10] *See, e.g.*, *Banks v. Unisys Corp.*, 228 F.3d 1357, 1359 (discussing an employer's ownership of an employee's invention under "employed to invent" principle which is a form of implied-in-fact contract).

[11] *See* Chisum *et al.*, *supra* note 1, at 485–88 (reviewing the law of inventorship and collecting sources).

[12] *Compare*, *Hess v. Advanced Cardiovascular Systems*, 106 F.3d 976 (Fed. Cir. 1997) (holding that person who provided "materials and suggestions" to inventors named in the patent did not establish conventorship because his contribution was not more than what a "skilled salesman would do in explaining how his employer's product could be used to meet the customer's requirements ... [and t]he extensive research and development work that produced the [invention] was done by [the named inventors]"). For more on ownership, see *infra*, Section IV.

to the system. We could change the law that governs the baseline rights each party brings to the transaction, or we could change the law that governs the transaction, or we could make changes, such as through collective action or education, that somehow influence the interaction without changes to the generally applicable legal regime.

Each of the above efforts for change assumes, of course, there is a reason for change. This begs the question about what is so wrong with these cases that some change is in order. If the concern is that the patients are not getting a share of the doctor's profits, then we should recognize that this is a distributional problem, not an efficiency problem. To be sure, distributional or fairness complaints are legitimate, but they also require a watchful eye to ensure that any effort to ensure equivalence in distribution of a resource does not dissipate the value of the resource. This means that, however you come out on the distributional issue, the one goal you must have is at least to keep the value of the patent high. If the concern is about the potential restrictive effects of patents in this field, then they will be considered later in the section discussing the impact of ownership on owners.[13]

Finally, the interaction between patent rights and another form of intellectual property rights must also be kept in mind. Patents can, to some extent, operate as an alternative to another form of intellectual property, called trade secret rights. Trade secret rights are relatively weak compared to patents in many respects because they only last as long as others do not otherwise discover the information that is the subject of the secrecy. To be sure, if independent discovery or reverse engineering are unlikely, then trade secret protection can be of greater economic significance than patent protection because it may last significantly longer. For example, the recipe for Coca-Cola remains a secret to this day, more than 100 years after a patent on the recipe would have expired. Nevertheless, trade secret protection requires the owner to keep the information secret and only gives protection so long as the information does not become generally known through some other means, such as independent creation or reverse engineering by a third party. As a result, patent rights may in some cases be an alternative to trade secrets that is more attractive to the owner because it can outlast independent origination or reverse engineering, and it is more attractive to the public because it leads to publication of the information through the publishing of the patent on the Patent Office web page. Moreover, reliance on trade secrecy can have other pernicious impacts on our social fabric. The need to keep information secret may induce individuals not to share information, or to use security measures that may be unpleasant.[14]

[13] *See infra* Section V.

[14] *Compare, Kewanee Oil Co. v. Bicron Corp.*, 416 U.S. 470, 486–7 (1974) (Burger, C. J.) (in the context of a discussion about the benefits of allowing trade secret protection, highlighting

Patents can also be useful as an alternative to some self-help mechanisms.[15] The ability to exclude use through a patent not only facilitates increased use, it also provides individual actors with a legal alternative to self-help approaches that may have some pernicious impact. Consider, for example, the form of self-help in the agriculture sector called "terminator" technologies, which many have feared might cause environmentally important plant species to die out. Terminator technology refers to seeds that have been genetically altered so as to yield crops whose resulting seed will be sterile.[16] The technology prevents farmers from harvesting seeds from crops they have grown using genetically engineered seeds, thereby forcing farmers to buy more of the original seed each planting season. Terminator technology can also be thought of as the agricultural equivalent of copy-protection technology in the software industry.

Such terminator and copy-protection technologies are each a form of self-help that can be used as an alternative to legal protection in a way that is likely more costly than legal protection. Consider a market for some modified form of seed that has been altered so as to make it especially valuable compared to other seeds. Since seeds generate plants that in turn produce more seeds, the sale of a seed must take into account the potential of vast progeny seeds that are themselves potent for germination. The seller must consider the risk that his buyer will generate maximal progeny, maybe even returning to the market to sell some progeny seeds in competition with the original seller. The price needed to cover for this risk will far exceed the price needed to cover a sale to a farmer who will only use the seed for production of a single crop, not for the generation of progeny seeds. Buyers seeking seed for the purpose of growing such a single crop will want to identify themselves convincingly to sellers. Sellers' willingness to sell to such buyers at the lower price will decrease to the extent the seller disbelieves that the buyer indeed intends to and will use the seed for a single crop. As a result, both pricing and consummation of that sale are frustrated. In contradistinction, terminator technology ensures that both sides of the sale will keep to its terms. Because both seller and buyer know the seed will only be of value for a single crop, pricing and consummation of that sale are facilitated.

---

social costs of the self-help measures that would be used by individual actors if legal forms of protection were not available).

[15] Indeed, the public accessibility of patents and the information they contain stands in stark contrast to what in some respects may be the alternatives—trade secrecy and scientific fraud. For more on these issues see the following chapters: 1. Ananda Chakrabarty, "Patenting Life Forms Yesterday, Today, and Tomorrow," 22. James F. Davis & Michele M. Wales, "The Effect of Intellectual Property on the Biotechnology Industry" 26. Horace Freeland Judson, "The Difficult Interface: Relations Between the Sciences and Law" and 23. Edward T. Lentz, "Are Real Business People So Easily Thwarted?"

[16] One example of terminator technology is the Technology Protection System™ from Monsanto.

But technological self-help may not be attractive to individuals if a legal device will have the same effect, especially if more cheaply. One legal device may be a contract for sale having a restrictive term—such as a clause agreeing that the seed will only be used for a single crop. A problem with such a contract may be that it will have enforcement problems. The ordinary contract remedy of expectation damages is likely to underdeter breach.[17] In addition, contract remedies will have difficulty reaching any third-party transferees of progeny seeds. Patent law offers a convenient aid because patents can be licensed with restrictive terms and patent remedies include the right to an injunction against any infringer, including both third parties and those in contract privity with the patentee.[18] For this reason, courts have upheld patent licenses that restrict buyers to a single use.[19] Indeed, restrictive patent licenses have the added advantage of avoiding the potential risk of some harmful biological consequences that are feared to be associated with self-help devices like the terminator technology—such as the potential for its accidental spread through cross-pollination to other plants for which germination is otherwise desired.[20] Therefore, patents—especially when used with enforceable restrictive patent licenses—may operate to help avoid some of the problems with self-help technologies like the terminator gene. To be sure, these benefits that patents may provide that other forms of protection such as trade secrets and some forms of self-help do not, must be weighed against the costs of patents that are explored at the end of this chapter.

Therefore, as explored previously, patents are more limited than other property rights along a number of dimensions; and both patents and other forms

[17] *See, e.g.*, Fred S. McChesney, *Tortious Interference With Contract Versus "Efficient" Breach: Theory and Empirical Evidence*, 28 J. LEGAL STUD. 131 (arguing that so-called efficient breaches of contracts that are viewed as efficient from a static perspective often are not efficient when viewed from the dynamic perspective).

[18] *See, e.g.*, Chisum, *et al.*, *supra* note 1 at 1223–1308 (discussing patent remedies including the right to an injunction and, where infringement is willful, to attorney fees and treble damages).

[19] *See, e.g.*, *Mallinkcrodt v. Medipart*, 976 F.2d 700 (Fed. Cir. 1992) (single-use restriction on patented catheter held enforceable).

[20] One feared mechanism by which such pernicious spreading of terminator technologies might take place is discussed in Kojo Yelpaala, *Owning The Secret of Life: Biotechnology And Property Rights Revisited*, 32 MCGEORGE L. REV. 111, 209:

> Terminator Seed technology has the potential for serious environmental damage. Through cross-pollination, the Terminator Seed technology could spread from farm to farm and into other varieties of seeds. Given that the Terminator Seed technology can be combined with ordinary non-patented seeds and with other genetically engineered technologies such as the herbicide-resistant plant technology, the spread of the Terminator Seed technology would be virtually unstoppable. Imagine the thousands of different varieties of maize in Mexico being exposed to the Terminator Seed technology from a few farms. With time the technology could threaten the bio-diversity of seeds in Mexico.

of property rights often come into contact with each other. This review of these differences and these interactions will inform the exploration of the actual object of the ownership interest in a DNA patent in the next section, and the exploration of the comparative impact of patents on non-owners in the following section.

III. THE OBJECT OF OWNERSHIP IN PATENTS

The object of the ownership interest in a DNA patent, as in any patent, is set forth in the claims. While the meaning of the words used in a claim is often the focus of much litigation, whether the issue is one of validity or infringement,[21] "*the claims of the patent are the measure of the patentee's right to exclude*."[22] Patent claims in this country are said to be peripheral in nature, not central, because the claims set forth the outer metes and bounds of the right to exclude rather than the "gist" or "heart" of the invention. As Judge Giles Rich often said about patents, "*the name of the game is the claim* ... [and] the function of claims is to enable everyone to know, without going through a lawsuit, what infringes the patent and what does not."[23]

A. The Rules about What is Not Owned

Before trying to understand what a patent claim covers, it is first important to recognize what it does not cover. More specifically, the patent right to exclude, often labeled with the term "monopoly," is often not able to give the patentee much market power. Patents often are not correlated with monopolies because there is rarely a one-to-one correlation between a patent and a market and the term monopoly only has functional meaning—as opposed to mere rhetorical force—when applied to an entire market. In addition, the patentee may face market threats from old technologies, alternative noninfringing technologies, and future technologies.[24]

[21] *See*, *SmithKline Diagnostics, Inc. v. Helena Labs.*, 859 F.2d 878, 882 (Fed. Cir. 1988) ("The claims must be interpreted and given the same meaning for purposes of both validity and infringement analyses").

[22] Giles S. Rich, Foreword, in Chisum *et al.*, *supra* note 1, at iii–vi (emphasis in original).

[23] *See, e.g.* Giles S. Rich, *The Extent of the Protection and Interpretation of Claims—American Perspectives*. 21 INT'L REV. INDUS. PROP. & COPYRIGHT L. 497, 499, 501 (1990) as quoted in *Hilton Davis Chem. Co. v. Warner-Jenkinson Co.*, 62 F.3d 1512, 1539 (Plager, Circuit Judge, with whom Chief Judge Archer and Circuit Judges Rich and Lourie join, dissenting) (emphasis in original).

[24] *See* Kieff *supra* note 8, at 729–31 (collecting sources).

The coverage of a patent claim is most strenuously cabined by the rules that govern patentability. While the *Chakrabarty* case held that living things can be patented just like other things, the case was more specifically dealing with a perceived exclusion for living things from patentable subject matter.[25] The Supreme Court in that case held that patentable subject matter includes "anything under the sun made by man," and it rejected any *per se* exclusion for certain technologies, instead holding that patents on living things are to be tested against the same requirements for patentability generally applicable to all fields of technology. As the Court put it when rejecting the argument that the patent system envisioned by the framers of the Constitution was not intended to apply to such new areas of technology as recombinant DNA:

> This Court frequently has observed that a statute is not to be confined to the "particular application[s] . . . contemplated by the legislators." This is especially true in the field of patent law. A rule that unanticipated inventions are without protection would conflict with the core concept of the patent law that anticipation undermines patentability. Mr. Justice [William] Douglas reminded that the inventions most benefiting mankind are those that "push back the frontiers of chemistry, physics, and the like." Congress employed broad general language in drafting § 101 precisely because such inventions are often unforeseeable.[26]

Thus, the requirements for patentability are applicable to all technologies, and each must be satisfied for the claimed invention to be covered by a valid patent. They include the requirements of novelty, nonobviousness, and disclosure.

The novelty requirement of patent law exists to prevent patents from being issued on products or processes that anyone is otherwise using. Under the novelty requirement, which is codified in Section 102 of the Patent Act,[27] a claimed invention that is expressly or inherently disclosed in any single item of prior art is said to be "anticipated," and is considered to be not patentable for lack of novelty.[28]

[25] The requirement of statutory subject matter stems from Section 101 of the Patent Act. 35 U.S.C. § 101 (utility and statutory subject matter). For more on the statutory subject matter requirement, including the *Chakrabarty* cases and its progeny, *see* Chisum *et al.*, *supra* note 1, at 728–828.

[26] *Diamond v. Chakrabarty*, 447 U.S. 303, 315–16 (internal citations and footnotes omitted).

[27] 35 U.S.C. § 102 (novelty and statutory bars). The novelty requirement is measured from the time of invention. The statutory bars, which may also be viewed as a grace period, to patents from issuing on inventions that have been disclosed in printed publications or in public use more than one year before the filing date of the patent application.

[28] For more on the novelty and statutory bars, see Chisum *et al.*, *supra* note 1, at 323–513.

Similarly, the nonobviousness requirement of patent law exists to prevent patents from being issued on products or processes that anyone is otherwise about to use. Under the nonobviousness requirement, which is codified in Section 103 of the Patent Act,[29] a claimed invention is considered to be not patentable for obviousness if: 1) it is fully disclosed among several items of prior art; 2) those items of prior art provide some teaching, motivation, or suggestion that they be combined to form the claimed invention; *and* 3) there would be a reasonable expectation that such a combination of prior art references to form the claimed invention would be successful in yielding the claimed invention.[30]

The novelty and nonobviousness requirements work together. They each consider all verifiable items of prior art.[31] They also ensure that each item of prior art must be evaluated under both requirements: novelty and nonobviousness.[32]

The impact of these prior art requirements of the patent system reveals the illusory nature of at least one of the problems some worry might be raised by cases like the patient/doctor examples explored earlier in the section on the types and rights of ownership.[33] The prior art rules of patent law require that the patent in the Canavan's case, for example, will in no way prevent patients who happen to have been born with the gene from continuing to possess their own gene because such a claim would cover the prior art. As discussed later, patents like this one only cover isolated versions of the gene, or test kits using that isolated version.[34] To the extent there is a concern that patents on such test kits prevent the patient from looking into her own body, then we must compare such test kit patents with patents on other exploratory tools in medicine like an Xray machine or an NMR machine.

[29] 35 U.S.C. § 103 (nonobviousness).

[30] For more on the nonobviousness requirement, see Chisum *et al.*, *supra* note 1, at 514–706.

[31] *See, e.g.*, Chisum *et al.*, *supra* note 1, at 441–451 (describing evolution of case law treating 35 U.S.C. § 102(g) as a provision under which prior use may count as prior art even if not public, as long as it is not abandoned, suppressed, or concealed, and discussing the amount of evidence needed for a piece of prior art to qualify under that provision).

[32] *Id.*, at 554–584 (prior art that triggers any subsection of § 102 is available for analysis under § 103); *see also In re Foster*, 343 F.2d 980 (CCPA 1965) (*reversing In re Palmquist* 319 F.2d 549 (CCPA 1963) to hold that despite plain meaning of the statute, art qualifying only under § 102(b) may support an analysis under § 103). For the reasons discussed more thoroughly by Parchomovsky and Lichtman *et al.*, the result in *Foster* is important to mitigate the costs associated with strategic disclosure. Douglas Lichtman *et al.*, *Strategic Disclosure in the Patent System*, 53 VAND. L. REV. 2175 (2000); Gideon Parchomovsky, *Publish or Perish*, 98 MICH. L. REV. 926 (2000).

[33] *See generally supra* Section II. *See also* text accompanying note 13.

[34] *See infra* text accompanying notes 39–46.

B. The Rules about Disclosing What is Owned

The disclosure requirements of patent law exist to meet a number of needs. Under the disclosure requirements, which are codified in Section 112 of the Patent Act,[35] the inventor must provide a sufficiently detailed written description of the claimed invention so as to enable those in the art to practice the claimed invention, including the best way known to the patentee, and so as to enable those reading the patent to know as definitely as possible what will infringe the claims and what will not.

Patent law's disclosure requirements decrease social costs by serving to give clear notice about the property rights, and to decrease the chance of duplicative efforts towards the same invention.[36] The Federal Circuit's strong reading of the written description requirement to put the public on clear notice of what will infringe and what will not makes sense because the patentee as the drafter is the least cost avoider of such ambiguities. This legal development was controversial to be sure—it has been the subject of extensive criticism by many considered to be propatent and even some considered to be antipatent—yet it marks an important weapon in the patent system's arsenal for fighting social cost.

Therefore, propatent arguments that are against this development because it leads to the invalidation of particular patents must be weighed against the fact that this requirement helps to minimize the social cost of the system. In this connection, the costs of drafting a more complete application and of sacrificing patents whose disclosures are held insufficient must be weighed against the alternative social costs of transacting and litigating over, and transacting around, such patents.

In addition, antipatent arguments against particular patents—such as those on gene fragments, for example—must similarly be weighed against the fact that under present case law such patents are much less likely to cause the pernicious clogging of downstream innovation than originally feared[37] because

[35] 35 U.S.C. § 112 1–2 (setting forth the disclosure requirements of patent law: 1) written description; 2) enablement; 3) best mode; and 4) definiteness, which is also stated as the requirement that the claims particularly point out and distinctly claim).

[36] As discussed below, they also increase the cost of drafting each patent. The net effect of these different impacts is explored below.

[37] *See* Arti Kaur Rai, *Regulating Scientific Research: Intellectual Property Rights and the Norms of Science*, 94 NW. U. L. REV. 77, 126–29 (1999) (suggesting that patents on multiple gene fragments, such as ESTs, could block the use of a larger DNA sequence of which they are a part, and citing Michael A. Heller & Rebecca S. Eisenberg, *Can Patents Deter Innovation? The Anticommons in Biomedical Research*, 280 SCIENCE 698 (1998) (arguing that patents can deter innovation in the field of basic biological research)). This argument and its implications are explored in depth in the other important works by Eisenberg. *See, e.g.*, Rebecca S. Eisenberg, *Property Rights and the Norms of Science in Biotechnology Research*, 97 YALE L. J. 177 (1987) (exploring potential negative

under this case law many such downstream activities would not infringe most such valid claims for a number of interrelated reasons.[38] The intuition behind this view is not new. Indeed, according to Judge Rich, claims present a fundamental dilemma for every patentee in all fields of technology because "the stronger a patent, the weaker it is, and the weaker a patent, the stronger it is."[39] He meant that a broad patent claim is strong on offense because it covers more and therefore is more likely to be infringed, but it also is weak on defense because it may cover something in the prior art or fail to contain a sufficiently detailed disclosure and therefore it is more likely to be invalid. On the other hand, a narrow claim is weak on offense, as it covers less and therefore is less likely to be infringed, but it also is strong on defense as it is less likely to cover something in the prior art or fail to contain a sufficiently detailed disclosure, and therefore also is less likely to be invalid.[40]

Consider the examples of some older claims that actually have been litigated and that might colloquially be described as covering natural substances, such as adrenalin, Vitamin B, strawberry flavoring, or erythropoeitin (EPO). On the one hand, if those claims actually did cover such compounds, without any

---

impact of patent rights on scientific norms in the field of basic biological research); Rebecca S. Eisenberg, *Patents and the Progress of Science: Exclusive Rights and Experimental Use*, 56 U. CHI. L. REV. 1017 (1989) (exploring an experimental use exemption from patent infringement as a device for alleviating potential negative impact of patent rights on scientific norms in the field of basic biological research); Rebecca S. Eisenberg, *Public Research and Private Development: Patents & Technology Transfer in Government-Sponsored Research*, 82 VA. L. REV. 1663 (1996) (offering preliminary observations about the empirical record of the use of patents in the field of basic biological research and recommending a retreat from present government policies of promoting patents in that field).

[38] F. Scott Kieff, *Facilitating Scientific Research: Intellectual Property Rights and the Norms of Science—A Response to Rai & Eisenberg*, 95 NW. U. L. REV. 691, 699–700 (2000) (showing why a patent claim directed to a gene fragment like an EST cannot be construed to cover a larger DNA sequence, such as a substantial portion of an entire gene, and citing Kieff, *supra* note 8 at 721–22 (noting that if the patentee attempts to argue that the claim to the smaller fragment covers the fragment within the environment of the larger DNA, then the claim is likely to be held invalid over the prior art or for lack of adequate disclosure because, for the claim to be valid, the claimed subject matter must be new and nonobvious, and the patent application must disclose the metes and bounds of the claimed subject matter with physical and chemical detail as well as how to make and use it; and alternatively pointing out that since ESTs exist in nature in the company of the other DNA of the genome, a typical EST claim must be limited in order to overcome this prior art to a version of the EST in some specific environment other than its natural one, such as isolated from all other DNA or inserted into an artificially engineered piece of DNA, and the details of the degree of isolation or of the engineered piece of DNA must also be provided so as to satisfy the disclosure requirements)).

[39] *See, e.g.*, Giles S. Rich, *The Proposed Patent Legislation: Some Comments*, 35 GEO. WASH. L. REV. 641, 644 (1967) (responding to proposed legislation S. 1042 and H.R. 5924, 90th Cong., 1st Sess. (1967) and Report of the President's Commission on the Patent System (1966)).

[40] *Id.* (explaining patentee's dilemma, or in his words, "puzzle").

additional requirement, then the claims would cover the existence of those compounds in their natural state. Because each has existed in the natural state for quite some time, the claims would be woefully invalid over the prior art for lack of novelty as discussed earlier. According to the old maxim of patent law: "That which infringes, if later, would anticipate, if earlier."[41] On the other hand, if those claims were valid, we must suspect—and indeed it turns out to be the case—that the claims were substantially more limited. Indeed, in each case, the claim was directed to a purified form of the compound. As a result, although perhaps acceptable to colloquially describe these claims as covering adrenalin, Vitamin B, strawberry flavoring, or EPO without any additional requirement, it is absolutely crucial to understand that when read in their entirety and when applied by the courts, those claims only covered the purified forms of the compounds. Moreover, for those claims to be valid under the disclosure requirements, those patents must also have provided detailed instructions on how to purify, as well as specific "blaze marks" along the purification trail, so as to give notice as to which particular levels of purity would infringe.[42]

C. Applying These Rules to DNA

The same rules and practices apply to modern DNA claims that relate to the discovery of a specific gene or even a fragment of DNA. For the claim to be valid over the prior art, it cannot cover the sequence as it exists in each cell in each one of us. A claim directed to an isolated version, or a version on a vector, might overcome the prior art as long as that version has not been previously isolated. But the breadth of such a claim might be somewhat limited. Under the disclosure requirements, a claim that simply recited an isolate or vector might face problems if stretched to cover isolates or vectors comprising allelic variants, or even species variants, absent some detailed teaching about how to find those variants and how to recognize whether they infringe. While some patentees, including in the EPO[43] case, as well as in the beta-interferon[44] and insulin[45] cases, have tried to assert claims that endeavor to capture such variants with functional language—such as a claim to any sequence that encodes a protein having the same function—the Federal Circuit has made quite clear that such claims do not satisfy the disclosure requirements.

[41] *Peters v. Active Mfg. Co.*, 129 U.S. 530, 537 (1889).

[42] *See In re Ruschig*, 379 F.2d 990, 994–5 (CCPA 1967) (analogizing the written description requirement of notice to the practice of marking trails though a forest by making blaze marks on certain trees).

[43] *Amgen, Inc. v. Chugai Pharm. Co.*, 927 F.2d 1200 (Fed. Cir. 1991) (EPO).

[44] *Fiers v. Revel*, 984 F.2d 1164 (Fed. Cir. 1993) (beta-interferon).

[45] *Regents of the Univ. of Cal. v. Eli Lilly and Co.*, 119 F.3d 1559 (Fed. Cir. 1997) (insulin).

Instead, a patentee interested in trying to cover such variants must provide a crisp set of physical or chemical tests for determining which variants are covered. For DNA, this is not impossible, but it requires some gambling. Because DNA is double-stranded and because binding between strands is a function of their similarity, which can be tested via binding experiments under different levels of stringency, a claim can be written that is directed to a specific sequence—call it SEQ ID #1—or any heterologous piece of DNA that binds to it under conditions of a certain level of stringency. In such a claim, the patentee is gambling that whatever sequence that happens to be commercially significant will so bind, while whatever sequences that were in the prior art will not. Meanwhile, the claim can at least be in compliance with the disclosure requirements as long as the patentee has provided detailed instructions on how to test for stringency—such as temperature, pH, and annealing time—and which level of stringency is required.[46]

A final matter on the patentability requirements as they relate to DNA claims is worth considering. The criticism of claims directed to DNA fragments such as ESTs is often phrased as a challenge to the patent's "utility."[47] But an attack on utility grounds is difficult to understand. If the charge is that the claim is useless, there should be no concern because none will infringe a useless patent claim.[48] In contrast, a charge that the claim is too useful would only support the claim as complying with the patent system's constitutional purpose, which is to promote the progress of the useful arts.[49] If the utility is uncertain, the patentee has an incentive to license it broadly, so as to increase the chance of being able to extract some part of whatever utility is later uncovered.[50] To be sure, the patent system could be amended to require the patentee to stake out a utility at the time of filing and then limit the patent to only that utility. But such a rule would have to be explored for its impact on the central purpose behind the present patent system, which is to provide incentives for investment in the complex and risky process of commercializing inventions after they have been made.[51] Efforts to identify a metric for judging utility would also be required. In addition, a utility limitation would require a consideration of whether the patentee would have to prove that added limitation, which might be refuted

[46] *See* S. Leslie Misrock & Stephen S. Rabinowitz, *The Inventor's Gamble: Written Description and Prophetic Claiming of Biotechnology Inventions* in Chisum *et al.*, *supra* note 1, at 319.

[47] *See, e.g.*, Heller and Eisenberg, *supra* note 37.

[48] Kieff, *supra* note 8 at 721–22 (showing why the utility requirement is itself useless).

[49] *See* U.S. Const. art I, § 8, cl. 8.

[50] Kieff, *supra* note 8 at 712–714 (discussing the powerful incentive to license broadly that is caused by risks of commercialization, such as those that would obtain where commercial utility is uncertain).

[51] *See* Kieff, *supra* note 8 (showing how patent system operates, by design, as an economic tool for promoting commercialization of new technologies). *See also*, Chisum *et al.*, at 58–90 (reviewing incentive to commercialize and other incentive theories of the patent system).

by subjective evidence of the infringer that the intended use was otherwise. Indeed, it is for this reason that patentees typically avoid including such intent-based elements in their claims under the present regime.

IV. THE OWNER OF OWNERSHIP IN PATENTS

As mentioned earlier, patents are owned at least in the first instance by the inventor or inventors.[52] While express or implied contractual relationships may transfer ownership, ownership begins with the person or persons who made a significant contribution to the conception or reduction to practice of the actual invention itself.[53] To best understand how patents on DNA may operate in practice, it may help to look to see who are the inventors—and, as a result, the owners—and who are not. Consider the account given in a recent article that is somewhat critical of the present use of patents on DNA:

> For example, in the case of ESTs, companies [that are] focused on the development of end products have taken an attitude towards patentability [that is] very different from that of companies focused on upstream research. While HGS and Incyte have filed patent applications on ESTs (and have also maintained post-filing secrecy), the pharmaceutical firm Merck & Co. has put into the public domain the results of an EST identification project it sponsored at Washington University. In making these results widely available, Merck hopes to take advantage of the efforts of those who will use the results to do fundamental research. In other words, Merck believes that its own comparative advantage lies in using the fundamental research of others to do downstream work directed towards the formulation of particular drugs. Similarly, in the context of SNPs, a number of pharmaceutical companies have formed a non-profit collaboration, known as the SNP Consortium, that aims to identify and place in the public domain information about the approximately 300,000 SNPs in the human genome. The pharmaceutical companies, whose comparative advantage lies in developing basic research about SNPs, are racing to catch up with various biotechnology companies, such as Incyte, Millenium Pharmaceuticals Inc., and Genset, that have already filed patent applications on this upstream research.[54]

[52] *See infra* notes 10–12 and accompanying text. In the case of joint inventors, each may use or license use of the claimed invention without the consent or and without an obligation to account to the other. 35 U.S.C. § 262 (joint owners).

[53] *Id.*

[54] *See* Rai, *supra* note 37, at 134. (footnotes omitted).

This story is remarkably reminiscent of an earlier public policy battle over patents in yet a different field of technology that was considered to be too new for patent protection: the software industry. Today, computer software is recognized to be patentable subject matter under case law that began a year after the *Chakrabarty* case in the biotechnology field, such as the Supreme Court decision in *Diamond v. Diehr*.[55] But before the *Diehr* decision, the law was quite different. Beginning with the decision in *Gottschalk v. Benson*[56] in 1972, the law seemed to indicate that software was not patentable subject matter. Writing two years later about how to view the holding in that case, Judge Rich provided this description:

> I find it more significant to contemplate the identities of the troops lined up for battle in *Benson* and observe which side obtained the victory. On the one side was the Government, against patenting programs or software, supported by the collective forces of major hardware (*i.e.*, computer) manufacturers and their representative associations who, for economic reasons, did not want patents granted on programs for their machines. On the other side was Benson *et al.* and their assignee and assorted lawyers and legal groups who were in favor of patent protection for programs or software. The anti-patenting forces won the victory. . . .[57]

Both of the above accounts suggest that a decision to eliminate the particular types of patents at issue would be beneficial to a large company like

[55] 450 U.S. 175 (1981) (holding the process for molding rubber involving a computer program using the Arrhenius equation to calculate an estimate of the mold temperature was patentable subject matter). More recent cases include *In re Alappat*, 33 F.3d 1526 (Fed. Cir. 1994) (in banc) (holding a program for drawing a smooth wave form on a computer screen to be patentable subject matter); and *State Street Bank & Trust Co. v. Signature Fin. Group Inc.*, 149 F.3d 1368 (Fed. Cir. 1998), *cert. denied*, 119 S. Ct. 851 (1999) (holding a computer-driven hub and spoke mutual fund accounting system to be patentable subject matter).

[56] 409 U.S. 63 (1972).

[57] *In re Johnston*, 502 F.2d 765, 774 (CCPA 1974), *reversed by Dann v. Johnston*, 425 U.S. 219 (reversing on other grounds than those discussed in the excerpt). Indeed, the majority opinion in *Benson* relies heavily on the 1966 report by a Presidential Commission on the patent system that has been described by former Patent Office Commissioner Gerald Mossinghoff.

> The 1966 report of the President's Commission on the Patent System was largely a battle between AT&T, which strongly supported the patenting of software, and IBM, which bitterly opposed it. IBM's position as a mainframe manufacturer and seller was that software should be unpatentable and should be given away free of charge. AT&T, as primarily a software developer, felt precisely the opposite.

Gregory J. Maier & Robert C. Mattson, *State Street Bank in the Context of the Software Patent Saga*, 8 GEO. MASON. L REV. 307, 336 n.63 (1999) (citing interview with the Honorable Gerald J. Mossinghoff, former Commissioner of Patents and Trademarks, in Arlington, Va. (Sept. 13, 1999)).

IBM or Merck. If helping these companies is the public policy objective, then such elimination is consistent with the objective. But if the stated concerns about patents on DNA relate to their impact on individual biological scientists,[58] then it is not clear that weakening or eliminating these patents is consistent with stated goals. This is especially true if it is these individuals who are likely to find themselves as patentees.

This discussion about who is likely to be an owner may raise questions about what business models we might want operating in this sector. If the list of appropriate business models is determined to include entities that sell or license scientific reagents, such as new genetic constructs for use in studying genetic diseases or therapies, then patents on such DNA fragments might be quite useful. Moreover, patents might be especially useful in helping individual or underfinanced inventors to compete against larger commercial enterprises in this market for reagents.[59] While the characterization of the provision of genetic constructs as a business model may sound too "commercial," or in some ways incongruent with the academic norms of the basic biological science community, it must at least be recognized that exchanges of such materials do take place. Therefore, the next section on the implications of ownership for nonowners considers different types of legal regimes to help determine which will best facilitate such arrangements.

V. THE IMPLICATIONS OF OWNERSHIP IN PATENTS FOR NONOWNERS

The patent law literature teaches that the right to exclude that is the core of the patent system's enforcement rules can actually operate to increase use by facilitating *ex ante* investment in the complex, costly, and risky commercialization activities required to turn nascent inventions into new goods and services.[60] This right to exclude competitors who have not shared in bearing the initial costs of commercialization provides incentives for the holder of the invention and the other players in this market to come together in an organized way and incur the costs necessary to facilitate commercialization of the patented invention.[61] The drafters of our present patent system—the 1952 Patent Act—had

[58] Concern about the impact of patents on basic biological scientists seems to be the predominant theme in the literature that is critical of these patents. *See infra* note 37.

[59] Kieff, *supra* note 8 at 744 (discussing the ability for patents to serve as the crucial slingshot in any David Co. v. Goliath Inc. type of competition).

[60] Kieff, *supra* note 8, at 707–710 (explaining how the right to exclude use promotes commercialization by facilitating the social ordering and bargaining around inventions that are necessary to generate output in the form of information about the invention, a product of the invention, or a useful embodiment of the invention). *See also*, Chisum *et al.*, at 58–90 (reviewing incentive to commercialize and other incentive theories of the patent system).

[61] Kieff, *supra* note 8, at 707–710.

precisely this concern for commercialization in mind when drafting the statute and were motivated by the specific fear that, for example, the handicapped in need of a new wheelchair might not find one to buy if the patent system did not provide an incentive for it to be brought to market in the first instance.[62]

A. Ownership Operates *Ex Ante* to Bring Commercialization

In the field of biotechnology, the need for such *ex ante* commercialization incentives may be particularly strong because the costs and risks of commercialization are not only exceptionally high, but they also are likely to be borne only by a first mover. As I note more fully elsewhere:

> Costs of commercialization in the biotechnology industry are exceptionally high—it is estimated to take over ten years and several hundred million dollars to bring a single drug to market. Risks of commercialization in the biotechnology industry are also exceptionally high—only a minute fraction of all compounds that reach clinical trials successfully complete trials and make it all of the way to market. Moreover, these costs and risks are unique to the first-mover because marginal cost in this industry is also uncommonly low, when compared to initial commercialization costs.... In fact, since the changes in applicable patent law beginning around 1980 that are the focus of patent critics, such as the availability of patent protection for living organisms and gene fragments, the United States biotechnology community has enjoyed particularly rapid and large advances in technology and overall prosperity, especially compared with the biotechnology communities of other countries that did not readily adopt such changes in applicable patent law as quickly as the United States, if at all.[63]

Thus, the use of the *ex ante* incentives provided by patents seems to at least correlate with the remarkable recent success of the biotechnology industry.

To be sure, a separate concern is whether the patent system has developed a set of rules about licensing transactions that operate *ex post* to maximize the likelihood that all those wanting such use will get it. The important work by Eisenberg, Heller, and others has shown how the transaction costs imposed by property rights in such tools for genetic research might lead to a suboptimal level of subsequent, or downstream, research.[64] Because the

[62] *Id.*, at 736–746 (showing how the drafters of the 1952 Patent Act were motivated by the commercialization theory and specifically contemplated such a wheelchair example).

[63] *Id.* at 724–25 (collecting sources) (footnotes omitted).

[64] *See supra* note 37.

transaction-costs problem may be solved without eliminating property rights in patents, it is first important to explore the extent of the social costs suggested by this literature in light of the tools the patent system has evolved for their mitigation. This is the topic explored in Part B.

B. Transactions Operate *Ex Post* to Increase Output

Both putative licensees of a patented technology who place a high value on such use and those who place a low value on such use are attractive targets to a patentee as long as the patentee is allowed to set a different price for different users—a practice called price discrimination. Patent law allows patentees to price discriminate among such licensees because this gives patentees a strong financial incentive to ensure all those desiring use get use—even a monopolist who can price discriminate will push output to the full competitive output level.[65] Such beneficial price discrimination can take place because patent law (and contract law) allows for the enforcement of the restrictive licenses needed to prevent arbitrage between low-value and high-value users.[66] In the presence of such a system, a patentee is rationally motivated to avoid posting an excessive price because to do so would scare away would-be paying customers, a money-losing venture. Indeed, the strong likelihood that those obtaining and maintaining patents are concerned about ordinary business issues—as opposed to more esoteric personal concerns—increases the likelihood that they will respond to such pressures.[67]

Even where the user is not able to pay any price, the patentee may be rationally motivated to grant a license for free. The granting of a free license may provide the patentee with an inexpensive way to preserve the legal force of the patent property right for use in other transactions with paying customers.[68] The patentee may also be able to derive advertising benefits from such uses, as long as they are successful uses and their low price does not cause

[65] Kieff *supra* note 8, at 727–732 (showing how the patent system's facilitation of tie-ins and other forms of price discrimination, where technological and economic factors alone might prevent price discrimination, provides incentives for the patentee to elect to keep output at competitive levels).

[66] *Id.* The prevention of arbitrage is essential for price discrimination to work. For example, those obtaining senior-citizen discounts could sell their low price tickets to patrons who would otherwise have to pay full price if movie theatres did not require some proof of age on admission, which may be as simple as looking at the ticket holder.

[67] *See supra* text accompanying note 7.

[68] Kieff, *supra* note 38 (discussing a property owner's rational decision to allow free users so as to avoid the cost of monitoring low value uses while preserving the full scope of the property right for other high value uses).

customer-relations harm with the high-paying customer base.[69] Thus, even very low value users are likely to be able to obtain licenses from the patentee. For example, the patents that cover some of the most basic technologies in the field of modern basic biological science—such as hybridomas and calcium phosphate transfection—are widely licensed for free to academic scientists. Other patents, such as the one covering the process of PCR, are licensed to anyone who buys from the patentee a machine for performing the process.

While patentees may be rationally motivated to sell permission to each user, and users may be rationally motivated to buy permission from patentees, such sales may not be consummated because of various market failures.[70] In response to these concerns, some commentators have argued that patents should be protected by a liability rule[71] instead of a property rule.[72]

It is first important to recognize that some of the causes of the transaction costs identified in the literature are likely also to exert some mitigating effect on those costs. One type of transaction cost explored in the literature is the valuation problem caused by an inability for parties bargaining over a license to accurately predict potential uses. But this type of uncertainty may also have a positive impact because broad licensing may be a way to increase the chance that at least some licensee generates some value from which the patentee can extract a share.[73]

[69] The price discriminating patentee will face a difficult customer-relations balance here. High paying users may be offended to learn of the availability of a lower price, but they may also be motivated to buy when learning about the successes of the patented technology. While the net impact of these competing forces is uncertain, the patentee's desires to preserve value while avoiding transaction costs discussed *infra* at note 68 will likely tip the net balance of incentives to be towards the use of such free licenses in certain cases.

[70] *See supra* note 37.

[71] An entitlement enjoys the protection of a property rule if the law condones its surrender only through voluntary exchange. The holder of such an entitlement is allowed to enjoin infringement. An entitlement has the lesser protection of a liability rule if it can be lost lawfully to anyone willing to pay some court-determined compensation. The holder of such an entitlement is only entitled to damages caused by infringement. *See* Guido Calabresi & A. Douglas Melamed, *Property Rules, Liability Rules, and Inalienability: One View of the Cathedral*, 85 HARV. L. REV. 1089 (1972); *see also* Jules L. Coleman & Jody Kraus, *Rethinking the Theory of Legal Rights*, 95 YALE. L. J. 1335 (1986).

[72] *See*, e.g., Ian Ayres and Paul Klemperer, *Limiting Patentees' Market Power Without Reducing Innovation Incentives: The Perverse Benefits of Uncertainty and Non-Injunctive Remedies*, 97 MICH. L. REV. 985 (1999) (criticizing recent increases in certainty in patent law and suggesting the use of liability rules instead of property rules for patent enforcement).

[73] *Cf*, Kieff *supra* note 8, at 726:

> Furthermore, the large risks of commercialization in the biotechnology industry also provide a particularly strong incentive for patentees in this industry to license broadly as a method for reducing risk. Licensing to hedge risk makes sense because once a firm becomes competent in pursuing one avenue of development, the firm may have difficulty keeping

It is also important to recognize that there are already some liability rule provisions in patent law today. Otherwise infringing uses that are by or for the federal government enjoy sovereign immunity protection that effectively results in a compulsory licensing regime.[74] In addition, the high costs of litigation under the present rules of civil procedure and the ability for an infringer to be kept effectively judgment-proof through the corporate and bankruptcy laws may also operate as a form of liability rule gloss on the present property rule regime.[75]

In addition, the political process provides several solutions for would-be licensees on a case-by-case basis. They may prevail on the government simply to provide such use in particular cases.[76] They may alternatively prevail on the government to subsidize their ability to pay.[77]

But before adopting broader liability rule treatment, it is worth at least considering whether the market power ensured by a property rule may have the beneficial effect of inducing even more new technologies. To the extent that some would-be licensees may not be able to obtain permission for use, despite manifesting some willingness to pay some positive price,[78] the presence of such potential customers and the potential for an independent patent each provide incentives for others to bring to market some alternative noninfringing substitute.[79]

---

track of other potential avenues. Indeed, the use of joint ventures and other licensing strategies to reduce commercialization risk in the biotechnology industry is well recognized.

[74] 28 U.S.C. § 1498 (providing limited waiver of sovereign immunity for acts of infringement by or for the federal government and instead allowing suits against the government in the United States Court of Federal Claims for a reasonable royalty).

[75] Kieff *supra* note 8, at 734, n. 154. ("Concerning procedure, litigation costs may be high enough to prevent the patentee from seeking court intervention against an infringer. Concerning substance, the limitations on liability that are available to a would-be infringer through the use of the corporate form or bankruptcy laws, for example, may encourage acts of infringements that are essentially judgment proof").

[76] See *supra* discussion accompanying note 74. The recent public demand for the patented drug Cipro to treat anthrax infection provides an example from the healthcare arena of just such behavior. *See, e.g.*, Terence Chea, *Vaccines Are Hot Topic, But Not Hot Investment*, The WASH. POST, E01 (Dec. 13, 2001) ("At the height of the anthrax crisis, government officials considered overriding German drugmaker Bayer AG's Cipro patent to purchase pills at a better price. Under threat of losing its patent, Bayer agreed to sell the government the antibiotic at half price").

[77] *See*, Douglas Gary Lichtman, *Pricing Prozac: Why the Government Should Subsidize the Purchase of Patented Pharmaceuticals*, 11 HARV. J. L. & TECH. 123, 124–25 (1997) (arguing that the government should offer a cash subsidy to any consumer who values a patented good above marginal cost but is unwilling or unable to pay to such a price). *Compare* Kieff, *supra* note 8, at 716, n. 91 (noting that such proposals face the distortion and implementation concerns generally raised against subsidies).

[78] *See Id.*, at 731 (discussing possibility that some licensees may not be able to obtain permission to use the patented invention).

Ensuring use through a switch in the patent system towards overall liability rule treatment may also frustrate the initial ability of the patent system to provide the *ex ante* incentives discussed earlier to commercialize new technologies. The use of liability rules may lead to a net increase in social cost and frustrate the very efforts for ordering and bargaining around patents that are necessary to generate output of patented inventions in the first instance, thereby decreasing overall social access to new technologies.[80] As recognized by Merges, it is precisely because private parties have a comparative advantage over courts in valuing patents and patented inventions that a property rule is likely to work better than a liability rule, according to the established test for choosing between the two types of regimes.[81] Again, the strong likelihood that those obtaining and maintaining patents are concerned about ordinary business issues—as opposed to more esoteric personal concerns—increases the likelihood that they will indeed be comparatively better able to make such valuations.[82]

But even assuming this literature has correctly identified the nature and magnitude of such social costs, it is necessary to compare the potential pernicious impact of breakdowns in transactions over such genetic materials and techniques under two legal regimes: one in which patents are available and one in which they are not.[83] Under a regime in which patents are not available, basic biological scientists will still be functioning in a market in which the central currency comprises scientific kudos—such as publication in prestigious journals, citation by peers, general prestige, the award of research grants, academic appointments and tenure, and salary.[84] Under a regime in which patents are available, the currency will also include patents.[85] Because patents will fatten out the market by bringing immense amounts of funding from diverse sources and other resources to the basic biological research community, it is likely that the regime under which patents are available is likely to suffer less pernicious impact of these social costs.[86] Indeed, it is precisely this increase in wealth and diversity of resources that has been recognized as a critical factor in the great success the community has enjoyed since the 1980 *Chakrabarty*

[79] *See* Chisum *et al. supra* note 1 at 75–76 (discussing incentive to design around patented inventions).

[80] Kieff *supra* note 8, at 732–36 (showing how the potential infringements induced by a liability rule will discourage investment in the commercialization process *ex ante* and may even result in a net destruction of social wealth if the collective costs of entry and exit across infringers exceeds the social surplus otherwise created by the invention).

[81] *Id.*, at 734, n. 152. (citing Robert P. Merges, *Of Property Rules, Coase, and Intellectual Property*, 94 COLUM. L. REV. 2655, 2664 (1994)).

[82] *See supra* text accompanying note 7.

[83] Kieff, *supra* note 38 at 698 (discussing the importance of performing such a comparative institutional analysis).

[84] *Id.* at notes 24, 34–35, and 68, and accompanying text.

decision and the subsequent explosion of patents in the field.[87] Simply put, the comparison between a market for kudos and a market for kudos-plus-cash shows that although the patent regime will impose social costs on the basic biological research community, they will be less than they would be if patents were not available.

VI. CONCLUSION

As a final prescription, it may be well worth trying to gather some empirical evidence to better understand what types of breakdowns are actually occurring.[88] Subjective perceptions of the individual legal and business strategies for market participants could be probed to ascertain some of the reasons why price discrimination is not being used to its fullest extent. The costs of monitoring and individualized pricing may surface as candidates. Another candidate might be concerned about the enforceability of the restrictive contract terms that are essential for preventing arbitrage.[89] Similarly, subjective perceptions and objective behaviors of individual members of the basic biological science community might be probed to ascertain actual mechanisms, circumstances, and frequencies of breakdowns in transactions over basic biological materials or protocols. For example, although scientific norms suggest that absent the patent right to exclude there will be sharing of information—at least the exchange of actual protocols and constructs discussed in published articles—it may turn out that requests for such sharing are not satisfied. In this connection, the failure to comply with such a request—even for seemingly innocent reasons like being "too busy"—might have the same impact as the trade secrecy problems discussed earlier in the section on the types and rights of ownership.[90] If it turns out that the materials or protocols that are exchanged happen to be less effective versions, then these exchanges may resemble the self-help problems discussed earlier in the section on the types and rights of ownership.[91] Conversely, it might turn out that when patents are in play, the uses receive no express

[85] *Id.* at 703.

[86] *Id.*

[87] *Id.*

[88] Indeed, this is a project of mine that is under way. I thank participants in the Boston Area Intellectual Property Law Roundtable Series and the Washington University Empirical Research Group for their many helpful comments on draft survey instruments and strategies for collecting data.

[89] There is indeed some evidence of the willingness for some courts to not enforce such provisions by relying on a myriad of doctrines, including antitrust, patent misuse, contract unconscionability, and even preemption. For more on these theories see Chisum *et al. supra* note 1 at 1066–1096.

permission but yet are not litigated, at least not right away for immediate injunctive relief.[92] In this case, there may actually be implied licenses being given quite regularly as discussed earlier in the section on the implications of ownership for nonowners.[93] Depending upon the results of such empirical research, it may turn out that legal changes are in order—such as the softening of the patent right to exclude or the strengthening of the enforcement of contract terms; or it may turn out that behavioral changes are in order—such as better education of the participants in this market about the actual legal rules and available techniques.

Acknowledgments

I gratefully acknowledge the financial support of the John M. Olin Foundation and the Washington University School of Law. I also gratefully acknowledge contributions from participants in the "Conference on Intellectual Property and The Human Genome Project" held at Washington University in St Louis on April 12–13, 2002, as well as more detailed comments provided by Michael Abramowicz, Chris Braccy, Richard Epstein, Doug Lichtman, Chuck McManis, Rob Merges, Troy Paredes, and John Witherspoon.

[90] *See infra* text accompanying note 14.
[91] *See infra* text accompanying notes 16–20.
[92] *See infra* text accompanying notes 66–69.
[93] *Id.*

8

Steady the Course: Property Rights in Genetic Material

Richard A. Epstein
James Parker Hall Distinguished Service Professor
The University of Chicago Law School
Chicago, Illinois 60637

Peter and Kirsten Bedford Senior Fellow
The Hoover Institution
Stanford, California 94305

0065-2660/03 $35.00

I. ALL-OR-NOTHING ON PROPERTY RIGHTS

Few issues today seem to generate more passion than the question of property rights in the human genome. The entire topic has become enmeshed in a multi-front war that takes place in successively narrower theaters.[1] The battle begins with broad questions about attitudes that one has to property rights, writ large, whether it be in human beings, body parts, tissues, cells, or molecules. Do we think that these elements, or some large portion of them, are, by nature, inappropriate candidates for reduction to private ownership?[2] For those—and there are many—who think that it is immoral or worse to reduce living substances to private ownership, the debate is over.[3] But for those who find these moral objections either insufficient or misplaced, negotiating this initial hurdle requires that others be surmounted as well. Within the traditional economic framework, is the case made for creating private forms of property when "the commons" is useful for delivering at least some kinds of goods and services? That question, in turn, quickly leads to a discussion that is more focused on matters associated more generally with intellectual property law and patent law in particular. The first inquiry asks: Why it is appropriate for the law to adopt a system of patent protection instead of some other form of intellectual property protection, such as copyright or trade secrets? Or more generally, why the use of private property systems at all, as opposed to state systems of bounties for the production and dissemination of this information? But once it is assumed that patent protection counts as the dominant alternative, we must ask how the protection of the genome in its many phases comports with the broader objectives of the patent law on the one hand, and the particular doctrinal requirements of current patent law on the other.

Let me state my conclusions quickly at the outset. I am not persuaded by categorical objections to creating property rights in human or other living substances. But once that question is left behind, we do have to face hard questions about the structure of property relations. For that question, I favor all-or-nothing solutions. There are some human and genetic substances that should be left in the public domain, and there are many that should be governed

[1] For a comprehensive account of all the arguments, pro and con, see generally, Arti Kaur Rai, *Regulating Scientific Research: Intellectual Property Rights and the Norms of Science*, 94 NW. U. L. REV. 77, 126–29 (1999); F. Scott Kieff, *Facilitating Scientific Research: Intellectual Property Rights and the Norms of Science—A Response to Rai & Eisenberg*, 95 NW. U. L. REV. 691 (2001); Donna M. Gitter, *International Conflicts Over Patenting Human DNA Sequences in the United States and the European Union: An Argument for Compulsory Licensing and a Fair-Use Exemption*, 76 N.Y.U. L. REV. 1623 (2001); Molly A. Holman and Stephen R. Munzer, *Intellectual Property Rights In Genes and Gene Fragments: A Registration Solution for Expressed Gene Tags*, 85 IOWA L. REV. 735 (2000).

[2] For this argument in another context, see Lawrence Lessig, THE FUTURE OF IDEAS: THE FATE OF THE COMMONS IN A CONNECTED WORLD (2001).

[3] *See, e.g.*, Jeremy Rifkin, THE BIOTECH CENTURY (1998); Jeremy Rifkin, *Genes Ought to Belong to Us All—Not Just to "Bio-Prospectors,"* HOUSTON CHRON., July 2, 2001, at 5C.

by the ordinary regime of patent protection, with some marginal adjustments for the distinctive problems of dealing with genetic substances and the biochemical agents used to treat them. I am suspicious of programs that seek to tweak the system with a variety of complex arrangements that rely on subsidies, tax breaks, rewards, honors, and, most notably, compulsory licensing to outperform this simpler set of rules.[4] This position leaves a number of important degrees of freedom. All-or-nothing solutions can be altered by varying the length of the patent term or the scope of the patent. They can hold that a patent applies with respect to some uses, *e.g.*, commercialization, but not with respect to others, *e.g.*, basic research conducted on a not-for-profit basis. The argument here is not meant to bias the shift between private and public domain property and strongly endorses the traditional practice of treating the inventions contained in expired patents to fall into the public domain.

So much for what the all-or-nothing approach does. But what it does not do is welcome those clever schemes designed to split the difference by weakening the normal rights of exclusion attendant on property rights and subjecting them to either schemes of government condemnation on the one hand, or compulsory licensing schemes on the other. It is tempting to think that the problems of monopoly inherent in any scheme of patent protection could be addressed by some system of forced exchanges, such as those that are involved under the law of eminent domain. But they will not work in this area. There is no coherent scheme to decide which patents should be condemned, or how the government should deal with them once condemnation has taken place. Nor do we have the enormous institutional capabilities to make all patent holders into public utilities who are obliged to do business with all comers under prices that are established and administered by the state. The current system is imperfect on any number of points. It is easy to find examples where current institutions lead to unhappy results. It is surely possible to envision intelligent forms of tinkering about the edges of established doctrine. There is a real question, for example, about the optimal scope of a patent. But the current legal structure, whatever its shortcomings, has fueled the enormous surge in biotechnology in the past generation. It is best to leave well enough alone. The all-or-nothing solutions—either its private property or in the public domain—dominate any and all efforts to carve out some elegant but inoperable middle position.

The creation of a property rights system in genetic substances is part and parcel of this general debate. In order to place some order on the discussion, the chapter proceeds as follows: Section 1 addresses the moral arguments against property rights in human materials; Section 2 addresses the choices between common and private property; Section 3 addresses the various schemes for

[4] F. Scott Kieff, *Property Rights and Property Rules for Commercializing Inventions*, 85 MINN. L. REV. 697 (2001), for a detailed criticism of the shortfall of these alternative allocation rules.

forced reassignment of property rights, first by condemnation and then by compulsory licensing; Section 4 addresses the issue in connection with genetic material. It concludes first that all expressed sequence tags (ESTs) should be left in the public domain, and second that the usual rules of patent protection work about as well as can be expected for most genomic products. We should not let the best become the enemy of the good.

II. MORAL ARGUMENTS AGAINST PROPERTY RIGHTS IN THE GENOME AND ELSEWHERE

One of the systematic objections to the creation of property rights in genetic material represents the vocalization of a powerful and deep sentiment that these technical activities are inconsistent with our considered judgments of what it is to be a human being. The point has been raised in many contexts, such as the cloning debate.[5] It surely has had an extensive influence on the way in which philosophers and bioethicists have attacked other problems in the health care system. The point usually starts with some invocation of the proposition that all things that matter in society should not be controlled by the market, which for these purposes can be defined as a system of voluntary exchange, protected and enforced by the state, of a known set of property rights in labor and external things. Those property rights, the argument generally goes, are more easily conferred to persons on things that are external to the self than they are on constituents of one's own being. Here the language of ownership (how does one own himself?) is always regarded as awkward if only because it is impossible to see how the agent (who does the owning) can be identical with the "thing" or "body" that is subject to ownership. Personhood is thus regarded in some sense as a unique category of experience, to which the usual rules on autonomy and voluntary exchange do not apply in any straightforward style.

The articulation of this general position has had payoff in the history of ethical and legal thought. Such notables of the English liberal tradition as Henry de Bracton, John Locke, and William Blackstone believed both in the autonomy of the will and in the prohibition against suicide.[6] Their views helped

[5] For a representative sample of views, see CLONES AND CLONES: FACTS AND FANTASIES ABOUT HUMAN CLONING (ed. Martha C. Nussbaum & Cass R. Sunstein, 1998).

[6] *See, e.g.*, 2 BRACTON ON LAWS AND CUSTOMS OF ENGLAND 423 (f. 150) (G. Woodbine ed., S. Thorne transl., 1968). "[J]ust as a man may commit felony by slaying another so may he do so by slaying himself." John Locke, SECOND TREATISE OF GOVERNMENT, Chap. II, § 6: "But though this be a state of liberty, yet it is not a state of license: Though man in that state have an uncontrollable liberty to dispose of his person or possessions, yet he has not liberty to destroy himself, or so much as any creature in his possession, but where some nobler use than its bare preservation calls for it." The last phrase seems to be an oblique protection for animals against cruelty. 4 W. Blackstone, Commentaries* 189. "The law has ... ranked [suicide] among the highest crimes."

influence the Supreme Court to rebuff any challenge to the prohibitions against assisted suicide.[7] The prohibition against suicide, for example, rested on the belief that self-preservation was the norm for all individuals, so that no person could deprive himself of the agency to reverse his course of action in the future. The objections to voluntary slavery took much the same line: The sale of self was not the same as the sale of any external object, given this unity between the subject and the object of the legal rights. The great fear here was that the sale of self into slavery would result in the permanent loss of legal capacity, so that it would become impossible for that person to be able to reverse the transaction.

None of these concerns, of course, has anything to do with the immediate issues surrounding the patenting of various manifestations of genetic material. In these contexts, the concern with autonomy is, if anything, urged in precisely the opposite direction. The now classic case on the subject is *Moore v. Regents of California.*[8] At issue in *Moore* was whether the individual patient who had supplied an invaluable cell line was entitled to some ownership interest in the patent that proved to be of tremendous commercial value. The case itself revolved around two potential theories of liability. The first involved a claim for the conversion of the plaintiff's genetic material by the action of the defendant —a claim that necessarily presupposed that the defendant had a property interest in his genetic material in the first place—and that it was not "abandoned" during the course of surgery. The second theory was based on an analogy to the law of informed consent in medical malpractice and claimed that the defendants breached their fiduciary duty when they did not disclose the reasons why they continued to call the plaintiff back for more and more tests, all of which increased their available supplies of his critical cell lines, without disclosing to him their ulterior purpose.

The mere fact that a plaintiff could bring up the conversion claim shows how far this particular protest against the ownership of human beings is from matters of suicide or slavery. The California Supreme Court rejected that claim, though it accepted a version of the disclosure obligation. But on a forward-looking basis, the differences between the two positions disappear. In the simplest scenario under the disclosure obligation, a plaintiff who knows the value of his cell line can refuse the doctor's service and look elsewhere, perhaps under a contract with some other physician, to undertake the extraction of all needed materials by himself and then sell those materials to other individuals as part of an ordinary commercial transaction. Once they have been removed from the body, they become a form of tangible property subject to exclusive ownership like any other chattel. The sales or licenses of that material can then be used to give the owner a contractual right to share in the proceeds of any patent

[7] *Washington v. Glucksberg*, 521 U.S. 702 (1997) and *Vacco v. Quill*, 521 U.S. 793 (1997).
[8] 249 Cal. Rptr. 494 (Cal. App. 1988), *aff'd and rev'd in part in* 793 P.2d 479 (Cal. 1990).

that the statutory inventor can claim under the system. It thus becomes a business judgment as to whether the genetic material is sold for a lump sum or a variable payment, contingent on its therapeutic usefulness. On this view of the matter, Moore was stymied only because of his ignorance at the time of the initial surgery and the subsequent visits.

The real question in *Moore* relates to the long-term implications that flow from the creation of self-ownership of tissues and organs. If the conversion theory had been accepted, as was the case in the Appellate Court, then the critical issue of fact is abandonment, which should not be presumed with respect to material to which both the original owner and, as it were, the subsequent finder attach extraordinary value. Requiring any form of disclosure thus will quickly eliminate any dispute over the abandonment question, because it eliminates the mistake of fact that gives the assertion of abandonment its tenuous plausibility. At that point, the property rights would have been robust.

California Supreme Court, however, refused to accept the conversion claim and thus never had to face the abandonment question. In deciding against the property rights approach in this context, the California Supreme Court anticipated many of the concerns in this area when it noted any claims of private ownership of genetic materials could dull the pace of medical innovation in much the same way that interlocking patents, subject to different holders, could block research on the genome.[9] The point lies certainly at the core of any debate over the scope of patent protection in the genome, but it has little consequence in any forward-looking world once the Court held that the treating physicians had a definitive obligation to disclose the fact that the cell line could come from the plaintiff's tissue. Once that disclosure takes place, astute individuals who are aware of the unique value of their own cell lines will market them to the highest bidders.

Now, suppose the physician decides to deliberately breach that duty to disclose. Here, the rejection of the conversion claim means that any third person who takes the patient's genetic material will have clear title to its use. But, although the point is not nailed down in *Moore*, it appears that the physician (or his institutional employer) could be liable in damages for the same amount that could be recovered from the third party if the conversion claim had been allowed. Punitive damages could easily be added into the mix. The upshot is that nondisclosure becomes costly for the institutional defendant,

[9] *See* Michael A. Heller & Rebecca S. Eisenberg, *Can Patents Deter Innovation? The Anticommons in Biomedical Research*, 280 SCIENCE 698 (1998). The term anticommons adds little to the debate. It only refers to a system in which multiple vetoes stand between an idea and its realization. It is thus a restatement of how the holdout problem can reduce value. For a general proof of the proposition that multiple rights of exclusion have parallel (negative) consequences with multiple rights of access, see James Buchanan & Yong J. Yoon, *Symmetric Tragedies: Commons and Anticommons*, 43 J. LAW & ECON. 1 (2000). For a systematic evaluation of this claim in the patent area, see *infra* at 158–164.

even if the tort of conversion is banished from the area. Going forward, the remedial differences between the conversion and the nondisclosure regime should not, in most cases, influence the pattern of control. The better view is conversion because it creates cleaner property rights in those cases in which individuals do enter into various kinds of business transactions. But that point is not a first-order consideration. Either way, the person with the distinctive cell lines gets to control them himself.

The above analysis presupposed that the initial choice between conversion and disclosure theories determined the ultimate distribution of rights in this cell line version of the eternal triangle. But like all property and tort solutions, this one is subject in principle to contracting out. These contracts could be designed to achieve one of two objectives: The first is to exclude the individual patient from the fruits of the use of his or her genetic materials. That result may well make sense if genetic materials gathered from a large cohort of treated individuals is used to concoct some new genetic compound that depends on assembling the genetic material of hundreds of different individuals. At that point, either no compensation or nominal compensation seems appropriate. And the entire subject is largely moot because the compensation could easily come in a small credit on the bill for services, which may not be separately acknowledged at all.

The situation is much more complex where the cell line from a single individual packs as much punch as Mr. Moore's. Here the treating institution could insist on a generalized waiver of all genetic material before any treatment is undertaken. But that waiver could be declined, or itself be subject to attack on the ground of insufficient disclosure of the size of the stakes involved with the transaction. Hence, in today's climate, this contractual waiver could be undermined by a disclosure obligation that imitates the underlying doctrine of informed consent. Waivers are very hard to come by, even when normally required in routinized transactions.[10] It would not work well here. Nor should it be tried. The better approach in these situations is to have a provision that allows the institution to use the extracted material in exchange for a value to be determined by some preset formula or by arbitration. That agreement could also set the ground rules for further cooperation between the parties, which is critical when other body tissue and fluids contain the key cell line. They could provide for additional insurance to the patient or do a whole host of other things that make joint sense. An agreement that has that larger objective is far less likely to be attacked as illegitimate because of its intrinsic substantive balance. Nor should one fear that markets will result in these skewed alternatives. The usual view of nineteenth century employment relations in England was that workers waived their rights of recovery against an employer for accidental injury on

[10] *See, e.g., Obstetrics & Gynecologists Ltd. v. Pepper*, 693 P.2d 1259 (Nev. 1985).

the job. But in many dangerous industries, such as mines and railroads, this is not what happened. Instead, these contracts typically put in place a pioneer equivalent of the workers' compensation system, which allowed for limited compensation for all accidents that arose out of and in the course of employment.[11] It may be that transactions costs block some transactions, but those that do survive will, in general, lead to improvements for both parties.

Viewed more broadly, the issue in *Moore* is in some sense orthogonal to that found in the genome cases, for it seeks to track down unique genetic material while most genomic research finds value in the genetic material shared by all persons in common so that, as a first approximation, it is the genomic material or the derivatives, such as complementary DNA (cDNA), that have medical and economic value. But *Moore* does serve the valuable function of showing that the strong objections against patenting human material cannot rest on any variation of the philosophical tradition of personal autonomy.

III. THE COMMON VERSUS PRIVATE PROPERTY

A. A Mixed Equilibrium

The next line of argument against the patenting of genes or other genetic material looks less to theories of individual autonomy and more to theories of overall social welfare. The question is: What form of property rights in genetic materials will best promote human resources? The potential variations on the theme of property ownership are not unique to intellectual property; they apply to all sorts of resources. In dealing with this question in its most general form, the optimal system of property rights requires us to negotiate our way between Scylla and Charybdis. Any system of exclusive ownership in distinctive issues runs the risk of hold-ups. Any system of common (by which I mean open)[12] ownership runs the risk of governance and coordination. The more individuals who fall under the tent, the less likely it is that they will be able to find ways to rule themselves effectively. The divergence of views on what uses to make of what assets will place genuine strains on their alliance, which is why most

[11] Richard A. Epstein, *The Historical Origins and Economic Structure of Workers' Compensation Law*, 16 GA. L. REV. 775 (1982).

[12] An open commons is one in which all may enter. It differs from closed commons, in which a large number of identified people have rights of entry. For the various permutations, see Clifford G. Holderness, *The Assignment of Rights, Entry Effects, and the Allocation of Resources*, 18 J. LEGAL STUD. 181 (1989). On the commons generally, see Elinor Ostrom, GOVERNING THE COMMONS (1990); Carol Rose, *The Comedy of the Commons: Custom, Commerce, and Inherently Public Property*, 53 U. CHI. L. REV. 711 (1986). The plasticity of human arrangements is quite large. For a discussion of the semicommons in which land is used privately for growing grain and collectively for grazing cattle, see Henry E. Smith. *Semicommon Property Rights and Scattering in the Open Fields*, 29 J. LEGAL STUD. 131 (2000).

commons tend to have single uses, such as grazing or traffic. Where heavier levels of investment are needed for particularized uses, the sensible social allocation calls for good neighbors separated by good fences. In other cases, we want to knock down those fences and create some form of commons.

This tension has been with us from the earliest times. The creation of the commons carries with it the risk of tragedy, at least if Garrett Hardin is to be believed.[13] But it is, in principle, quite difficult to draw any inference about the relative strength of private and public property without some real knowledge of the physical characteristics of the particular resource—land, chattel, water, oil and gas, spectrum, intellectual property—and the details of the legal regime, either common or private, that governs its use. It is very difficult, for example, to find strong regimes of private property that govern flowing water. The standard Roman treatment of property rights noted that air and water were to be held in common, and that no one had the right to exclude others from their use.[14] The "going concern" value of the resource depends not only on its chemical formula but also on its ability to flow from site to site, to support multiple human uses (bathing, recreation), and to support multiple forms of animal and plant life. The system, therefore, is one in which private and common uses uneasily *but necessarily* coexist.[15]

There are strong reasons for the creation of the limited commons in water. If each riparian owner could stop its flow, or at least the navigation along his portion of the river, could lead to the loss of its collective use altogether. The (limited) commons removes that blockade right. But all systems of property rights involve implicit economic trade-offs, such that open access creates the reciprocal risk of congestion, which also reduces the value of navigation. The task is to identify the smaller risk in this setting. Placing intellectual property in a commons, however, does not give rise to the same risk, for the use of an idea or invention by another does not crowd out its use by a competitor. It is for this reason that placing certain forms of intellectual property within the public domain has always had an enormous attraction and remains an indispensable part of any property rights regime.[16] The cost of running the system is virtually zero. No one needs to purchase or license the property in question from anyone else because it is always there for the taking. The difficulties in running a system of private contract and private property rights thus are neatly sidestepped, and the use and absorption of the information takes place at a rapid rate. When

[13] Garrett Hardin, *The Tragedy of the Commons*, 162 SCIENCE 1243 (1968).

[14] *See* JUSTINIAN'S INSTITUTES, Book II, Title I.

[15] For discussion, see Richard A. Epstein, *On the Optimal Mix of Common and Private Property*, 11 SOC. PHIL. & POL. (No. 2) 17 (1994).

[16] For my views on the heterogeneity of intellectual property, see Richard A. Epstein, *Intellectual Property: Old Boundaries and New Frontiers (Addison C. Harris Lecture)*, 76 IND. L. J. 803 (2001).

patents and copyrights are created for limited periods of time, the reversionary interest to the public is in the form of public domain (as opposed to government-owned) property, that has these same welcome characteristics. As will become clear, we work best in the world of extremes, where we have either a system of strong private property rights on the one hand, or public domain property on the other. What should be avoided at all costs are mixed regimes that feature forced exchanges, be it by state condemnation of patents or by compulsory licensing agreements.[17]

In order to understand how these property rights regimes apply to patent issues generally and to the genome in particular, it is useful to return to the water situation. Suppose that a river (or a bridge) is in need of some improvement in order to become navigable, and this improvement requires some form of private investment. How should this be handled? One possibility allows a single regulated monopolist to charge some kind of toll sufficient to thin out the traffic but not so high as to lead to underutilization of the resource. That response, in turn, gives rise to the famous marginal cost controversy.[18] The bridge costs money to build but nothing to maintain. Any positive charge on its operation involves the exclusion from the bridge of someone who attaches positive value to its use that is lower than the price charged. But reduce the cost of the bridge to zero and then there is no reliable way to determine whether it should have been built in the first place.

Patents raise the same issue. The marginal cost of an additional dosage of a new drug, for example, may be virtually zero, so that any positive price precludes new users at the margin. But even if it is not, there is always a gap between the competitive and monopoly price, such that monopoly pricing always produces dead-weight losses. The decision to price it at zero (or at marginal cost) requires some other institution to make the initial decision as to whether and to what extent investment should be made in the development of this drug and none other. We cannot avoid some real-world inefficiency; we can only trade in one problem for another.

B. Multiple Monopolies

However difficult that last choice might be, it is easy to think of a worse situation: sequential monopolies. One grim possibility (though one, thankfully, never entertained at common law) is to allow each riparian owner to extract a toll of his own choosing, and this combination of obstacles could reduce the value of the river to virtually nothing. That could not happen within a single

[17] See *infra* at 167–175.

[18] See Ronald H. Coase, *The Marginal Cost Controversy*, 13 ECONOMICA (N.S) 169 (1946). *Reprinted in* Ronald H. Coase, THE FIRM, THE MARKET, AND THE LAW 75 (1988).

jurisdiction where property owners are subject to the rule of one sovereign. But it could well happen when two independent sovereigns each run their own portion of a longer waterway. The best way to think of this problem is as an illustration of the so-called double marginalization question: When the dust settles, two unrelated monopolists, acting independently, reduce the social gains from the utilization of their resources.[19] The basic intuition is that each of the monopolists will ignore the harms that his decision imposes on any other monopolist, so that acting separately the riparians do far worse than they would if they coordinate. This problem arises only, however, when the two monopolists stand in an upstream and downstream relationship to each other, a term that is descriptive, not metaphorical, in the water case. So understood, the problem of uncoordinated exclusion is the mirror image to the problem of uncoordinated overuse of many rivers or other rivalrous commons, where again the problem arises because each of the potential users of the common resource ignores the costs that his use will impose on others.[20]

The double marginalization problem presents the opposite of the normal situation, where the (horizontal) monopoly is created when two competitors combine their operations. To continue the water metaphor on this case, it is as though the river forked into two branches at the only point where tolls were allowed, such that the owners of each branch could each charge tolls. So long as they operated separately, they formed a Cournot duopoly whose price lies between the monopoly and competitive price.[21] In contrast to the upstream/downstream situation, their combination reduces social welfare by driving the price further from the competitive solution. The combination that is welcome in one context (upstream/downstream) becomes dangerous in another (sidestreams, to coin a phrase). The spatial relationship really matters; the terms vertical and horizontal are not just blackboard conventions, but are defined by the elemental forces as gravity.

One great difficulty in connection with patent law generally, and with the genome in particular, is that we cannot draw easy spatial diagrams to indicate the relationship that one invention has to any other within the sphere. The first and most obvious feature of all intellectual property is that it does not face the problem found with water usage, where congestion effects do matter. In principle a thousand individuals could use (or, for that matter, prove) the Pythagorean theorem before dinner and it remains as available to everyone else as it did before. That characteristic is not characteristic simply of those ideas or

[19] *See, e.g.*, Dennis Carlton and Jeffrey Perloff, MODERN INDUSTRIAL ORGANIZATION, 398–410 (3rd ed, 2000). *See also* material, *supra* note 8.

[20] See Lessig, (*supra*) at 19–23.

[21] For a general exposition, see Gibbons, etc (1992); for its application to patents, see Carl Shapiro, NAVIGATING THE PATENT THICKET: CROSS LICENSES, PATENT POOLS, AND STANDARD SETTING, *in* INNOVATION POLICY AND THE ECONOMY 119 (A. Jaffe, J. Lerner, and S. Scott (2000)).

laws of nature that have long been held to fall outside the scope of patent protection. They are equally true of any invention (or writing under copyright law) as well. The point is almost beyond dispute because one of the tests of patentability is the requirement that the inventor describes his invention with specific particularity so that others with ordinary skill in the art could make it from his plans.[22] It is here that we face the rub. In the case of water, that is a resource that is (at least in the simplest scenarios) given by nature, so that the key questions are those of its effective utilization. But nature does not gush forth ideas or inventions, and the question then arises as to what system of human relations will best direct both the creation and utilization of ideas and inventions.

In the case of a single stand-alone patent—which is most emphatically not the case with the genome—the issue of legal protection tends to gravitate to the well-known "bargain." The inventor gets the exclusive right to use his invention for a limited number of years so long as he makes public information on how it is created, so that at the expiration of the term it falls into the public domain. The monopoly is created in order to accelerate the creation of inventions. Here it should never be forgotten that the *social* benefits from the new invention come *during* as well as *after* the patent period. First, it is better that people have the opportunity to purchase a good at a monopoly price than they have no opportunity to purchase the good at all. Thereafter the deal gets even sweeter because the resource comes at zero marginal cost to all those who will use it. As everyone can use the resource, its subsequent exploitation takes place within a competitive framework. Second, even during the period of monopoly ownership, the patent disclosures could easily provide information that could encourage others to develop complementary or supplemental goods.

It is, however, one thing to identify the existence of the trade-off between monopoly today and public domain tomorrow. It is quite another to figure out whether this deal is good or bad from the point of view of the inventor and society. The position of a single present inventor, at least when considered in isolation, does not give rise to any pause. Without the monopoly, the inventor might have chosen some other line of work; he would be better off with limited monopoly protection than with none at all. Alternatively, the inventor could have relied only on the other forms of protection for his invention available to him; again he would be unambiguously worse off than he would have been with the additional choice of the patent monopoly. So, at first blush, the deal looks good from his side against the feasible set of alternatives.

[22] *See* 35 U.S.C. § 112. "The specification shall contain a written description of the invention, and of the manner and process of making and using it, in such full, clear, concise, and exact terms as to enable any person skilled in the art to which it pertains, or with which it is most nearly connected, to make and use the same, and shall set forth the best mode contemplated by the inventor in carrying out his invention."

The situation, however, becomes more clouded in a dynamic sense in light of the multiple roles occupied by inventors as a class. No inventor works in a void; all build off a set of ideas and techniques that were always in the public domain or have lodged there after the expiration of other patents. It may be that this inventor could never have reached the point of invention if the intellectual commons were to be stripped bare. No inventors wants a system that gives patent protection that is too broad or too long, for that would strangle his own opportunities for innovation in two ways: It would reduce the amount of public domain information from which he could rely, and it would reduce the field of action into which he could move. The case for open use of inventions is even more insistent for society as a whole. It does not consist solely of inventors but of a large number of individuals who use inventions in all sorts of direct and indirect ways but do little, if anything, to create them (even if they own shares in companies with extensive patent portfolios). Yet even noninventors are aware that they cannot utilize today inventions that were not made yesterday.

How, then, do we make the appropriate balance between protection and access? Debates over this subject are legion. The first question that must be asked is what the location of any new patent is relative to all others in the patent universe. Speaking at the most general level, patents do not self-identify as upstream, downstream, or sidestream. Everything depends on the network of existing patents into which they are thrust. It is not even clear that the directional element implicit in these descriptions applies across the board, so that the term "thicket" has been used to capture the potentially unruly and nondirectional formations that patents can take.[23] Yet, in particular situations, patents might well line up as either strict complements or substitutes, at which point the analysis follows the argument made previously about water rights. The integration of complements avoids the double marginalization problem. The integration of substitutes facilitates cartels. We need to be aware of that difference.[24]

The angular distribution of patents in hot subject matter areas is not lost on patent prosecutors, who take enormous care in structuring patent claims for their inventors: Make patent as large as possible to secure an effective scope for future use.[25] But make it too large and inventors might run into difficulties with infringement or prior art. Finding that balance requires a detailed knowledge of the particular landscape for the inventor and poses genuine problems of analysis for the rest of us. Thus, in some cases, the new patent could be quite benign in its application. It may introduce a new treatment for, say, intestinal

[23] See Shapiro, *supra* at note 21.

[24] On which more, see *infra* at 176.

[25] *See generally*, Robert P. Merges & Richard R. Nelson, *On the Complex Economics of Patent Scope*, 90 COLUM. L. REV. 839, 866 n.118 (1990).

disorders that at first has surgery as its only competitor. But shortly thereafter someone else comes up with a rival drug over which it has a monopoly that is usable for roughly the same end. Here the legal patent monopoly is demoted to an economic duopoly in relatively short order, and it is, in turn, subject to erosion by the invention of further substitutes for the proposed treatment, even from remote technologies such as lasers. In this case, the creation of side-by-side legal monopolies hastens the emergence of a quasi-competitive situation long before the expiration of the patent period. It is as though there were two or more forks in the river, with no prospect of any form of unified control. The patent system thus appears to foster a great social deal insofar as the second patent competes away many of the gains of the first patent long before expiration. The rash of new entrants undercuts the idea of any systematic patent blockade. Some patents undermine blockades that other patents seek to create.

Within this confused array, however, any two patents could operate in upstream/downstream configuration. That possibility has already been suggested in connection with Human Genome Sciences' recently patented CCR5 receptor gene, where drug companies could be required to get releases from multiple patent holders if it turns out that other inventors obtain patents adjacent to HGS's patent.[26] But this is not the only possible outcome for that one gene. The presence of so many other patents in this imaginary space may also create rather different configurations. Thus, at each patent gate, we could face a choice between two duopolies instead of two monopolies. The social outcome is likely to be better in the first setting than in the second because the double marginalization problem should be mitigated by the absence of a single monopolist at either stage. Of course, the patterns could be more complex and more like a neural network than a river. For some uses of Patent A, it is necessary to take advantage of Patent B, but for other uses of Patent A, Patent B is quite irrelevant, but the choice between Patents C and D is critical.

More generally, it is easy to think of thousands of different patent pathways through which some new conception may travel in order to crystallize into a commercial application. These patents are not always thickets;[27] they frequently have a clear directional bias that thickets lack. But usually there is no unique path so the holdout value at any key pressure point is difficult to determine. It is hard to make a universal judgment about the applicable patent regime at so general a level when we do not know which configuration of patents dominates the landscape. It is interesting to note that the professionals in the field are, in general, optimistic about their ability to transact in this business. At a conference at Washington University in 2002 it was said publicly

[26] Gitter, at 1669 citing, Paul Smaglik, *Could AIDS Treatments Slip Through Patents Loophole?* 404 NATURE 322, 322 (2000).

[27] For the use of the term, see Carl Shapiro, *supra* at note 21.

and more than once that experienced operators in the field did not know of a single worthwhile product that had been kept from the market because of blockade difficulties.

That conclusion has been fortified by a recent survey published in *Science*,[28] whose three authors found that the strong patent protection that surrounded the use of research tools had done little to thwart the rate of innovation. The techniques that were involved included "licensing, inventing around patents, going offshore, the development and use of public databases and research tools, court challenges, and simply using the technology without a license (*i.e.* infringement)."[29] This list is not dispositive on the question, but it is suggestive. The strategies of licensing, inventing around patents, and using public databases and research tools are wholly unexceptionable because they only take advantage of the opportunities that the patent system intends to create. The use of offshore research devices is trickier to evaluate because that option would be removed if the patent system offered worldwide protection of the sort that is contemplated under the Agreement on Trade Related Aspects of Intellectual Property, or the TRIPS agreement. In any event, that device is then not available to use in domestic markets.

More puzzling, if not troublesome, is any decision to work without getting a license. To be sure, such uses do amount to infringement, but the social arrangements are more complex, for many patent holders may turn a blind eye to certain infringements. Their actions need not constitute disinterested generosity; they could rest on the self-interested calculation that the use of these tools, especially in universities for research purposes, could produce results that could find new commercial applications for their patented goods.[30] Thus, even without the benefit of an explicit research exception to patent protection, practices along those lines appear to be well ingrained in the United States unless universities start to act with commercial concerns, at which point the amnesty is off and the litigation begins.[31]

The full range evidence is, as it always seems to be, incomplete, but the burden seems now to be on those who claim that the patent system builds blockades against innovation through the creation of an anticommons, as Heller and Eisenberg have suggested.[32] The vast and sustainable increase in the rate of

[28] John P. Walsh, Ashish Arora, Wesley M. Cohen, *Work Through the Patent Problem*, 299 SCIENCE 1021 (2003).

[29] *Id.*

[30] The European Community recognizes a formal research exemption from the patent protection, *see Commission of the European Communities, Proposal for a Council Regulation on the Community Patent*, art. 9 (August 1, 2000) 43 *Off. J. Eur. Communities* 278 (2000); available at europa.eu.int/eurlex/pri/en/oj/dat/ 2000/ce337/ce3370001128en02780290.pdf

[31] *Madey v. Duke University*, 307 F.3d 1351 (4th Cir., 2002), on which see Rebecca Eisenberg, *Patents, Swords and Shields*, 299 SCIENCE 1018 (2003).

[32] Heller & Eisenberg, *supra* note 9.

genomic patent filings is not consistent with the view that new patents strangle innovation; in fact, it points to the opposite conclusion. The lawyers and businesspeople who work in this area are built evidently for speed, not stubbornness, and the gnawing sense that inaction leads to slow starvation and death keeps the rate of transactions humming. Of course, one could dispute this characterization of the field, but I should be reluctant to belittle or disregard this information. Transactions are what define this field. Patent positions are what make these possible. The burden is on those who want to explain why the property rights regime should be displaced in particular contexts.

IV. FORCED TRANSFERS OF PATENT RIGHTS

The situation, moreover, is not unique to intellectual property. It can arise, albeit in less dramatic fashion, with respect to land where separate assets stand in fixed relation with each other. Any individual who owns Blackacre has the exclusive right to that territory, and so, too, does his neighbor over at Whiteacre. But, like patents, we cannot tell simply from their location how these two plots of land will interact with each other. It could easily happen that each will be used for a single family residence, in which case the side-by-side legal monopolies are preconditions to a competitive market. But if both of those plots are needed to complete a key road, then their sequential use in an unregulated market creates the double marginalization problem.[33] The legal responses are predictable insofar as no one thinks that use of the condemnation power is inappropriate to assemble land for a highway, but this problem is easy in comparison with the patents because our objectives are far more modest and the two-dimensional physical layout of the land gives us a relatively clear guide as to which parcels must be included once the (political) task of route selection is accomplished. The question now arises as to what social responses could be introduced to deal with the admitted problems that some patents may create on some occasions. In this section I deal with, and generally reject, two such responses: condemnation and compulsory licensing arrangements.

A. Condemnation

One way to deal with the patent-monopoly is to adopt the condemnation option used to assemble land in highway cases. As with the state subsidy of the bridge, the purpose of condemnation is to eliminate the dead-weight loss that is

[33] *See*, for further discussion, Richard A. Epstein, BARGAINING WITH THE STATE 36–37 (1993), where I used the electrical parallel to circuits that are organized in series and parallel.

associated with monopoly pricing but to preserve the pecuniary incentive for invention that the current patent law provides. The theory is that the government acquires the patent for a price that compensates the inventor for his labors. It then turns around and licenses the invention, or sells products produced with it, at their marginal cost, or places it the public domain. The lump sum payment does not distort future use, so we get optimal innovation and optimal utilization as part of one package. The position in question is one that reflects the new-found fondness for the use of liability rules (which allow buyouts at collective valuations) over strong property rules by which entitlements (patents included) are only transferred with the consent of their owner.[34]

A moment's reflection should show why this general approach to the patent problem should be dismissed as a hopeless pipe dream. The best way to make the point is that condemnation does not work very well in those settings for which it is most suitable: the condemnation of land for public works. In the land context, the target of condemnation is frequently raw land for use as a highway. That land normally has an easily determined market value. Subjective values, moreover, are far more likely to inhere in the family home than in the back forty acres, which may be already leased anyhow. But the simplicity is deceptive, even in the easiest cases. The first question is *which* land should be targeted for condemnation. In some cases, local citizens vie to have the road built near them; in other cases to keep it far away. These costs are worth bearing because roads (at least most roads) have to be built, so that the costs of route selection are part of the cost of doing business, even if not linked to the acquisition of any parcel once the program is in place.

The problems do not cease, however, with the identification of the targeted land. Valuation is a constant headache. Often only fractional interests are taken for roads, raising problems with severance and spillover values. The residual land retained by one party may be enhanced in value owing to the condemnation and this could give rise to an offset against liability. Alternatively, the residual land could be rendered functionally barren because of the loss of some key component (the land is no longer large enough to be zoned commercial, for example) so that some severance damages should be added to the mix so as to leave the owner indifferent to the government imposition. Yet that compensatory ideal is systematically frustrated because the landowner does not get moving expenses or because of appraisal fees, loss of goodwill, and the like.[35] Neighbors whose value diminishes in consequence of the taking are also

[34] For one notable exception to the general trend, see generally Robert P. Merges, *Contracting into Liability Rules: Intellectual Property Rights and Collective Rights Organizations*, 84 CAL. L. REV. 1293 (1996), noting that in the copyright area, the firms themselves were able to design a distinctive licensing system that was superior to any that the state could impose of its own will. For discussion, see *infra* at 175–176.

ignored under the mix. Goodwill is ignored; subjective value is ignored; costs of appraisal and legal defense are ignored. Every bias in land acquisition cases favors the state. These biases are, moreover, only accentuated when the condemnation is directed at homes and business, where the valuation issues become still more complex.

The use of condemnation (or prizes) is a far more complex issue for patents. The rich network that surrounds any individual patent raises at the very least all of the contextual questions found with land. The initial question is which patents should be subject to the buyout and which ones should not. Here, one alternative is to condemn all patents once their validity is determined, at which point the system will be overwhelmed with hundreds if not thousands of complex condemnation proceedings. All of these could drag on for years, often for inventions that will never be able to turn a profit. In the interim, exploitation of the patent will lag out of uncertainty as to whether the additional investments will receive at least adequate compensation from the government.

Alternatively, the government could engage in selective condemnation, choosing only certain key inventions for this process. But that approach will unleash immense political pressures over the exercise of the power. Will some people angle to have their inventions condemned because they think that they will not retain value in the long run? Will they resist the condemnation of a competitive drug for fear that the government will then underprice their own product? Will they encourage the condemnation of the products of competitors in order to enmesh them in controversy?

Once the process was set in motion, it would raise in acute form all the valuation questions that dog land condemnations, only worse. If some patents within a cluster are taken over while others are not, then an accurate system of condemnation has to take into account the influence on retained interests, positive or negative, to be true to the original ideal of full compensation for the economic losses brought on by the condemnation. But as the identification of "neighboring patents" is far from self-evident, these calculations will be far from routine events. If 13 other patents have their values altered by the condemnation of the one patent from the suite, how are those collateral losses assessed? What duties of mitigation are incumbent on the private owner?

Even self-sufficient patents will produce their share of knotty condemnation difficulties. The net value of any patent depends on a complex set of judgments about its exploitation, sale, and pricing, all of which are constantly revised in response to new market conditions and technical innovation. It is relatively easy (but still very hard) to predict future cash flows when assigning a value to land, which is done all the time in voluntary transactions. But there are

---

[35] For an examination of these biases, see Richard A. Epstein, TAKINGS: PRIVATE PROPERTY AND THE POWER OF EMINENT DOMAIN 182–194 (1985).

few naked cash sales of patents precisely because it is so hard to discount their future worth to present value. Within the condemnation framework, their valuation at the very least depends on complex calculations as to the allocation of joint costs and the possibilities of entry of novel competitors. The fiscal pressures to lowball compensation will lead to systematic expropriation of patent values in any individual case. The condemnation process cannot work as an answer to the monopoly problem.

B. Compulsory Licensing

1. Basic Proposal

One variation of the eminent domain position is to abandon the notion of state takeover of patents but to preserve the idea that state coercion can overcome the blockade effects of patenting through a system of compulsory licensing.[36] In part, the argument for government intervention rests on the sense that individual biotech firms will be reluctant as a general matter to forfeit the exclusive control over their patents by entering into various cross-licensing and patent pool arrangements. There is, indeed, some limited precedent for federal intervention to mandate cross-licensing in specific contexts relating to the preservation of health, the development of nuclear energy, and the preservation of clean air.[37] There is no doubt that just as government intervention is a useful, if limited, antidote to the holdout problem, so, too, mandatory cross-licensing could achieve that same result on select occasions. But it is a far cry to assume that this device should be imposed (as opposed to, used) as a general practice in patent law, or even that fraction of the patent law that is confined to biotechnology.

My uneasiness here does not rest solely on the predictable opposition from within the biotech industry,[38] although that should serve as a warning bell to the aggressive implementation of these schemes, which will flounder unless they can rely on the cooperation of the relevant parties. But the opposition to this proposal runs deeper. In this context, as in others, the strong presumption should be in favor of a regime of strong property rights (defeasible only with owner's consent) unless and until there is some clear showing of necessity that justifies the switch to a regime that requires forced purchases.[39] The simple

[36] *See, e.g.*, Gitter, at 1679–1691.

[37] *Id.* at 1682.

[38] *See* W. R. Cornish, INTELLECTUAL PROPERTY: PATENTS, COPYRIGHT, TRADE MARKS AND ALLIED RIGHTS 291 (3d ed. 1996). Philippe G. Ducor, PATENTING THE RECOMBINANT PRODUCTS OF BIOTECHNOLOGY AND OTHER MOLECULES 157 (1998). Gitter at 1681–1682.

[39] I have said my piece on this issue on other occasions, see Richard A. Epstein, PRINCIPLES FOR A FREE SOCIETY: RECONCILING INDIVIDUAL LIBERTY WITH THE COMMON GOOD (1998); *A Clear View of The Cathedral: The Dominance of Property Rules*, 106 YALE L. J. 2091 (1997).

complications of negotiating deals under uncertainty does not meet that threshold; indeed it describes a condition that is applicable to just about every dynamic industry under the sun. To be sure, there may be exceptional circumstances that call for a compulsory licensing program. That was proposed, for example, for Cipro during the anthrax scares in the aftermath of 9/11. But even then, circumstances moved so quickly that this problem fell to the back burner as soon as a voluntary deal was completed. We should take warning, given the extensive negotiations over a single use of a single drug, which was known to go off patent in 18 months. There are good reasons for the admitted "rarity" of the practice.[40] It takes a vast leap of faith to posit the social and technical infrastructure that could make it worthwhile to implement on a mass basis.

All this is not to say that the current system is perfect. The monopoly problem still remains in some form, such that the ostensible allure of the proposal comes from the vision that intransigent holdouts would become happy royalty holders. Professor Donna Gitter offers one description of the proposal:

> Congress should enact a compulsory-licensing system, which would require an owner of patent rights in a DNA sequence to license that sequence to any and all scientists pursuing commercial research related to that sequence in return for a reasonable licensing fee. The licensing fee would not be established by the individual licensor, but would instead depend on the commercial value of the product developed as a result of the research. Thus, potential licensees would not be dissuaded from making an initial investment to pursue research, because the amount of the royalty payment would be tied to the success of the product they develop. This system is also fair to the licensor, who would receive adequate compensation from licensees who achieve financial success through their use of the patented sequence.[41]

Her proposal does not stand alone, but builds on a similar proposal for the compulsory licensing of "research tools," advanced earlier by Janice Mueller.[42] Mueller's compulsory licensing scheme would embrace, in addition to ESTs, "cell lines, monoclonal antibodies, reagents, animal models, growth factors, combinatorial chemistry, libraries, drugs and drug targets, clones and cloning tools (such as PCR [polymerase chain reaction]) methods, laboratory equipment and machines, databases and computer software."[43] Thereafter, the

[40] *Dawson Chem. Co. v. Rohn & Haas Co.*, 448 U.S. 176, 215 & n. 21 (1980) (noting the Congressional proposal in the Patent Act).

[41] Gitter, at 1679.

[42] Janice M. Mueller, *No "Dilettante Affair": Rethinking the Experimental Use Exception to Patent Infringement for Biomedical Research Tools*, 76 WASH. L REV. 1, 58–66 (2001).

proposal calls for a "reach-through" royalty structure that targets the products sent to market by means of the research tools, even if sold after the expiration of the underlying patent.[44] Unlike the usual voluntary license, Mueller believes that the compulsory licensee need make no disclosure of the nature or the details of the use as long as he makes a declaration of "an intention to use" that puts the patentee on notice "so that it can police any subsequent introduction of new products into the marketplace by the tool user."[45] The stated royalty structure "is computed at twenty-five percent of the licensee's pre-tax profit rate on his sales," subject to "fine-tuning in individual cases."[46]

In one sense, these proposals might be defended for being more modest than they seem. Far from being command and control mechanisms that force parties to do business on terms that they do not like, they simply invite them to renegotiate their arrangements in ways that better serve their joint interests. But compulsory licenses have more teeth than this optimistic account suggests because they change the threat positions of the parties to any negotiation. The patent owner who cannot get the terms he wants is no longer allowed to walk away from the deal. He must continue to do business with someone who will not budge from the compulsory license unless he gets terms that he finds better than the ones guaranteed under the law. At the very least, therefore, these proposals deny a patent holder the right to choose the parties with whom he will do business in the first place. Necessarily, they make exclusive licenses a dead letter, even though these are often consummated in practice. In order for this use of compulsion to be a sensible social procedure, one has to have some confidence in the licensing rules themselves. Yet too much optimism lurks behind these compulsory licensing proposals, which underestimate at every turn the contractual density and sophistication of licensing transactions.

To start from the beginning, all compulsory licensing schemes show a decided preference for some kind of "liability" as opposed to "property" rule, whereby the holder of any given entitlement is subject to its purchase at some for some fair valuation determined by some independent party.[47] When introduced

---

[43] *Report of the National Institutes of Health (NIH) Working Group on Research Tools 3*, at http://www.nih.gov/news/researchtools/index.htm (June 4, 1998) [hereinafter NIH Research Tools Report], referred to in Mueller, *Dilettante Affair*, at 11.

[44] This is a perfectly sensible way of doing business in voluntary or statutory markets. Clearly the use for which the royalty was paid took place during the period of patent protection. Only the compensation is deferred, which could take place with fixed as well as variable sums. What will not happen is a royalty, absolute or contingent, for work done after the time of patent expiration.

[45] Mueller, (*No "Dilettante Affair"*), at 59.

[46] *Id.* at 64.

[47] For the initial distinction, see for one defense of the superiority of liability rules, Guido Calabresi & A. Douglas Melamed, *Property Rules, Liability Rules and Inalienability: One View of the Cathedral*, 85 HARV. L. REV. 1089 (1972). Ian Ayres & Eric Talley, *Solomonic Bargaining: Dividing a Legal Entitlement To Facilitate Coasean Trade*, 104 YALE L. J. 1027, 1036–1037 (1995).

by Guido Calabresi and A. Douglas Melamed in 1972, protection by liability rules only helped explain why injunctions need not necessarily be awarded in all nuisance cases. In some situations, the defendant could continue to pollute, presumably at some defined levels, so long as some provision was made for the payment of permanent or temporary damages.[48] In practice, the law of nuisance has not been as receptive to the damages-only approach as Calabresi and Melamed's characterization suggests. The Courts prefer in most cases to award the injunction as a right unless there is some manifest imbalance between the trivial advantages to the plaintiff and the massive dislocations to the defendant.[49] In some other cases, it has been suggested that liability rules are better ways to decide which of two people will be able to make better use of a divided asset (*e.g.*, land in which a determinable fee is subject to a possibility of reverter) by inducing the parties to reveal the intensity of their preferences for the unitary asset.[50]

This thumbnail sketch shows that, at their inception, liability rules were meant at most to serve the modest role of adjudicating boundary disputes; somewhat later it was proposed that they take a transitional role in moving assets between persons. Assuming the soundness of these conclusions, for the sake of argument, the traditional accounts did not contain any hint that liability rules should sanction the deliberate expropriation or use of the property of another in complex and cooperative business transactions. One sense of the limits of these forced commercial exchanges comes from the law of contract. The all-too fashionable theory of "efficient breach" posits that a promisor can break his promise, pay expectation damages, and pocket the surplus.[51] The appeal here is for the substitution of a liability rule for the stronger property rule protection from, say, a remedy of specific performance. The argument is that the promisee is not worse off (because of the expectation damages) while the promisor is better off. The unilateral breach with compensation amounts to a Pareto improvement. But the entire scheme flounders because the computation of expectation damages is next to impossible whenever the breach by promisor creates ripple effects that are hard to quantify. The usual reaction to efficient breach among businesspeople is to gasp in genuine horror about its universal application. Most trades work on the assumption that the expectation measure of damage is permissible only for want of a better, *i.e.*, in those cases in which the defendant is not in a position to perform a contract. The usual response is expulsion from a trade association when the defendant engages in a deliberate breach of contract in order to sell the goods to some third party above the

[48] *See* Calabresi & Melamed at 1116.

[49] For discussion, see Richard A. Epstein, TORTS, § 14.7 (1999).

[50] *See* Ayres & Talley, *supra* 1030–1035.

[51] On the question of efficient breach, see Daniel Friedmann, *The Efficient Breach Fallacy*, 18 J. LEGAL STUD. 1 (1989); Richard Craswell, *Contract Remedies, Renegotiation, and the Theory of Efficient Breach*, 61 SO. CAL. L. REV. 629 (1988).

contract price.[52] As with the other case, the liability rule tends to be rejected even when the breach disrupted all relations between the parties. This commercial response should offer a warning about any widespread substitution of liability rules for property rules in contexts that call for future interactions. Even if liability rules have some niche uses in necessity cases, they represent an impossible way to organize complex social interactions if they are construed to mean that any individual is entitled deliberately to take the property of another individual as long as there is some assurance that he will be required to pay the appropriate social price. The level of instability is enormous, for no remedy can calculate the full consequential damages that follow from the loss of entitlements.

In this context, it bears stressing that the purpose of a license agreement is not simply to place the patent in the hands of one party or another. It is to institute at a conscious level some system of divided control over the use of information in the ordinary course of business. These shared arrangements require continued trust and cooperation, which is hardly to be expected when one person is able to foist himself on another. It is instructive to see just how difficult it is to organize a system of compulsory licenses when divided, not unitary, control is the end of the process.

2. Mechanics of Compulsory Licenses

Recall that the Gitter proposal applies only to DNA sequences, while the Mueller proposal extends to a wide array of other products, such as computer software. But the danger of creeping compulsion cannot be overlooked in assessing this proposal in either its broad or narrow manifestations. The first question is the simple one of who initiates the license. In the ordinary voluntary transactions, the licensor, as owner of the patent, can control the stable of licensees and pick those who he believes will have the greatest chance of producing some invention or product that will yield a high return from the relationship. Resources of monitoring and supervision can thus be effectively conserved. That form of control is wholly lost once the patent holder is converted into a de jure common carrier that must make licenses available as a right to all who wish to work with some research tool or DNA sequence. Unlike in other situations, this sequence information is in the public domain and it provides information that allows the new licensee to begin work on the project without notification to the patent holder. In order to curtail that risk, it will become necessary to mandate that licensees post on some neutral site a notice of its "intention to use" the covered materials of another company, and

[52] *See* Lisa Bernstein, *Merchant Law in a Merchant Court: Rethinking the Code's Search for Immanent Business Norms*, 144 U. PA. L. REV. 1765 (1996).

to subject that party to heavy damages in the event that they proceed without supplying the requisite notice.[53]

Unfortunately, an enormous gap remains between a notice provision on the one hand, and a completed contract on the other. So the first task that the licensor must face is how to respond once it learns of the notice. Must the patent holder now enter into some detailed investigation of the financial position and technical abilities of the putative licensee to determine whether it is worthwhile to incur the expenses necessary to police the activities of the licensee in order to preserve some future royalty stream? Voluntary licenses, of course, contain provisions whereby the activities of the licensee are policed, often through the use of elaborate "milestone" provisions, in which future work under the license is tied to previous progress. The cost of overseeing these operations, however, is generally taken into account in setting out the overall deal. Otherwise, the licensor would have a negative expected value for the transaction, which would then not ordinarily go forward. But with these compulsory licensees, no front-end payment is contemplated, and security arrangements and guarantees are virtually impossible to devise, especially if the licensee is permitted to conduct its work under a shroud of secrecy, so that the firm has to go out of pocket on monitoring, knowing of the serious risk that the licensee could be subject to reorganization, takeover, or bankruptcy, leaving its ultimate collection rights very much in danger. The point here is no small one for a firm that has multiple patents, each of which could become the subject of hundreds of mandatory licenses. It makes matters only worse that the licensing firm may use multiple patents to get its own product to market. This effort to solve the holdout problem could create, in short order, massive administrative burdens over which licensed patent contributed what to the final outcome.

Matters do not look any better when one considers the possible terms and conditions of the licenses so imposed. As a benchmark for this analysis, it is useful to consult the voluntary licensing agreements right now in place for various forms of technology. These agreements are, of course, complex affairs that must track the wide range of permutations that frequently arise in any ongoing relationship. The terms of particular licenses are confidential, but even a quick look at a specimen template of an exclusive license shows the complexity inherent in deals of this sort. The exclusive license agreement used by Harvard University[54] is only a few pages—before the particulars are filled in. It contains detailed provisions that govern not only the conduct of licensee but also of its affiliates,[55] lest there be confusion as to what activities are covered.

[53] On which see Mueller, *No "Dilettante Affair"* at 58–59.

[54] Harvard Exclusive License Agreement as of August 27, 2003, at http://www.hms.harvard.edu/otl/form_exclusive.html

[55] *Id.* at 1.1

It contains detailed description of the materials supplied[56] and covered the products and processes.[57] It contains a four-part definition of what counts as "net sales" on licensed product.[58] It deals with such matters as patent rights[59] and sublicensee income.[60] It contains various reservations so that its exclusive license does not stop noncommercial research work by Harvard or other licensees. It contains, as it must, a best efforts clause, so that the exclusive license does not remain fallow.[61] Further, it has complex provisions to determine the royalties owed under the license.[62] And it includes provisions that deal with reporting[63] and record keeping[64] to backstop the agreement.

It is easy to see how this list can be expanded: Are there specific procedures for the modification of terms? What are the rights for termination or cancellation? Assignment and sublicensees? Are payment provisions lump sum, periodic, or contingent? Is there security for payment provisions? Cooperation provisions? Sharing of information between the parties or with third parties? Compliance with regulatory norms? Remedies for breach of contract before, during, and after the formation of the agreement? Coverage of payments after the expiration of the patent? For its international use? It seems highly unlikely that any compulsory license could cover enough of these issues to be sustainable in the long run.

The payment issue alone shows the magnitude of the problem. The only system that is practicable in compulsory licensing is a percentage of revenues, along the lines of most voluntary agreements. But there is a subtle difference. It is not enough to say that consequential losses don't matter under compulsory licenses because ordinary licensing agreements don't call for them.[65] True enough, but that is because the patent holder will *not* enter into any voluntary licensing arrangement if any consequential losses from the new entrant exceed the expected royalty stream. A selection effect weeds out some transactions that will now take place under the compulsory licensing stream. These additional transactions are loss transactions to the patent holder, so that over the full range of deals, his economic position will be eroded by compulsory licensing. That downward bias in returns will still be greater if the return is

[56] *Id.* at 1.2
[57] *Id.* at 1.5 & 1.6.
[58] *Id.* at 1.8.
[59] *Id.* at 1.11
[60] *Id.* at 1.10
[61] *Id.* at 3.2 (d) & (e).
[62] *Id.* at Art. IV.
[63] *Id.* at Art. V.
[64] *Id.* at Art. VI.
[65] Nor should they. *See* Richard A. Epstein, *Beyond Foreseeability: Consequential Damages in the Law of Contract*, 18 J. LEGAL STUD. 105 (1989).

posited at some "competitive" level, *i.e.*, lower than the voluntary market would yield for this particular transaction.

Other problems with transactional mechanics will undermine this scheme. Mueller's proposes a standard of a "reach-through" 25% of pretax profits. But there is little reason to think that any single number will capture the wide variation in voluntary deals. One key element in the royalty rate involves the amount of downstream work that is left to be done once licensing is complete. If that amount of work is extensive, then the license royalty is likely to be lower. If the work required is slight, then the rate would increase. Picking a presumptive standard that is too high would render the entire licensing system inoperative because no potential licensee will seek to exercise its rights under the compulsory scheme. In the limit, one way to repeal the compulsory license scheme is to mandate that the licensee pay 100% of its revenues to the licensor. But alternatively the figure set could be below the market royalty rate, at which point the licensor may not be able to recoup its expenditures with a reasonable rate of return. Here the product, if made at all, will be gobbled up by a large number of nonexclusive licenses which might yield too little in revenue to encourage systematic investment over time. There is, in short, no single number which makes sense as a presumptive target for licenses, and no set of ad hoc exceptions to the rule that any public body can administer. It makes sense neither to allow the use to be made today subject to a royalty to be determined later, nor to hold off the license until the question of particular rates receives some public validation.

But even if we could set optimal royalty levels that work for the full range of covered products, this system will fail as an administrative matter if the licensor is not allowed to monitor the books to determine these profits, and will likely fail even if that monitoring is allowed. Most percentage leases in real estate work off gross revenues, not net profits, precisely because the latter figure is so elusive in contested settings. Even if some voluntary transactions calculated royalties off net profits, it is doubtful whether that practice would make sense when the number of transactions increases, and the dollar stakes of the transactions, in at least some cases, go down, and in which the level of access and trust is far lower. Transplanting isolated terms of business arrangements to new contexts simply does not work. What is needed is a keen appreciation of how each individual term fits into the large mosaic.

In sum, the negotiations over any particular agreement present an uneasy mix between standard provisions on the one hand, and dickered clauses on the other. Someone has to engage in fine-tuning, but legislatures cannot do it for contracts any more than they can do it for violins. More precisely, owing to the wide range of licensed and derivative products and processes, there is little hope that all will be able to gravitate to the same standardized agreement, or whether certain kinds of clauses will be relevant for certain classes of arrangements. All products are not created equal; more narrowly, all cDNA sequences may not be created equal, either. All that is known from the outside is that every business agreement involves trade-offs between the ideal incentives on the one hand, and the administrative costs of enforcement on the other. We rely on voluntary agreements not because we (collectively) know the right answers but because we know that transacting parties in sophisticated transactions have both the incentives and the knowledge to get it right.

Yet, what can be said when the terms of the compulsory license are stacked up against those which are part and parcel of every voluntary agreement? Initially, we can be confident that the terms of these agreements should, if anything, be more exhaustive than their voluntary counterparts. When contracting parties select each other they can rely on a full set of informal, reputational, or relational sanctions to help keep each other in line. As the level of trust diminishes, the need for explicit monitoring between the parties increases at a time when the sheer numbers of these agreements strains all the resources available for monitoring. Where net licenses are used, moreover, some procedures must be developed to allow the patent holder to determine whether the licensee has padded expenses in order to reduce the net profits owed under the agreement, or reallocated the income to some other product to the same customers, perhaps as part of a complex package. The inquiries require the parties to delve into the intricacies normally associated with rate-of-return regulation used for regulated public utilities, such as the phone company, where suits drag on for eons, and the term "reasonable" does not begin to unpack the host of interpretive and factual difficulties. The entire history of rate regulation for network industries should lead one to doubt that anyone could price these licensee fees in such a manner that leaves the licensing companies intact. This is one road down which no one should travel.

V. CONTRACTING STRATEGIES

In light of the difficulties with any set of state-mandated forced licenses or exchanges, the sound collective approximation is to stick with the patent system and to rely on other devices to make needed adjustments at the edges. Virtually all these devices rely on some form of contract that avoids the holdout problems that arise when distinct interlocking arrangements are held by unrelated parties. One common device is the humdrum system of assignments. Only inventors are entitled to file for patents, and corporations do not invent. To avoid the excessive Balkanization of the process, a corporate employer typically requires its employees by contract to assign all patents that he receives to his employer. The full arrangements will, in most cases, contain some return compensation for the inventor above and beyond the usual salary, as a spur to invention.[66] The automatic assignment provisions allow the institution to organize a suite of coherent inventions, often from multiple workers within the same discipline. The alternative scenario allows each inventor to keep his initial invention, which in turn could create difficulties where related inventions have different combinations of inventors drawn from a pool of related workers. These arrangements have been attacked from time to time under contract law on grounds of exploitation, unconscionability, and the like. But their overall efficiency-enhancing characteristics should immunize them from such challenges. Nor should the antitrust law apply to the division of labor within the firm, but only to connections made across firms.

The next consolidation device is more tricky precisely because it does involve cooperation between firms. Cross-licensing agreements give each firm access to the technology of the other but carry with them the risk of coordinated pricing-arrangements that could violate the antitrust prohibition against horizontal arrangements. Patent pools involve a procedure whereby two or more firms create a package of patents that they then license en masse to third parties. In both situations, the critical distinction is between economic *substitutes* and economic *complements*, for pooling arrangements are far more attractive in the latter situation than in the former.[67] With complements, the effect of the pool is to allow coordinated licensing which would eliminate some of the inefficiencies of the Cournot duopoly. Often these are royalty-free, much like the restrictive covenants and restrictions imposed on the residents of planned-unit developments. Nonetheless, if the commercial value of the respective patents suites are unequal, some overall transfer payment might be added to ensure that the transaction is win/win. But the key point here is that the licensing takes place in bulk so as to avoid the tedium of measuring value for individual patents.[68] But with substitutes, the effect of the pooling arrangement could be the formation of a cartel in restraint of trade, as the Department of Justice has in fact ruled.[69] But whatever those difficulties are, the one solution that seems doubtful is a per se ban on all patent pools when the alternative is to magnify the holdout problems that frequently loom so pervasive in the area.

[66] The University of Chicago, for example, has such a policy with regards to inventions that arise out of employment, which stipulate percentages of the payment that go to the inventor, to his division, and his department. The university undertakes all the expenses of bringing the invention to market.

[67] *See* Shapiro, *Navigating the Patent Thicket*, at 5–6.

[68] *Id.* at 12

[69] *See* http://www.usdoj.gov/atr/public/press_releases/1997/1173.htm Press release of 6/26/97, which approved a "one-stop" patent pool involving nine firms and one university for "essential patents," and created a mechanism whereby and "independent patent expert" would pass on whether patents should be added or removed from the pool. *Id.* at 3. An "essential" patent refers to those which are complements. *See also* the press release of 12/17/98, http://www.usdoj.gov/atr/public/press_releases/1998/2120.htm

What, then, is the correct response to this array of interlocking, overlapping, and independent monopolies in the patent world? In light of all that has been said, the effective choice boils down to either a general commons or the current regime. Writing on a blank slate, Congress could create a new commons by refusing to exercise its patent power at all. The hope might be that the lower rate of invention would be more than offset by the higher rate of utilization of the fewer inventions that are developed in the first place. But that said, the crude empirical judgment of just about everyone is that the gains from increased incentives to create at some stage outweighs the losses from restricted dissemination, so the patent system should be preserved.[70] One way to reach this conclusion is that monopoly profits are never that great to begin with (say in the order of 20%), and that owing to the imperfect correlation between legal and economic monopolies, many of those losses are competed away. All this dislocation is better than the total wipeout that would result if innovation grinds to a halt. Contestable empirical judgments are required on all dimensions, but we must make these nonetheless. The patent system should stand more or less in its current configuration.

VI. ON TO THE GENOME

The next portion of this chapter deals with the question of how these general conclusions should operate with respect to the genome. In order see how this works out, the chapter will proceed at two levels. The first asks how the patent bargain plays out without reference to the particular terms of the statute, that is, under some idealized version of the patent law that keeps to the property system and asks only one question: Is this set of transactions one for which it is worthwhile to create a private monopoly? The second level then examines the issues related to the patentability of genetic materials under the current law. The question then arises: How do these theories play out when applied in a more systematic fashion to the genome? Here the problem for genetic patents is in a sense simpler than that for the patent law in general. The range of genetic patents is narrower than those for all inventions, so that it becomes possible to

[70] For a listing, see Michael Abramowicz, *Perfecting Patent Prizes*, mss, at note 4. Included on his list are, Fritz Machlup, *An Economic Review of the Patent System* in Subcomm. on Patents, Trademark, and Copy Right of the Senate Comm. on the Judiciary 85th Cong. 44–45 (Comm. Print 1958) (exhaustive review of patent system); Brian Peckham, *Should the U.S. Patent Laws Be Abolished?* 11 J. CONTEMP. PROBS. 389, 421 (1985) (declining to advocate immediate abolition of the patent system); James Bessen & Eric Maskin, *Sequential Innovation, Patents, Imitation* (Massachusetts Inst. of Technology Working Paper No. 11/99, 1999). For an interesting compendium of quotations denouncing the patent system, see Gordon Irlam, Coalition for Networked Information Forum, http://www.cni.org/Hforums/cni-copyright/1994-04/0648.html

hazard more confident generalizations about the operation of the patent system in this proper subset of its overall domain, even before we examine the specific conditions of patentability under the current law.

A. The Patent Law

In order to set the stage for the discussion that follows, it is useful to run through the normal requirements needed to ensure patentability. The inquiry is important because nothing in the language of the patent act itself makes explicit reference to the overall patent bargain between the inventor and the state. But the individual elements required for patentability do, for the most part at least, seek to identify the terms and conditions under which the grant of a patent is likely to produce the desired results.

This process is, in general, divided into two stages. The first of these, which addresses the question of patent eligibility, simply asks the question of whether the subject matter of the proposed invention is a machine, process, manufacture, or composition of matter.[71] If it is not one of the above, then it just does not qualify for any form of patent protection. The implicit economic logic of this approach is that candidates for patent protection that fall outside these various elements do not deserve case-by-case examination to see if the patent bargain makes sense from the public's point of view. Mathematical ideas and compounds of nature are always in a sense there for the taking, so that no one should be allowed to claim an idea or an element just because he or she got it first. Madame Marie Curie has no claim for the exclusive use of radium, even though she isolated and purified it first. She might be able to get a process patent, and she could certainly keep the radium that she purified. But beyond that, she is not able to prevent others from finding different ways (*e.g.*, chemical reactions) to isolate radium.

In some cases, it is an open question of whether a particular claim falls on one side of the line or another. Supreme Court Justice William Douglas is generally regarded as wrong in *Funk Bros. Seed Co. v. Kalo Inoculant Co.*[72] to the extent that he rejected as a product of nature a mixture of six strains of bacteria that worked together as an effective inoculant for leguminous plants. It took a fair bit of ingenuity to sort out which strains of bacteria could coexist with other strains, and it was that array, rather than any of the natural strains of bacteria, that constituted the composition of matter (or manufacture) that would be eligible for patent protection if all the other hurdles could be overcome. Even though each of the separate strains were found in nature, their juxtaposition was

[71] 35 U.S.C. §101.
[72] 333 U.S. 127 (1948).

not. The 1980 watershed case of *Diamond v. Chakrabarty*,[73] which allowed the patentability for man-made microorganisms, could be distinguished from *Funk* on the ground that it involved a new composition of matter rather than a simple admixture of existing items. But the usual view, which I share, is that *Chakrabarty* has moved the boundary line between natural products and human creations, so that the latter occupies a larger scope than before.

Even when all this is done, however, the patent law contains an implicit prohibition that no person can obtain a legal monopoly over ideas on the one hand, and natural substances on the other. The history of that proposition has been exhaustively documented in a recent article by Linda Demaine and Aaron Fellmeth.[74] The clear impulse for this general position is that no system of rewards should be created for people who just find something that others could find at some later point in time for themselves. If it was created, it would be bad from a social point of view because the exclusive grant could cover matters of great value to which the patentee has added nothing. One prominent variation on this basic theme holds that the same prohibition against patenting ought to apply to natural substances that have been isolated and purified by human invention. In this case, the easiest and most sensible response is to allow someone to patent the *process* under which the purification and isolation has taken place as long as the other requisites of the patent law have been met, but not to patent the *substance* to which that process has been applied. Just this solution was articulated as early as 1874 by Justice William Strong in *American Wood-Paper Co. v. Fibre Disintegrating Co.*,[75] in which the patent claim covered refined cellulose that had been extracted from wood, straw, and other fibrous materials. The plaintiff sought to use the materials in the preparation of various forms of paper, but his patent application was rebuked in the following words:

> There are many things well known and valuable in medicine or in the arts which may be extracted from divers substances. But the extract is the same, no matter from what it has been taken. A process to obtain it from a subject from which it has never been taken may be the creature of invention, but the thing itself when obtained cannot be called a new manufacture.[76]

Strong then illustrated the point by noting that a new process for the creation of prussic acid could be patented even if the prussic acid itself could not

[73] 447 U.S. 303 (1980).

[74] *See* Linda J. Demaine & Aaron Xavier Fellmeth, *Reinventing the Double Helix: A Novel and Nonobvious Reconceptualization of the Biotechnology Patent*, 55 STAN. L. REV. 303 (2002).

[75] 90 U.S. 566 (1874).

[76] *Id.* at 693.

be.[77] The decision here clearly rested on the notion that the extracted material was not patent eligible. The case did not concern any of the individual elements relating to the nonobviousness or utility of this particular claim. In principle, this same approach can be carried over to the new wave of biotechnology, including, of course, genetic substances. In addressing this issue, John J. Doll, the director of biotechnology at the Patent and Trademark Office (PTO), has urged that these "isolated and purified" sequences must be protected in order to supply and reward structure for work in this area.[78] But in this statement he did not respond, as Demaine and Fellmeth have noted,[79] to the obvious objection that the standard rule that denied patent protection to isolated and purified substances carries over without missing a beat to recombinant DNA technology, as the traditional case law required.

The question then arises whether the Doll pronouncement just misses the boat or whether the common position that supports genomic patents can escape this charge. In order to see how this works out, it is necessary to set out briefly some of the basics of recombinant DNA. Human DNA is an elaborate code that instructs each cell how to combine different amino acids into proteins. Those instructions are contained in codons, which are triplets of three nucleotides, or bases: adenine (A), guanine (G), cytosine (C), and thymine (T). These four nucleotides generate 64 (4^3) combinations. But they code for only the 20 amino acids that the body synthesizes. The code therefore is degenerate or redundant in the sense that there is no bijective (or one-to-one) correspondence between a given triplet and the associated amino acids. Each codon codes for one nucleotide, but many different triplets code for a single amino acid. The coding, however, does not take place directly. The DNA works through another molecule called messenger RNA, or mRNA. The ability to manufacture these proteins depends on the manipulation of mRNA, which serves as a template to generate complementary DNA, or cDNA. The cDNA produced will code for the same protein, but it is an *artificial* molecule that need not have the same configuration as the DNA found in nature. At this point, the patent process involves more than the isolation and purification of prussic acid, or indeed any other natural element or substance.

This point is important because it suggests that genomic patents should be regarded not merely as process patents but as subject matter patents that meet the test for a new composition of matter set out in the patent law. Nonetheless, in their recent broadside against genomic patents, Demaine and Fellmeth downplay this structural difference in order to claim that genomic patents should not be issued at all. They write:

[77] *Id.*

[78] John J. Doll, *The Patenting of DNA*, 280 SCIENCE 689 (1998).

[79] Demaine & Fellmeth, 55 STAN. L. REV. at 305.

> That this position has swayed circuit courts and the PTO is surprising in light of the strong theoretical and scientific counterarguments against it. The argument both misstates the meaning of "purification" as defined by the Supreme Court and misrepresents the difference between natural DNA transcription and scientific recombinant DNA cloning. In the first place, the contention that recombinant DNA is not equivalent to purified, natural DNA is scientifically unsound. The notion of removing introns, regions of a DNA sequence that do not code for proteins, is not a human invention. In fact, during the process of natural DNA transcription, an mRNA molecule is created as a copy of a gene in preparation for protein synthesis. During the creation of the mRNA, only the exons are reproduced. Thus, by merely matching the complementary nucleotides to those in a naturally occurring mRNA molecule, an "isolated and purified" sequence of nucleotides is created. Recombinant DNA technology produces, in effect, a kind of imitation of a naturally transcribed gene. Under well-established patent principles, a product that is an imitation of a preexisting product (naturally occurring or not) is not patentable per se. The practical implementation of recombinant DNA technology (a process) may be new outside of a living body, but it is hardly a new concept. The identity of naturally transcribed DNA to recombinant DNA provides an independent basis to find a lack of invention.[80]

There are several difficulties with this position. In the first place, it renders largely beside the point Demaine and Fellmeth's long and tortuous defense of the rule that natural substances cannot be patented as long as the patentee only purifies and isolates what is found in nature. In this case, the cDNA is a product synthesized by human intelligence. To call recombinant DNA technology a "kind of imitation" is to give a backhand admission that the molecule in question is not an exact duplicate of what nature itself provides, so that the traditional prohibition against patenting natural substances no longer applies. There may well be questions whether this or that genetic substance meets the second-tier questions on patentability, but the roundhouse blow to the entire system of patent protection, now in place for well over a decade, does not seem well advised.

Nor is there any reason to be alarmed at this result. To see why, suppose that in *American Wood-Paper* there was but one and only one method for the isolation and purification of prussic acid. At that point the process patent would give its holder the same economic advantage as a patent over the substance

[80] *Id.* at 408–409.

itself. Frequently exclusive holders of secret processes prefer to hold them as trade secrets, for if the product is one of standard manufacture, the mere appearance of that product on the marketplace offers no indication that the rival seller has turned the secret to his own advantage. He could have used some alternative process to reach the same result. But once the process is shown to be exclusive, then it can be patented, for if anyone else produces the purified and isolated product, then it is no more difficult to sue for infringement of the process patent than a substance patent. It is only when an alternative process is developed that the prior process patent loses its exclusive force and the questions of enforcement (whose patent was used and why?) come to the surface.

The situation here is no different. There is no reason to think that any naturally occurring genetic substance, now and ever more, can be made only by constructing a cDNA molecule through recombinant techniques. This substance patent can be undermined in exactly the same way as the exclusive process patent. All it takes is human ingenuity to find an alternative pathway to the naturally occurring substance. The situation has, of course, analogies with common and private property in land. Suppose that X owns the only ski lift to a ski run at the top of a mountain that is open to the public. The charge for the use of the lift is identical to what he could charge if he owned both the lift and the ski run. There is no double marginalization problem because no one can be excluded from the slope. Yet we do not treat this situation as any different from one in which a single person owned both the ski lift and the ski run (which operates in competition with other ski lifts and run in the region). But the moment a second operator builds a ski lift to the top, the monopoly hold is broken, which would not be the case if the first owner had owned the slope as well. In most cases, it is a process patent that holds the key to public domain property. In this case, it is a substance patent that produces that result.

Nor is it difficult to see why Demaine and Fellmeth take so restrictive a view on the subject. Elsewhere in their paper they give their express approval of the restrictive interpretation of patent eligibility adopted in *Funk* and their disapproval of the more expansive definition of that term found in *Chakrabarty* in large because that decision countenanced the patenting of various human life forms.[81] It is one thing to attack the use of genomic patents on the restrictive views that dominated before the more expansive readings of *Chakrabarty*, but it is a good deal more difficult to reach that result once it is recognized that the decision counts a good law that helped to spur the huge biotech boom.

This analysis of the problem helps to explain and justify the outcome in the extensive dispute over the patenting of a gene that codes for erythropoietin (EPO) for the treatment of human diseases. The key case for this short

[81] *Id.* at 315–19.

demonstration is *Amgen v. Chugai Pharmaceuticals.*[82] The most salient points about this dispute was that no one claimed the right to patent EPO as found in nature, but only the purified and isolated cDNA that allowed for its production and the "procaryotic or eucaryotic host cell" that had been transformed to allow expression of this gene. These two limitations are critical because it means that anyone who developed a different recombinant DNA sequence could have a fair shot at patenting that gene.[83] Certainly anyone who produced EPO by the older and inefficient technique of "concentration and purification of urine" would be free to do so as well.

Nor was this exercise in recombinant technology a romp in the park. Dr. Edward Fritsch, who headed the project for Genetics Institute, Inc. (GI), had labored over the program in a consistent fashion for 3 years, from 1981 to 1984, until he was able to reduce it to practice. He had no idea when he began whether he would succeed in the end. His claim lost out to Dr. Fu-Kuen Lin's early reduction to practice because the Court held (rightly) that Fritsch's initial strategy to use "two sets of fully degenerate cDNA probes" was not tantamount to having a clear visualization of the molecule that he wished to create. The case at no point raised the possibility that the cDNA that coded for EPO belonged in the public domain. Indeed, for any of the three participants of the dispute to argue for that point hands a victory on a platter to any outsider who then commences to produce the EPO without having incurred the costs needed to develop the techniques for its generation. It is also deeply ironic that all of the arguments about the patentability advanced by the parties were directed to the processes by which EPO was purified and isolated, and none was directed toward the molecule which was (as a "composition of matter") the subject of the patent in the first place. As John Doll has stated: "In order for DNA sequences to be distinguished from their naturally occurred counterparts, which cannot be patented, the patent application must state that the invention has been purified or isolated or is part of a recombinant molecule or is now part of a vector."[84] The first phrase on purification and isolation may well be in question, but the stricter requirements in the last two elements were satisfied in *Amgen* where the key claim "was the novel *purified and isolated* sequence which codes for EPO and neither Fritsch [the rival claimant] nor Lin [Amgen's scientist] knew the structure of the physical characteristics of it and had a viable method of obtaining that subject matter until it was actually obtained and characterized."[85]

[82] 927 F.2d 1200 (Fed. Cir. 1991).

[83] Here the ambiguity arises out of the potential application of the doctrine of equivalents, which would, in my judgment, block the patenting of a new gene that simply represented the substitution of one or more codons in the isolated genes that coded for the same protein as the codons that it displaced. But what do I know?

[84] John J. Doll, *The Patenting of DNA*, 280 SCIENCE 689, 689 (1998).

[85] Amgen, at 1206.

What was so admirable about the *Amgen* opinion is that it ran through the full set of obstacles that any application must run through in order to become a patent, and demonstrated quite clearly that this application satisfied them all without any special kind of tugging and hauling. It hardly works to deny the utility of EPO given the readily identifiable uses for it from the date of its synthesis. It could hardly be said that the synthesis was obvious given that three teams of crack scientists worked so hard to get to this solution. Nor could the "best mode" objections to this particular patent (which were rightly dismissed as groundless) ever be a serious issue on the larger structural question of whether to allow genetic patenting at all. The requirement only demands that the applicant give the clearest account within his knowledge as of the time of patenting, and, in principle, that requirement can always be satisfied with some fuller reckoning of the process of invention and production.

So in the end, what is striking about the cases of genomic patents is that they suggest that the traditional analysis of patentability offers us a reasonable proxy through the patent thicket.[86] It is not that individual problems will not arise. Clearly borderline cases are part and parcel of the patent business whether we deal with pharmaceuticals, computers, or biotech. But the point is that marginal problems require marginal responses, not social revolutions. Here it looks as though the law has got it right. Once we remember that the best is always the enemy of the good, we should strive to leave well enough alone. It is not credible to assert that the efforts, cost, and magnitude could take place without the shelter of the patent system.

B. Express Sequence Tags

We are now in a position to contrast the treatment in *Amgen* with the well-publicized controversy over express sequence tags (EST), where the analysis is quite different. ESTs are a kind of gene fragment that works as a useful identifier or probe for a particular gene.[87] The EST is a subset of cDNA that represents only one randomly selected segment of the gene from which it was derived. As Molly Holman and Stephen Munzer explain "an EST is generally about 400 to 500 nucleotides in length and encodes about 130 amino acids, whereas full-length genes are generally between 2000 and 25,000 nucleotides . . . ."[88] In most cases, it is not possible to infer the full structure of the gene from the EST involved, nor to speculate about its function. As Holman and Munzer observe:

[86] That patent law basically gets the balance right by using the combined forces of the various validity requirements to keep patents from having a scope that is "too broad" is the subject of the work by Kieff. *See, e.g.*, Kieff, *supra* note 1, at 699–700 (citing F. Scott Kieff, *Property Rights and Property Rules for Commercializing Inventions*, 85 MINN. L. REV. 697, 721–22 (2001)).

[87] For a fuller description, see Holman and Munzer at 741–750.

[88] Holman and Munzer at 749.

> ESTs can be useful as tools in isolating full-length genes, locating coding regions on genomic DNA, in identifying patterns of express in tissues other than the tissue of origin of the ESTs, or in the same tissue under different condition, and in other applications where unknown DNA fragments can be used. But, with a few exceptions, these uses are primarily intermediate. They do not themselves result in a "product," but rather allow one to continue down the path to a useful end result.[89]

The main controversy arose over the patentability of these ESTs, which was pursued and then dropped by the National Institutes of Health under the leadership of Dr. Craig Venter. The application was eventually withdrawn,[90] but thereafter Incyte Pharmaceuticals filed on a mass production basis what appears to be 400,000 patent applications for EST tags.[91]

I first encountered the issue of EST some years at a symposium on intellectual property held at the University of Chicago. Professor Rebecca Eisenberg had written a general evaluation of the patent position of these ESTs,[92] to which I added a short response,[93] which, in my usual guarded style, opined that "I could hardly conceive of a weaker case for patent protection than this one." That bald conclusion was based solely on my own sense as to the desirability of the patent bargain from the point of view of the public. The basic intuition for the position runs as follows: Once the techniques had been developed for the isolation of these ESTs, it was (and is, even more so) an easy task to identify them in buckets, as the deluge of Incyte patent applications demonstrates. In virtue of the fact that the ESTs are almost always "intermediaries," once identified and so owned, their use lies largely in limiting the access that other investigators could have to the genes for which these tags supplied, as it were, a port of entry. No one thought, or thinks, that ESTs in and of themselves were a valuable pharmaceutical product. They form simply one link in the chain of processes that must be traversed in order to create useful pharmaceuticals out of the balky materials that nature had given us.

More to the point, perhaps, there did not seem to be any reason to pay the social price of giving patent protection to these large libraries of gene tags because private companies, most notably Merck, were prepared to fund efforts

[89] *Id.*

[90] *Id.* at 750–754. *See also* Christopher Anderson, *NIH Drops Bid fro Gene Patents*, 263 SCIENCE 909 (1994).

[91] Holman & Munzer, *Id.* at 750–754; Elliot Marshall, *Patent Office Faces 90-Year Backlog*, 272 SCIENCE 643 (1996).

[92] Rebecca S. Eisenberg, *Intellectual Property at the Public Private Divide. The Case of Large-Scale Gene Sequencing*, 3 U. CHI. L. SCH. ROUNDTABLE 557 (1996).

[93] Richard A. Epstein, *Property Rights in cDNA Sequences: A New Resident for the Public Domain*, 3 U. CHI. L. SCH. ROUNDTABLE 575 (1996).

at Washington University in St. Louis to isolate and publish these ESTs in ways that brought them immediately into the public domain.[94] In these circumstances, it would be mistaken to assume that this decision was made out of a disinterested sense of the public good. Rather, the motivation for the decision was that Merck was better off if the ESTs were unilaterally placed in the public domain than it would have been if these were kept as private property. It made that decision in the knowledge that other firms would be able to free ride on its decision to engage in unilateral publication of the information. The only reason for making this judgment is that the blocking value of the ESTs (at least at the time these decisions were made) was far greater than their use value. It was worth, in a word, privately creating some form of a public good. The quiescence with the EST cases suggests that other firms share this vision. No individual firm could simply pull its patent application for ESTs with the knowledge that other firms might prevail on their own. So the applications remain in place, even when submitted by firms who think that the first best solution is for all ESTs to fall within the public domain. As long as no one succeeds, everyone is better off. But if one firm succeeds then the usual logic of the prisoner's dilemma exerts its corrosive effect: All will want to obtain blockade positions if one does.

Nor need this effort be confined to ESTs, for the same logic could extend to any information about the genome that facilitates the creation of new commercial products without itself being a commerical product. Systematic efforts are now under way to ensure that basic information about the genome is placed into what may be termed quasi-public domain databases available for general use. For example, the object of the International Genomic Consortium[95] (IGC) is to assemble databases that can be protected by patents, which in turn can be subject to open licenses for all. That system differs from a pure public domain system in that the IGC could use its patent position by licensing its materials to groups that make all or part of their own information available under similar licenses. It is a most difficult question to decide whether these sorts of arrangements are better than pure public domain, which other firms may incorporate without restriction into their own patentable inventions.

On a separate front, NATURE also has worked hard to make sure that all the gene sequences referred to in its publications are made a part of the public domain.[96] Other scientists have sent out the same call or alarm. Once ESTs or

[94] Holman & Munzer, at 755

[95] http://www.intgen.org

[96] *Editorial, Human genomes, public and private*, 409 NATURE 745 15 (February 2001): "As indicated in our 'Guide to Authors,' we require the results of genome sequence analyses, as with protein structure coordinates, to be immediately available from an appropriate data-base without restriction. This supports an unwritten contract with our readers that what they see described is what they can use, without obstacles, whether they work in the commercial or academic sector (an increasingly blurred distinction)."

other information is placed in the public domain, any private firm may provide annotated libraries of the raw data, which it could market under contract as trade secrets. Presumably, the same could be done with information that is obtained under some general public license. That revenue stream would depreciate over time, but it could never be cut short by the patenting of the information that they had assembled. Looked at simply as a structural bargain, the patenting of these ESTs did not, and does not, seem to be a good bargain from the point of view of the public at large.

That conclusion is, as a matter of theory, only strengthened by the considerations addressed in this paper. Whatever the abstract difficulties in working out the interaction among different patent monopoly in the general case, that problem does not exist in this situation. The function of ESTs is largely invariant, as is their mode of operation. Each EST is a gateway to some gene on which some useful work could be done. No EST has any end use. Structurally, patent protection for ESTs seems like a bad idea because it adds on a downstream link in the chain of production. Sometimes this difficulty could be moderated, for two or more ESTs serve to identify a gene, so that the nice question comes up whether the second EST can be patented over the first if the two contain some but not all common elements. But even if there are two or more ESTs that give entry into a given gene, access will be more restricted under the duopoly than it would under any open access system—a system that can be effortlessly created by leaving all ESTs in the public domain. We know that there are ample incentives to create these gene tags wholly without patent protection. Why, then, should any protection be offered if it provides at best only small marginal stimulus for more rapid isolation and identification of these tags in addition to real danger of entry blockade? If this analysis is correct, then the ESTs should be denied patent protection even if they represented some new chemical entity obtained by the intelligence of human beings. The simple fact that others put them into the public domain suggests that the public comes out the loser if they are made the subject to the patent bargain.

What is instructive about the current battle over the genome is that patent eligibility is not an effective stop against, for example, the patenting of ESTs (a point that eluded me in my 1996 denunciation of the practice). These are not like radium in the sense that they are creations through the rote manipulations of DNA and mRNA. Doctrinally, their exclusion from protection has to come for some other reason. Those reasons are contained in the usual litany of other tests that are applied on a case-by-case basis to see, ultimately, whether it is a good bargain from the point of view of the public to extend patent protection to this cases. It is on these second-tier grounds that one has, within the confines of the patent statute, to seek (in a result-oriented sense, perhaps) ways to knock out patent protection for the ESTs. One possibility for this result is to read the "utility" requirement of the law more strictly than is done with ordinary machines and processes, where the ability to commercialize a given invention is thought to go to the value but not the patentability of products.[97] But it is all too easy to imagine broad classes of chemical compounds that can be identified, and perhaps even synthesized, without any real sense of their worth. The patent law, on its face, looks as though all four classes of patent-eligible inventions should be treated as a piece, but this requirement has been tighter for chemicals[98] relative to other areas. One of the attacks on the ESTs has been a proposal to read the utility requirement so tightly that it requires the patent applicant to demonstrate specific, substantial, and credible utility in ways that are explicitly meant to make it more difficult to patent ESTs.[99] The impulse is surely welcome, but the conclusion should be done in a categorical fashion, which is the only way to deal with the millions of applications of this sort that are before the PTO. Perhaps if utility does not work, then the sheer volume of the ESTs indicates that all of them (except perhaps the first batch) were so "obvious" that they should likewise be denied protection under patent law, which only allows patents to be issued for nonobvious inventions.[100]

ESTs have generated a somewhat different response in a long and thoughtful article prepared by Holman and Munzer. They rightly chide me for my hyperbolic denunciation of the patent claims for ESTs and insist that these tags have enough useful applications so as to meet the "utility" prong needed to secure patentability, unless, of course, that requirement is made more stringent in this context.[101] Nonetheless, they shy away from giving these tags full patent protection and urge as a conscious legislative reform a proposal of patent registration that offers lesser protection for ESTs but at the same time eases the path toward their receipt of legal protection. Their proposal for registration requires the state to register the patent on a showing that it is "novel," *i.e.*, that is has not been identified before, without a showing of any of the other requisites normally required for patentability: nonobviousness, best mode of production, etc. The consequence of registration is to give the owner exclusive use for a 5-year period, then to give the owner royalties under some compulsory licensing system for the next 5 years before the EST finally falls into the public domain.

The first question is whether this proposal addresses a problem that is in need of solution. In my view, it does not. So long as we have a private party that

[97] 35 U.S.C. § 101.

[98] *See Brenner v. Manson*, 383 U.S. 519 (1966).

[99] *See Patent and Trademark Office Utility Examination Guidelines*, 66 FED. REG. 1092, 1098 (Jan. 5, 2001). John J. Doll, the head of the biotechnology office of the PTO, is on record as noting this consequence. Martin Enserink, *Patent Office May Raise the Bar on Gene Claims*, 287 SCIENCE 1196, 1197 (2000).

[100] 35 U.S.C. § 103.

[101] Such has been suggested by John Doll, the head of the Office of Biotechnology of the Patent and Trademark Office.

is determined to place these ESTs in the gene tag, there is no functional justification for affording even this limited form of registration protection. In addition, it is not clear in the aggregate whether this protection is more or less valuable than ordinary patent protection. To be sure, the payoff from the registration is more limited than it is from patent protection. But, by the same token, the ease of registration means that individuals would spread their nets wider under this system than would be the case if they were outside the patent system altogether. Five years of restriction is a long time in terms of modern research, and the second 5-year period raises all the troublesome question of how to price ESTs under compulsory licensing. There seems to be no reason to disturb the current situation which in essence leaves these ESTs within the public domain.

VII. CONCLUSION

One of the most fundamental questions that arises in intellectual property law is the extent to which novel technology circumstances requires a displacement of traditional norms of property. This issue has been with us from earliest times. The rise of strong property interests in land, for example, only arose as a social response when agriculture became a viable mode of production in what had once been an exclusively hunter-gatherer society. The rise of cheap modes of printing gave birth to the law of copyright. The institutionalization of scientific research clearly formed a powerful spur for the creation of a viable law of patents. But just as one should not assume that legal rules can remain static through all forms of social change, so it is wise to remember that legal rules need not change, and often should not change, in response to each movement in technology and trade. On a number of previous occasions, I have argued that many of the fixtures of the earlier law serve us well in modern settings.[102] Much depends on how the law was structured in the first instance. With patent law, it seems clear that the full range of classical tests were designed to figure out when it was best to give someone the exclusive rights to a given invention even at the cost of limiting its usefulness once its production has been secured. That task is inherently messy and requires seat-of-the-pants judgments on issues for which everyone would prefer to have authoritative empirical evidence. But for these purposes, the critical point is that the fundamental trade-offs made in the traditional patent law calculations are exactly those that have to be made in connection with the genome. Once we are secure in the obvious fact that the fundamental conditions of patent law are unchanged, then the best way to meet

[102] *See* Richard A. Epstein, *The Static Conception of the Common Law*, 9 J. LEGAL STUD. 253 (1980).

some test of social utility is to stick with the traditional patent law until it is shown to be misguided in some basic judgment. Although many have clamored for a basic change, that case for revamping the system not yet been made. The innovation we need is today under the patent law, not of the patent law itself. "Steady the course" makes up in soundness what it lacks in novelty.

Acknowledgments

My thanks to Rebecca Eisenberg, Scott Kieff, and Randal Picker for a host of useful references and advice. Jenny Silverman and Dan Levine supplied valuable research assistance. The original version of this paper was first prepared for a Conference on the Intellectual Property Implications of the Humane Genome Project held at Washington University on April 12 and April 13, 2002.

9 Varying the Course in Patenting Genetic Material: A Counter-Proposal to Richard Epstein's Steady Course

Rochelle Cooper Dreyfuss
Pauline Newman Professor of Law
New York University School of Law
New York, New York 10012

I. INTRODUCTION

Richard Epstein's "Steady the Course: Property Rights in Genetic Material" makes a case for an all-or-nothing approach to patent protection for the fruits of biotechnology research, a case that rejects initiatives that would "tweak" the system to assure the availability of biomedical advances for use in basic research.[1] Epstein believes that we have stumbled into the best of all possible worlds and should instead stay, as his title suggests, on a steady course. Reading his paper makes one wonder why the literature questioning aspects of genomic patenting and proposing the sorts of interventions that he rejects—compulsory

[1] For an argument that goes even further than Epstein's in favor of property rules treatment for patents see F. Scott Kieff, *Property Rights and Property Rules for Commercializing Inventions*, 85 MpINN. L. REV. 697 (2001).

0065-2660/03 $35.00

licensing, experimental use defenses, condemnation proceedings—is growing so large so fast. I suggest that the disparity between Epstein's claim and what others working in the biotechnology field see arises from a disagreement on four underlying assumptions. I would like to explore these assumptions, which center around the nature of the patentee, the patent, the licensing relationship, and the goals of patent law. I end with a counter-proposal for protecting the public domain of science, which is intended to deal with some of the problems that I believe Epstein to be justified in raising.

II. FOUR ASSUMPTIONS

A. The Rational Patentee

The first of Epstein's assumptions lies at the bedrock of the traditional Chicago School. It is, of course, the notion of the rational actor. Limiting the patent right through government intervention is not necessary under this view because we can rely on the patentee to engage voluntarily in whatever activity these interventions are thought to encourage. If there is unfilled demand for a patented invention, the patentee can be counted on to expand production; if the invention has other applications, the patentee will enter these other fields. Even if the patentee cannot, for some reason, engage in such activities, the patent holder's economic interest lies in finding licensees who can.

In support of this proposition, Epstein cites the "anthrax" incident, noting that even before it was necessary to fashion a compulsory license to assure adequate distribution of Cipro, the patentee, Bayer, had already agreed to supply pills at dramatically lower prices than it previously charged.[2] Bayer's action was not unique: The AIDs epidemic inspired similar behavior on the part of GlaxoSmithKline. More recently, Pfizer, Eli Lilly, and GlaxoSmithKline announced price discounts on all their drugs for low-income Medicare patients, and Geron relinquished some of the rights it claimed to have licensed from the University of Wisconsin in response to the limitations the Bush administration has placed on stem cell research.[3] But as encouraging as these initiatives are, they do not amount to proof that patentees can be counted on to satisfy society's

[2] The federal government was supplied at a price of 95 cents per tablet, substantially less than the $1.75 previously charged, *see* BAY NEWS INTERNATIONAL, October 25, 2001, http://www.news.bayer.com/news/news.nsf/ID/NT0000E15E

[3] *Burroughs Moves To Lower Prices*, N.Y. TIMES, February 13, 1992, Section D; Page 4; Column 2; *Aspen Pharmacare Receive Voluntary License from GlaxoSmithKline on Anti-retroviral Patents in SA*, http://www.aspenpharmacare.co.za/showarticle.php?id=135; Melody Petersen, *Prescription Card to Link Discount Programs is Planned*, N.Y. TIMES, March 11, 2002, at Section C, p. 7, col. 5; Andrew Pollack, *University Resolves Dispute On Stem Cell Patent License*, N.Y. TIMES, Jan. 10, 2002, at Section C, p. 11, col. 1.

needs for their inventions. Not every company will be as altruistic as Bayer, GlaxoSmithKline, Eli Lilly, or Pfizer. Indeed, an equal number of anecdotes can be cited to support the opposite proposition, which argues there are risks that patented innovations will be suboptimally utilized. Consider, for example, the genesis of one of the few public-regarding provisions of the current Patent Act—the one that limits the liability of medical practitioners for performing infringing medical activity.[4] That provision was enacted after a surgeon began to assert a patent on an incision against other surgeons.[5] There is also the Myriad Genetics story. Myriad holds the patent on the genetic tests that screen for mutations of the BRCA 1 & 2 genes, which are correlated with breast cancer. Myriad has been threatening to bring infringement suits against medical researchers who use the test only to learn more about breast cancer.[6]

Epstein supports his claim by noting that "at a conference at Washington University in 2002 it was said publicly and more than once that experienced operators in the field did not know of a single worthwhile product that has been kept for the market because of blockade difficulties."[7] Of course, the difficulty of proving a negative *is* known; in fact, there were also conferees who said they knew of problems. Moreover, the absence of blockades may, in part, be attributed to a long-standing assumption that the conduct of basic research at universities is within a common-law experimental use defense.[8] Since the conference, the Federal Circuit has made it clear that universities (and presumably, other basic research institutes) enjoy no such exemption.[9] Accordingly, frustrated research agendas may soon become more common: Patentees now have more power to insist on licenses, and private research institutions may be less willing to countenance unauthorized use, knowing there is a risk of treble damages for willful infringement.[10] In this context, it is important to remember

[4] 35 U.S.C. § 287(c)(2).

[5] *See* Todd Martin, *Patentability of Methods of Medical Treatment: A Comparative Study*, 82 J. PAT. & TRADEMARK OFF. SOC'Y 381, 401–02 (2000).

[6] *See* Donna M. Gitter, *International Conflicts over Patenting Human DNA Sequences in the United States and the European Union: An Argument for Compulsory Licensing and a Fair Use Exemption*, 76 N.Y.U. L. REV. 1623, 1650–51 (2001).

[7] Epstein, at 166–167.

[8] *Peppenhausen v. Falke*, 19 Fed.Cas. 1048, 1049 (C.C.S.D.N.Y. 1861).

[9] *Madey v. Duke University*, 307 F.3d 1351 (Fed. Cir. 2002) ("Our precedent does not immunize any conduct that is in keeping with the alleged infringer's legitimate business, regardless of commercial implications. For example, major research universities, such as Duke, often sanction and fund research projects with arguably no commercial application whatsoever. However, these projects unmistakably further the institution's legitimate business objectives, including educating and enlightening students and faculty participating in these projects. These projects also serve, for example, to increase the status of the institution and lure lucrative research grants, students and faculty." *Id.* at 1362).

[10] Public universities are in a somewhat different situation by virtue of *Florida Prepaid Postsecondary Education Expense Board v. College Savings Bank*, 527 U.S. 627 (1999) and *College Savings Bank v. Florida Prepaid Postsecondary Education Expense Board*, 527 U.S. 666 (1999), which

that there is an important difference between many of the entities that engage in basic research and purely commercial researchers: The former do not always have the resources—or even the expectation of resources—to pay licensing fees for the technologies they need. They also lack an inventory of existing licenses that could be used to barter for access (through, for example, cross-licensing and pooling arrangements).

More generally, not all patentees will act rationally all the time. There is increasing work in behavioral economics on deviations between observed behavior and what might be expected of the rational actor.[11] Although the law and economics crowd is starting to pick up on this literature,[12] it has not, so far, used it to analyze patent licensing. There is reason to believe such an examination would be worthwhile. For example, Rebecca Eisenberg has shown that academic researchers who regard their work as at the cutting edge and the university technology transfer officers who represent them often hold out for inappropriately large royalty agreements.[13] More prosaically, patentees often enter into exclusive licensing deals. Economic theory suggests that patentees usually do better with nonexclusive licenses (because the licensees are then forced to compete against each other). Nonetheless, exclusive licensing arrangements are relatively common. The obstacles they pose to innovation are nicely illustrated by the problems that CellPro had in trying to utilize its technology in the face of a patent that Johns Hopkins University had exclusively licensed to Baxter Healthcare and which Baxter was not fully exploiting.[14]

---

held that state actors are not monetarily liable for past royalties. State actors can, however, presumably be enjoined from continuing to infringe.

[11] *See generally*, Sendhil Mullainathan and Richard H. Tahler, *Behavior Economics*, http://papers.ssrn.com/sol3/papers.cfm?abstract_id=245733

[12] *See, e.g.*, Christine Jolls, Cass R. Sunstein, and Richard Thaler, *A Behavioral Approach to Law and Economics*, 50 STAN. L. REV. 1471 (1998).

[13] Rebecca S. Eisenberg, *Bargaining Over the Transfer of Proprietary Research Tools: Is the Market Failing or Emerging?* in EXPANDING THE BOUNDARIES OF INTELLECTUAL PROPERTY 223 (Rochelle Dreyfuss, *et al.*, eds. 2001). *See also*, Arti Kaur Rai, *Regulating Scientific Research: Intellectual Property Rights and the Norms of Science*, 94 NW. U. L. REV. 77 (1999) (showing how patents can frustrate exchanges among members of the basic biological science community). Compare, F. Scott Kieff, *Facilitating Scientific Research: Intellectual Property Rights and the Norms of Science—A Response to Rai & Eisenberg*, 95 NW. U. L. REV. 691 (2001) (showing how patents can facilitate these exchanges, especially when viewed as an additional tool beyond the otherwise available tools of academic kudos, such as citations, publications, grants, and tenure).

[14] *See, e.g.*, *Johns Hopkins University v. CellPro, Inc.*, 978 F.Supp. 184 (D.Del. 1997). Admittedly, Baxter offered CellPro a license. However, the offer was on very unfavorable terms: $500,000 upfront, applied against royalties "not to exceed 30% of the value of the kit" CellPro was developing—terms well above what Baxter had offered others. Significantly, the alleged reason for the disparity was that Baxter viewed CellPro as its leading competitor and ahead of it in development. *See generally*, Avital Bar-Shalom and Robert Cook-Deegan, *Patents and Innovation in Cancer Therapeutics: Lessons from CellPro*, 80 THE MILBANK QUARTERLY, 637 (2002).

B. The Benign Patent

It should also be noted that not every failure to license is irrational. This brings me to the second assumption that I want to examine: the notion that patents are, on the whole, benign in that they can be invented around. Epstein gets this idea across by dwelling on several examples of what he calls "sidestream" relationships, where alternative patents can be licensed with the result that the patentees compete down the terms of public access.[15] But that is not the piece of the puzzle that the biotechnology industry is actually worried about. Instead, it is concerned (as Epstein acknowledges) with the upstream/downstream problem. The phenomenon of patents covering inventions with substantial research use is actually relatively new to patent law. In the classic case of *Brenner v. Manson*,[16] the Supreme Court drew a line between end uses, which it considered patentable, and research uses, which it held to be unprotectable. A patent, the Court said, "is not a hunting license... not a reward for the search, but compensation for its successful conclusion."[17] That distinction, which released discoveries of mainly research significance to the public domain, has broken down in recent years, particularly in biotechnology, which is much more science-intensive than the innovation industries prevalent at the time patent law was developing.[18] In biotechnology, inventions can have immediate commercial applications as diagnostics or treatments and thus qualify for patent protection, and yet they can be crucially important to researchers. As a result, there are patents that now confer exclusivity not only in product markets[19] but also in innovation markets; by the same token, there are patentees who now regard the research opportunities presented by their patents as part of their reward.

[15] Epstein, at 165.

[16] 383 U.S. 519 (1966).

[17] 383 U.S. at 536. *See also Funk Bros. Seed Co. v. Kalo Inoculant Co.*, 333 U.S. 127 (1948)(holding that products of nature are not statutory subject matter); *O'Reilly v. Morse*, 56 U.S. (15 How.) 62 (1853)(abstract principles not statutory subject matter).

[18] *See, e.g.*, Francis Narin and Dominic Olivastro, *Status Report: Linkage Between Technology and Science*, 21 RESEARCH POLICY 237 (1992)(using citations in patents to journals publishing basic science as a measure of the relationship between basic science and patentable subject matter; demonstrating that the tie between science and technology is becoming closer over time and is more pronounced in drugs, medicine, chemistry, and computing than in fields such as machinery and transportation). Admittedly, it may be that something different is going on: That the notion (often attributed to Vannever Bush) of a linear progression from basic science to applied science to technology was always inaccurate and its deficiencies are now better understood (or, its deficiencies are more salient as science has changed). But even under this view, there is a strong argument that the law should be modified. *See generally*, Donald E. Stokes, PASTEUR'S QUADRANT: BASIC SCIENCE AND TECHNOLOGICAL INNOVATION passim and 21 (1997).

[19] In this context, product market means the market for products, processes, and the products of processes.

This, in turn, has serious consequences. One doesn't have to believe in the irrational patentee to worry about holdouts. Patentees may refuse to license patent rights to those who might develop superceding inventions on which the licensed patent does not read.[20] If a patent protects sequential invention along a quality ladder, a rational strategy for the patentee may well be to develop products slowly, milking each market before progressing to the next one.[21] Furthermore, in cases where the invention's potentials are difficult to evaluate, the risk-averse patentee may prefer to wait to license until the significance of the patented contribution is clarified. Many of the molecular structures of interest to biomedicine are, at bottom, information products, valued for what they say and not—as with chemicals—for what they do. It is thus highly significant that copyright law, which has long protected information products, contains a complex array of compulsory licensing requirements, some of which are specifically targeted at the problem of cumulative development.[22]

To put this another way, while it is surely the case that a patentee's private interest *may* coincide with the interest of the public, there are also circumstances when the patentee's interest is inimical to that of the public. Epstein expresses some willingness to deal with such problems by, for example, varying patent term or patent scope.[23] However, clashes between public and private interests are not easily resolved with such variations. Changing patent duration would help only if the term were made short enough to accommodate the rapid pace of innovation in this field. But significantly shortening the term might undermine the incentive effects of the patent reward. Broadening scope likewise would not help. Broader patents would cover a wider swath of development opportunities, making it possible for the patentee to control more lines of research; they may also slow down the rate of effective invention. Narrower patents would cover fewer end-products, giving the patentee even more reason to refuse to license developers. Moreover, narrow patents produce thickets of rights that increase transaction costs and may pose barriers to entry. Thus, as long as patenting continues to move upstream, it behooves us to consider interventions, such as the compulsory licensing patents of social significance, to reduce the costs associated with blocked innovation markets.

[20] An example is a drug developed from a patented target.

[21] For examples, see *Special Equipment Co. v. Coe*, 324 U.S. 370 (1945)(allegation that patent applicant would suppress patent on pear canning process); *Rite-Hite Corp. v. Kelley Co., Inc.*, 56 F.3d 1538 (Fed.Cir. 1995)(refusing to incorporate invention into fully-automated product). This problem has also arisen in the context of mergers, where the FTC has been concerned that the merged companies may cut off research lines in order to protect products from cannibalization, see Dennis A. Yao & Susan S. DeSanti, *Innovation Issues Under the 1992 Merger Guidelines*, 61 ANTITRUST L. J. 505 (1993).

[22] 17 U.S.C. §§ 107–121; see especially § 107 (fair use).

[23] Epstein at 155.

Epstein's categorical rejection of these kinds of approaches is, in some ways, curious. First, he says it is not possible to deal with the upstream/downstream issue because it is difficult to "make some universal judgment about the applicable patent regime... when we do not know which configuration of patents dominates the landscape."[24] But it is hard to see how his proposals—to remove inventions from the category of patentable subject matter when their blocking value is greater than their use value, or to recognize "use" patents that cover only commercial uses[25]—are any easier to implement. They, too, require an understanding of the position of patents in the research stream; they also require complex assessments of value and ignore the fact that many research projects have both commercial and noncommercial goals. Further, there is nothing that says we need to make a "universal judgment." Arti Rai and Rebecca Eisenberg, for example, have suggested a limitation that turns on the source of funding;[26] I end this essay with a proposal that relies on self-identification by downstream users. As long as it is clear that a landscape contains a stream, that property owners are arrayed along it, and that their relationships can adversely affect each other or third parties, it is worth making sure that their actions do not systematically cause harm. If the scope of operation of the chosen method is made clear, there is little need to worry about affecting relationships among property owners whose holdings are differently configured. Certainly, there are countless rules and institutional arrangements surrounding the rights of upsteam and downstream owners of riverside property, and sparse evidence that these rules have adversely affected the value of other property relationships.

C. The Malignant Compulsory License

Epstein is additionally concerned about the cost and quality of relationships forged by force of law. Thus, he paints a rather a bleak picture of the cost of compulsory licensing: searching for the right party, providing notice, conducting negotiations, setting prices, monitoring performance, and sharing control in the absence of "trust and cooperation."[27] This brings up the third assumption I want to look at, namely the idea that there is a disparity between a voluntary and a compulsory license with which we need to be concerned. My main point here is that compulsory licensing requirements do not operate in the manner Epstein contemplates. Indeed, they are often conceived of as correcting market

[24] Epstein at 166.

[25] Epstein at 155, 186.

[26] Arti K. Rai and Rebecca S. Eisenberg, *Bayh-Dole Reform and the Progress of Biomedicine*, 66 LAW & CONTEMP. PROBS. 289 (2002).

[27] Epstein at 175.

imperfections.[28] Some simply shift burdens of initiating negotiations or proof; some create off-the-shelf arrangements that reduce transaction costs. But most important, the core significance of compulsory licensing requirements is that they act as credible threats, not as actual business deals. Knowing that arrangements will be imposed if they do not act voluntarily, patentees are pushed to the bargaining table. In this regard, it is significant to note that developed countries that have compulsory licensing requirements in their laws find that potential research licensees rarely need to resort to court to enforce them.[29]

Furthermore, even if there were cases requiring compulsory licensing, there is little reason to think the licenses would be significantly harder to maintain than purely voluntary agreements. To be sure, pricing is a problem, but it is also a problem in negotiated arrangements. As to maintaining relationships, potential licensing partners are hardly the strangers that Epstein posits. Biotechnology is a sophisticated field; publishing research results and making the rounds of conferences and seminars are *de rigueur*. As important, the research community is highly networked. As Woody Powell has documented, individual firms and researchers are enmeshed in a host of cross-ties through joint ventures, research partnerships, strategic alliances, equity investments, and licensing arrangements.[30] Because mobility and the capacity to attract partners are crucial to success in this environment, it is essential for researchers to maintain their reputations as cooperators and as honest dealers. Since that requires adherence to social norms irrespective of the origins of relationships, it is unlikely that anyone will see the need for the special arrangements envisioned by Epstein.

[28] *See, e.g.*, Wendy J. Gordon, *Fair Use as Market Failure: A Structural and Economic Analysis of the Betamax Case and its Predecessors*, 82 COLUM. L. REV. 1600 (1982).

[29] *See, e.g.*, Gianna Julian-Arnold, *International Compulsory Licensing: The Rationales and the Reality*, 33 IDEA: J.L. & TECH. 349 (1993). Even Robert Merges, a classic transactional optimist, concedes that the " 'visible hand' of government" may sometimes be needed "to prod or force parties into [pooling] transactions," Robert P. Merges, *Institutions for Intellectual Property Transactions: The Case of Patent Pools*, in EXPANDING THE BOUNDARIES OF INTELLECTUAL PROPERTY, *supra* note 13, at 123, 165.

[30] Walter W. Powell, *Inter-organizational Collaboration in the Biotechnology Industry*, 151 J. INSTITUTIONAL & THEORETICAL ECON. 197 (1996); Walter Powell et al., *Network Position and Firm Performance: Organizational Returns to Collaboration in the Biotechnology Industry*, 16 RES. SOC. ORG. 129 (1999). Similar networks are found in other industries, *see, e.g.*, David Teece, *Capturing Value from Knowledge Assets: The New Economy, Markets for Know-How and Intangible Assets*, 40 CAL. MGMT. REV. 55 (1998). *Cf.* Ronald J. Gilson, *the Legal Infrastructure of High Technology Industrial Districts: Silicon Valley, Route 128, and Covenants Not to Compete*, 74 N.Y.U. L. REV. 575 (1999)(documenting the reliance of the software industry on employee mobility and other forms of shifting alliances).

D. The Goals of Patent Law

Of course, what Epstein is really concerned about is not maintaining the licensing relationship. It is the price-leveling effect of imposing it. As to that, he is surely right: A patentee who cannot walk away from a negotiation will likely extract a lower return than one who can. However, this concern raises the fourth assumption requiring analysis, the assumption that patentees are entitled to maximize the returns on their investment. Is this assumption true? It certainly isn't reality: Economic studies repeatedly show that the social benefits produced by innovation outstrip the private returns available under the patent system.[31] Yet no one suggests strengthening patents so that spillovers can be fully captured. One reason is that the patent system is aimed at encouraging the *optimum*, not the *maximum*, level of innovation. Another is that, in a sense, it is spillovers that "make" the deal—they are an important part of what society gets in exchange for suffering the costs of exclusivity.

Encouraging the optimal level of innovation may not even require that patent holders receive as much of a reward as they currently do. Admittedly, we are in an era in which the notion of allowing rights holders to capture every scrap of surplus has flourished. But it is important to see that, like upstream patents, this, too, is a new idea. I have addressed this notion in other places,[32] and will not go into a full critique here, except to note that it resonates not at all with our constitutional scheme. Intellectual property rights are not antecedent to the Constitution in the way that real property rights seem to be. Rather, Congress is empowered to create intellectual property rights.[33] The rights created must be limited. Most significantly, they can be created only for specific purposes—in the case of patents, for the purpose of promoting the progress of the useful arts. I would not go so far as to say that patents on upstream inventions—inventions that are not purely useful arts—are unconstitutional. I think they are constitutional. As other contributors at this conference point out, upstream patents can promote progress in specific, crucial sectors of the biotechnology industry. But I also think that the returns on this new kind of patent deserve careful scrutiny. If returns from product markets were once enough to stimulate the optimum level of innovation, then rewards from innovation markets may not be needed at all. Or, only a portion of them may be necessary. If so, then the remainder could be foregone in favor of facilitating research in other ways, such as by reducing the costs associated with licensing

[31] *See, e.g.*, Zvi Griliches, *The Search for R&D Spillovers*, 94 SCAND. J. ECON. S29 (1992); Richard C. Levin, *Appropriability, R&D Spending, and Technological Performance*, 78 AM. ECON. REV. 424, 427 (1988).

[32] Rochelle Cooper Dreyfuss, *Expressive Genericity: Trademarks as Language in the Pepsi Generation*, 65 NOTRE DAME L. REV. 397 (1990).

[33] U.S. CONSTIT., art. I, sec. 8.

research-related inventions. The fact that so many researchers in biotechnology are advocating this position, and especially the fact that some of the advocates are, like Merck, among the major developers of end-use products, suggests that a mixed approach toward promoting progress is worth considering. To that end, I offer the following counter-proposal.

III. A COUNTER-PROPOSAL

Epstein does usefully call attention to a set of issues that anyone wishing to depart from the steady course should consider. He is right that it would be difficult to decide which inventions are problematically upstream, that pricing and monitoring within compulsory licensing arrangements could be tricky, and that complex government involvement should be disfavored. He does not, however, compare these problems to those engendered by his own scheme, which include suffering holdout costs in cases where the blocking value of the patent, although high, is less than the use value; or sacrificing the incentive and signaling effects of patents when the use value, though high, is less than the blocking value. In my view, inventions that have both research and end uses present a case where the current system could use tweaking.

I would do it by analogizing to the remedies provision that was instituted in 1996 to limit liability for certain uses of patented surgical and medical processes. As noted earlier, this provision was enacted because of concerns that there would be inadequate access to important developments.[34] In the context of patents on upstream inventions that are not available in the market on reasonable terms, the analogue to this provision might read as follows:

> With respect to a basic researcher's performance of a research activity that constitutes infringement under section 271(a) or (b) of this title, the remedy provisions of the Patent Act shall not apply against the basic researcher or against the institution in which the work takes place.

[34] Public Law 104–208, 35 U.S.C. § 287(c)(2)("With respect to a medical practitioner's performance of a medical activity that constitutes infringement under section 271(a) or (b) of this title, the [remedy provisions] of this title shall not apply against the medical practitioner or against a related health care entity with respect to such medical activity." Sections 271(a) and (b) cover making, using, selling, offering for sale, importing, or actively inducing infringement, when these occur within the United States; the remedy provisions included are §§ 281 (general provision), 283 (injunctions), 284 (monetary damages), and 285 (attorney fees)). *See generally*, Gerald J. Mossinghoff, *Remedies Under Patents on Medical and Surgical Procedures*, 78 J. PAT. & TRADEMARK OFF. SOC'Y 789 (1996).

The term "basic researcher" would require a definition. Given that one of the problems in this area is that researchers are often engaged in projects with mixed goals, a definition that relied on the commercial nature of the activities would not be effective. One approach would be to define basic researchers as those who work solely in nonprofit organizations, or—somewhat more restrictively—in nonprofit educational organizations. A problem here is that many of those engaged in the kind of research this provision is intended to further will nonetheless be excluded because they have ties to industry. Another problem is that nonprofit educational organizations are active patentees themselves. The availability of a research exemption would give them a competitive advantage over commercial enterprises engaged in similar work.

A better alternative is to have those seeking an exemption self-define: to identify themselves as basic researchers working for noncommercial purposes.[35] This could be done with waivers. A researcher who wished to benefit from the special liability rule would be required to sign a waiver in which he or she agreed to publish findings promptly and to refrain from patenting the discoveries made in the course of using the invention for which the waiver was sought. For example, if a researcher wished to use a patented technique for identifying the BRCA 1 and 2 genes, which are linked to breast cancer, he or she could sign such a waiver and then use the techniques without authorization and without paying royalties. Should the researcher, in the course of working with the patented technique, discover another gene involved in breast cancer, nothing associated with that discovery could be patented. The work would, however, be published.[36]

This approach has several advantages. It avoids the need to characterize an invention as upstream or downstream at the time of patenting, or to decide whether a particular use is socially desirable (or even whether it is sufficiently noncommercial). Because exclusive rights are not denied even when they protect inventions of research significance, these patents would continue to be available to serve as signals, as facilitators of technology transfer, and as vehicles for appropriating end-use (as well as commercial research-use) value. As important, the proposal eliminates all need to price the usage made of the patented invention. Indeed, the proposal recognizes the differences in resources

[35] To circumscribe the reach of the proposal (that is, to assure a greater market for patented technologies), researchers eligible for the exemption could be limited—for example, to researchers at noncommercial institutions or at universities.

[36] If a venue for publication could not be found, the work could be the subject of a statutory invention registration, 35 U.S.C. § 157. Such work is disseminated through the patent system, but there is no exclusive right. Nor is the application examined in detail. I would also consider as a friendly amendment Richard Nelson's suggestion that patents on the results of this work would be permissible, as long as they were licensed on a nonexclusive basis, see Richard Nelson, *The Market Economy, and the Scientific Commons*_RESEARCH POLICY_(forthcoming 2003).

available to basic research (as opposed to commercial) institutions. The waiver element has interesting effects of its own. Most prominently, it serves to enrich the public domain because all resulting work is published and not patented. In the university setting, it may also soften the effect of the Bayh Dole Act, which is often incorrectly treated as requiring (rather than permitting) the recipients of federal funds to patent their inventions. In a sense, the effect of the waiver is to partially monetize and internalize the public benefit of allowing work to fall into the public domain.[37] Providing basic researchers with an incentive to forego patenting may also help to restore the Mertonian ethos that once dominated basic scientific work. And it does not require special government involvement other than maintaining a registry for waivers.

Admittedly, there are concerns with this approach. Patentees may complain that it reduces the size of product markets and licensing markets. For example, the market for research tools would be smaller because those who sign waivers would not need to license them. Obviously, the extent of this problem depends on the number of people who seek the benefit of the exemption. The number is likely to be small, given the necessity of waiving the right to patent developments that can be commercialized.[38] Moreover, because these are likely to be people who were not sued under the assumed common-law defense, the actual market for the tools will not shrink much. Finally, since the exemption applies only to technologies that are not offered on reasonable terms, patentees can avoid this loss by making licenses available.

The waiver approach also devalues patents by decreasing the size of the innovation market: Discoveries made by basic researchers deplete the research opportunities available to the holder of the rights the researchers are utilizing. If the research makes available at competitive prices substitute products that do not fall within the scope of the patent, subsequent product markets could also be affected. There are several responses. First, as noted above, the need for a waiver means that there will not be many products developed in this way. Second, because the waiver gives notice of what lines are being pursued non-commercially, duplicative efforts can be avoided and the patentee (and its licensees) will not waste resources pursuing work that researchers will be making

[37] The Bayh Dole Act envisions the possibility that the federal government will elect to pursue patent rights in inventions that are not patented by the funded party, 35 U.S.C. § 202. The government generally licenses on a nonexclusive basis, which means that even if it elected to patent, the information would be accessible to all potential users. However, it may be desirable to change the Act to assure that information invented with the benefit of a waiver finds its way into the public domain.

[38] Small size could be further assured by confining the availability of the exemption to research conducted at nonprofit (or educational nonprofit) institutions. Further, the waiver approach could be limited to cases in which the patentee refused to license on reasonable terms, see Nelson, *supra* note 36.

public.[39] Third, it is highly likely that the waiver provision will make the exemption attractive mainly to those whose work is truly basic—that is, not expected to yield readily commercialized output. If that is so, then the opportunities removed from the innovation market would be rather low on the patentee's own priority list—they would, in fact, be squarely within the area where inefficient "hunting" is especially likely to occur. The waiver approach, in other words, is partly attractive because of its capacity to identify the situations where a chill on research is a potential problem. Fourth, while it is true that this exemption essentially requires patentees to subsidize basic research, the waiver also requires basic research to subsidize patentees. This is because the work that comes out of research under this exemption will be published and thus open new opportunities, which the patentee is free to pursue because they are in the public domain. In addition, basic researchers provide educational benefits to those who later hire the researchers' trainees (such as students and post-docs).

The waiver proposal also raises some practical difficulties, but none is insurmountable. Institutions will have to determine who can authorize waivers; the need for them may create friction between administrators and researchers. These problems are substantial, but they are not very different from those raised by congressional proposals for requiring states to waive sovereign immunity in the wake of the *College Savings Bank* cases.[40] Under these proposals, waivers will be required before states (and state universities) can patent their work. Presumably, such procedures would also require the identification of authorized agents and could similarly engender disagreements over what is best for the state, the institution, and the inventors.

Because research is often serendipitous, it may sometimes be the case that a scientist who signed a waiver makes a commercially significant discovery with the patented technology. Because such an invention may require patent protection to promote and facilitate further development, buyouts should be permitted. The researcher would then negotiate directly with the patentee, in much the same way that improvers now negotiate with those who hold blocking patents. Because the price exacted would compensate the patentee for uses the researcher made in both innovation and product markets, the patentee would ultimately capture the commercial value of its own lost research opportunities. Of course, questions might then arise as to whether a particular invention was

[39] It might be argued that the intent of the Bayh Dole Act is to make sure that research results are the subject of exclusive rights. But that argument cannot be squared with the optional nature of the Act, or indeed, of patent law generally.

[40] *See supra* note 10. For an example of an attempt to abrogate sovereign immunity for intellectual property infringement, see Property Protection Restoration Act of 1999, S. 1835, 106th Cong. (1999). *See generally*, Mitchell N. Berman, R. Anthony Reese, and Ernest A. Young, *State Accountability for Violations of Intellectual Property Rights: How to "Fix" Florida Prepaid (And How Not To)*, 79 TEX. L. REV. 1037 (2001).

the product of the research subject to the waiver. Answering this question will require courts to examine the scope of the research project and determine the role that the patented technology played in making the new discovery. That could be a complex inquiry, but it is not one that is very different from the examination made when inventorship is called into question, or when infringement is alleged.[41]

At bottom, problems will surely arise if we stray from the "steady course." The hard question is whether the steady course is taking us where we want to go. The patent law with which we are familiar was developed to respond to the legal requirements of an industrial age. But we now see the relationship between basic and applied science in a new light; we need to vary the course so that our law reflects the needs of the information society in which we now live.

Acknowledgments

I would like to thank Graeme Dinwoodie and Katherine Strandberg for helpful discussions and the DePaul University College of Law and its Center for Intellectual Property Law & Information Technology for giving me an opportunity to discuss these ideas at its Symposium on Ownership and Control in the Academic World.

[41] See 35 U.S.C. §§ 111, 116, 271. All of the proposals discussed in this and Epstein's article would also need to be tested against the requirements of the TRIPS Agreement; that issue is the subject of another project, see Graeme B. Dinwoodie and Rochelle Cooper Dreyfuss, *Preserving the Public Domain of Science Under International Law*, in INTERNATIONAL PUBLIC GOODS AND TRANSFER OF TECHNOLOGY UNDER A GLOBALIZED INTELLECTUAL PROPERTY REGIME. (Keith Maskus & J. H. Reichman, eds., forthcoming 2004).

10

Reaching through the Genome

Rebecca S. Eisenberg
Robert and Barbara Luciano Professor of Law
University of Michigan Law School
Ann Arbor, Michigan 48109

ABSTRACT

Advances in genomics research have called forth new strategies for patenting DNA sequences. Gene patenting, which began inconspicuously in the early days of the biotechnology industry in the 1970s and 1980s, did not generate significant public controversy until the advent of high-throughput DNA sequencing in the 1990s.[1] By this point, it was such a well-established practice[2] that categorical

[1] The absence of public controversy over the allowance of patents on DNA sequences stands in marked contrast to public reactions to the allowance of patents on microorganisms, *Diamond v. Chakrabarty*, 447 U. S. 303 (1980), animals, Ex parte *Allen*, 2 U.S.P.Q.2d (BNA) 1425 (Bd. Pat. App. & Interf. 1987), computer software, *Diamond v. Diehr*, 450 U.S. 175 (1981), and business methods, *State Street Bank & Trust v. Signature Financial Group*, 149 F.3d 1368 (Fed. Cir. 1998). The issuance of patents in each of these areas promptly provoked outspoken opposition, along with media commentary and Congressional hearings. Similar controversy did not attend the patenting of DNA

challenges to the patentability of DNA seemed quaint and out of touch. Yet something was plainly different. In the early days, patenting genes looked like patenting drugs. By the early 1990s, it looked more like patenting scientific information. We have a reasonably clear story about why we should issue patents on drugs; the case for issuing patents on scientific information is less clear.

Patents on research discoveries arising far upstream from end-product development threaten the interests of research scientists, who fear impediments to the free use and dissemination of new discoveries, and of downstream product developers, who fear that they will be foreclosed from pursuing certain research and development (R&D) pathways or that their profits will be diluted by the claims of upstream predecessors. At the same time, it is not obvious how upstream patent owners might use patents to capture the value that their discoveries contribute to downstream product development, particularly in the face of concerted resistance to sharing the wealth. This strategic challenge is leading upstream innovators to pursue novel patent claiming and licensing approaches that raise unresolved doctrinal and policy questions.

I. THE OLD MODEL: GENES AS PRODUCTS

The first generation of DNA sequence patents, directed toward genes encoding proteins of potential therapeutic value, covered tangible materials used to produce therapeutic products. Such patents typically claimed an "isolated and purified" DNA sequence corresponding to the amino acid sequence for the protein, along with the recombinant materials that incorporated that DNA sequence for use in making the protein in cultured cells.[3] The effect was similar

sequences until the public outcry over NIH patent applications on gene fragments identified in the laboratory of Dr. Craig Venter in the early 1990s, in the early days of the Human Genome Project. *See* Reid G. Adler, *Genome Research: Fulfilling the Public's Expectations for Knowledge and Commercialization*, 257 SCIENCE 908 (1992); Rebecca S. Eisenberg, *Genes, Patents, and Product Development*, 257 SCIENCE 903 (1992); Bernadine Healy, *Special Report on Gene Patenting*, 327 NEW ENG. J. MED. 664 (1992); Thomas D. Kiley, *Patents on Random Complementary DNA Fragments?* 257 SCIENCE 915 (1992).

[2] *See, e.g., Hormone Research Found., Inc. v. Genentech, Inc.*, 904 F.2d 1558 (Fed., Cir. 1990); *Scripps Clinic & Research Found. v. Genentech, Inc.*, 927 F.2d 1565 (Fed. Cir. 1991); *Amgen, Inc. v. Chugai Pharm. Co.*, 927 F.2d 1200 (Fed. Cir. 1991).

[3] *See, e.g.*, U.S. Patent No. 4,757,006 (July 12, 1988), which claims, inter alia:

1. An isolated recombinant vector containing DNA coding for human factor VII:C, comprising a polydeoxyribunucleotide having the [following] sequence:

. . .

4. A non-human recombinant expression vector for human factor VIII:C comprising a DNA segment having the [following] sequence:

. . .

5. A transformed nonhuman mammalian cell line containing the expression vector of claim 4.

to a patent on a drug, although the gene patent was directed to the recombinant materials used in the production process rather than to the pharmaceutical end product itself. The Patent and Trademark Office (PTO) and the courts treated these patents the same way they treated patents on new chemical compounds.[4] The analogy may never have been perfect, but it did the job of providing commercially effective patent protection that motivated investment in the development of new products.

Neither the PTO nor the courts expressed any significant hesitation about the patent eligibility of these first-generation DNA sequence claims, as long as the inventions were claimed in a form that distinguished them from naturally occurring products.[5] The convention of using the words "isolated and purified" in the claim language to distinguish claimed inventions from "products of nature" had already been established before the advent of commercial biotechnology in cases involving purified Vitamin B12,[6] adrenalin,[7] and aspirin.[8] Most lingering doubts about the patentability of biological discoveries were effectively laid to rest by the Supreme Court ruling in *Diamond v. Chakrabarty*,[9] which held that a genetically modified organism was eligible for patent protection. Using very broad language, the Court said that Congress intended to extend patent protection to "anything under the sun that is made by man."[10]

This was important because patents matter to investors in the biopharmaceutical industry. In many other industries, firms report that patents are not very important to R&D investment decisions. They say that other factors matter

[4] *See, e.g.*, *Amgen Inc. v. Chugai Pharmaceutical Co. Ltd.*, 927 F.2d 1200, 18 USPQ 2d (BNA) 1016, 1021 (Fed. Cir. 1991) ("A gene is a chemical compound, albeit a complex one . . .").

[5] *See, e.g.*, *Amgen Inc. v. Chugai Pharmaceutical Co.*, 13 USPQ.2d (BNA)1737, 1759 (D. Mass. 1989) ("The invention claimed in the '008 patent is *not* as plaintiff argues the DNA sequence encoding human EPO since that is a nonpatentable natural phenomenon 'free to all men and reserved exclusively to none.' . . . Rather, the invention as claimed in claim 2 of the patent is the 'purified and isolated' DNA sequence encoding erythropoietin.").

[6] *Merck & Co. v. Olin Mathiesen Chemical Corp*. 253 F.2d 156 (4th Cir. 1958).

[7] *Parke-Davis & Co. v. H.K. Mulford & Co.*, 189 F.95 (S.D.N.Y. 1911), *aff'd*, 196 F.496 (2d Cir. 1912).

[8] *Kuehmsted v. Farbenfabriken*, 179 F. 701 (7th Cir. 1910).

[9] 447 U.S. 303 (1980).

[10] *Id.* at 309 (citing S. Rep. No. 82-1979, at 5 (1952); H.R. Rep. No. 82-1923, at 6 (1952)). Even in the early 1980s, when the courts were still wary of protecting information technology, *e.g.*, *In re Grams*, 888 F.2d 835 (Fed. Cir. 1989); *In re Meyer*, 688 F.2d 789 (CCPA 1982), they viewed tangible DNA as a chemical, or "composition of matter" in the language of the patent statute, 35 U.S.C. §101, rather than as information. Perhaps if the PTO and courts had understood DNA to be a storage medium for information, they would have hesitated before bringing it within the patent system's expanding tent. Instead, paradoxically, for awhile it was far easier to patent *nature*'s information technology than it was to patent *human-made*, electronic information technology. *See* Rebecca S. Eisenberg, *Re-examining the Role of Patents in Appropriating the Value of DNA Sequences*, 49 EMORY L. J. 783 (2000).

more in determining the profitability of innovation, such as being first to market, and that patents are primarily trading currency to use in cross-licensing negotiations to get other patent-holders to leave them alone.[11] In the pharmaceutical industry, one hears a very different story. Firms report astronomical (and growing) costs of new product development, with many costly failures for each successful product.[12] If competitors could enter the market and drive down prices on the successful products without incurring R&D costs on the full range of successful and unsuccessful candidates, pharmaceutical R&D would quickly become unprofitable.

Early biotechnology firms saw themselves as "high tech" pharmaceutical firms developing therapeutic protein products,[13] and they, too, wanted patents to prevent free riders from destroying their profits. Patents on genes promised to provide that protection and allowed these new firms to raise capital in the financial markets, or in some cases, to get major pharmaceutical firms to collaborate with them to develop new products. For these firms, the case for patent protection of DNA sequences was essentially the same as the case for patent protection of drugs.

Some firms still follow the basic business model pioneered by firms like Genentech and Amgen. These firms seek to identify, in the deluge of genomic information emerging from the Human Genome Project, those sequences that encode new therapeutic proteins that might become pharmaceutical products. But these nuggets represent only a small portion of the value of genomic information in contemporary biomedical research and product development, and private firms have accordingly become much more diverse in their research and business strategies. Some of these firms are deploying new patent strategies in an effort to appropriate the value of genomic information for use in future research by others.

[11] *See* Wesley Cohen, *et al.*, *Protecting Their Intellectual Assets: Appropriability Conditions and Why U.S. Manufacturing Firms Patent (or Not)*, National Bureau of Economic Research, Working Paper No. 7552, 2000 (discussing the importance of patents relative to other mechanisms of appropriation across various industries and concluding that patents are particularly important in the pharmaceutical arena); Richard C. Levin *et al.*, Appropriating the Returns from Industrial Research and Development, in 3 BROOKINGS PAPERS ON ECONOMICS ACTIVITY 783 (Martin N. Baily, *et al.*, eds., 1987).

[12] This story is elaborated with facts and figures on the Web site of the Pharmaceutical Research and Manufacturers of America, www.phrma.org. *See also* J. A. DiMasi, R. W. Hansen, & H. G. Grabowski, *The Price of Innovation: New Estimates of Drug Development Costs*, 22 J. HEALTH ECON. 151–85 (2003).

[13] *See generally* Regional Oral History Office, University of California, Berkeley, *Robert A. Swanson: Co-Founder, CEO, and Chairman of Genentech, Inc., 1976–1996* (1996–97), available from the Online Archive of California; http://ark.cdlib.org/ark:/13030/kt9c6006s1/

II. THE NEW MODEL: GENES AS RESEARCH TOOLS

Research that builds upon genomics and bioinformatics platforms can contribute to the development of products that are several steps removed from the information base that helped researchers along the path to discovery. Although the patent system has been generally receptive to these platform discoveries,[14] it is not obvious how to use patents to capture the value that they contribute to future products.

From a business strategy perspective, the challenge for patent owners is to reach beyond their patented research platforms to seize a share of revenues from future end-product sales. Future drug sales are a particularly attractive target, as new drug development has been the most conspicuously profitable endpoint for bioscience research. Perhaps this is why the pharmaceutical industry, although generally favoring a strong patent system, has been notably ambivalent about patents on platform technologies.[15] Although patents on *drugs* make drug developments profitable, patents on the prior "upstream" inventions that explain disease pathways and mechanisms and identify potential drug targets impose costs on drug development. They threaten to become so many siphons at the feeding trough of new drugs, draining away profits in different directions.

From a public policy perspective, the challenge is how to allocate intellectual property rights along the complex course of cumulative innovation in biomedical research. Too much protection for upstream innovations could dilute incentives to develop new products, while too little protection could dilute incentives to develop the platform technologies from which future products will emerge. Although there is a rich theoretical literature on the law and economics of cumulative innovation,[16] it may be illuminating to examine the available doctrinal and policy choices in the contemporary context of genomics research. In this patent-sensitive research and commercial environment, institutions are

[14] Lawrence M. Sung, *The Unblazed Trail: Bioinformatics and the Protection of Genetic Knowledge*, 8 WASH. U. J. L. & POL. 261 (2002); Jorge A. Goldstein, *Patenting the tools of drug discovery*, DRUG DISCOVERY WORLD (Summer 2001) at 9.

[15] Rebecca S. Eisenberg, *Intellectual Property at the Public-Private Divide: The Case of Large-Scale cDNA Sequencing*, 3 U. CHI. L. S. ROUNDTABLE 557 (1996); Rebecca S. Eisenberg, *Will Pharmacogenomics Alter the Role of Patents in Drug Development?* 3 PHARMACOGENOMICS 571 (2002).

[16] *See, e.g.*, Ted O'Donoghue, *A Patentability Requirement for Sequential Innovation*, 29 RAND J. ECON. 654 (1998); Jerry R. Green & Suzanne Scotchmer, *On the Division of Profit in Sequential Innovation*, 26 RAND J. ECON. 20 (1995); Howard F. Chang, *Patent Scope, Antitrust Policy, and Cumulative Innovation*, 26 RAND J. ECON. 34 (1995); Richard R. Nelson, *Intellectual Property Protection for Cumulative Systems Technologies*, 94 COLUM. L. REV. 2674 (1994); Suzanne Scotchmer, *Standing on the Shoulders of Giants: Cumulative Research and the Patent Law*, 5 J. ECON. PERSP. 29 (1991).

motivated to pursue their own intellectual property claims aggressively while resisting the claims of others. These ongoing strategic moves and countermoves highlight the policy levers that are in play and offer a window into bargaining between actual upstream and downstream innovators.

III. REACH-THROUGH STRATEGIES

A recurring motif for these strategic interactions begins with the identification of a novel gene or set of genes that appears to be involved in a disease pathway, or that appears to be a potential drug target. The researchers who identify these genes and their relevance seek to establish patent positions that will allow them to claim a share of revenues from future products that engage the mechanism they have described or that bind to the target they have disclosed. There are a number of ways that they might do this, and the law might either facilitate or impede these strategies on grounds of public policy.

A. Reach-through Licenses

Perhaps the best established of these strategies, although still somewhat controversial, is the use of "reach-through" license agreements that provide access to patented research platform technologies in exchange for either royalties on future products that would not otherwise infringe the upstream patent (reach-through royalties) or a promise of an exclusive or nonexclusive license under future patents on inventions to be made by the user (grant-backs).[17] Owners of platform technologies often seek reach-through royalties when the user is a pharmaceutical firm or other institution engaged in product development.[18] Grant-backs are more typical when the user itself is a nonprofit institution, such as a university, that does not develop products but nonetheless generates further intellectual property.[19] Both are primarily contract strategies in which the role of patent law is largely limited to determining background rights. Patent law might nonetheless be deployed either to promote these strategies or to retard them. For example, doctrinal choices that facilitate the patenting of upstream research discoveries, such as minimizing the "utility" standard for patent

[17] For a more extended discussion of negotiations over reach-through license terms, see Rebecca S. Eisenberg, *Bargaining over the Transfer of Proprietary Research Tools: Is This Market Failing or Emerging?* in R. Dreyfuss, H. First & D. Zimmerman eds., EXPANDING THE BOUNDARIES OF INTELLECTUAL PROPERTY: INNOVATION POLICY FOR THE KNOWLEDGE SOCIETY (Oxford University Press 2001): 223–249 (hereinafter *Bargaining Over Research Tools*).

[18] *Id.* at 243. *See, e.g., Bayer AG v. Housey Pharmaceuticals, Inc.*, 228 F. Supp. 2d 467 (2002).

[19] *Bargaining Over Research Tools*, *supra* note 17, at 244.

protection,[20] impel more users of research tools to negotiate for licenses. Conversely, patent law could strengthen the hand of tool users in their negotiations with patent owners by excluding the activities of researchers from patent infringement liability,[21] or by forbidding the use of reach-through licenses as a form of patent misuse.[22]

Although reach-through terms are increasingly common in agreements proposed by research tool owners, many tool users object to these terms. The pharmaceutical industry has been particularly stalwart in its resistance to this strategy, preferring to pay fixed, up-front fees for access to research tools rather than agreeing to share the wealth in potential future products that might generate billions of dollars in sales.[23] Pharmaceutical firms will go to some trouble to avoid incurring reach-through royalty obligations, including recreating unpatented materials, using a slightly different sequence to avoid infringing a narrow patent, or even going offshore to screen drugs against patented targets.

Cash-poor biotechnology firms are more likely than large pharmaceutical firms to agree to reach-through terms.[24] This may be because they lack the resources to pay up front or to invent around patents, or because they have a more free-wheeling culture that tolerates the infringement of patents on proprietary research tools, only pausing to seek licenses for platform technologies that play a conspicuous role in their research. Perhaps firms that have not yet grown accustomed to living off of rents on patented drugs find it easier to agree in the abstract to divert some of those rents to the coffers of other institutions, or perhaps they expect to transfer the payment obligation to a major pharmaceutical collaborator if and when they identify a promising product.

[20] *See, e.g., In re Brana*, 51 F.3d 1560 (Fed. Cir. 1995) (reversing a rejection for lack of utility of a patent claim directed to compounds for use as antitumor substances, notwithstanding the failure of the specification to show that the claimed compounds would work for this purpose and noting that "[u]sefulness in patent law, and in particular in the context of pharmaceutical inventions, necessarily includes the expectation of further research and development.").

[21] In fact, the Federal Circuit took a restrictive approach to the common law "experimental use" defense in its recent decision in *Madey v. Duke University*, 307 F.3d 1351 (2002). By holding ineligible for the defense noncommercial academic research in pursuit of a university's mission to "further the institution's legitimate business objectives, including educating and enlightening students and faculty participating in these projects," the Federal Circuit effectively strengthened the hand of patent owners and promoted the licensing of proprietary research tools. But *cf. Integra Lifesciences v Merck KGaA*, 331 F.3d 860, 872 (opinion of Newman, J. concurring in part and dissenting in part) (endorsing research exemption for "research conducted in order to understand or improve upon or modify the patented subject matter").

[22] *See, e.g., Zenith Radio Corp. v. Hazeltine Research, Inc.*, 395 U.S. 100 (1969) (holding that conditioning the grant of a patent license upon payment of royalties on unpatented products is patent misuse). *Cf. Bayer AG v. Housey Pharmaceuticals, infra* notes 28–30 and accompanying text (reach-through royalties permissible on a voluntary basis for the mutual convenience of the parties).

[23] *Bargaining Over Research Tools, supra* note 17, at 243 and notes 61–62.

[24] *Id.* at 245 note 69.

If reach-through royalties are anathema to pharmaceutical firms, grant-backs are anathema to universities when they review agreements for access to proprietary research tools.[25] University scientists, like scientists in biotechnology firms, get away with infringing many patents that are unlikely to be enforced against them, and sometimes believe (generally incorrectly) that their activities are exempt from infringement liability.[26] They nonetheless sometimes need to sign agreements to get access to proprietary materials and databases that they cannot duplicate with their own resources, and they rarely are willing or able to make the kind of up-front payments for these platform technologies that pharmaceutical firms use to avoid reach-through obligations. Reach-through royalties make little sense as an alternative payment system for university researchers, who are unlikely to develop products that will generate a payment stream. Instead, owners of proprietary research platform technologies may propose grant-back terms that promise the tool owner a license under future patents. Such precommitted licenses are a source of great dismay to licensing specialists within universities. From their perspective, university researchers who promise to grant future licenses to providers of proprietary research tools are doing "free research" for the tool owners.[27] Moreover, *ex ante* promises of licenses to use future discoveries compromise the stewardship of the university over these discoveries once they are made. Exclusive licenses under future patents are the reward that universities offer to sponsors that pay the full costs of research, not to providers of mere research tools. Provisions for options to take exclusive licenses under future patents quickly collide with each other if the same researcher uses resources from more than one source offered on similar terms. Even promises of nonexclusive licenses under future patents tie the hands of the university, preventing it from offering more lucrative (and often more commercially viable) exclusive licenses to firms that might otherwise sponsor further research or invest in developing technologies emerging from university-based research.

So far, users of research tools have failed to persuade the courts that the use of reach-through provisions in licenses should be impermissible. In *Bayer AG v. Housey Pharmaceuticals, Inc.*[28] a pharmaceutical firm argued that the owner of patents on screening methods that identify potential pharmaceutical products was misusing its patents by licensing them on terms that required the payment of reach-through royalties on future products that were not themselves covered by the patent claims, but were identified through the use of the patented screening methods. The district court concluded that, although it would be

[25] *Id.* at 244–45.
[26] *See* note 21, *supra*.
[27] *Bargaining Over Research Tools*, *supra* note 17, at 237–38.
[28] 228 F. Supp. 2d 467 (2002).

patent misuse for a patentee to "condition" a license on the payment of royalties on unpatented products and activities,[29] reach-through royalty terms are nonetheless permissible for the mutual convenience of the parties, where the evidence indicates that the patent holder was willing to consider other payment options.[30]

In sum, for a variety of reasons, patent owners who seek to use reach-through licenses to leverage patents on upstream research tools into a share in the value of downstream products have encountered significant resistance in the market for licenses. So far, however, the courts have not precluded tool owners and users from voluntarily using reach-through terms in license agreements.

B. Reach-through Remedies

A related, but distinct, question is whether courts should award reach-through royalties on products developed through the unlicensed use of patented research tools as a remedy for patent infringement. This approach has been favored by some commentators[31] but not yet sanctioned by the courts, although patent owners are pursuing various ways of framing the argument.[32]

Professor Janice Mueller has proposed changing the law to provide a reach-through royalty as a remedy for "developmental use" of patented research tools by users who give notice to patent holders.[33] The goal of this proposal is to bypass the transaction costs associated with *ex ante* licensing and to facilitate prompt access to research tools while ensuring an appropriate return to the patent holder. In effect, the proposal amounts to compulsory licensing of research tools, with a payment obligation measured by a reach-through royalty.

If reach-through royalties become common in licenses to use patented research tools, then such royalties would seem to be an appropriate measure of compensation for past infringement under current law, as an approximation of a

[29] *Bayer AG v. Housey Pharmaceuticals, Inc.*, 169 F. Supp. 328 (2001) (denying patent holder's motion to dismiss claim of patent misuse on the pleadings).

[30] 228 F. Supp. at 470–71.

[31] Janice Mueller, *No "Dilettante Affair": Rethinking the Experimental Use Exception to Patent Infringement for Biomedical Research Tools*, 76 WASH. L. REV. 1 (2001); James Gregory Cullem, *Panning for Biotechnology Gold: Reach-Through Royalty Damage Awards for Infringing Uses of Molecular Sieves*, 39 IDEA 553 (1999).

[32] *See, e.g.*, *Bayer AG v. Housey Pharmaceuticals, Inc.*, 169 F. Supp. 328 (D. Del. 2001) (rejecting argument that sale in the U.S. of a pharmaceutical product identified abroad through use of U.S.-patented drug screening method infringes the U.S. patent on the drug screening method). *Cf. Trustees of Columbia University v. Roche Diagnostics*, 150 F. Supp 191, 203 (D. Mass. 2001) (rejecting argument that importation into U.S. of noninfringing product made abroad through use of U.S.-patented materials infringes patent under "fruits of the poisonous tree" theory).

[33] Mueller, *supra* note 31, at 58–59.

hypothetical license negotiation between a willing licensor and licensee.[34] But this and other arguments for reach-through royalties as an appropriate measure of compensation to patent holders are hard to sustain in the face of concerted resistance to reach-through royalties in the market for licenses. The compensatory goal of patent damages is not served by an *ex post* measure of damages that research tool users actively resist *ex ante* in negotiating the terms of license agreements.

C. Reach-through Claiming

A more powerful strategy for allowing developers of research tools to capture the value that their tools contribute to future product development is "reach-through claiming."[35] This strategy focuses on the patent itself rather than on license terms, using claim language that is broad enough to cover future products directly. If tool developers can get patents that cover not only the tools themselves but also future products developed through use of the tools, they need not persuade tool users to agree in advance to pay future royalties on product sales, nor do they need to persuade courts to adopt innovative reach-through remedies for past acts of infringement. The claim language will provide the basis for a remedy against ongoing infringement at the time of commercial sale, allowing tool owners to keep future products off the market unless product developers get licenses. If users refuse or neglect to sign licenses at the research stage, they will still have to deal with tool developers before they can profit from product sales.

Patent claims that reach beyond the technological accomplishments of the patent holder are by no means unprecedented. Because the principal constraint on the scope of patent claims is the content of the prior art in the pertinent field of technology,[36] it is common for pioneering inventions that open up new markets to receive broad patents that dominate future advances, including future products that require significant further R&D.[37] An example of an

[34] *Cf. Georgia-Pacific Corp. v. U.S. Plywood*, 318 F. Supp. 1116 (SDNY 1970).

[35] For a description of reach-through claiming and a careful analysis of the doctrinal issues it raises under the patent laws in the U.S., the European Union, and Japan, see *Report on Comparative study on biotechnology patent practices Theme: Comparative study on "reach-through claims,"* (Trilateral Project B3b), posted on the Internet at www.uspto.gov/web/tws/B3b_reachthrough.pdf (visited Sept. 9 2003) (hereinafter "Comparative Study").

[36] 35 U.S.C. §§ 102, 103.

[37] The principal doctrinal constraint on the scope of such patent claims has traditionally been the requirement of an enabling disclosure, 35 U.S.C. § 112, ¶ 1, that teaches those of ordinary skill in the field how to make and use the invention across the scope of the claims without the need for "undue experimentation." *Genentech, Inc. v. Novo Nordisk* A/S, 108 F.3d 1361, 1365 (Fed. Cir.), *cert. denied*, 522 U.S. 963 (1997).

early advance in biotechnology that received broad "reach-through claims" is the Cohen-Boyer gene-splicing technique patented by Stanford University.[38] The patent claims covered not only the enabling technology, subsequently put to use in many different academic and industrial laboratories in a broad range of R&D projects, but also any recombinant organisms created through use of the technology. The claims to recombinant organisms reached through the disclosed technology to cover later-developed starting materials used in recombinant production of proteins, giving the patent owner a dominant claim over a whole generation of biotechnology products. Because the relevant patent rights were licensed widely and nonexclusively on reasonable terms,[39] their validity was never tested in the courts.

In a series of decisions over the past decade, the Federal Circuit has taken a parsimonious approach toward claim scope on a variety of doctrinal fronts, especially in the biotechnology field. Of particular relevance to reach-through claiming strategies, the Federal Circuit has developed a robust "written description" requirement and a skeptical standard for establishing "possession" of claimed inventions that reach beyond an inventor's tangible laboratory accomplishments.[40] Although at least one member of the Federal Circuit has called for the court to reconsider decisions expanding the reach of the written description requirement,[41] this line of cases continues to stand as a formidable obstacle to reach-through claiming strategies. Meanwhile, reach-through claims have become sufficiently common to prompt the patent offices of the United States, the European Union, and Japan to issue a joint report on the legal issues they raise.[42]

[38] U.S. Patent No. 4,237,224 (Dec. 2, 1980). *See* Margaret Young, *The Legacy of Cohen-Boyer*, SIGNALS MAGAZINE (June 12, 1998), posted on the Internet at http://www.signalsmag.com/signalsmag.nsf/657b06742b5748e888256570005cba01/b7367c099e624afe8825662000609d01 (visited Sept. 9, 2003); Joan O'C. Hamilton, *Stanford's DNA patent 'enforcer' Grolle closes the $200M book on Cohen-Boyer*, SIGNALS MAGAZINE (Nov. 25, 1997), posted on the Internet at http://www.signalsmag.com/signalsmag.nsf/657b06742b5748e888256570005cba01/2d348d68e91004988825655b000b4862 (visited Sept. 9, 2003).

[39] Niels Reimers, *Tiger by the Tail*, 17(8) CHEMTECH 464–471 (1987).

[40] *See, e.g.*, *Regents of the University of California v. Eli Lilly*, 119 F.3d 1559, 1566–69 (Fed. Cir. 1997); *Fiers v. Revel*, 984 F.2d 1164 (Fed. Cir. 1993); *Amgen v. Chugai Pharmaceutical Co.*, 927 F.2d 1200 (Fed. Cir. 1991); U.S. Dep't of Comm., Pat. & Trademark Off., Guidelines for Examination of Patent Applications Under the 35 U.S.C. § 112, ¶ 1, "Written Description" Requirement, 66 FED. REG. 1099 (Jan. 4, 2001). *See also* Janice M. Mueller, *The Evolving Application of the Written Description Requirement to Biotechnological Inventions*, 13 BERKELEY TECH. L.J. 615 (1998); Mark D. Janis, *On Courts Herding Cats: Contending with the 'Written Description' Requirement (and Other Unruly Patent Disclosure Doctrines)*, 2 WASH. U. J. L. & POL'Y 55 (2000); Arti K. Rai, *Intellectual Property Rights in Biotechnology: Addressing New Technology*, 34 WAKE FOREST L. REV. 827 (1999).

[41] *Enzo Biochem, Inc. v. Gen-Probe Inc.*, 296 F.3d 1316 (Rader, J., dissenting from denial of rehearing en banc); *Moba, B. V. v. Diamond Automation, Inc.*, 325 F.3d 1306, (Fed. Cir. 2003) (Rader, J. concurring).

Consider a typical scenario in which upstream researchers identify a novel gene. By searching in databases for homologous sequences, they might recognize the gene as encoding a new member of a previously characterized family of receptors that play a role in a particular disease pathway, suggesting that the newly identified gene might be useful in screening for drugs to treat that disease. A patent that claims the novel receptor itself (or the DNA sequence encoding it), perhaps including additional claims covering the use of the receptor in a drug screen, will give the inventor a bargaining chip for negotiating with firms that want to use the receptor in screening for new drugs. But pharmaceutical firms bargain hard to limit their future obligations to owners of such patents. Moreover, they might avoid the patents entirely by completing their screening before the patent issues, or by going offshore to do the job. The patent owner would be in a much stronger position if he or she could obtain claims covering drugs that are identified through the use of the receptor in drug screening.

The principal obstacle to obtaining a claim that would cover these as yet unidentified drugs is the disclosure requirements of the patent laws. Section 112 of the U.S. Patent Act requires that a patent application "shall contain a written description of the invention, and of the manner and process of making and using it, in such full, clear, concise, and exact terms as to enable any person skilled in the art ... to make and use the same...."[43] The Federal Circuit has stressed the distinctness of the "written description" requirement from the "enabling disclosure" requirement, holding that it is not enough to provide an enabling disclosure of how to make a product that is not described in the specification.[44] In a series of cases involving biotechnology product claims, the Federal Circuit has said that the "written description" standard, which serves to ensure that the inventor was "in possession" of the invention as of the patent filing date, requires disclosure of information about the structure of products covered by the claim, not just a description of their function.[45] Claims to ligands of a disclosed receptor, identified only by

[42] *See* Comparative Study, *supra* note 35.

[43] 35 U.S.C. § 112.

[44] *Amgen Inc. v. Hoechst Marion Roussel, Inc.*, 314 F.3d 1313, 1330 (2003); *University of California v. Eli Lilly & Co.*, *supra*, 119 F.3d 1559 at 1568; *Fiers v. Revel*, *supra*, 984 F.2d at 1170–71. Judge Rader, joined by Judges Gajarsa and Linn, has argued that the written description requirement should be coextensive with the enabling disclosure requirement except in cases involving priority disputes. *Enzo Biochem, Inc. v. Gen-Probe Inc.*, 42 Fed. Appx. 439, 445 (Fed. Cir. 2002) (dissenting from denial of re hearing en banc). *See also Moba, B.V. v. Diamond Automation, Inc.*, 325 F.3d 1306, (Fed. Cir. 2003) (Rader, J. concurring). *Cf. id.* at 1327 (Bryson, J. concurring) (noting that nothing in the language of § 112 would justify construing written description and enablement as distinct requirements only in cases involving priority disputes).

their function of binding the receptor with no description of structure, seem to fail this standard. Such a description does not permit visualization of the compounds,[46] but is merely "an attempt to preempt the future before it has arrived."[47]

But research has advanced to the point where some firms have a strategy for addressing this problem. Researchers studying a new receptor might crystallize it and determine the Cartesian coordinates corresponding to its three-dimensional structure using Xray crystallography. Through these methods it might be possible to visualize the receptor on a computer screen, creating a three-dimensional model of the drug target for use in designing drugs.[48]

Information about the size and shape of binding sites on receptors might help patent applicants satisfy the written description requirement for reach-through claims covering potential drugs that fit within such binding sites, even if they have not actually made and screened the compounds. Arguably the disclosure of crystal coordinates and binding sites provides enough structural information about the size and shape of the future drug products, linked to the function of binding the target, to permit visualization of molecules falling within the scope of the claim.[49] In other words, the applicant is not simply claiming any molecules that have the desirable function of binding the receptor with no

---

[45] *University of California v. Eli Lilly & Co., supra,* 119 F.3d 1559 at 1568 ("In claims to genetic material . . . a generic statement such as 'vertebrate insulin cDNA' or 'mammalian insulin cDNA,' without more, is not an adequate written description of the genus because it does not distinguish the claimed genus from others, except by function. . . . It does not define any structural features commonly possessed by members of the genus that distinguish them from others. . . . A definition by function . . . does not suffice to define the genus because it is only an indication of what the gene does, rather than what it is."); *Fiers v. Revel, supra,* 984 F.2d at 1171 ("Claiming all DNA's that achieve a result without defining what means will do so is not in compliance with the description requirement; it is an attempt to preempt the future before it has arrived."). *Cf. Enzo Biochem, Inc. v. Gen-Probe Inc.*, 296 F.3d 1316, 1324–25 (Fed. Cir. 2002) (approving of PTO Guidelines indicating that "functional descriptions" might satisfy the written description requirement "when coupled with a known or disclosed correlation between function and structure").

[46] *See University of California v. Eli Lilly & Co.*, 119 F.3d 1559, 1568 (Fed. Cir. 1997) (generic claim to DNA failed to satisfy written description requirement in that disclosure did not permit one skilled in the art to "visualize or recognize the identity of the members of the genus").

[47] *Fiers v. Revel, supra,* 984 F.2d at 1171.

[48] Patent claims have been issued for methods of identifying candidate inhibitor compounds that involve introducing crystal coordinates for a target into a computer program and superimposing models of inhibitor test compounds to identify those that fit spatially into a binding site of the target. U.S. Patent 6,083,711. *See* Jorge A. Goldstein, *supra* note 14.

[49] *See* Dep't of Comm., U.S. Pat. & Trademark Off., Guidelines for Examination of Patent Applications Under the 35 U.S.C. 112 ¶ 1, "Written Description" Requirement, 66 FED. REG. 1099, 1106 (Jan. 5, 2001) (stating that written description requirement may be met by disclosure of "functional characteristics when coupled with a known or disclosed correlation between function and structure").

disclosure of structure, but is further describing structural features necessary to do the job.

This strategy is fraught with hazards for the patent applicant, both scientific and legal. Biology is full of surprises, and informed speculation about the size and shape of molecules that will bind a receptor based on protein crystal coordinates might prove to be wrong. Proteins fold and unfold in response to different environmental stimuli, and the shape of a receptor on the surface of a cell in the environment in which it encounters a drug may be quite different than the shape found for the crystallized protein in the laboratory. If skilled practitioners would have reasonable doubt about the likely success of the remaining work necessary to identify compounds falling within the scope of the claim, the application will fail the enablement standard, which requires that the disclosure enable those skilled in the art to make and use the full scope of the claimed invention without "undue experimentation."[50]

Even if the patent issues and the inventor's informed speculation about the structure of functional ligands proves correct, prior art may later come to light that renders the patent invalid. Many compounds exist in the prior art, but relatively few have been defined by size and shape in a way that would permit straightforward searches to identify those that fall within the scope of claims defined in terms of fit within a described binding site. A broad claim to a genus of compounds fails to meet the novelty standard if even a single member of the genus was disclosed in the prior art, even if the properties of the prior art compound that make it fall within the scope of the claim were merely inherent and not disclosed.[51] If a prior art compound in fact has the shape and size described in the claim, and therefore would bind the receptor, even though no one knew it, the patent is invalid. Broad claims make big targets.[52]

Although it may be possible for clever lawyers to identify arguments for the patentability of reach-through claims that are not technically inconsistent with precedent, the trend of authority over the past decade is against them.

[50] *Genentech, Inc. v. Novo Nordisk A/S*, 108 F.3d 1361, 1365 (Fed. Cir.), *cert. denied*, 522 U.S. 963 (1997).

[51] *Titanium Metals Corp. of America v. Banner*, 778 F.2d 775 (Fed. Cir. 1985).

[52] This is, of course, a risk for all broad claims, not just for reach-through claims. But reach-through claims to compounds having a specified shape as determined through computer-visualization of a target receptor are particularly vulnerable to this risk because an inventor who is drafting claim language on the basis of informed speculation without any confirming results in the lab is unlikely to have narrower claims to fall back on. By contrast, an inventor who has made several members of a family of compounds and seeks a broad generic claim to the entire family might still be able to retain valid claims to particular disclosed species even if subsequently identified prior art renders the broad claims invalid.

A recent district court decision in the case of *University of Rochester v. G.D. Searle & Co.*[53] may be a harbinger of likely judicial reception of reach-through claiming strategies. The patent at issue arose out of the discovery by scientists at the University of Rochester that stomach irritation associated with non-steroidal anti-inflammatory drugs is caused by the inhibition of a protective enzyme (PGHS-1 or Cox-1) that is distinct from a similar enzyme (PGHS-2 or Cox-2) that causes inflammation. They hypothesized that molecules that selectively inhibit Cox-2 only, without inhibiting Cox-1, might provide relief from pain and inflammation while avoiding these side effects. Without identifying or testing any such molecules, the university obtained a patent on a "method for selectively inhibiting PGHS-2 in a human host, comprising administering a non-steroidal compound that selectively inhibits activity of the PGHS-2 gene product to a human host in need of such treatment."[54]

As soon as the patent issued, the university brought a patent infringement action against pharmaceutical firms that had developed selective Cox-2 inhibitor products. The defendants moved for summary judgment of patent invalidity for failure to comply with the written description and enablement requirements, and the district court held the patent invalid on both grounds. The court noted that "the claimed method depends upon finding a compound that selectively inhibits PGHS-2 activity." Having failed to describe such a compound other than by its function, the plaintiff failed to show that they were in possession of their claimed method and thus failed to meet the written description standard: "[I]t is, as defendants argue, akin to 'inventing' a cure for cancer by utilizing a substance that attacks and destroys cancer cells while leaving healthy cells alone. Without possession of such a substance, such a 'cure' is illusory, and there is no meaningful possession of the method."[55]

Given the lack of detail in the patent as to how to select suitable compounds, the court concluded that one would have to engage in undue experimentation, with no assurance of success, in order to practice the invention, and the patent thus also failed to meet the enablement standard:[56] "At most, its description will enable a person of ordinary skill in the art to *attempt to discover* how to practice the claimed invention. That is not enough."[57]

Although this district court decision does not carry the authority of decisions of the Federal Circuit, it is carefully reasoned and broadly consistent with prior decisions of that court.

[53] U.S. Dist. LEXIS 3030 (W.D.N.Y. 2003).

[54] U.S. Pat. No. 6,048,850 (issued April 11, 2000).

[55] 2003 U.S. Dist. Ct. LEXIS 3030, 31–32.

[56] The court did not reach the separate argument that the patent failed to enable the claimed method of treating humans because it does not sufficiently disclose specific dosages, formulations, or routes of administration. 2003 U.S. Dist. LEXIS 3030 at 45 n.10.

[57] *Id.* at 53.

IV. NORMATIVE ASSESSMENT OF REACH-THROUGH MECHANISMS

Reach-through licensing, reach-through remedies, and reach-through claims are each mechanisms by which upstream inventors might capture some of the value of downstream products. Each of these mechanisms depends upon a constellation of legal rules (contract, patent, antitrust) that might be fine-tuned to promote, permit, discourage, or prohibit its use, depending upon how the legal system approaches these reach-through practices as a normative matter. Cogent arguments can be and have been advanced both in favor of these practices and against them.

Reach-through proponents might argue that, in the absence of mechanisms for sharing the value of downstream discoveries, the stand-alone value of upstream innovations may be too low, leaving too little motivation for investing in technology platforms for use in further research.

In theory, owners of upstream inventions could recover value from users through pay-as-you-go licenses without any reach-through provisions, just as owners of patented building tools recover value from users who buy tools in a hardware store without seeking a share of the value of future construction projects. Some patented reagents are available in anonymous market transactions that are analogous to the purchase of a hammer in a hardware store, and pharmaceutical firms bargain for similar pay-as-you-go terms in negotiations over access to research tools. But it is often difficult for owners and users to agree upon a fair value for the use of tools in a research project, and some users who have promising research plans may lack the resources to make substantial up-front payments. Reach-through provisions might be useful both as a valuation formula and as a financing mechanism, permitting transfers of research resources when the parties otherwise would be unable to settle on mutually agreeable terms.

Disagreements about value are pervasive in negotations over the transfer of research tools. By its nature, research involves investigation of the unknown, and exchanges of research tools thus necessarily involve imperfect information. This problem is aggravated for new research tools, yet prompt dissemination of cutting-edge research platforms may be critical to the pace of scientific advance. Owners and users may differ in their assessments of the likely outcome of a proposed project using a new tool. The best information about the value of a tool to a particular user is likely to be divided between the owner of the tool (who has better information about the capabilities and limitations of the tool itself) and the prospective user (who has better information about the contemplated research project). These parties may be reluctant to share information with each other and skeptical of each other's purported assessments. Even after the conclusion of a successful R&D project, the parties may disagree about whether or to what extent the outcome was accelerated or enhanced by access to a particular research tool. Reach-through provisions might be a way of resolving these differences by making payment obligations contingent on agreed markers of proven value, such as future product sales.

The biomedical research community has become quite diverse, and its members differ greatly in the resources at their disposal to pay for research tools. Pharmaceutical firms and some large biotechnology firms are relatively prosperous, while universities and start-up firms are often cash-poor. These differences in ability to pay make it difficult to standardize terms across transactions and caution against prohibiting terms that some tool users might prefer given their resource constraints. Reach-through royalty provisions might be attractive to some cash-poor users as a way of deferring payment until they begin earning money from a successful research project.

Grant-back provisions in licenses might facilitate exchanges of research tools between scientific or commercial rivals or other parties that do not trust each other. Such provisions protect the provider of a research tool from future intellectual property claims that might otherwise be asserted by the user, in effect forcing mutual forbearance from future assertion of patent rights. Nonexclusive grant-back provisions might even be used to create a form of commons among institutions that share research tools, analogous to the "copy left" strategies of the open source movement in the software field. If users of proprietary research tools commit themselves to making their future inventions nonexclusively available to prior tool providers, they thereby disable themselves from granting exclusive licenses to these inventions in the future, promoting widespread dissemination of incremental advances in the course of cumulative research. Indeed, once exclusive licensing is precluded, subsequent innovators may have little incentive to bother filing patent applications on future discoveries.

Reach-through mechanisms that do not require an *ex ante* agreement, such as reach-through patent claims that dominate future products or reach-through royalties as a remedy for the past use of a tool in developing a product, may be a way of reducing transaction costs that impede prompt dissemination of new research tools, permitting research to proceed without the need for constant negotiations over access to each proprietary input.[58] These mechanisms conserve on transaction costs by making it unecessary to reach agreement before research proceeds, confining the need for bargaining to research that yields commercially successful products. At that point, product-developing firms must agree to share the wealth with prior contributors or face infringement claims.

Each of these reach-through mechanisms may also be criticized on a number of grounds. In the context of biomedical research, giving reach-through royalties on blockbuster drugs to the many upstream inventors whose discoveries have contributed to understanding disease pathways might overreward upstream

[58] Professor Mueller has elaborated this point. *See* J. Mueller, *supra* note 31.

innovators at the expense of those who make more substantial investments in carrying research forward to the point of developing a useful product. More generally, reach-through royalties impose a tax on product development, diluting incentives of downstream innovators to reward upstream innovators who may have no continuing involvement in the project. Such an allocation might undermotivate downstream R&D.

Reach-through licenses might raise antitrust concerns by permitting the leveraging of patents on early discoveries into control of future inventions. This concern seems more compelling for grant-back provisions than for reach-through royalties. Grant-back provisions do not merely divide the wealth, but they could also extend the monopoly power of an early innovator by allowing it to capture the patent rights of subsequent innovators who might otherwise become competitors. Grant-back provisions could also aggravate "anticommons" problems by continuously augmenting the number of rights-holders at the bargaining table as cumulative research proceeds.[59]

How one balances these competing concerns depends on how one views the relative need to fortify incentives at different points in the course of cumulative innovation. Those who worry more about the adequacy of incentives for early stage innovation and less about the adequacy of incentives for later stage innovation are likely to favor reach-through mechanisms that help early innovators capture some of the value of later innovations. Conversely, those who worry more about the adequacy of incentives for downstream research and product development are likely to view reach-through mechanisms as unwise.

One reason to worry more about the adequacy of incentives for upstream research is that upstream research results are less likely to have a high stand-alone value than downstream research results,[60] which by definition are closer to market. Issuing broad patent rights to upstream research performers, or otherwise permitting them to use their intellectual property to reach through to the profits from downstream product development, ensures that they recover the full social value of their inventions, including the value that they contribute to subsequent inventions that might be more directly profitable. This argument supports the time-honored approach of the patent system to reward pioneers in a field with broad patent rights, while giving only narrower claims to those who make follow-up improvements. Although the improvements may have more stand-alone commercial value than the primitive versions of the invention developed by the pioneer, the pioneer has taken greater risk and shown more

[59] *See* Michael A. Heller & Rebecca S. Eisenberg, *Can Patents Deter Innovation? The Anticommons in Biomedical Research*, 280 SCIENCE 698 (1998).

[60] This analysis is cogently elaborated in the work of Suzanne Scotchmer. See sources cited in note 16, *supra*.

ingenuity by opening up a new field. Giving pioneers broad patents allows them to force subsequent improvers to negotiate for licenses, thereby capturing for pioneers some of the follow-on value created by those who merely tweak inventions to make them more marketable.

In the context of biomedical research, however, competing intuitions tug in the other direction. Upstream biomedical research tends to be relatively cheap and gets more support from public funding, while downstream product development—especially drug development, which is subject to costly regulation and characterized by many product failures—is relatively costly and risky and relies heavily on private funding. Many of the path-breaking, pioneering discoveries in upstream biomedical research that paved the way for lesser (but more financially viable) discoveries were paid for by National Institute of Health (NIH), calling into question the need for heightened intellectual property incentives to motivate this upstream research.[61] In recent years, some private firms have tried to figure out business models for generating biomedical research information to provide platforms for downstream drug discovery, especially in genomics and bioinformatics. But often by the time private firms see such an opportunity, the upstream research has become relatively routine and mechanical and is no longer pioneering.

For example, when Celera decided to embark upon a private sector venture to complete the human genome sequence, the publicly funded Human Genome Project had already completed much of the path-breaking work that brought this goal within reach.[62] Considerable work remained, but there was little question that the job could be done. In this context, the "upstream" work of sequencing the genome looked relatively routine, low risk, and mechanical compared to the "downstream" work of figuring out the significance of the information in understanding biology and disease and using it to develop new diagnostic and therapeutic products.

Concerns that the patent system may overreward routine upstream DNA sequencing are aggravated by the apparent collapse of the nonobviousness standard for determining the patentability of DNA sequences in the Court of Appeals for the Federal Circuit.[63] (Indeed, the fact that private investors chose

[61] Although since passage of the Bayh-Dole Act of 1980 Congress has generally promoted patenting the results of government-sponsored research, the original purpose of this policy initiative was not to improve incentives to do upstream government-sponsored research, but to improve incentives to transfer resulting inventions to the private sector for downstream commercial development. *See* Rebecca S. Eisenberg, *Public Research and Private Development: Patents & Technology Transfer in Government-Sponsored Research*, 82 VA. L. REV. 1663 (1996).

[62] Justin Gillis & Rick Weiss, *Private Firm Aims to Beat Government to Gene Map*, The Washington Post, May 12, 1998, § 1 at 1 (LEXIS News Library); Nicholas Wade, *Scientist's Plan: Map All DNA Within 3 Years*, N.Y. TIMES, May 10, 1998, Section 1, p.1, col. I (Lexis); J. Craig Venter *et al.*, *Shotgun Sequencing of the Human Genome*, 280 SCIENCE 1540 (1998).

[63] *In re Bell*, 991 F.2d 781 (Fed. Cir. 1993); *In re Deuel*, 51 F.3d 1552 (Fed. Cir. 1995).

to race with an ongoing government-sponsored upstream research initiative rather than simply welcoming the upstream subsidy and reserving their own resources for the pursuit of downstream research goals is itself an indication that the patent system may have been overrewarding the upstream research.) These factors suggest greater cause for concern about motivating and rewarding downstream research than upstream research.

Other factors arguably call for greater attention to the incentives of upstream biomedical researchers. First is the brute fact that the biotechnology industry, already in its third decade, remains, for the most part, unprofitable, while the pharmaceutical industry has been extremely profitable over the same time period. Perhaps the pharmaceutical firms are smarter about business than the biotechnology firms, but perhaps the gross disparity in bottom lines reflects in part a failure of the biotechnology industry to capture for its own shareholders the social value that it has contributed to the pharmaceutical industry. Although the limited profitability of the biotechnology industry has not—so far—prevented new firms from attracting investment capital, the apparent gap between the social value of biotechnology and the profitability of firms that invest in this technology cautions against further constraining the intellectual property strategies available to these firms. Reach-through license agreements might help biotechnology firms that specialize in the development of upstream research platform technologies to keep more of the value that they are creating.

Second, although not all observers agree,[64] some believe that the biomedical research community suffers from persistent bargaining failures over the terms of access to research tools. To the extent that the community finds it difficult to arrive at mutually agreeable terms of exchange for research tools, it makes little sense for the law to foreclose options that might help the parties get to "yes." Reach-through provisions might help the bargaining parties resolve disputes about valuation and finance the license through deferred payments that come due if, and only if, the research yields a successful product.

The deferred financing aspects of reach-through royalties are likely to be particularly beneficial to cash-poor biotechnology firms that are otherwise at a considerable disadvantage in competing against pharmaceutical firms for access to research tools on a pay-as-you-go basis. In effect, reach-through royalty provisions allow upstream and downstream biotechnology firms and universities to share the risks and rewards of research without draining cash in the upstream stages of R&D.

[64] *See* John Walsh, Ashish Arora, and Wes Cohen, *Patenting and licensing of research tools and biomedical innovation*, in S. Merrill, R. Levin and M. Myers, eds. INNOVATION IN A KNOWLEDGE-BASED ECONOMY. (forthcoming National Academies Press).

Reach-through provisions in the form of license grant-backs—*i.e.*, precommitments to license future discoveries back to the provider of an upstream research platform—are more troubling, but they may also be a valuable contract option. Although they raise potential antitrust concerns by allowing early innovators to exercise continuing control over subsequent research and perhaps to suppress new innovation, grant-back provisions may be a necessary defensive maneuver to permit early sharing of research tools without worrying about facilitating future domination by a competitor.

In sum, although reach-through terms are currently encountering some resistance in the market for licenses to patented research tools, particularly from pharmaceutical firms and universities, sometimes the parties have good reasons for agreeing to such terms. In this evolving market, it seems unwise to stretch the limits of antitrust law in order to foreclose the use of such terms in license agreements.

On the other hand, market resistance to reach-through royalties in the terms of license agreements cautions strongly against adopting reach-through royalties as a remedy for patent infringement. Such royalties dilute incentives for downstream product development. Moreover, reach-through royalties calculated in the course of an infringement action after the research has yielded a commercial product may tend to be higher than royalties calculated in the course of negotiations between the parties before the research has taken place when its outcome is highly uncertain. The law should follow, not lead, actual contracting practices, awarding reach-through royalties as a measure of damages if and when they become common as a license term for research tools. At that point, the terms of actual reach-through licenses negotiated *ex ante* will provide a benchmark to guard against overvaluation *ex post*.

Reach-through claims that give upstream research performers dominant patent rights over future products raise all of the hazards of reach-through damage awards and more. Approved by PTO examiners in the course of *ex parte* prosecution, they are not tied even to a hypothetical agreement between the parties about the value of a tool in future product development, nor does the examiner have the benefit of hindsight in determining that the tool was in fact critical to downstream R&D. Downstream product development increasingly draws on multiple upstream research tools, raising the prospect that future pharmaceutical products will be dominated by multiple overlapping claims if reach-through claiming strategies become more commonplace. Such claims would dilute incentives for product development and jeopardize the introduction of new products by giving multiple patent owners hold-up rights.

So far, the PTO and the courts have shown a sensible skepticism toward reach-through claims sought for genomic discoveries. Although the written description requirement has emerged in recent years as a principal obstacle to reach-through claims in this setting, and the limits of this requirement are not yet clear, skepticism toward such claims is broadly consistent with a long-standing antipathy of the courts toward patent claims that dominate fields in which considerable inventive work remains to be done.[65] The enablement standard tends to limit the reach of patent claims to those variations and enhancements of an inventor's actual accomplishments that can be achieved through future work that is reasonably routine and predictable. This is a wise limitation that permits patent owners to share the value of incremental enhancements while preserving incentives for future innovators to profit from more ambitious (and nonobvious) advances. The contemporary relationship between advances in genomics and the development of future pharmaceutical diagnostic and therapeutic products does not call for a departure from this cautious approach.

V. CONCLUSION

The primary commercial value of recent advances in genomics and bioinformatics probably lies in the contributions that these advances will make to the development of future diagnostic and therapeutic products. It is not obvious how to use patents on the upstream advances themselves to capture a share of the value of these downstream products, nor is it clear as a matter of policy how this value should be allocated. Reach-through terms in license agreements for the use of proprietary research tools have had a mixed reception in the market so far, but owners and users of research tools should not be precluded from using this mechanism to apportion the value of incremental innovations. Reach-through strategies that are endorsed in the market in the terms of consensual license agreements are more likely to reflect an appropriate balance between rewards and incentives to upstream and downstream innovators than reach-through remedies awarded by courts for past acts of infringement or reach-through claims allowed by the PTO in the course of *ex parte* patent examination. Legal provision of reach-through rights should follow indications in the market that such an allocation of rewards is appropriate, not lead the way.

[65] *O'Reilly v. Morse*, 56 U.S. 62 (1854) (telegraphy); The Incandescent Lamp Patent, 159 U.S. 465 (1895) (electric light bulbs).

11

The Human Genome Project in Retrospect

Michael Abramowicz
Associate Professor of Law
George Mason University
Arlington, Virginia 22201

0065-2660/03 $35.00

I. INTRODUCTION

In his will dedicating funds to the prizes that would bear his name, Alfred Nobel specified that the prizes would reward the best research conducted in the preceding year.[1] The absence of delay may seem odd to those familiar with the modern operation of the Nobel Prizes,[2] but Nobel may have hoped that the prospect of an immediate prize would help stimulate investments in research. Even if that is so, though, why would Nobel seek to offer rewards at the end of each year instead of providing funding at the beginning of a year? Modern institutions funding scientists, after all, generally fund those with the strongest proposals for future research, not those whose work in the preceding year was the strongest. The answer may be that Nobel was not thinking about maximizing research, or he was recognizing that the Prizes' prestige would dwarf their dollar value. Regardless of Nobel's thoughts, however, I wish to use the topic of this symposium, the Human Genome Project, to advance and consider a proposal for the funding of science that is a variant of Nobel's approach. In particular, I will propose scientific grants that depend primarily or even exclusively on researchers' performance in the period immediately prior to awarding the grants.

Some have argued that scientific funding should have more of a retrospective element.[3] After all, grant institutions do consider the reputation of scientists to some degree.[4] To show how the approach that I consider would differ from proposals focusing on reputation, let us consider the design of this symposium. F. Scott Kieff and Charles McManis have done an admirable job of assembling a distinguished group of lawyers and scientists to discuss the Human Genome Project, and I am honored and delighted to be included. Presumably, the selections were made largely on the basis of the participants' reputation to the selectors, and we have all been wooed with honoraria, good company, and a swank hotel. A different approach might have been to issue a call for paper proposals and to select the best among them. But the approach that I am considering would be different from both of these, and it would work only if the symposium were a repeated game. Under this approach, the honorarium that I would receive for the next symposium would depend on my performance at this

[1] *See* Code of Statutes of the Nobel Foundation § 1 (June 29, 1900), *reprinted in* Elisabeth Crawford, THE BEGINNINGS OF THE NOBEL INSTITUTION: THE SCIENCE PRIZES, 1901–1915, at 221 (1984).

[2] The statutes implementing the Nobel Foundation interpreted Nobel's stipulation so liberally as to make it a nullity. *Id.* at 222–23.

[3] *See, e.g.*, Rustum Roy, *Peer Reviewed Productivity-Based Formula for Funding University Research*, 22 MINERVA 316 (1984) (offering a formula that would be used to determine funding of university departments, based on factors such as past publications).

[4] For example, the National Science Foundation has long considered researcher competence as one of many factors in grant proposals. *See, e.g.*, NATIONAL SCIENCE FOUNDATION, GUIDE TO PROGRAMS FY 1991, at ix (1991).

one, and my performance at that one would determine my funding for the one after that. The symposium, of course, is not a repeated game, but scientific research often is.

Perhaps the most obvious benefit of a retrospective grant institution is that it would save transactions costs; no proposals need to be prepared, and only those who have performed research will need to explain the significance of their accomplishments. While addressing this briefly, I will focus on a different point, that the proposal will increase innovation in scientific research. There are two foundations in the literature for my quest to identify an institution that can accomplish this task. First, the ongoing debate on the appropriateness of patenting the human genome,[5] which is prominent in many of the discussions at this symposium, emphasizes the tension between the incentive benefits of intellectual property regimes and the provision of research for the public domain. Second, one legal scholar, Thomas McGarity, has offered a detailed administrative law assessment of agencies that award grants through peer review.[6] The most significant problem that McGarity finds with peer review is that it often closes off funding to mavericks who wish to challenge accepted orthodoxies, thus limiting the possibility of significant scientific advances.[7] The retrospective grant institution that I describe is a cross between a patent regime and a traditional grant regime, involving centralized provision of funds while providing incentives for original and risky approaches.

Perhaps the strongest rhetorical attack against my argument would be to note the danger of a slippery slope. If it is advisable to provide grants on the basis of past performance, why not switch all the way to a reward system?[8] Indeed, my support for retrospective grants is based in part on the observation that grant money can be used as a way to reward private organizations for taking financial risks. Perhaps the strongest positive explanation for the absence of

[5] Significant contributions to this literature include, Rebecca S. Eisenberg, *Patents and the Progress of Science: Exclusive Rights and Experimental Use*. 56 U. CHI. L. REV. 1017 (1989); F. Scott Kieff, *Facilitating Scientific Research: Intellectual Property Rights and the Norms of Science—A Response to Rai & Eisenberg*, 95 NW. U. L. REV. 691 (2001); Michael A. Heller & Rebecca S. Eisenberg, *Can Patents Deter Innovation? The Anticommons in Biomedical Research*, 280 SCIENCE 698 (1998); and Arti Kaur Rai, *Regulating Scientific Research: Intellectual Property Rights and the Norms of Science*, 94 NW. U. L. REV. 77 (1999).

[6] Thomas O. McGarity, *Peer Review in Awarding Federal Grants in the Arts and Sciences*, 9 HIGH TECH. L.J. 1 (1994).

[7] *Id.* at 39–42. McGarity considers what he terms "radical alternatives to peer review," including review based on publication history. *See supra* note 3 and accompanying text.

[8] Proposals for reward systems include Steve P. Calandrillo, *An Economic Analysis of Intellectual Property Rights: Justifications and Problems of Exclusive Rights, Incentives, and the Alternative of a Government-Run Reward System*. 9 FORDHAM INTELL. PROP. MEDIA & ENT. L.J. 301 (1998); Michael Kremer, *Patent Buyouts: A Mechanism for Encouraging Innovation*, 113 Q.J. ECON. 1137 (1998); Michael Polanvyi, *Patent Reform*, 11 REV. ECON. STUD. 61 (1944); and Steven Shavell & Tanguy van Ypersele, *Rewards Versus Intellectual Property Rights*, 44 J.L. & ECON. 525 (2001).

prize and retrospective grant institutions is an assumption that scientific organizations cannot be subject to risk, that they must know their funding in advance. This assumption is bizarre in a world in which private parties routinely invest resources without assurance of profit, and I will argue against the notion that science is unique in needing government to pick winners in advance. Given this argument, a prize system is just one obvious step beyond the retrospective grant institution, with money being used to reward past science with no requirement that it be used to further future science.

I would be willing to slide down the slippery slope. Indeed, I have argued in a separate paper that prize or reward systems might sometimes serve as complements to a patent system.[9] One virtue of a reward system relative to the retrospective grant institution is that decisions on rewards need not be made at frequent intervals. Thus, a project can be judged at its completion or even later, with extra time serving to improve the prize givers' abilities to assess the significance of a particular contribution.[10] A strike against a prize system is that it is quite different from any existing legal institution. The ultimate purpose of this paper is to show that several of the ideas underlying my advocacy of a prize system could be implemented without complete adoption of a retrospective prize system.

The essay will proceed as follows. Part I will recount the administrative history of the Human Genome Project, focusing on the debates about how the Project should be structured. My task is primarily descriptive, but my intent in offering this account is not to determine what the relevant questions are. To the contrary, I hope to show that those engaged in the debates considered a relatively narrow set of options and made questionable assumptions about what goals the administrative structure of the project should strive to promote. Part II sketches how a retrospective grant institution might work, considers objections, and assesses the extent to which this institution advances the postulated goals. Part III concludes.

II. THE HISTORY, STRUCTURE, AND FUNDING OF THE HUMAN GENOME PROJECT

At first glance, the Human Genome Project might appear to be the perfect model of "big science." This might seem so both in the sense that the Project was successful—the genome was sequenced, after all—and in the sense that the Project provides data on government involvement in big science generally. The

[9] *See* Michael Abramowicz, Perfecting Patent Prizes (2002) (unpublished manuscript, on file with author).

[10] *Id.* at 86–93.

ultimate success of the Project, however, remains unclear. The government's decision to invest in the Project reflected not just a belief that the gene sequence was intrinsically valuable, in the same way that any scientific or artistic project may increase our self-awareness, but also the view that the sequence ultimately would produce valuable remedies and medications. These are mostly in the future and even when they arrive, the question of how much, if at all, the Human Genome Project advanced them will remain. With legal institutions, in contrast to hard science, there are rarely controlled experiments.[11] As isolated data points go, however, the Human Genome Project has much to offer. Let us consider a brief history of the Project, with initial attention to occasions in which intellectual property approaches appeared to pose a challenge to the government-orchestrated Project.

A. A History of Private Challenges to the Human Genome Project

In 1987, as government agencies dithered over whether to undertake the major scientific endeavor of sequencing the human genome or to embark on a less ambitious gene mapping project, Dr. Walter Gilbert decided to break away from the debate.[12] He announced that he would form a private corporation that would sequence the genome, obtain a copyright on the data, and then sell it to researchers.[13] This sparked debate over whether anyone could claim copyright protection for the sequence of the human genome. While his announcement created controversy over both intellectual property rights in genome research[14] and the roles of private and public research,[15] the fervor subsided when he failed to raise funds to launch his private effort.[16] Many believed anyway that the federal government should fund and lead such a daunting exercise in basic scientific research,[17] citing the potential public benefits and the prospect of

[11] Michael Dorf and Charles Sabel have urged a form of government that is inherently experimental. *See, e.g.*, Michael C. Dorf & Charles F. Sabel, *A Constitution of Democratic Experimentalism*, 98 COLUM. L. REV. 267 (1998).

[12] Leslie Roberts, *Controversial From the Start*, 291 SCIENCE 1182, 1184 (Feb. 16, 2001).

[13] Leslie Roberts, *Who Owns the Human Genome*, 237 SCIENCE 358 (July 24, 1987).

[14] *Id.* at 359. At a June 26, 1987, conference titled "Issues of Collaboration for Human Genome Projects," much of the discussion centered on whether Dr. Gilbert could legally copyright the human genome sequence. Those who agreed with Gilbert that copyright protection was appropriate analogized the genome sequence to a computer program or to a series of letters like those in a book. *Id.* Others argued that the copyright would be not only inappropriate, but also unconstitutional. *See* Ira H. Carmen, *Letters: Ownership of the Human Genome*, 237 SCIENCE 1555 (Sept. 25, 1987).

[15] *See* Lorance L. Greenlee, *Letters: Ownership of the Human Genome*, 237 SCIENCE 1555 (Sept. 25, 1987).

[16] Roberts, *supra* note 12, at 1184.

[17] OFFICE OF TECHNOLOGY ASSESSMENT, U. S. CONGRESS, MAPPING OUR GENES—GENOME PROJECTS: HOW BIG? HOW FAST? 93 (1988) [hereinafter OTA REPORT]; *see also* Greenlee, *supra* note 15.

underinvestment by the private sector.[18] But the debate over whether genome research should be public or private did not end with the inception of the publicly funded Human Genome Project.[19]

Dr. Gilbert's early frustration with what would soon be called the Human Genome Project foreshadowed the later actions of J. Craig Venter, who left the National Institutes of Health (NIH) to head a private effort to sequence the human genome using an alternative approach.[20] Venter's exit from NIH spurred the public Human Genome Project to reevaluate its methods and compete with his private effort.[21] This competition resulted in the release of a sequenced human genome by both the Human Genome Project and Venter's Celera Genomics in June, 2000. Though controversy remains over the extent to which Celera relied on public data,[22] there is no doubt that competition led to completion of the draft sequence far ahead of the original schedule.[23] Part of Venter's motivation in breaking away and creating a private endeavor was the prospect of being able to patent the genes he sequenced and gain profits from those patents.[24] This abbreviated history suggests that competition and methodological innovation ultimately benefited the Human Genome Project, but the competition and innovation largely occurred outside of the project itself. That is because the Project itself offered no structural mechanisms to encourage competition and innovation, as we will now see.

B. Structural Options Considered at the Outset of the Human Genome Project

Ten years before the Human Genome Project officially began, academics, government researchers and private researchers entered into nationwide discussion about the possibility of mapping and sequencing the human genome.[25] As it

[18] Lewis D. Soloman & Suzanne E. Schoch, *Developing Critical Technologies: A Legal and Policy Analysis*, 9 SANTA CLARA COMPUTER & HIGH TECH. L.J. 153 (1993).

[19] *See, e.g.*, Mark D. Adams & J. Craig Venter, *Should Non-Peer-Reviewed Raw DNA Sequence Data Release Be Forced on the Scientific Community*, 274 SCIENCE 534 (1996); Keith Aoki, *Authors, Inventors and Trademark Owners: Private Intellectual Property and the Public Domain*, 18 COLUM.-VLA J.L. & ARTS 191 (1994).

[20] Roberts, *supra* note 12, at 1186–88.

[21] *Id.*

[22] *See, e.g.*, Scott Hensley & Antonio Regalado, *Scientists Publish Critique of Celera's Work—Rivals Charge Firm Recycled Public Data in Genome Map*. WALL ST. J., Mar. 5, 2002, at A2.

[23] Eliot Marshall, *Rival Genome Sequencers Celebrate a Milestone Together*, 288 SCIENCE 2294 (June 30, 2000).

[24] Venter has since departed from Celera as the company moves away from basic research provision to pharmaceutical development. Andrew Pollack, *Scientist Quits The Company He Led in Quest For Genome*, N.Y. TIMES, Jan. 23, 2002, at C1.

[25] Leslie Roberts *et. al.*, *A History of the Human Genome Project*, 291 SCIENCE 1195 (Feb. 16, 2001) (noting that the first proposal of a method to map the entire human genome was made in 1980 by scientists at the Massachusetts Institute of Technology).

became clear that mapping and sequencing the genome were possible, the debate shifted to how it should be done. Many advocated a traditional "big science" approach that the Department of Energy (DOE) would head.[26] There had never been such an approach to microbiology and many resisted it, pressing instead for a smaller and decentralized effort, similar to NIH's ongoing support of investigator-initiated research in genome mapping.[27] Still others pressed for a cooperative effort between the DOE and the NIH.[28] The National Resource Council (NRC) and the Office of Technology Assessment (OTA) both presented reports to Congress regarding the structure and funding of a government project to sequence the human genome.[29] These reports recognized the ultimate goals of the Human Genome Project as rapid accumulation of knowledge about the genome, efficient storage and distribution of that information, and conversion of the information gained into both general theories and specific products.[30]

The NRC and OTA together considered five basic types of structures for the Human Genome Project: one agency, single agency leadership, interagency agreement and consultation, interagency task force, and consortium.[31] The creators of the Human Genome Project wanted a structure that would organize communication in the scientific community and would facilitate planning a multi-faceted research program, fostering effective partnerships and effectively allocating funding. OTA recommended that Congress choose a structure for the project based on "perceptions of necessary patterns of authority, of quality and scope of experience in research and development, and of fiscal and economic priorities."[32] As these priorities suggest, participants in the debate seemed to assume that an advantage of relatively more centralized structures is that they would facilitate the establishment of a consistent research agenda. Perhaps this emphasis on consistency makes sense for NASA; if two different scientific teams were to build two different parts of the spacecraft, it is important that the parts be compatible with each other. As Ronald Coase argued in *The*

[26] Leslie Roberts, *Agencies Vie over Human Genome Project*, 237 SCIENCE 486 (July 31, 1987).

[27] James D. Watson, *The Human Genome Project: Past, Present and Future*, 248 SCIENCE 44, 45 (Apr. 6, 1990). Although NIH came later to the game, there was competition between DOE and NIH for control of any government-funded human genome research. Both pursued their own genome-related projects during the debate over genome mapping. Roberts, *supra* note 26, at 487–88.

[28] Roberts, *supra* note 26, at 488.

[29] Roger Lewin, *Genome Projects Ready to Go*, 240 SCIENCE 602 (Apr. 29, 1988); Robert Mullan Cook-Deegan, *Origins of The Human Genome Project*. FRANKLIN PIERCE L. CTR. GENOME PAPERS, *at* http://fplc.edu/risk/vol5/spring/cookdeeg.htm, at 5.

[30] OTA REPORT, *supra* note 17, at 11; COMMITTEE ON MAPPING AND SEQUENCING THE HUMANE GENOME, NATIONAL RESEARCH COUNCIL, NATIONAL SCIENCE FOUNDATION, MAPPING AND SEQUENCING THE HUMAN GENOME (1988) [hereinafter NRC REPORT].

[31] OTA REPORT, *supra* note 17, at 115.

[32] *Id*, at 116, NRC REPORT, *supra* note 30, at 93.

Nature of the Firm,[33] sometimes transactions costs are sufficiently high that it is cheapest to organize production in a single entity. But there is little reason to think that such transactions costs and compatibility issues are relevant for genomics research. The significance of the sequencing of one gene does not depend on the method used for sequencing another. And if different approaches may lead in quite different directions, including, for example, some DNA sequencing and some analysis of proteins, the project is likely to produce greater benefits than if it follows a linear path, because the most valuable work on a subsidiary task may be more valuable than the least valuable work on a principal task. Those arguing about different proposed structures, however, failed to take these considerations into account.

1. One Agency

In the one-agency model, a single expert administrative agency would organize all genomic research.[34] The agency would set all research policy and execute all of the research using government-employed scientists, either within the agency or at the national laboratories. Proponents maintained that this would promote a consistent research policy, without explaining why that would be valuable. A problem with the approach is that once the expert agency chose a method and specialized in it, there would be no incentive for it to find better methods or, if better methods were available, to abandon the original choice. The one agency model failed, however, not for these reasons, but because of significant political drawbacks. Several agencies had already developed expertise in specific areas of research. Projects such as the GenBank and DNA repositories were already inter-agency.[35] The one-agency approach would have eliminated the involvement of all but one of those agencies and consolidated the interagency projects. Because of these problems, NRC, OTA, and Congress never seriously considered the one-agency model. It was quickly "dismissed as unnecessary and politically unworkable."[36]

2. Single-Agency Leadership

Under a single-agency leadership model, Congress would dedicate one agency to act as the coordinating and monitoring agency for all genomic research.[37] Unlike the one-agency model, this structure would not restrict the funding to

[33] Ronald H. Coase, *The Nature of the Firm*, 4 ECONOMICA (n.s.) 386 (1937), *reprinted in* RONALD H. COASE. THE FIRM, THE MARKET, AND THE LAW 33 (1988).

[34] OTA REPORT, *supra* note 17, at 115.

[35] *Id.*

[36] *Id.*; NRC REPORT, *supra* note 30, at 86, 88.

[37] OTA REPORT, *supra* note 17, at 116; NRC REPORT, *supra* note 30, at 94.

one agency, though it would designate one agency to administer the entire project,[38] with other agencies able to compete for funding.[39] The NRC proposed two modified versions of the pure single-agency model. In the one recommended by the majority of the NRC committee, a single agency would lead but a scientific advisory board would advise, appropriately.[40] In the second model, an interagency committee with a scientific advisory board would guide the project, but a single agency would retain control of the daily administration of the project.[41] Those who backed the single-agency leadership models stressed their clear designation of authority and the ability of a single entity to determine a coherent theme and focus for all genome research.[42] Detractors of this approach pointed out not that such consistency might be unnecessary, but that a single agency would not be able to marshal the resources of other agencies and might have a commitment limited to specific areas of genome research.[43]

Perhaps the fatal problem, however, was once again that if this approach were adopted, a particular agency would have had to be selected to take the lead role: DOE or NIH.[44] Neither NRC nor OTA was willing to propose which particular agency should be the one chosen to lead the project.[45] The reports' assessments of the strengths of the two agencies help reveal the apparent priorities. DOE's greatest strength as a choice for lead agency was seen to be its ability to administer large, focused research programs and its demonstrated strong interest in genome research.[46] DOE's apparent weakness was its minimal experience in molecular biology, as its primary interest in genomic research had been limited to energy-related issues.[47] The reports thus seemed to assume that considerable expertise was necessary for an agency to conduct a project, ignoring the possibility that an agency could rely primarily on outside scientists. Meanwhile, although the reports applauded NIH's commitment to traditional

[38] NRC REPORT, *supra* note 30, at 94.

[39] *Id.*

[40] NRC REPORT, *supra* note 30, at 3, 93.

[41] *Id.* at 97.

[42] OTA REPORT, *supra* note 17, at 116.

[43] *Id.*; *see also* Lewin, *supra* note 29, at 603.

[44] Both OTA and NRC included the National Science Foundation (NSF) in their list of potential lead agencies, but neither gave much attention to the possibility. The NRC Report commented simply that the NSF was "involved in the development of technology and instrumentation relevant to the human genome project." NRC REPORT, *supra* note 30, at 94.

[45] OTA REPORT, *supra* note 17, at 118; NRC REPORT, *supra* note 30, at 94.

[46] OTA REPORT, *supra* note 17, at 118; NRC REPORT, *supra* note 30, at 94. DOE had been responsible for major projects in physics including the Manhattan Project. In addition, DOE had aggressively pursued its own human genome project while other agencies expressed less commitment to the idea of such a project.

[47] OTA REPORT, *supra* note 17, at 118. The genesis of DOE's genomic research came from administrators' desire to understand the effects of radiation from things like "dirty bombs" on humans, particularly on their genetic make-up. *Id.* at 6.

research through grants,[48] and although NIH was viewed as a leader in DNA research and had acted as overseer in some larger projects,[49] critics complained that because NIH focused on grants, it might not be able to support a focused effort as massive as the Human Genome Project.[50]

3. Interagency Agreement and Consultation

This type of arrangement "eschews any formal creation of authority and relies on the good will of the participants to exchange information freely," which reflects the "flexible, decentralized organization that are [the] strengths of American science."[51] Unlike the other options, this arrangement did not require new legislation.[52] Each agency involved would theoretically work in its own "institutional area of interest," and principals of the agency would communicate about research goals and developments.[53] Detractors warned that the "interests" of varying agencies would overlap and create duplication and that the potentially divergent priorities of various agencies would lead to a breakdown in communication.[54] There were also concerns that such an informal arrangement would not be appropriate to an expensive long-term effort such as genome mapping, which detractors argued yet again needed clear authority, accountability, and mechanisms to resolve disputes between agencies.[55]

4. Interagency Task Force

The Interagency Task Force model, also called the Interagency Committee model, combined aspects of the single-agency leadership model and the informal agreement and consultation model.[56] A task force made up of principals from participating agencies would have the authority to design and direct the overall genome project.[57] It would also formally gather information, prepare reports, and formulate recommendations and future plans for the project. A task force would be the central point of contact for media and politicians. Detractors worried that the selection of the task force's membership and chairperson might become lengthy political battles.[58] Delegation of research projects themselves

[48] *Id.* at 116–17; NRC REPORT, *supra* note 30, at 94.
[49] NRC REPORT, *supra* note 30, at 94.
[50] OTA REPORT, *supra* note 17, at 117.
[51] *Id.* at 119.
[52] *Id.* at 15.
[53] *Id.* at 118.
[54] *Id.* at 119.
[55] *Id.* at 118–19.
[56] *Id.* at 119; *see also* NRC REPORT, *supra* note 30, at 96; Lewin, *supra* note 29, at 603.
[57] OTA REPORT, *supra* note 17, at 119; *see also* NRC REPORT, *supra* note 30, at 96.

would lead to political infighting among the agencies over what distribution would be equitable. Some argued that a task force would create yet another level of meaningless bureaucracy, serving to do nothing more than further remove the administration of science from those performing the research.[59] Though this point comes close to recognizing that innovation is likely to come bottom-up rather than top-down, the focus seemed to be on the distance between the administrators and the scientists, not the locus of decision-making authority.

5. Consortium

The consortium model was the only model discussed that involved the private sector in the organization of the project.[60] Based on smaller consortia between states and their universities,[61] the consortium model involved government research facilities and agencies as well as private companies who would be interested in the commercial applications of research. Generally, consortia involve government funding at the outset of basic research and private funding for development projects that stem from basic research.[62] The theory is that government and universities allow basic research to move forward, while firms use the developing technologies to create new products.[63] The main problem with the consortium model was seen as preventing the setting of a research agenda.[64] Thus, while the debate did consider the possible use of the private sector, it did not consider that the private sector might advance basic research itself, and it assumed once again that project coherence was necessary.

6. The Solution

Ultimately, NRC recommended that Congress adopt the single-agency leadership model, and OTA recommend a collaborative effort between agencies.[65] Congress allocated funds to both NIH and DOE to participate in a coordinated

---

[58] Drafters of the NRC report seemed less bothered by this than drafters of the OTA report. The NRC report suggested that the committee would be made up of members from NIH, DOE, NSF and other interested agencies, and that the chairmanship would rotate among NIH, DOE, and NSF, NRC REPORT, *supra* note 30, at 96.

[59] OTA REPORT, *supra* note 17, at 121.

[60] *Id.*

[61] An example is the Midwest Plant Biotechnology Consortium. Cited specifically by OTA, this consortium was created to increase America's position in agriculture by encouraging basic plant biotechnology research. *Id.* at 122.

[62] *Id.* at 121.

[63] *Id.* at 122.

[64] *Id.* at 123.

[65] Cook-Deegan, *supra* note 29.

Human Genome Project.[66] In 1988, Congress appropriated twelve million dollars to DOE for its initiative and slightly over seventeen million dollars to the National Institutes of General Medical Sciences (NIGMS), a part of NIH, earmarked specifically for "genome studies."[67] NIH promptly created an advisory committee to decide its genome research priorities.[68] Concerned about how their programs would relate to each other, DOE and NIH signed a "Memorandum of Understanding" that created a joint subcommittee whose members came from the genome research advisory committees of each agency.[69] The subcommittee set to work creating the first 5-year plan for the Human Genome Project. In the meantime, both NIH and DOE increased their 1990 budget requests for genome research. In response, Congress appropriated twenty-eight million dollars to DOE's program and just under sixty million dollars to the newly created National Center for Human Genome Research (NCHGR), which later became the National Human Genome Research Institute (NHGRI) at NIH.[70] Congress recognized the interagency commitment to a Human Genome Project and promised three billion dollars over 15 years. Using this funding, DOE and NIH would finance research with the goal of finding all the human genes and sequencing the DNA base pairs.[71] Although the amount of funding and the breadth of the Human Genome Project have changed over the years, NIH and DOE remain the principal agencies charged with heading the effort.[72]

C. Funding Mechanisms

Of greater consequence than the allocation of authority among agencies are the mechanisms used to support scientific research. Both DOE and NIH support genome research through a combination of internal work, contracts with national laboratories and research centers, and grants to individual researchers, research institutions, and private companies.[73] NIH supports more extramural research than it does intramural research, while DOE uses most of its Human Genome Project budget to support intramural research at the national

[66] Watson, *supra* note 18.

[67] *Id.* at 46.

[68] *Id.*

[69] Leslie Roberts, *NIH and DOE Draft Genome Pact*, 241 SCIENCE 1596 (Sept. 23, 1988).

[70] Cook-Deegan, *supra* note 20, at 6.

[71] U.S. Human Genome Project, *5-Year Research Goals 1998–2003: Time Table Accelerated on U.S. Human Genome Project*, available at http://www.ornl.gov/TechResources/ Human_Genome/hg5yp/

[72] In 2001, the United States Government spent $394.8 million on Human Genome Project research. Human Genome Project Information, Human Genome Project Budget, *at* http:// www.ornl.gov/hgmis/ project/budget.html

[73] OTA REPORT, *supra* note 17, at 93.

labs.[74] Private companies conduct research in conjunction with or similar to the Human Genome Project with the hope of profiting from that research through the use of intellectual property protections including patents.[75]

1. Direct Appropriations and Government Contracts

Both NIH and DOE receive funds for the Human Genome Project through congressional appropriations.[76] While the vast majority of these appropriations are earmarked only for the Human Genome Project, some appropriations carry more specific directions.[77] An appropriation may direct DOE to use a certain laboratory or may indicate a specific area within the Human Genome Project for which specific monies are meant to be used.[78] Such congressional earmarks are the least flexible mode of financing, offering no incentive by private parties or the government to innovate. Otherwise, NIH and DOE can use the appropriated funds to do intramural research or to employ contractors to perform specific research directed by the agency. NIH's intramural research is done by salaried government employees and recipients of NIH research fellowships.[79] Such funding generally is not conditional on any kind of performance review. When an agency contracts for specific research projects with national laboratories or other entities, these contracts are for specific research tasks or projects.[80] These contracts span a period of years and are not regularly subject to competitive bidding, as they are usually mere addenda to existing operational contracts.[81]

[74] OFFICE OF TECHNOLOGY ASSESSMENT, U.S. CONGRESS, FEDERAL TECHNOLOGY TRANSFER AND THE HUMAN GENOME PROJECT 3, 12 (1995) [hereinafter OTA TECH. TRANSFER REPORT].

[75] *Id.* at 3; *see also* Eliot Marshall, *The Company That Genome Researchers Love to Hate: Human Genome Services, Inc.*, 266 SCIENCE 1800 (Dec. 16, 1994); Eliot Marshall *et al.*, *In the Crossfire: Collins on Genomes, Patents, and Rivalry*, 287 SCIENCE 2396 (Mar. 31, 2000); Elizabeth Pennisi, *Genomics Comes of Age*, 290 SCIENCE 2220 (Dec. 22, 2000).

[76] *See* Watson, *supra* note 27, at 45–47; *see also* Jeffrey Mervis, *R&D Budget: Growth in Hard Times*, 263 SCIENCE 744 (Feb. 11, 1994).

[77] Often specific research programs receive funding based on either the agency's specific request or on the priorities of particular politicians. *See* Andrew Lawler, *Easing the Squeeze on R&D: Growth in the Economy Frees Money for Research*, 278 SCIENCE 1390 (Nov. 27, 1990); *see also* National Institutes of Health, *FY 2002 Budget Request*, Statement of Francis S. Collins available at http://www.nhgri.nih.gov/About_NHGR1/Od/Admin/Budget/fy02congStmnt.html

[78] *See, e.g.*, 147 CONG. REC. S1139 (daily ed. Nov. 1, 2001) (statement of Sen. Boxer) (discussing the Department of Energy appropriations bill and correcting the name of the University of Southern California program to be supported with federal funds). Regardless of whether an appropriations bill actually spells out what lab funding should go to, members of Congress often know, at least in the case of intramural and contract research, who is likely to receive the funding.

[79] *See* Division of Intramural Research, *About the DIR*, at http://www.nhgri.nih.gov/Intramural_research/about_DIR.html

[80] For example, Lawrence Livermore, Lawrence Berkeley, and Los Alamos national laboratories have contracted with DOE to sequence chromosomes 5, 16, and 19. *See* The Joint Genome Institute, *Decoding the Human Genome*, at http://www.llnl.gov/str/Branscomb.html

2. Grants to Individuals and Institutions

When giving grants, the agencies conduct peer review on independent researchers' grant proposals, and committees decide whether to offer grants and how much money to give.[82] The agencies award grants to both academic centers and commercial ventures, with private entities increasingly winning grants over academic centers.[83] Both DOE and NIH evaluate grant applications on a variety of factors, of which the competency of an applicant's personnel is one.[84] Although past performance will be considered at least in the DOE for those who seek to renew existing grants, the process does not give this consideration overriding significance. Grants thus provide for a type of competition, but the competition is primarily in the writing of grant proposals rather than in the performance of scientific research.

3. Cooperative Research and Development Agreements

Cooperative Research and Development Agreements (CRADA's), which Congress first created under the Federal Technology Transfer Act of 1986,[85] allow national laboratories to partner with private firms to develop products based on their research.[86] Each agreement can last for up to but no more than four years. Under a CRADA, "the government provides personnel, services, facilities, equipment or other resources (but not funds) and the nonfederal partner provides funds, personnel, services, equipment or other resources toward the conduct of specific research or development efforts consistent with the missions

[81] *See* Lawrence Livermore, NATIONAL LABORATORY, INSTITUTIONAL PLAN FY 2001–2005 (2000) (noting that Livermore's work for the U.S. Government is performed under general contract W-7405-Eng-48).

[82] *See* Center for Scientific Review, *A Straightforward Description of What Happens to Your Research Project Grant Application After It Is Received for Peer Review*, http:///www.csr.nih.gov/REVIEW/peerrev.htm (Aug. 24, 2001).

[83] *See* Elliot Marshall, *Commercial Firms Win U.S. Sequencing Funds*, 285 SCIENCE 310 (1999) (reporting on $15 million in grants to Genome Therapeutics Corp. and Incyte Pharmaceuticals, Inc.).

[84] U.S. Department of Energy, Office of Science, *Grant Application Guide*, at http://www.er.doe.gov/production/grants/process.html The National Human Genome Research Institute, *Guide for Applicants: Genome Research Resource Grants (P41)*, at http://www.nhgri.nih.gov/Grant info/Funding/Statements/p41 guide.html The factors considered by NIH for resource development grants include the potential for significant impact of the new technology on genome research, the scientific or technical significance and originality of the proposal, the appropriateness and adequacy of the experimental or engineering approaches to be used, documented community need for the resource, appropriateness and adequacy of plans for technical support, qualifications and experience of the principal investigator and the staff, reasonableness of budget and duration proposed and the adequacy of dissemination or commercialization plans.

[85] 15 U.S.C.A. § 3701 et seq.

[86] OTA TECH. TRANSFER REPORT, *supra* note 74, at 8, 47.

of the federal facility."[87] Both the government and its partner provide resources for research and development that is "consistent with the missions" of the national lab.[88] In 1995, surveys of CRADA participants revealed views that these agreements had facilitated the sharing of resources and provided an economic benefit to the private entities, but not necessarily to the government agencies.[89] Although CRADAs offer a variant on grants, if anything they are less likely to invite competition and innovation, because partners do not have to compete regularly in offering new proposals.

D. Goals of Structure and Funding

So far I have argued that the Human Genome Project founders placed too much emphasis on consistency and developed mechanisms for distributing funds unlikely to promote innovation. I shall now offer a list of goals that an ideal system would meet. Each of these goals ultimately is subservient to a larger goal, maximization of the production of discoveries leading to technological progress that individuals would value. This is a broad goal, but it does rule out two possibilities. First, it rules out the possibility that the goal of governmental science should be to maximize knowledge irrespective of its usefulness. Knowledge may be intrinsically valuable,[90] but if we assume that the goal of science is knowledge for knowledge's sake, then there is no obvious metric by which we can compare the success of one institution with that of another. Scientific experiments often turn out to be useful in unexpected ways,[91] but this does not mean usefulness is an inappropriate criterion. Rather, it means that if some areas of scientific inquiry are more likely to lead to surprising byproducts than others, we should, all else being equal, seek to encourage more research in the former areas than in the latter. The focus on usefulness is, in any event, more important for the assessment of how the government implements its commitment to scientific advance generally than it is for the assessment of possible programs by which the government might seek to advance scientific knowledge in some area, whether that area is genomics or theoretical physics.

Second, the criterion of maximization of useful scientific advances assumes that more science is better than less. Maximization of science, however, need not necessarily improve human well-being. One reason is that science

[87] *Id.* at 14.

[88] *Id.* at 14.

[89] *Id.* at 3, 20.

[90] For a discussion of the distinction between pure and applied science, and the claim that scientists make to autonomy from outside regulation, see Barry R. Furrow, *Governing Science: Public Risks and Private Remedies*, 131 U. PA. L. REV. 1403, 1415–17 (1983).

[91] The classic example is Alexander Fleming's fortuitous discovery of penicillin. *See, e.g.*, L. Ludovici, FLEMING: DISCOVERER OF PENICILLIN 131–34 (1952).

could produce discoveries with negative consequences, perhaps catastrophic ones.[92] Another is that society conceivably could spend too much on science, at the expense of satisfying other human wants, from immediate health care needs to national security to entertainment.[93] Nonetheless, I assume away these issues for analytical simplicity. If there is too much government science, a remedy is not to create a government program, and if there is too much private science, a remedy is to shorten patent terms.[94] My interest is in how government should design an administrative program to promote science, given the assumption that more science is better than less.[95]

Given the ultimate goal of maximizing scientific advances, I will offer five criteria: flexibility, competition, nonexclusivity, nonredundancy, and cost minimization. My purpose for now will be to explain the criteria, without detailed comparison of different institutions' performance against the criteria, a project that I will turn to later in this essay.[96]

Flexibility. Organizations often change their goals and the means by which they seek to meet those goals in response to new information. The stock market is perhaps the institution in our society that is most conducive to flexible decisionmaking. Equity flows to corporations that seem to have the best prospects, and if new information contradicts high expectations for a company, its activities will be contracted. Recent events like the telecommunications crash and the Enron fiasco may fairly be used to point out that stock markets are

[92] *Cf.* David Whitehouse, *Not the End of the World*, BBC NEWS (July 22, 1999), *available at* http://news.bbc.co.uk/hi/english/sci/tech/newsid399000/399513.stm (discussing controversy over whether a certain experiment might cause the creation of a black hole that would devour the earth). While we often dismiss such possibilities, perhaps finding them paranoid or even amusing, that may be a reflection of the human cognitive tendency to pay too little attention to small probability, high magnitude costs.

[93] *See* Matthew Erramouspe, Comment, *Staking Patent Claims on the Human Blueprint: Rewards and Rent-Dissipating Races*, 43 UCLA L. REV. 961 (1996) (considering this possibility). Government funding of science also could lead to reduced advancement in science, if inefficient government science crowds out efficient private sector science. *See* Terence Kealey, THE ECONOMIC LAWS OF SCIENTIFIC RESEARCH (1996) (arguing that government investments in science have generally been welfare-reducing).

[94] An alternative remedy might be to make patent rights less certain. *See* Ian Ayres & Paul Klemperer, *Limiting Patentees' Market Power Without Reducing Innovation Incentives: The Perverse Benefits of Uncertainty and Non-Injunctive Remedies*, 97 MICH. L. REV. 985 (1999) (exploring the tradeoff between patent certainty and term length).

[95] My own instinct is that the benefits probably are worth the cost for anything the government reasonably might be expected to spend in the genomics area. Whether the space program receives excessive funding is a more difficult question. *Cf.* Bonnie E. Fought, Comment, *Legal Aspects of the Commercialization of Space Transportation Systems*, 3 HIGH TECH L.J. 99, 107–10 (1988) (discussing the cost efficiency of NASA).

[96] *See infra* Part II.D.

not perfect, but they also show that once poor prospects become manifest, the market responds. The flexibility of the stock market allows shareholders to cut their losses. When a project is to be funded outside of traditional capital markets, some alternative way of ensuring flexibility is desirable. A flexible project would be one that first, could be expanded or contracted as the project turns out to be more or less successful than imagined, and second, can move research in more profitable directions. On the former score, the Human Genome Project had no explicit mechanism that would lead to contraction if the project proved to be wasteful. In theory, Congress might have stopped funding, but the creation of a government bureaucracy devoted to the program makes such a decision less likely. The latter aspect of flexibility presents the more serious problem for the Human Genome Project. The premise of the project was that sequencing of the entire genome would be useful. There is some evidence that this premise has turned out to be correct, as even the sequencing of "junk" DNA has led to important discoveries.[97] But the Project lacked any mechanism that would have redirected funds to other research projects if sequencing junk itself turned out to be junk.

Competition. Just as competition in markets for goods and services is likely to lead to the highest quality of such goods and services, so too is competition in approaches likely to improve science. The possibility of private gains—whether reputational or financial—is the strongest inducement to innovation. The preliminary evidence from the Human Genome Project supports this observation, as the competition from the Celera group led to an earlier-than-expected completion of the initial sequencing. The competition that occurred, however, was more happenstance than a product of any specific structural mechanism intended to produce competition. Moreover, there is no guarantee that private competition will materialize for future big science projects, given the uncertain state of intellectual property rights in the area. Perhaps the absence of any specific structural mechanism can be attributed to a view that sequencing seemed to be a methodical project, requiring more brute force than innovation. That fact that the innovation turned out to be important even for such a project highlights the need for encouraging competition.

Nonexclusivity. If the results of scientific research are made available to all, the number of future innovations will be greater. Availability of data was certainly a goal of the Human Genome Project, as the resulting sequences

[97] *See, e.g.*, Kristen Philipkoski, *DNA Junkyard Yielding Gold*, WIRED NEWS (Feb. 12, 2001), *available at* http://www.wired.com/news/technology/0.1282.41750,00.html Some scientists, however, argue that even if junk DNA has some function that sequencing helps elucidate, it is still not the best use of resources to sequence the complete genome. *Id.* ("[Bill] Haseltine believes studying the whole genome is a waste. 'It's clear we should focus on genes and not the genome,' he said.").

were to be placed and in fact were placed in the public domain.[98] The designers of the Project, however, gave little consideration to whether private researchers could obtain patents on sequences. Perhaps that absence of attention ultimately proved beneficial, by providing a means by which private parties were able to compete. This highlights that the possibility of property rights may provide a useful incentive for private innovation. My point is simply that if we can produce innovation without exclusive property rights, that is likely to be better than producing innovation with them. An important caveat is that property rights may help encourage commercialization of the fruits of government research;[99] such property rights, however, need not necessarily be on the government research itself.

Nonredundancy. If different sets of scientists work on the same problem at the same time, some of the work they produce might be redundant, and there will be fewer scientific advances than if scientists worked on different projects. There may be some situations in which redundancy is worthwhile if the successful completion of one goal is so much more important than the marginal project that the additional competition outweighs the redundancy. Perhaps one virtue of the Human Genome Project as originally designed was that it was nonredundant. Instead of having different teams of scientists sequencing the same genes simultaneously, scientists would sequence each gene once. Because the Project did not prevent separate private research, however, it failed to stop redundancy. Once again, Celera's entry may have been socially worthwhile on the whole, but science would be advanced further if it were possible to harness the benefits of competition without redundancy.

Cost minimization. This consideration includes not only the cost of the administrative apparatus itself, but also costs incurred by private parties in connection with the administrative scheme. For example, if a particular design for big science, such as reliance on government contracts with ambiguous terms, will lead to litigation, that litigation is a social cost that offsets the benefits of the project. The Human Genome Project should be credited with inducing relatively little litigation,[100] but the involvement of multiple agencies may have increased more traditional types of costs. One type of cost that is particularly significant is political rent-seeking costs,[101] such as the costs associated with lobbying government officials. These are

[98] *See* http://genome.ucsc.edu/

[99] *See* F. Scott Kieff, *Property Rights and Property Rules for Commercializing Inventions*, 85 MINN. L. REV. 697, 717–36 (2001).

[100] None of the nine published cases mentioning the Human Genome Project involves litigation over the Project itself. Search of Westlaw, ALLCASES database (Mar. 13, 2002).

[101] *See* Dennis C. Mueller, PUBLIC CHOICE II at 229–38 (1989) (providing an overview of rent seeking).

likely to be particularly high to the extent that Congress appropriates money for particular organizations, or to the extent that there is a chance that Congress might do so. These costs are of concern not only directly, but also because they may distort government decision-making.

III. A PROPOSAL FOR ADMINISTRATION OF SCIENCE

I have argued that competition and innovation in genomics research happened in spite of the Human Genome Project, rather than because of it. One might argue that this is unfair, because the developers of the Human Genome Project created it against a background of legal institutions, and the creators of the structure for the Human Genome Project cannot be faulted for implicitly taking advantage of those institutions' strengths, whether consciously or not. It would be a mistake, however, to assume that we have achieved just the proper combination of policies so that the strengths of one will diminish the weaknesses of another. Moreover, it would be foolish to assume that a multipronged approach is necessarily the best strategy. A multipronged approach might be worse than a single-minded one if the negative aspects of a particular funding strategy manifest themselves even when that strategy is inducing only a relatively small amount of investment. For example, if patent law induces a relatively small percentage of sequencing work, its benefits may be negligible, yet the anticommons problem suggests that its costs might be quite high.[102]

Thus the combination of patents, grants, contracts, and intramural research that the Human Genome Project exemplifies may not be optimal. The most obvious alternative would be to choose one of the above. We might well be better off, for example, if we left further genomics work (and if we had left past genomics work) entirely to the patent process, perhaps a modified version with either stronger property rights or a fair use doctrine to weaken them.[103] Although there might still be holdup problems, economies of scale might lead to the establishment of a relatively small number of dominant players, and thus the conditions for an effective private rights-management organization might have formed.[104] And similar arguments might be offered for the primacy of either of the other solutions. I recognize the importance of the

[102] *See* Heller & Eisenberg, *supra* note 5 (discussing the problem); *see also* Michael A. Heller, *The Tragedy of the Anticommons: Property in the Transition from Marx to Markets*, 111 HARV. L. REV. 621 (1998) (introducing the anticommons problem in a general property context).

[103] *See* Maureen A. O'Rourke, *Toward a Doctrine of Fair Use in Patent Law*, 100 COLUM. L. REV. 1177 (2000).

[104] *See* Robert P. Merges, *Contracting into Liability Rules: Intellectual Property Rights and Collective Rights Organizations*, 84 CAL. L. REV. 1293 (1996) (emphasizing that private organizations often emerge to manage and license rights).

debate over which of the approaches would be the best if we were forced to select one, because perhaps we should be forced to make just such a choice. Nonetheless, none alone will be able to meet all of the goals we would like to achieve in funding big science, and thus the choice among these alternatives must be made while holding one's nose.

Instead of advocating for any of these, I will sketch out an alternative model. The model adheres more closely to the grant process than to its alternatives, in part based on a practical recognition that any conscious further government effort to encourage innovation in big science will probably take that form. The approach, however, draws on the virtues of all three strategies, seeking to create a regime that involves some centralized government control but considerable private competition. I recognize that one should address the task of building a better structure with some trepidation, both because what works on paper may not work in practice, and because what sounds radical on paper will never make it into practice. While the overall institution that I aim to describe would be different from what we now have, it uses elements that are familiar. Where I offer new twists to old forms, I seek to do so in ways that the old institutions are still visible in the new.

A. A Sketch of a Retrospective Grant Institution

The essence of my proposal is that grants should be provided based on the success of previous research results on the same project rather than on the basis of future plans. Peer review, of course, typically does involve consideration of an applicant's past efforts, at least implicitly, except in situations in which peer review is blind.[105] Ordinarily, however, reviewers select grants on the basis of anticipated future performance, rather than past performance, even when taking into account the success of past results or the reputation of the applicants. This practice presumably reflects the common sense intuition that it is easier to predict how successful a project will be with knowledge of what the project will entail. More information, it might seem, is better than less information, and if the government is to select those projects that have the greatest likelihood of success, common sense suggests that its decision makers are far better off if they can assess the project itself.

Common sense, however, misses that the question is not *which* information will be considered, but *who* will consider it. There are, in essence, two decisionmakers: the grant giver and the scientist (a shorthand that I will use to encompass also private, academic, and other organizations pursuing scientific research). Even if the grant giver does not directly consider what project is more

[105] *See generally How Blind is Blind Review?* AM. J. PUB. HEALTH. July 1991, at 843 (discussing blind review and the contention that it eliminates bias).

likely to be successful, a scientist provided with a grant but concerned about receiving a grant in the next period as well will make an assessment among alternative projects. Individual scientists are likely to be better situated than government grant givers to make this decision. In part, this is true because scientists are likely to confront practical research questions most directly and thus are most likely to have the knowledge and information needed to answer those questions. Equally importantly, unlike government decisionmakers, individual researchers have a stake, professional if not financial, in the quality of the decision and thus strong incentives to make it well.

That scientists rather than reviewers will make the fundamental choices in a retrospective grant system points to several advantages of such a regime. First, it encourages scientists to pursue those research strategies that they themselves believe are most likely to be productive. In a more typical grant regime, the focus of scientists will be on preparing a proposal likely to appeal to those making grants, such as members of the administrative agency, and subsequently will be on completing the grant as proposed. Retrospective grants do not eliminate the scientist's incentive to cater to others' tastes, because scientists will still seek to achieve results that grant givers are likely to find important. Nonetheless, a scientist who believes that her approach is more likely to produce a desired result than an alternative approach will be able to prove it.

Second, the system will encourage risk-taking. A scientist who otherwise would not expect to receive large grants might adopt an unconventional approach even if this approach is unlikely to succeed. If the project fails, the scientist will expect to receive less than if the scientist had taken an unconventional approach, though perhaps not zero, as the demonstration that a particular approach is ineffective itself can be a valuable contribution. And if a project thought unlikely to be productive does succeed, the scientist will receive considerably larger grants. The possibility that upstarts will take risks in turn may encourage leaders, scientists, or laboratories receiving relatively large grants to do the same.[106] Of course, more risk-taking is not always better, as sometimes it might be undesirable for scientists to pursue a low expected value, high upside project. The experience of the Human Genome Project, however, suggests that risk-taking may lead to substantial advances. Moreover, even with retrospective decision making, scientists are likely to accept less risk than is socially desirable because individuals are generally more risk-averse than society at large.[107]

[106] An absence of risk-takers is similarly likely to dampen innovation in corporate law. *See* Marcel Kahan & Ehud Kamar, The Myth of State Competition in Corporate Law (2002) (unpublished manuscript, on file with author). (arguing that there is not much competition among state corporate laws, thus relieving Delaware of much of the burden of competition).

[107] *See* Kenneth J. Arrow & Robert C. Lind, *Uncertainty and the Evaluation of Public Investment Decisions*, 60 AM. ECON. REV. 364 (1970). (demonstrating that if the number of taxpayers is large, then public planners can ignore the risk of any particular project).

Third, the system will make science more flexible and better able to respond to new information about the approaches that are likely to produce useful advances. This is true both within a period in which grants are being performed and across periods. Within a research period, a scientist who believes that a project is not going well or who recognizes that an abrupt change in plans would be desirable will have an incentive to switch gears. With traditional grant institutions, a scientist might prefer to finish a project that is unlikely to be useful, and thus complete the grant, than switch to a new project. Sometimes, a scientist might conclude that completion of an unsuccessful project will provide more social value (and thus a larger future grant) than a late switch to a new project, but at least the scientist has the incentive to make the appropriate comparison. If one scientist's efforts change other scientists' assessments about which approaches are most likely to produce useful results, the other scientists will have an incentive to follow the first scientist's lead.

The most serious potential disadvantage to grants that are based on *ex post* performance is that they might lead to redundant research, as with patent races. Suppose, for example, that the benefits of one research project seem likely to be more than twice as great as those of the next research project. Then two different firms might perform the same research, even if they expect that the ultimate grant determinations would give each only half credit for discoveries. Even with less of a gap, failure among firms to coordinate with each other or a game of chicken in which neither party wants to give up the preferred research project might lead to redundancy. A virtue of *ex ante* grants is that decision makers ordinarily will pick no more than one party to perform a particular project, thus eliminating the possibility of redundancy.

An agency providing grants on the basis of *ex post* performance, however, easily could perform the same function. Just because grant amounts are retrospective does not mean that the agency cannot make prospective assignments. If the agency dictates what a particular scientist should do, then the benefits discussed above would disappear, but there would be no such harm from the agency's designation of priorities on certain research projects. Doing so would still involve less work than screening proposals for ultimate merit, especially if the agency examines only proposals from institutions whose past work entitles them to funding. For example, if two different teams were sequencing genes, the agency could assign them different genes. Indeed, the agency ought to do so in the ordinary case even if the teams are using different techniques, unless there is reason to believe that any redundancy is worthwhile because the particular research project is so much more important than the marginal research project that would be ignored. Much of the overlap between Celera's work and that of the Human Genome Project could have been eliminated through the assignment of property rights to research different sections of the genome. (These property rights would not be patents, of course, but merely rights to research in a particular area, valid over other potential grant recipients.) The enforcement of such property rights could occur in the next grant cycle, with appropriate deductions made if a grant applicant had strayed onto another's jurisdiction.[108]

B. Refinements

I have as yet provided only the vaguest description of how a retrospective grant-giving institution would operate. While I do not intend to fill in regulatory minutiae, it may be worthwhile to consider some of the specific issues that likely would develop in the design of such an institution. The most obvious question is how the agency should handle the first and last periods in which grant money is being distributed. A simple solution for the first period would be to use prospective grants for that alone. Thus, the seed money would depend on an assessment of future plans, as well as factors like reputation and infrastructure, but performance in the first period would determine funding in the second period, so the benefits of retrospective decision making would not be lost. The last period, meanwhile, might be rewarded with a simple distribution of cash, or an open-ended grant. Taken together, this approach would lead to one more set of checks being issued than the number of periods, but this additional expenditure simply increases the incentive that scientists will have to participate in the process and engage in the desired research. If this were deemed undesirable, researchers might be required to put up their own seed money in the first period.

Perhaps the most important issue is who should make the decisions on grants. One concern might be that if the preferences of the decision makers is known, much of the value of a retrospective process might be lost. This objection can be overblown, as even a decision maker whose belief in a certain strategy is known may be willing to concede in making retrospective assessments, should research prove a different strategy to be effective. It should not, however, be underestimated. The problem may be particularly severe when the ultimate objective is far off. Suppose, for example, that a number of intermediate steps, each taking a research period, are necessary before any tangible product of research emerges, such as development and testing of a pharmaceutical drug. Then the retrospective grant institution might be a little different from a prospective one, because decision makers will assess the success of past work by their estimates of future success. Only if a scientist believes that success in an early step will sway decision makers will she be willing to take risks and pursue

[108] An alternative, though more speculative possibility, would be for the agency to auction off the right to pursue particularly attractive research prospects. *Cf.* Dean Lueck. *The Rule of First Possession and the Design of the Law*, 38 J.L. & ECON. 393, 403 (1995). (noting that auctions are the most common alternative to first possession rules in assigning property rights).

the path that she believes is most likely to be successful in the long term. Sensible steps are meaningful only if they are perceptible.

The key to finding a solution to this problem is the recognition that the problem is not so much the impossibility of proving the success of a stage of a research project, but the danger of bias in the officials deciding grants. If a biased decisionmaker decides grants for each period, then the decisionmaker might refuse to recognize that a daring project has exceeded expectations. If a decisionmaker is open-minded, however, the multi-stage nature of a research project is of little consequence. Such a decisionmaker should be willing to make honest assessments of progress and should take into account that some risky projects should be rewarded because of the possibility that they will lead to substantial advances. The problem is that even for honest and disinterested individuals, it may be impossible to place aside one's expectations of success from one round to the next, as decisionmakers may interpret equivocal evidence as reinforcing prior beliefs. Thus, even a grant-giver who believes he is willing to admit error may tend to believe that equivocal results reinforce his initial preferences.

Rotating the decisionmakers who dictate grant amounts may be sufficient to overcome this hurdle. If the individual who chooses a grant amount one period is different from the one who does so the next, then initial conceptions of the likely success of a project will not be eternal determinants of grant funding. Rather than rely on internal staff to make decisions about grants, an agency should rely on outside scientific experts, as in many peer review programs, with those consulted one period ineligible for at least several subsequent periods. One danger to be guarded against is that the choices of decisionmakers might be reflections of the preferences of agency decisionmakers, for then scientist will still cater to some particular decisionmaker's agenda. To avoid this problem, the agency's responsibility should not be to select the "best" decisionmaker, but to assess all applicants against a relatively objective set of minimum qualifying criteria. The decisionmakers in a particular period could then be chosen by lottery from among all those qualified. Although there is some chance that subjective considerations would influence whom the agency would deem sufficiently qualified, and thus who scientists would anticipate might set their grants, such an approach would be considerably more objective than the alternative.

This approach cannot eliminate the possibility that some decisionmakers will approach the grant task with preconceptions, and it may increase the risk that a decision on a particular grant will reflect an idiosyncratic preference. The important point, however, is that errors in grant decisions are of relatively little consequence, leading at worst to the occasional provision of resources to scientists who are not necessarily in the best position to make progress. The decisions that matter most are the decisions of scientists about what to research, made in anticipation of subsequent decisions on grants. And these decisions will be made not in the shadow of any particular decisionmaker, but in the shadow of a hypothetical average decisionmaker. Thus, even if decisions on grants are idiosyncratic—with a particular research team receiving more than enough to cover expenses one period, less the next—rotation ensures that decisions on what to research will not be. As long as a scientist believes that she can establish promise in an unconventional approach and that an average decisionmaker will recognize the value of diversified approaches, she will be willing to take that path.

Even with a rotation of decisionmakers, a grant could be based on some average of performance in previous periods rather than on the evaluation of performance in the immediately preceding period. There is a tradeoff here. On one hand, incentives for potential grant recipients will be greatest if the performance in a particular period is the sole determinant of the size of the grant. The difference appears primarily only in the latest periods, because in those periods performance will have only a small effect on future grant receipts. On the other hand, it may be problematic for scientists to have research support increase and decrease dramatically from period to period. I believe that this problem is not serious, for reasons I will explain below,[109] but at least this objection might make a system without some form of averaging politically infeasible. Perhaps an even more politically palatable system would be one in which performance grades would determine only whether there would be relatively small increases or decreases in grants from one period to the next. With such a system, a grant applicant's original reputation or proposal might have a substantial effect on subsequent grant payments, but there is still some incentive for innovation.

In discussing how the grant-making process should be structured, I have made one notable omission: I have not addressed the issues that dominated discussion at the onset of the Human Genome Project, how the agencies themselves should be structured. These decisions are of relatively little importance if research is funded through grants rather than being accomplished by internal agency scientists. Perhaps it is best to have just one agency running the project to minimize transactions costs, or perhaps it would be better for different agencies to control distribution to ensure that grants reflect a diversity of views. The difference will not likely be large. I do not expect that these questions will disappear, as government officials with an interest in increasing their own portfolio can be expected to seek jurisdiction over new science projects.[110] Of course, this recognition that government bureaucracies can be self-perpetuating only increases the normative case for a system in which a relatively small government agency would disburse money, because a small agency is less likely

[109] *See infra* Part II.C.

[110] *See* W. Niskanen, BUREAUCRACY AND REPRESENTATIVE GOVERNMENT 36–42 (1971). (noting the tendencies of members of bureaucracies to seek to increase their size.)

to be able to ensure its own preservation should a scientific project no longer be worth finding.

Perhaps the strongest objection to a system in which the agency officials' roles are administrative, coordinating the selection of grant decision-makers and disbursement of funds according to some formula, is that agencies might have specialized knowledge or expertise that could benefit the scientific process. Sometimes government scientists might be better than their private or academic counterparts, and a system in which grants are only made to non-governmental organizations may neglect them. This argument, however, is ultimately circular, for it fails to explain why the scientists work for the government rather than in the private or academic sectors. If they have developed expertise, they presumably can leave government work and compete for grants with private organizations. Perhaps there are occasions in which national security interests demand that the government rather than the private sector be responsible for scientific experimentation. Ordinarily, however, the benefits of having competitive forces stimulate research advances seem likely to be greater than any benefits of governmental organization. Indeed, my principal point is that government funding of science need not deprive us of vigorous competition.

C. Objections

Let me now consider two possible objections to a retrospective grants regime. The first objection points to my observation that even though the government agency will not consider prospective plans, individual scientists will have an incentive to do so. Perhaps agency officials are better able to develop objective assessments of scientists' ideas than are the scientists themselves. Such an argument might cite the cognitive psychology literature on the self-serving bias,[111] which here translates into individuals thinking more highly of their work than others do. This argument, however, could be made about any competitive regime. Perhaps we could have a government agency determine whether opening an ice cream parlor is a good idea, because individual entrepreneurs may overestimate the quality of their vanilla. Moreover, the retrospective grant system provides scientists incentives to pick the strongest project from among the various projects on which they might work, and there is little reason to think that the self-serving bias should distort an individual's choice of projects. Perhaps a scientist is more likely to favor an idea that he came up with than someone else's idea, but the ultimate quality rating at least will force a scientist to consider carefully various options.

[111] *See, e.g.*, David Messick & Keith Sentis, *Fairness and Preference*, 15 J. EXPERIMENTAL SOC. PSYCHOL. 418 (1979).

A second objection is that retrospective financing would be too chaotic. Perhaps scientists are most comfortable when their future financing is settled. The argument is analogous to a claim often made about academic tenure, that it provides professors with the freedom necessary for high-quality scholarship. Freedom of speech, however, is hardly relevant in most science projects,[112] and freedom to pursue projects is more likely to produce indifference to others' views about the usefulness of projects than to stimulate useful innovation. More fundamentally, freedom and security are products of the organizations that employ scientists, except in the case of the lone scientist. An academic institution or private organization can promise to provide salary and funding to a scientist regardless of whether an agency gives a grant at the expected level. That private organizations often do not provide complete security and freedom to scientists reflects that those who run such organizations believe that the costs of such provision would exceed the benefits and that the employees of such organizations prefer to work with such pressure than to work with less pressure but lower salary. A retrospective grant system would allow for such tradeoffs to be made by private actors, with academic institutions and for-profit companies presumably taking different approaches.

This response might seem to invite a variant of the objection, that uncertainty in funding may wreak havoc on the organizations that employ scientists. It is not always possible, the theory states, for an organization simply to scale up and down a project. Large-scale endeavours sometimes require organizational commitments to persevere through unexpected obstacles, but such perseverance may not be possible if the obstacles are an absence of funding. This argument seems to reflect a view that organizations producing scientific research are incapable of bearing risk. The patent system, though, proves that risk and science are not inherently incompatible.[113] Even academic and non-profit organizations can withstand some risk, taking losses on some projects while earning profits on others. Indeed, there is no reason that the government necessarily must require scientists to spend grants on further research. Grants used to reward past research will be more valuable the fewer strings attached, and so the government will maximize private investments in science the more flexibility scientists have in spending grants. Some organizations may respond by investing their own funds in early periods to build up a foundation for later, cheaper research. Grants used to reward such research ought to consider the quality of the results, not the investment in any given period.

[112] Such issues are potentially of greater significance for agencies like the National Endowment for the Humanities. *See* Alvaro Ignacio Anillo, Note. *The National Endowment for the Humanities: Control of Funding Versus Academic Freedom*, 45 VAND. L. REV. 455 (1992).

[113] *See, e.g.*, John F. Niblack, Ph.D., *Why Are Drug Development Programs Growing in Size and Cost? A View from Industry*, 52 FOOD & DRUG L.J. 151. (discussing the costs and risks associated with drug development.)

D. Assessment of the Proposal

Although I have provided only a sketch of a retrospective grant institution, no more is required to convey the theoretical point, and I have offered enough to allow for a preliminary consideration of the extent to which the institution meets the goals that I have set forth. First, consider flexibility. When scientists know that the compensation that they receive in the future depends on an *ex post* analysis of their work, the incentives to innovate and to change course in midstream will be greater than in cases in which compensation is tied only indirectly to performance. Thus the retrospective grant institution that I have described is certain to be more flexible than a traditional grant institution. It will be as flexible as a patent regime, but scientists in the two regimes will have flexibility to pursue different goals. With retrospective grants, scientists will have flexibility in the means they use to reach the goal of impressing subsequent government decisionmakers that they have contributed social value. With patents, scientists will have flexibility in the means they use to reach the goal of impressing customers or downstream users that their products will have private value to them. The advantage of the retrospective grant goal is that it emphasizes social value, which might be different from private value. The advantage of the patent goal is that private value is ultimately measured through revealed preferences rather than through a third-party assessment.[114]

The second criterion, competition, also recommends the retrospective grant institution over a more traditional alternative. The comparison is particularly favorable when the alternative is a government-provided intramural science, for the same reason that government generally has weak incentives when it is the provider of goods and services. Even against a traditional grant institution, the retrospective grant institution performs well. While both institutions involve competition, the traditional emphasizes competition in proposals, while the retrospective institution emphasizes competition in results. A potential disadvantage of the retrospective institution emerges in comparison to patent law. Competition in the retrospective grant institution is over a fixed sum, allocated by Congress or an administrative agency, and changeable only through the exercise of political power. The competition in patent law is scalable, so the number and size of the competitors will be proportional to the size of the projected rewards. The difference, however, will be of less significance if a broad approach is taken to defining the problems on which grant recipients might work. Perhaps defining the task of the scientific project as "sequencing" would entail a risk that more funds than necessary might be provided, but if the project

[114] "Patents at least let the market decide." H. I. Dutton, THE PATENT SYSTEM AND INVENTIVE ACTIVITY DURING THE INDUSTRIAL REVOLUTION, 1750–1852, at 26 (1984). (discussing nineteenth-century debates on patent protection.)

is genomics, there is a strong likelihood that there will be many projects producing positive social returns.

How the grant institutions, both traditional and retrospective, meet the third goal, which is nonexclusivity, depends on whether scientists are allowed to obtain patents on their discoveries. Some commentators have criticized policies that allow researchers to benefit from government largesse and then patent their discoveries as well,[115] but economically it is conceivable that a regime in which research is awarded from both patents and grants may be superior to one in which grant recipients cannot obtain patents.[116] Awarding grants shifts out the supply of innovation, but attaching a restriction preventing grant recipients from obtaining patents will not shift out the supply as much, and if patents are eliminated or restricted in the relevant field of endeavor,[117] whether net supply increases or decreases depends on the relative size of the grant and patent incentives. The question thus becomes whether we are better off with patent protection in a particular area in which the government promotes research through grants, and general arguments about the costs and benefits of the patent system apply.[118] At least, however, the grant institutions, unlike patents, may reduce the need for legal protections guaranteeing exclusivity. Neither the traditional nor the retrospective variety seems more likely than the other to promote this goal.

The grant institutions are also likely to be superior to the patent system in promoting nonredundancy. Patent races lead to redundant work before a patent is issued,[119] and inventing leads to redundant work after a patent is issued.[120] Grants can limit the number of races to develop a particular technology, without necessarily limiting races to make progress on parallel technology tracks. In addition, to the extent that grants allow for the promotion of

[115] *See, e.g.*, Rebecca S. Eisenberg, *Public Research and Private Development: Patents and Technology Transfer in Government-Sponsored Research*, 82 VA. L. REV. 1663 (1996).

[116] *See, e.g.*, Richard E. Romano, *Aspects of R&D Subsidization*, 104 Q.J. ECON. 863 (1989).

[117] The Patent and Trademark Office has signaled that it will restrict genomics patents by explaining how it will consider whether the patent application meets the utility requirement. *See Utility Examination Guidelines*, 66 Fed. Reg. 1092, 1098 (2001).

[118] *See, e.g.*, Fritz Machlup, AN ECONOMIC REVIEW OF THE PATENT SYSTEM, *in* SUBCOMM. ON PATENTS TRADEMARK AND COPYRIGHT OF THE SENATE COMM. ON THE JUDICIARY, 85th Cong. 44–45 (Comm. Print 1958). (summarizing such arguments.)

[119] For discussions of patent races, see Partha Dasgupta & Joseph Stiglitz, *Industrial Structure and the Nature of Innovative Activity*, 90 ECON: J. 266, 284–87 (1980); Donald G. McFetridge & Douglas A. Smith, *Patents, Prospects, and Economic Surplus: A Comment*, 23 J.L. & ECON. 197 (1980); and Gideon Parchomovsky, *Publish or Perish*, 98 MICH. L. REV. 926 (2000).

[120] *See, e.g.*, Louis Kaplow, *The Patent-Antitrust Intersection: A Reappraisal*, 97 HARV. L. REV. 1813, 1869 (1984); Donald F. Turner, *The Patent System and Competitive Policy*, 44 N.Y.U. L. REV. 450, 455 (1969).

nonexclusivity, they eliminate the need for inventing around—engaging in research solely to avoid collision with others' intellectual property rights—thus freeing resources from tasks that would offer only incidental social benefit. The traditional grant institution offers more natural protections against redundancy than the retrospective grant institution, for it is straightforward to select proposals that are not overlapping. We have seen, however, that it is possible to design a retrospective institution in which participants have property rights to particular areas of research.[121]

The retrospective grant institution also seems likely to reduce administrative costs relative to either of the alternatives. It may be easier (and thus cost less) to evaluate past achievements than to project achievements onto the future. More significantly, the retrospective grant institution greatly reduces the costs of seeking a grant, because prospective plans need be shown only in enough detail to allow the agency to prevent redundancy. Scientists presumably will write up the results of their experiments eventually with considerable care anyway, and drafting proposals may be a considerable burden.[122] Both grant institutions may help avoid the administrative costs associated with the patent process, particularly the costs of prosecuting patents and of patent litigation. These costs will be reduced as long as grant recipients are required to place any discoveries in the public domain. Meanwhile, with appropriate protections, all of these systems can be insulated from the most blatant of rent-seeking abuses,[123] which are most dangerous when Congress earmarks funds for particular purposes.

IV. CONCLUSION

Grant institutions may be sufficiently established to make it unlikely that Congress would adopt the change suggested here. Let me thus break up the argument that I have offered into three distinct points, none of which needs to be taken to an extreme to improve science policymaking. First, the more that grant decisions are based on immediate past performance rather than on general reputation or on anticipation of future performance, the better. Second, there is little need for a large administrative apparatus to run even so large a project as the Human Genome Project. Just because the government sponsors research

[121] *See supra* text accompanying note 108.

[122] For an anecdotal example of how investment in grant proposals can dissipate rents that otherwise might be used productively, see Roy, *supra* note 3, at 317. In Roy's example, 2200 proposals were submitted for grants, and Roy estimates that the process diverted "nearly 200 years of scientific work" from research. *Id.*

[123] Some simple steps would be a requirement that interested decisionmakers recuse themselves and a ban on *ex parte* contacts.

does not mean that the government itself must engage in research. And third, decisionmaking in assessing grants may be better if there is rotation of decision-makers instead of a fixed group. Except in a project that needs a high degree of coherence because different parts must fit together, consistency of vision is more a vice than a virtue. The goal of science is innovation, and its demands are sufficiently pressing that they cannot wait for natural selection to do all the work.

Acknowledgments

For exceptional research assistance, including drafting of sections of Part 1, I thank Jennifer Atkins.

12

Goat-Boy Roams the Halls?

Justin Hughes
Assistant Professor of Law
Benjamin N. Cardozo School of Law, Yeshiva University
New York, New York 10003

The papers presented for the first session of The Washington University Conference Series on Law and the Human Genome Project—which have since become chapters in this book—are rich and varied, addressing the issues surrounding genomic property at all sorts of levels—from the broad philosophical arguments arrayed in James Boyle's draft to precise issues like adequate incentives for research tool development (Rebecca Eisenberg) and patent claims to ESTs (Richard Epstein). The wealth of ideas that came over my electronic transom during those weeks leaves a commentator feeling like a poorly coordinated guy confronting that old video game "Galaxians" in which the invaders descend the screen at increasingly faster rates. Like that hapless fellow, I'll be satisfied if my contribution seems at least marginally better than random shots.

I. INFORMATION AND THE ADVOCACY OF REFORM

As an initial comment, I would point to the one level of discussion that doesn't appear in these papers: empirical work. I make this comment being someone who has written about intellectual property (IP) at some pretty abstract levels,

0065-2660/03 $35.00

only occasionally sprinkling the pages with real-world examples. Examples are better—or usually better—than no examples at all, but there are severe limits to such "anecdata."

The call for empirical work overlaps partially with, but is distinct from, the appeal for more economic work found in articles like Professor Eisenberg's *Analyze This*.[1] Fortunately, we are in a time when intellectual property scholars are beginning to address the dearth of empirical data about the subjects we study. Recent articles by Kim Moore, Mark Lemley, and John Allison are good examples.[2] Others of us may soon get in on the act. These are "first generation" (there may be a more appropriate technical term) because these are efforts in data *compilation* and *interpretation*—using extant data from the courts and the United States Patent and Trademark Office (USPTO).

Hopefully, this work will continue to be augmented with a deeper, richer wave of data *generation* research—the kind of work that Wes Cohen and Keith Maskus have done in different areas.[3] Data generation work includes all kinds of surveys about points raised in Rebecca Eisenberg's paper for this conference that are begging to be explored. What are the actual perceptions of investors, biotech execs, biotech researchers, etc. as to whether research tool innovations have insufficient yields? What is the actual incidence of research tools being underutilized because pharmaceutical companies and universities can't reach agreements with the tool owners on terms for use? To what extent are research tools transferred to pharmaceutical companies via the buyout of start-ups? Sure, some of these questions are not easy to study empirically because respondents have incentives to either (a) not respond or (b) respond in

[1] Rebecca Eisenberg, *Analyze This: A Law and Economics Agenda for the Patent System*, 53 VAND. L. REV. 2081 (2000).

[2] Kimberly A. Moore, *Judges, Juries, and Patent Cases—An Empirical Peek Inside the Black Box*, 99 MICH. L. REV. 365 (2000); John R. Allison and Mark A. Lemley, *Who's Patenting What? An Empirical Exploration of Patent Prosecutions*, 53 VANDERBILT L. REV. 2099 (2000); John R. Allison and Mark A. Lemley, *How Federal Circuit Judges Vote in Patent Validity Cases*, 24 FLORIDA STATE U. L. REV. 745 (2000).

[3] *See, e.g.*, Wesley M. Cohen *et al.*, *Protecting Their Intellectual Assets: Appropriability Conditions & Why U.S. Manufacturing Firms Patent (Or Not)*, National Bureau of Econ. Research Working Paper No. 7552, 2000 (concluding on basis of survey results that "patents are used in substantially different ways across different technologies, suggesting that policy and court decisions affecting the breadth of claims, applicable nonobviousness standards, likelihood of being upheld in court and other features of patents will likely have different impacts on invention and competition in different industries"). I'm also thinking of Keith Maskus's work studying the behavior of multi-national firms in different countries compared against the level of patent protection in those countries via a multi-national "patent index." *See*, Maskus, *Lessons from Studying the International Economics of Intellectual Property Rights*, March 28, 2000 draft, *presented at* TAKING STOCK: THE LAW AND ECONOMICS OF INTELLECTUAL PROPERTY RIGHTS, SYMPOSIUM OF THE VANDERBILT LAW REVIEW, Nashville, Tennessee, April 7–8, 2000; *See also*, Keith Maskus, *The International Regulation of Intellectual Property*, WELTWIRTSCHAFTLICHES ARCHIV 134, 186–208.

a self-serving manner. But even survey results that must discount X percent of respondents for being self-serving can produce meaningful, helpful results in formulating intellectual property policy. I am thinking of the late Edwin Mansfield's survey work producing interesting, albeit criticized, results on the relationship between corporate perceptions of patent protection and investment decisions in developing economies.[4]

The lack of such data nowadays is definitely surprising and might even be telling. The National Science Foundation (NSF) and the National Academies of Sciences have shown increasing interest in studying intellectual property and its impact on science. But to date the funding has been concentrated in conferences and what I'd call "think pieces."[5] Perhaps this is a necessary step in the process of engaging the scientific community on IP issues. To be fair, the Academies' National Research Council (NRC) has definitely taken steps in the direction of more empirically based research. If university access to propertized research tools is a major issue, it is one begging to be studied empirically by NSF and NRC—with the gloomy survey results going directly to the Chair of the House Science Committee and his immediate predecessor (who happens now to be Chair of the House Judiciary Committee, overseeing IP legislation).

In making his case against any compulsory license or condemnation proceedings against biomedical or genomic patents in Chapter 8, Professor Epstein concludes that the "strong presumption should be in favor of a regime of strong property rights ... unless and until there is some clear showing of necessity that justifies the switch to a regime that requires forced purchases."

Let me extend that to the broader range of issues that Professor Boyle covers—and in a way that I hope most will agree with: *In any situation where we are debating property rights to information, there should be a strong presumption in favor of the status quo—whether it is a regime of property or commons or some mix thereof—unless and until there is some clear showing to justify a switch in regimes.* This more generalized principle should apply where there are rights—patents over biomedical research tools—and where there are not rights—as in uncopyrightable scientific database or subpatentable innovations and advances (as ESTs may be).

What counts as a "clear showing"? We often talk about market failure as the thing that must be shown, but we should be supple enough in our thinking

[4] Edwin Mansfield, *Intellectual Property Protection, Foreign Direct Investment and Technology Transfer*, International Finance Corporation Discussion Paper 19, The World Bank (1994); Edwin Mansfield, *Intellectual Property Protection, Direct Investment and Technology Transfer: Germany, Japan, and the United States*, International Finance Corporation Discussion Paper 27, The World Bank (1995).

[5] A fine example would be the National Research Council's extensive study/discussion of the impact of database protection on science, NATIONAL RESEARCH COUNCIL, BITS OF POWER: ISSUES IN GLOBAL ACCESS TO SCIENTIFIC DATA (1997). This very thorough and helpful discussion is, nonetheless, reinforced by no more than what I call "anecdata."

to contemplate *market improvement* as a basis for a "switch" in regimes. What is important is the standard for a clear showing. To me, it should be either a *very good* empirical case or a *very, very, very good* theoretical case.[6] The reason for the difference is obvious: If the theory is right, the real-world evidence will eventually become apparent and we can make the correction then, with limited losses in social utility.

The problem of extra-copyright protection for databases is a good case in point: There is simply no good empirical evidence, let alone *very good* evidence, for additional legal protection on the level of the 1996 EU Database Directive.[7] And the best theoretical case is for much more limited protection in the spirit of misappropriation doctrine. The proposals about the efficiency of "reward" versus patent systems may be another example—F. Scott Kieff has quickly shown that the proposals are interesting but clearly don't make the case beyond a reasonable doubt.[8]

For sure, there are downsides to this standard for credible reform proposals. One downside is when our judges "discover" that the intellectual property law really is different than we all thought it was. Presto, chango, there is a new status quo. This is, of course, the *State Street Bank* decision and the resulting flood of business method patents. A lot of people would like to undo this handiwork by Judge Giles Rich—whether as a judge or a legislative staffer decades before—and roll back the clock on business method patents for the sake of returning to the *old status quo*. I'd caution against that and stick to a demanding standard for an justification to change the law. After all, just as *State Street Bank* solidified an ambiguous status quo in a way that seemed to radically alter the landscape, the *Feist* decision also solidified an ambiguous status quo in a way that seemed to radically alter the landscape. It's just that one court established IP rights and the other killed them off.

II. PROPERTY RIGHTS AND HYBRID REGIMES

Professor Kieff in Chapter 7 notes that those unhappy with the system have three roads they can travel in advocating reform: changes in the baseline rights; changes in the rules governing the transaction when someone wants rights they do not have; and changes in how people understand, use, and sometimes misuse

[6] Or, perhaps, the empirical case for a switch in regimes could be proven by clear and convincing evidence, while the theoretical case for a switch would have to be proven beyond a reasonable doubt.

[7] Stephen M. Maurer, P. Bernt Hugenholtz, and Harlan J. Onsrud, *Europe's Database Experiment*, 294 SCIENCE 789 (October, 2001).

[8] F. Scott Kieff, *Property Rights and Property Rules for Commercializing Inventions*, 85 MINN. L. REV. 697, 705–717 (2001).

their rights. One rarely hears proposals concerning the third category, the "socialization" of the participants. Perhaps that's because everyone concludes that improving behavior through social atmospherics is even less likely to succeed than curtailing behavior through the strong arm of the law. But a similar conference held in East Asia or Africa, where social concord is more highly prized, might conclude differently.[9]

Among these choices, Professor Epstein argues in Chapter 8 that what should be avoided are "those clever schemes designed to split the difference," usually by limiting the traditional property rights of patents. Professor Epstein's intuition is that such hybrids are more trouble than they are worth, and that for any given field, one should adopt either a public domain approach (barring propertization) or leave well enough alone and let propertization through the patent system run its course. As I understand it, he advocates the public domain model for express sequence tags (ESTs) and the unabridged patent system for (at least most) remaining genomic products.

I think that Professor Epstein is right about hybrid systems generally, but Professor Keiff's triumvirate of reform avenues tells us that there are distinct ways to "split the difference." In the first, one starts with a baseline of traditional patent rights and conditions those rights through some system of compulsory licensing or government taking—this is the approach that is the principal target of Professor Epstein. In the second, certain kinds of innovation can be marked off for a weakened property regime as a whole, often inspired by the "petit patent" systems in place in many countries. We should recognize at least one virtue of the latter: It should be easier to understand and calculate the *ex ante* incentive effects of a system of weak baseline rights than a system of much stronger baseline rights with a set of uncertain limitation to those rights. Professor Jerome Reichman's writing on subpatentable innovations lays out the direction of empirical work that could be undertaken in this area.

I agree with many of Professor Epstein's detailed comments in this area. For example, he notes that it make much more sense to propose compulsory royalty rates based on gross revenues than profits. That must be right. Too much creative bookkeeping can be done with pretax profits. I recall Arnold Schwarzenegger commenting that the film *Titanic* made so much money not even

[9] For an example of this, when Singapore adopted dispute resolution procedures to deal with claims of cybersquatting in the <.sg> space, the Singaporeans adopted ICANN's Uniform Dispute Resolution Policy almost whole cloth, but added a distinct mediation procedure. *See*, SINGAPORE DOMAIN NAME DISPUTE RESOLUTION POLICY (Version 1–6 November 2001)., available at http://www.nic.net.sg/pdf/SDRP.pdf (Article 4(e) provides that the parties "will be invited to consider whether they wish to the dispute mediated by the Administrative Panel before the Administrative Panel is called upon to decide the dispute," then sets out procedures for such mediation.)

Hollywood accountants could hide its profitability. Even among friends doing deals at Wolfgang Puck's, one asks for a piece of the gross. In a compulsory licensing regime, that's unquestionably the most sensible course.

He also seems correct in his concerns about adapting a condemnation system from real property to patent rights. But if his criticism on this topic hit its mark, I'm not sure it was ever a living target. When have we seen serious discussion in Washington of condemnation as a broad policy option for the "patent thicket"? The best example of a high-profile discussion of condemnation might be the talk in late 2001 of "breaking" the patent on Cipro. But even with the Cipro example, most of what was actually discussed—as Professor Epstein's paper recognizes—was a compulsory license regime. It's just that that point got muddied in lots of press reports worldwide.[10]

Yet perhaps the Cipro example does point to some real facts that might temper Epstein's concerns about condemnation: Situations where there will be public outcry to "seize" or "break" a patent are situations where a product has already been brought to market, *i.e.*, private actors have cleared the patent thicket enough to bring the product to market.[11] In other words, if the U.S. Government had taken the Cipro patent, would any other patent holders have stepped forward to argue that their patent claims covered Cipro? How would a court have treated such a claim, unmentioned in the years that Bayer had been producing the drug?[12]

More broadly, the tough question with an condemnation or compulsory license regime is usually hidden by the polite talk about "reasonable" or "adequate compensation" or a licensing fee tied to "the commercial value of the product developed as a result of the research" (quoting Professor Donna Gitter). Professor Epstein asks the hard question: "Must the royalty in question allow the firm to

[10] *See, e.g.*, Lesley Russell, *Fallout from bioterrorism: the US Government has changed it attitude toward the pharmaceutical industry*, THE CANBERRA TIMES, November 19, 2001 at A9 ("The US Government can invoke compulsory licensing, break the patent and buy cheaper generic versions of Cipro, something Canada has already threatened. The US Government can do this with no hearing, and the companies involved cannot sue to prevent it. They can sue only for compensation."); Irwin Stelzer, Politicians take too broad a view of broadband, SUNDAY TIMES (LONDON), November 11, 2001, business section (commenting that "America's recent assault on Bayer's Cipro patent suggests that governments are willing to seize intellectual property when it suits them. . .").

[11] For better or for worse. In August 2001, the Federal Trade Commission was studying whether drug companies are trying to block the launch of generic drugs but is also exploring if the generic firms are colluding with the big players to benefit. The FTC had investigated Barr for an arrangement with Bayer AG over a deal the two reached to end a patent dispute over antibiotic Cipro, which resulted in Bayer paying Barr $25 million a year. Theresa Agovino, "Barr Labs Ships Version of Prozac," AP (Aug. 2, 2001).

[12] In other words, political pressure for government action that amounts to a "taking" tends to occur when the property in question is widely seen as valuable—and a patent won't be seen as widely valuable unless it is already being successfully exploited by the private sector (with all the clearing of the patent thicket that entails).

receive the same revenues that it would have gotten had it been allowed to keep its initial patent monopoly?"[13] But I'm not sure that he asks this question harshly enough. If the owner/innovator does not receive "full and perfect" compensation, don't we have a basic problem with our whole incentive structure apparatus?

Discussions of compulsory license regimes for successful products or processes—the only ones for which anyone would want a compulsory license—usually gloss over the fact that successful products are typically the result of an investment process that products lots of dead ends. One aspect of intellectual property-protected commerce generally is that it appears that only a small minority of works/products is ever profitable. One in 10 is a figure commonly used for profitability of Hollywood films, pop music albums, and pharmaceuticals, making each industry into a sort of crapshoot occupation.[14] Other inventive areas may have even worse statistics. This low success ratio is often true for successful individuals inventors and artists, as well as the successful, large commercial concerns. For the IP incentive structure to work, the winning lottery ticket must be made to look very seductive. And the winning ticket looks seductive because the winner gets to "command" whatever price the market will bear.

In the end, doesn't that mean that the incentive structure of intellectual property rights depends on the *prospect* of some unfriendly, antisocial behavior on the part of the property owner? At the end of the day, the price garnered by the successful, propertized technology depends on the proprietor's ability to refuse the sale. If the inventor (or investor) is being rational in her *ex ante* analysis of whether to try to *become* a proprietor of a successful technology, that monopoly price is an input in her calculation. (This is implicit in much of what is said by Professor Kieff in Chapter 7.[15]) One principled way out of this conundrum would be to say that inventors and investors are sufficiently *irrational*, and that they won't notice that our compulsory licensing system is skimming some of the cream off the victorious patents. Irrational creatures that they are, they'll keep going for the golden ring.

[13] From Professor Epstein's original conference paper, "Steady the Course: Property Rights in Genetic Material," at 40. Available at http://law.wustl.edu/Academics/Faculty/Kieff/HGPIP/Epstein2.pdf

[14] In the case of feature films, the numbers really are not so grim. It appears that around six out of 10 films eventually recoup their original investment through all distribution channels. *See* www.mpa.org/anti-piracy/content.htm (last visited February 20, 2002).

[15] "Ensuring use through a switch in the patent system towards over-all liability rule treatment may also frustrate the initial ability of the patent system to provide the *ex ante* incentives discussed earlier to commercialize new technologies."

III. GOAT-BOY ROAMS THE HALLS

On the whole, it seems like we are well past what Professor Epstein called "[t]he categorical objections to creating property rights in human or other living substances." And yet there is still the problem I affectionately call "Goat-Boy"—what will we do with patent applications for complete, modified genome sequences which include some substantial amount of human DNA? What do we do when our chimeric creations are no longer monoclonal antibodies, but half humans/half goats?

Professor Boyle's draft both (1) reminds us that these huge problems are out there, and (2) prods us to think not just about our substantive position, but whether intellectual property scholars have any proper role—or any *privileged* role—in the discourse. As Richard Posner has recently told us, when academics speak outside their expertise, "quality tends to go to hell."[16] (I guess, as a judge, he is allowed to speak about anything.)

I have a simple thought on this: Congress and the USPTO are definitely going to be immersed in these debates—the opening shots have already been fired. So, if intellectual property academics stay out, they will feel left out. Certainly the view that you "can't patent a human being because it would be tantamount to slavery" needs to be addressed by people who are experts in human rights, but it also needs to be addressed by people who understand what patent claims can and cannot do.

IV. A FINAL ADAGE

In discussing some of these issues briefly with a friend at the NRC, he reminded me of the adage about the destiny of writers who stick to theory: A theorist has but two fates—the noble one is that his theories will be disproved; the more common is that they will simply be forgotten. With time, it would be nice to see many of our theoretical fears noted and disproved, either because we were wrong in theory or because our theories turned out to be self-denying prophesies, prompting changes that helped the future skirt the predicted dangers.

[16] Richard A. Posner, *The Professors Profess*, THE ATLANTIC MONTHLY, February 2001 at 28.

13

Comment on the Tragedy of the Anticommons in Biomedical Research

Edmund W. Kitch
Mary and Daniel Loughran Professor of Law
E. James Kelly, Jr.—Class of 1965 Research Professor
University of Virginia
Charlottesville, Virginia 22903

The patent laws are a continuing experiment. They enable an inventor to obtain an exclusive right to his invention if the application satisfies a small number of general principals. The invention must cover patentable subject matter;[1] must be novel[2] and nonobvious;[3] the inventor must be able to explain how the invention is made and used;[4] and the inventor must be able to describe the invention in claims.[5] There is no requirement, however, that the inventor be able to convince officials that the invention is important. This enables the patent laws to respond to inventions whose significance and importance is not understood at the time they are made. There are many theories about the patent system, but whether or not it works in practice is a matter of experience, not theory.

The structure of the claims of the particular patents that issue and the identity of the persons or firms that own them is the result of the nature of the technology, the degree of diffusion of the basic knowledge the inventions incorporate, the timing of the inventive effort, and the timing of the patent applications. One early inventor may succeed in obtaining a patent with broad claims because of a lack of prior art, yet fail to identify the critical steps necessary to make the invention a success. Another may do brilliant early work, yet end up with a patent whose claims cover nothing of subsequent applied

[1] 35 U.S.C.§ 101.
[2] 35 U.S.C. § 102.
[3] 35 U.S.C. § 103.
[4] 35 U.S.C. § 112.
[5] *Ibid.*

0065-2660/03 $35.00

importance. Patents on all the critical technology may end up with a single firm, or they may end up in many different hands. Because the patent application often precedes an understanding of the commercial significance and direction of the technology,[6] no one has an advanced understanding of an optimum allocation of patent rights.

The patent system is structured to overcome this basic problem because the rights conferred by a patent can be reallocated by contract. Thus, after the patents issue, the affected firms can reallocate the rights so as to maximize their value. Since the arrangements that maximize their private value will almost always be the arrangements that maximize their public value, this structure addresses the problem of the impact of issued patents on the social value of inventions.

In an important article, Michael A. Heller and Rebecca S. Eisenberg argue that patents might not work well in the area of modern biomedical research.[7] They argue that the increased role of private actors in the conduct and funding of the research, combined with developments in the technology and the patent system facilitating the grant of patents at an early stage of the research process may deter innovation. They describe this process as "The Tragedy of the Anticommons." A number of the papers presented at this conference echo their arguments.

However, the arguments are based on theory, not experience. The claim is that issued patents can block access to critical technologies and slow progress in a field. So it might be. It is easy to imagine that a patent might issue covering a fundamental technique whose use is essential to all researchers. One can imagine that the owner of such a patent might decide to keep the patented technology solely for his or her own use. One can imagine many plausible scenarios in which that sequence of events would slow progress.

It is true that there has been an unusually aggressive use of patenting by some actors in biomedical research, particularly in human genomics. Patents have issued on gene fragments whose function and use is unclear. But it is notable that no one who expresses these concerns points to particular patents or particular patent licensing policies that have caused problems. Patents on basic research techniques are licensed widely at license fees which the research community is prepared to pay. These patents, often owned by research institutions that made pioneer investments in the field, generate licensing fees that strengthen the finances of those research institutions. The field, meanwhile, continues to advance and the level of activity is high.

[6] I explained the reasons why this is so in Edmund W. Kitch, *The Nature and Function of the Patent System*, 20 J.L. & ECON. 265, 269–271 (1977).

[7] Michael A. Heller and Rebecca S. Eisenberg, *Can Patents Deter Innovation? The Anticommons in Biomedical Research*, 280 SCIENCE 698 (1998).

Identifiable biotechnology patents were issued by the patent office more than thirty years ago. At the time they were issued, few if any appreciated that they were part of the foundation for what would become a dramatic and important new field of technology. The early patents have already expired. Surely by now commentators concerned that the Tragedy of the Anticommons is an actual problem should be able to point to particular patents or the licensing policies of particular patent owners that have slowed progress in the field.

The failure to identify any patent, patents, or patent owners that have had this effect means that it is impossible for anybody, public or private, to construct a solution to the problem. At the outset of America's entry into World War I, the Secretary of War was so concerned about the conflict between the Wright Brothers patent on a method of maintaining lateral control of an airplane by warping the wing and the Glen Curtiss patent on wing flaps, that the War Department, with help from Congress (which threatened to condemn the patents), persuaded the parties to form an aviation patent pool.[8] Can anyone identify patents in the biomedical research area that create similar problems today? If they can be identified, numerous solutions are available.

So far, it appears that the Tragedy of the Anticommons in the area of biomedical research is something that could have occurred as a matter of theory. It is not as yet, however, a problem that has been shown to have actually occurred. At least so far, the patent system in this area appears to have been an experiment that has worked.

[8] Described in George Bittlingmayer, *Property Rights, Progress, and The Aircraft Patent Agreement*, 31 J. OF LAW & ECON. 227, 230–232 (1988).

14

An Outsider Perspective on Intellectual Property Discourse

David A. Hyman
Professor of Law
University of Maryland School of Law
Baltimore, Maryland 21201

I. INTRODUCTION

"In God We Trust, All Others Bring Data"[1]

Outsiders who are invited to conferences on specialty areas have several significant comparative advantages. As "innocents abroad," they are free to question the assumptions, rhetorical tactics, and doctrinal strategies that insiders simply accept as the rules of trade. Outsiders can also arbitrage applicable insights from their home discipline; it is usually more efficient to import applicable knowledge than to build it from scratch. When trade across disciplinary boundaries is unrestricted, the work goes faster, and cross-hybridization strengthens at least one (and potentially both) fields. Finally, outsiders are one-shot players with limited reputational assets at stake, so they are free to trample scholarly conventions, upset apple-carts, kick sacred cows, and generally make nuisances (albeit preferably attractive ones) of themselves. The result is that outsiders

[1] Charles M. Cutler, *Research Needs For Managed Care*, 15 HEALTH AFF. 93 (1996).

0065-2660/03 $35.00

can sometimes teach insiders profound things about themselves and their fields—think Alexis de Tocqueville's *Democracy in America.*

Other times, outsider efforts fall flat or are affirmatively misleading—think everything written by the French about America and Americans other than Democracy in America.[2] This problem arises because outsiders labor under significant comparative disadvantages. Specialties exist because there are gains from specialization—no sensible person picks a pathologist or convenience store clerk to perform neurosurgery. The complexities and subtleties that insiders spend years studying do not cease to exist merely because outsiders ignore them. Oliver Wendell Holmes concisely stated the danger of dismissing insider "knowledge": "ignorance is the best of law reformers. People are glad to discuss a question on general principles, when they have forgotten the special knowledge necessary for technical reasoning."[3] This reality also complicates the benefits of arbitraging insights across disciplines. Not all analogies are analogous, and the process of cross-disciplinary arbitrage frequently flattens and distorts important details on both sides of the disciplinary divide.[4]

With the hope that my comments will end up closer to the former than the latter, I offer the following observations, largely drawn from work I have done on narrative and legislation by anecdote.[5] The most striking thing about the conference papers is the mismatch between the presence of confident factual assertions about the state of the world, and the complete absence of empirical evidence supporting the same. Although this rhetorical tactic was pervasive, for the sake of simplicity, I focus on one issue that gave rise to much discussion at the conference.[6] On numerous occasions, it was stated that research and clinical treatment are being hampered by the existence of property rights in genes and DNA sequences. Speakers pointed to the exorbitant costs of performing screening

[2] *See* Andrew Sullivan, *America Knows Who Its Friends Are*, http://www.sunday-times.-co.uk/article/0,981-287105,00.html ("For most Americans, when the French call something simplistic, it's usually a sign that it's the right thing to do.").

[3] Oliver Wendell Holmes, THE COMMON LAW 64 (1963).

[4] This is a standard risk with "law and a banana" scholarship. *See* Mark A. Graber, *Law and Sports Officiating: A Misunderstood and Justly Neglected Relationship*, 16 CONST. COMMENTARY 293, 295 (1999) ("Casual legal interlopers into other disciplines risk making bald assertions that serious scholars in the non-legal field recognize as flatly wrong, if not downright silly . . . Relying exclusively on categories derived from the study of legal phenomena, law professors may miss the most interesting features of their interdisciplinary subject, features which generate entirely different models of interpretation, evaluation, and decision making.") *See also* Richard A. Posner, *Against Constitutional Theory*, 73 N.Y.U. L. REV. 1, 15 (1998) ("Analogies are typically, as here, inexact and often, as here, misleading.").

[5] *See* David A. Hyman, *Lies, Damned Lies & Narrative*, 73 IND. L. J. 797 (1998) [hereinafter *Narrative*]; David A. Hyman, *Do Good Stories Make For Good Policy?* 25 J. HEALTH, POLITICS, POL'Y & L. 1149 (2000) [hereinafter *Stories*].

[6] Lest I be thought to be picking on anyone in particular, the conference papers are replete with numerous examples of this phenomenon.

tests for genes which had been patented, the inconvenience and delay associated with having to send samples to a centralized location to have such tests performed, the chilling effects on the speed and direction of basic research when one must pay licensing fees or surrender future rights to obtain access to patented materials, the deleterious impacts on the social norms of science, including the demise of free and open sharing of research results and materials, and the like.

Leave aside all the obvious theoretical responses to such claims, such as wondering about the implications of such an analysis for the propriety of pharmaceutical patents more generally,[7] and the incentives for innovation in the face of an unduly casual attitude toward the scope and enforceability of intellectual property rights.[8] Leave aside as well the churlish suggestion that the golden age of science wasn't all that golden.[9] Focus instead on the evidentiary foundation (such as it is) for these claims. Those making these statements really didn't offer much in the way of supporting data, beyond an occasional highly salient example or first-person testimonial.[10] A charitable commentator would describe such information as "anec-data," and emphasize the good faith and exemplary reputation of those advancing such claims, and the prestigious

[7] The reasoning is exactly parallel and virtually syllogistic:

1. Being sick is a bad thing;
2. Drugs to treat illness are expensive because they are patented;
3. Patient access to drugs is restricted because of their cost;
4. Patents should not be granted on drugs or patent rights should be easily breakable, through mandatory licensing or expropriation if prices are unreasonable.
5. QED.

[8] I must confess a substantial degree of agnosticism on this point, because I lack the specialized knowledge to assess the trade-offs. However, the lesson from condemnation cases is that property-holders are invariably under-compensated by any regulatory regime designed to make them whole for expropriation, because a wide variety of property-owners' costs are not counted as costs (*e.g.*, relocation and demoralization costs), and regulatory takings are not typically counted as takings. Thus, even though the purpose of the takings clause is "to bar government from forcing some people alone to bear public burdens which, in all fairness and justice, should be borne by the public as a whole," ex post compensatory regimes do not, in practice, prevent such conduct. *Armstrong v. United States*, 364 U.S. 40, 49 (1960).

[9] *See, e.g.*, F. Scott Kieff, *Facilitating Scientific Research: Intellectual Property Rights and the Norms of Science. A Response to Rai and Eisenberg*, 95 NW. U. L. REV. 691 (2001). *See also* Arti K. Rai, *Evolving Scientific Norms and Intellectual Property Rights: A Reply to Kieff*, 95 NW. U. L. REV. 707 (2001).

[10] The designated "bogeyman" of the story seems to be Myriad Genetics Inc., which has patented the BRAC1 and BRAC2 gene sequences. For what it is worth, Myriad Genetics has apparently agreed to offer discount pricing for sequencing that is conducted exclusively for clinical research, and unflinchingly defends its rights to otherwise charge what the market will bear. *See* Kimberly Blanton, *Corporate Takeover Exploiting the Patent System*. BOST. GLOBE, Feb. 24, 2002, at 10. Not surprisingly, this has occasioned considerable complaint from governmental purchasers in other countries (*e.g.*, France and Canada), who have gotten used to using price controls and monopsony purchasing power to obtain pharmaceuticals at below-market rates.

schools with which they are associated. A less charitable commentator might use words like "anecdote" and point out that good faith, exemplary reputation, and a prestigious professional address are all charming attributes, but they fall far short of demonstrating that the assertion in question is accurate and representative. An uncharitable commentator might use words like "urban legend" and suggest that such "evidence" is unhelpful both in diagnosing the problem and in specifying the appropriate solution.

Still more interesting is the evident lack of concern (let alone embarrassment) about the dearth of empirical evidence on the subject in question. With one exception, this issue was simply not mentioned at the conference.[11] To be sure, the problem is not unique to this conference or this subject, but results from the selection and socialization processes that produce lawyers and law professors, along with the incentives under which these professions operate.[12] Professor Rosenberg neatly stated the problem (along with its causes and consequences) at a conference on civil procedure almost 15 years ago:

> The tendency of legally-trained minds to prefer thinking to counting is legendary. So is the lawyer's preference for learning by watching for the vivid case rather than tabulating the mine-run cases. The problem is not that watching this case or that is useless. A dramatic case or anecdote may be more informative and more memorable than a tubful of printouts. But the rub is that good anecdotes do not care if they are not representative; they can be badly misleading if generalized. Nor does the problem end with the misleading anecdote. No matter how carefully the facts or data are gathered to respond to the pivotal

[11] *See* Justin Hughes, "Goat Boy Roams the Halls," *supra* at 263.

[12] *See* Michael Heise, *The Importance of Being Empirical*, 26 PEPPERDINE L. REV. 807 (1999); Peter Schuck, *Why Don't Law Professors Do More Empirical Research?* 39 J. LEG. ED. 323 (1989).

On the other hand, asking law professors to do empirical research may be asking for trouble. *See* Lee Epstein & Gary King, *The Rules of Inference*, 69 U. CHI. L. REV. I (2002). *But see* Frank Cross, Michael Heise & Gregory C. Sisk, *Above the Rules: A Response to Epstein and King*, 69 U. CHI. L. REV. 135 (2002); Jack Goldsmith & Adriane Vermeule, *Empirical Methodology and Legal Scholarship*, 69 U. CHI. L. REV. 153 (2002); Richard L. Revesz, A *Defense of Empirical Legal Scholarship*, 69 U. CHI. L. REV. 169 (2002) (offering various defenses of empirical legal scholarship, and criticism of Professor Epstein & King's analysis). There is also a demand-side element; legislators and lobbyists generally do not pay attention to the studies that do exist, preferring to legislate on the strength of anecdote. *See* David A. Hyman, *Drive-Through Deliveries: Is "Consumer Protection" Just What the Doctor Ordered?* 78 N. C. L. REV. 7 (1999) (detailing how prohibition on drive-through deliveries was enacted on the strength of anecdotal complaints, even though available empirical evidence indicated practice was safe and effective); Michael Heise, *The Future of Civil Justice Reform and Empirical Legal Scholarship: A Reply*, 51 CASE W. RES. L. REV. 251, 251–54 (2000) (same, for civil justice reform).

> questions, there will be great trouble in penetrating made-up minds . . . [Lawyers, lawmakers, and judges] prefer anecdotes to tables.[13]

My criticism of prevailing practices in intellectual property discourse, and of legal academic discourse more generally, invites several obvious comebacks. What is the evidence that my complaints are anything more than anecdotal?[14] What's wrong with relying on anecdotal evidence when analyzing issues and proposing reforms? Isn't the development of the common law based on anecdotal evidence (*i.e.*, the facts of each individual case)? Doesn't anecdotal evidence crystallize and mobilize public, legislative, and scholarly opinion on even the most dull and arcane subject?[15] Don't good anecdotes simply speak for themselves? What is wrong with a form of proof that is both simple and transparent—particularly when math and statistics-phobic law professors are the ones dealing with the data?[16]

Despite these rhetorically appealing comebacks, there are, in fact, substantial risks associated with relying on anecdotal evidence (or "anec-data" for those determined to bootstrap their rhetorical position), which explain why it is shunned by scientists and medical researchers.[17] The problems with "anec-data" are usefully analyzed in terms of truthfulness/completeness, and typicality.

[13] Maurice Rosenberg, *Federal Rules of Civil Procedure in Action: Assessing Their Impact*, 137 U. PA. L. REV. 2197, 2211 (1989).

[14] Admittedly, my sample of intellectual property discourse is limited to attending a single conference, and reading some of the literature. As such, I am effectively drawing conclusions about the forms and modes of discourse in a field that I have not studied systematically. This "pot calling the kettle black" critique of my observations would have more merit had other intellectual property scholars not made similar observations. *See* Hughes, *supra* note 11, at 263–265.

[15] *See* Hyman, *Narrative*, *supra* note 5, at 800, note 14 (noting Senator Kennedy's use of consumer complaints about the abuse of pets shipped as cargo ("frozen dog" anecdotes) to bring issue of airline deregulation to life); *International Board of Teamsters v. United States*, 431 U.S. 324, 339 (1977) (anecdotal evidence brings "the cold numbers convincingly to life.").

[16] *See, e.g.*, Catharine MacKinnon, *Law's Stories as Reality and Politics*, in LAW'S STORIES: NARRATIVE AND RHETORIC IN THE LAW 232, 237 (1996) (Professor MacKinnon admits that "she bursts into tears at columns of figures."); Blake Fleetwood, *From the people who brought you the twinkie defense*, 19 WASH. MONTHLY 33 (1987) ("Many [judges] weren't that good at math or science or statistics [which is] why they went to law school.").

[17] *See* Michael J. Saks, *Do We Really Know Anything About the Behavior of the Tort Litigation System - And Why Not?*, 140 U. PA. L. REV. 1147, 1159–61 (1992) ("[A]necdotal evidence is heavily discounted in most fields, and for a perfectly good reason: such evidence permits only the loosest and weakest inferences about matters a field is trying to understand. Anecdotes do not permit one to determine either the frequency of something or its causes and effects . . . and have the power to mislead us into thinking we know things that anecdotes simply cannot teach us."). In particular, anecdotal evidence cannot distinguish causation from coincidence, reporting error, self-deception, observer bias or intentional fraud.

II. TRUTHFULNESS/COMPLETENESS

People lie. When they aren't affirmatively lying, they often shade the truth, downplaying some facts and emphasizing others, to enhance the persuasiveness of the story they are telling. As such, "uncomfortable" facts, the context necessary to appreciate the reasons for existing institutional arrangements and the adverse consequences associated with proffered reforms, are likely to be omitted entirely from the anecdotes that are offered.[18] As my colleague, Bob Condlin has written, "such inclusion and exclusion of data is the story-teller's prerogative, presumably because it is not relevant to the message she wants to convey (in other words, to the story she wants to tell). But that is the problem with stories. They are always an advocacy move, used as much to make a point as to discover one, even if the storyteller does not think so."[19] This ability to "load the evidentiary dice" suggests that anecdotes should be approached with considerable skepticism—and the more egregious the described conduct, the greater the degree of skepticism required.

Heightened skepticism is also required because these anecdotes do not emerge into public view at random or by accident. Instead, they are sought out, packaged and "spun" by policy entrepreneurs and advocacy groups, who use them to further their agenda. If the "spin" sometimes overtakes the facts, most advocates can doubtless convince themselves that they have committed no great sin, since they *know* they are on the side of the angels.[20] Indeed, with sufficiently diligent effort, advocacy groups can even offer simultaneous anecdotes which support diametrically opposed positions.[21]

[18] *See* Martha Minow, *Stories in Law*, *in* LAW'S STORIES, *supra* note 16, at 24, 31 (noting problem of selectivity in storytelling, and conscious refusal to include "additional stories which convey unattractive features of the community that I was trying to paint in a sympathetic light.").

[19] Robert J. Condlin, *Learning From Colleagues: A Case Study in the Relationship Between "Academic" and "Ecological" Clinical Legal Education*, 3 CLIN. L. REV. 337, 339–40 n. 29 (1997) (commenting on narrative account of relationship with client which omits details of horrific crime for which client was convicted and sentenced to death.).

[20] *See, e.g.*, Katharine Dunn, *Fibbers: The Lies Journalists Tell*, THE NEW REPUBLIC, June 21, 1993, at 18 ("Of all the lies that are swallowed and regurgitated by the media, the ones that hurt the most come from the Good Guys, the grass-roots do-gooders, the social work heroes, the non-profit advocacy groups battling for peace, justice and equality . . . a lot of reporters don't check facts provided by non-profit organizations because they assume non-profits don't have anything to gain by lying . . . The well-meaning grow desperate for results and stoop to the tactics of their enemies. It happens all the time"); Daniel Koshland, *Scare of the Week*, 244 SCIENCE 9 (1989) ("Each group convinces itself that its worthy goals justify oversimplification to an 'ignorant' public.")

[21] *See* Hyman, *supra* note 5, at 804–6 (presenting competing anecdotes on tort reform and property rights/environmentalism). *See also* Gina Kolata, *Ethicists Struggle Against the Tyranny of the Anecdote*, N.Y. TIMES, June 24, 1997, at C4 (competing anecdotes struggle to define global appropriateness (or lack thereof) of physician-assisted suicide).

When the only source of information we have is someone complaining about the conduct in question, it is fair to wonder whether we are getting the truth, let alone "the whole truth and nothing but the truth."[22] The result is that anecdotes frequently mis-frame, if not completely misrepresent the costs and benefits of the status quo and its alternatives. It is no accident that the legal system generally declines to take action on the say-so of one party[23] and looks with considerable disfavor on limitations on the right to confrontation and cross-examination.[24] These questions about truthfulness and completeness obviously call into question the utility of anecdotal evidence.

III. TYPICALITY

Scrupulously accurate and complete narratives can still be unrepresentative. Atypical narratives can lead to the adoption of policies which make the

[22] As Professor Tushnet noted in criticizing news accounts of political correctness, "[t]he victim's account of the incident is the only source of evidence. The reports never note that victims have a perfectly understandable desire to present what happened to them in a way that makes them appear best. When the reports are offered by people with a political ax to grind, one can fairly wonder exactly what happened." Mark Tushnet, *Political Correctness, the Law, and the Legal Academy*, 4 YALE J. L. & HUMAN. 127, 131 (1992).

[23] *See Fuentes v. Shevin*, 407 U.S. 67, 83 (1972) ("Because of the undertandable, self-interested fallibility of litigants, a court does not decide a dispute until it has had an opportunity to hear both sides - and does not generally take even tentative action until it has itself examined the support for the plaintiff's position.").

[24] *See Coy v. Iowa*, 487 U.S. 1011, 1016, 1019–1020 (1994) ("[T]he Confrontation Clause guarantees the defendant a face-to-face meeting with witnesses appearing before the trier of fact.... It is always more difficult to tell a lie about a person 'to his face' than 'behind his back.'"); Christopher B. Mueller & Laird C. Kirkpatrick, EVIDENCE 6.29 (1995) ("If the calling party's opponents cannot subject the witness to cross-examination for reasons that are not his fault, some remedy is necessary... If cross-examination is permanently blocked, the direct testimony usually should be stricken in both civil and criminal cases, or a mistrial declared if the direct testimony is critical and striking it would not be effective.").

It is these attributes that differentiate the common law approach from pure regulation by anecdote, since common law courts have at least some tools with which to attempt to get at the truth of the matter. It is far less clear that common law courts do a good job formulating efficient rules of decision when their raw materials are atypical/unrepresentative fact situations. *See* Andrew Morriss, *Bad Data, Bad Economics, and Bad Policy: Time To Fire Wrongful Discharge Law*, 74 TEX. L. REV. 1901, 1914 (1996) ("Courts created wrongful discharge law on a foundation of anecdotes drawn from the peculiar sample of cases that reach state courts. In general, anecdotes are a poor basis for public policy. Anecdotes gathered by surveying people in lawsuits are even worse. The grim picture of the workplace those anecdotes paint is contradicted by the evidence that does exist about the extent of the problems employees face in the workplace.") *But see* George L. Priest, *The Common Law Process and the Selection of Efficient Rules*, 6 J. LEGAL STUD. 862 (1982); Paul Rubin, *Why is the Common Law Efficient?* 6 J. LEGAL STUD. 51 (1977) (arguing that the common law is efficient).

underlying problem worse, or cause other unintended consequences. The problem was nicely framed by Professor Saks:

> the trouble with legislation by anecdote is not just that some of them are false or misleading. Even if true and accurate, anecdotes contribute little to developing a meaningful picture of the situation about which we are concerned. It makes a difference if for every ten anecdotes in which an undeserving plaintiff bankrupts an innocent defendant, one, ten, one hundred, or one thousand equal and opposite injustices are done to innocent plaintiffs. The proportion of cases that results in one or the other error, and the ratio of one kind of error to the other, ought to be of greater interest to serious policy-makers than a handful of anecdotes on either side of the issue. Reforms are intended to change that ratio and the tens of thousands of anecdotes the ratio summarizes.[25]

Absent proof of representativeness/typicality, a single narrative in support of or opposition to a particular policy could be just that: singular. Even if a single anecdote is representative of a larger reality, one must know how frequent that larger reality actually is before deciding what, if anything, to do about it.[26] More generally, context (*i.e.*, how the mine run of cases are handled) matters a great deal more than the facts—however good or bad they may be—of any given case in assessing the overall merits of the system.[27] Unfortunately, disregarding this point can result in "reforms" which are intended to address the small percentage of transactions that go poorly, but end up disrupting the majority of the market, which works tolerably well.[28]

[25] *See* Saks, *supra* note 18, at 1161. *See also* Richard A. Epstein, *Legal Education and the Politics of Exclusion*, 45 STAN. L. REV. 1607, 1619–620 (1993) (issues of truthfulness, frequency, and typicality preclude generalization).

[26] *See* Richard Posner, *Legal Narratology*, 64 U. CHI. L. REV. 737, 742–44 ("The significance of a story of oppressiveness depends on its representativeness. In a nation of more than a quarter of a billion people all blanketed by the electronic media, every ugly thing that can happen will happen and will eventually become known; to evaluate policies for dealing with the ugliness we must know its frequency, a question that is in the domain of social sciences rather than of narrative... The risk of narratology to which MacKinnon herself succumbs in her writings on pornography is that of atypicality. MacKinnon is a magnet for the unhappy stories of prostitutes, rape victims, and pornographic models and actresses. Even if all these stories are true (though how many are exaggerated? Does MacKinnon know?), their frequency is an essential issue in deciding what if anything the law should try to do about the suffering that the stories narrate.")

[27] *See* Richard A. Epstein, *Discussion*, 45 STAN. L. REV. 1671, 1678 (1993) ("[If] you are trying to understand the way in which social reality works then the important thing to remember is that the prosaic and the boring is often far more important in the way in which the world organizes itself than is the exotic and profane.").

[28] Legal academics are particularly prone to this tendency. Any deviation from absolute perfection in the performance of a system is typically taken as a license to tinker, if not up-end the

As if matters were not already complicated enough, anecdotes are most persuasive when they appeal to our passions and prejudices—and the better they are at doing so, the more likely they are to be credited as representative, whether they are or not.[29] Even if a story is highly representative, other considerations may dictate a policy diametrically opposed to the one suggested by the story.[30] Finally, by its very nature, anecdotal evidence worsens the tendency in policy debates to privilege identifiable lives over statistical lives—hardly a recipe for sensible and cost-effective policies.[31]

IV. FROM THEORY TO PRACTICE

Let me recast my analysis from the meta-theoretical to the concrete and offer a highly representative anecdote about the risks of using anecdotal evidence. During 1997, Congressional hearings were held on the performance of the Internal Revenue Service (IRS). A handful of taxpayers offered testimony about their mistreatment at the hands of the IRS. The cases included a taxpayer who committed suicide because of a long-running bitter dispute with the IRS, and

entire system. Such attitudes empower legal academics to second-guess those who must actually live with the system, but it is a classic example of "nirvana fallacy" reasoning in action. *See* Harold Demsetz, *Information and Efficiency: Another Viewpoint*, 12 J.L. & ECON. 1, 1 (1969) ("The view that now pervades much public policy economics implicitly presents the relevant choice as between an ideal norm and an existing 'imperfect' institutional arrangement. This nirvana approach differs considerably from a comparative institution approach in which the relevant choice is between alternative real institutional arrangements.").

Worse still, when things are working tolerably well, they are, almost by definition, not reducible to salient anecdotes, and so the tolerable performance of the system never registers in the (exclusively anecdotal) scheme of things. A comparative institutional analysis helps moderate these academic tendencies, because it makes it clear that "bad is often best, because it is better than the available alternatives." Neil K. Komesar, IMPERFECT ALTERNATIVES: CHOOSING INSTITUTIONS IN LAW, ECONOMICS, AND PUBLIC POLICY 204 (1994).

[29] *See* David A. Hyman, *Regulating Managed Care: What's Wrong With A Patient Bill of Rights*, 73 S. CAL. L. REV. 221, 241 (2000) ("the more compelling the anecdote, the less likely we are to consider issues of typicality and frequency—meaning the risk of being led astray is a direct function of the persuasiveness of the anecdote.")

[30] *See* Hyman, *Stories*, *supra* note 5, at 1153 (2000).

[31] *See* Clark C. Havighurst, James F. Blumstein, and Randall Bovbjerg, *Strategies in Underwriting the Cost of Catastrophic Disease*, 40 LAW & CONTEMP. PROBS. 122, 140–41 (1976) (contrasting willingness of society to sacrifice identifiable lives v. statistical lives; "it is difficult to improve significantly on the commonplace observations that human beings cannot empathize with faceless abstractions and that 'squeaking wheels'—the complaints of known victims, such as the very vigorous lobbying of kidney-disease patients—not the silence of statistical unknowns will get the government grease. Spending 'millions to save a fool who has chosen to row across the Atlantic has external benefits' lacking from highway safety spending.").

another taxpayer who alleged that IRS agents invaded his home, knocked his colleague's 12-year old son to the ground and held him there at gunpoint, and forced his 14-year old daughter to get dressed in front of them.[32] The picture that was painted was of an agency that was, at best, inflexible, and, at worst, out of control.[33] The Clinton Administration attempted to finesse the issue by having the IRS Acting Commissioner apologize for his agency's conduct in any given case, while the President simultaneously defended the IRS' behavior as a general proposition.[34] However, Congress was sufficiently outraged that they passed a law which placed new strictures on the IRS, including ten offenses (the "ten deadly sins") for which the sanction was immediate termination.[35] Subsequent review demonstrated that there were serious concerns about the truthfulness and completeness of some of the proffered testimony,[36] and no evidence to indicate the conduct in question was representative.[37] As one commentator noted, "horror stories happen but they are relatively rare. Mistakes are inevitable at an agency with more than 100,000 employees who process more than 200 million returns every year."[38] The legislation has also had unintended consequences; the IRS Commissioner has admitted that "trying to balance customer service and compliance created an 'excruciating dilemma,' "[39] and enforcement actions have declined precipitously as taxpayers threaten to report agents for violation of one of the ten deadly sins.[40] Thus, the perils outlined previously are quite real and cannot be dismissed on theoretical grounds.

[32] *See* Leandra Lederman, *Of Taxpayer Rights, Wrongs, and A Proposed Remedy*, 87 TAX NOTES 1133 (2000).

[33] Editorial, *Reforming the Tax Collector*, N.Y. TIMES, Sep. 26, 1997, at A26 ("for many in Congress the anecdotes reflect a hostile, corrupt agency that cannot correct itself"); John M. Broder, *Demonizing the I.R.S.*, N.Y. TIMES, Sep. 20, 1997, at D1 (recounting anecdote of taxpayer who committed suicide because of IRS harassment; "Committee staffers say that while the bleak conclusion of Mr. Kugler's tale is obviously more extreme than most, it is emblematic of the way the I.R.S. operates—inflexible, insensitive, intrusive and, ultimately, ineffective.").

[34] Peter Baker, *Clinton Defends IRS*, WASH. POST, Oct. 01, 1997, at A6.

[35] *See* IRS Restructuring & Reform Act of 1998, Pub. L. No. 105–206, 112 Stat. 685 (1998).

[36] *See* General Accounting Office, Tax Administration: Allegiations of IRS Employee Misconduct GAO/GGD-99-82 (indicating that several taxpayer assertions at hearing of IRS misconduct could not be substantiated).

[37] *See* Michael Hirsh, *Behind the IRS Curtain*, NEWSWEEK, Oct. 6, 1997, at 29, 30 ("if the IRS hearings were high drama, they failed to shed much light on how widespread such abuses are, where they occur or who's really responsible.") To the extent there is a problem, Congress should not escape its share of blame, since they wrote the laws the IRS is trying to enforce. *See* Paul Glastris, *Lien on Congress*, U.S. NEWS & WORLD REP. Oct. 6, 1997, at 32.

[38] Paul Wiseman, *IRS May not be the Monster Critics Say It Is*, USA TODAY, Nov. 5, 1997, at 17A.

[39] *See* Cindy Zirkle, *IRS Decline in Enforcement Activity Draws Congressional Interest*, CCH Business Owners' Toolkit, http://www.toolkit.cch.com/columns/taxes/00–289irsenforce.asp

[40] *See* Brain Friel, *Treasury asks Congress to Make IRS' Ten Deadly Sins Less Deadly*, GOVERNMENT EXECUTIVE MAGAZINE, http://www.govexec.com/dailyfed/0202/020702b1.htm ("IRS

V. CONCLUSION

Unfortunately, empirical legal scholarship is the paradigmatic "dog that doesn't bark in the night."[41] However, the absence of such scholarship is particularly problematic when the stakes are the regulation of intellectual property.

I recognize that many people will be inclined to dismiss the risks I have outlined in this piece, and proceed full-steam ahead with anecdote-driven legislation and regulation. Before doing so, they should have to first explain why anecdotal evidence will give a more accurate and complete picture of the world of intellectual property and its imperfections than it did with the alleged campaign to torch black churches; the supposed rampant child abuse in day-care centers; the purported rape crisis on college campuses; the alleged poisoning of the country by alar, saccharin, cyclamates, and electromagnetic forces emanating from high-voltage power lines; the supposed problem of domestic violence on Super Bowl Sunday; and, of course, the purportedly pervasive problem of IRS misconduct.[42] In each of these instances, anecdotes gave a highly misleading impression of the existence and frequency of the "problem." The sequence of events is as unmistakable as it is invariant: "call it the 'whoops factor,' a phenomenon that starts with shoddy research or the misinterpretation of solid research, moves on quickly to public outcry, segues swiftly into the enactment of new laws or regulations, and often ends with news organizations and some public policy mavens sounding like the late Gilda Radner's character Emily Litella, as they sheepishly chirp, "Never mind!"[43]

What then should be done? When dealing with anecdotes, one should be exceedingly skeptical and substantially discount the value of the information in light of its provenance and questionable generalizability. Indeed, in the absence of good data, "don't just do something, sit there" turns out to be good advice—perhaps the best of all possible advice.[44] It is unlikely the regulation of intellectual property will turn out to be an exception to this general rule.

---

revenue agents have complained that some recalcitrant taxpayers have used the Ten Deadly Sins to stall IRS action against them. Agents have also said that the list has made them more hesitant to pursue tax violators. IRS enforcement activity has fallen in recent years.")

[41] David A. Hyman, *A Second Opinion on Second Opinions*, 84 VA. L. REV. 1439, 1462 (1998).

[42] *See* Steven A. Holmes, *It's Awful! It's Terrible! It's . . . Never Mind*, N.Y. TIMES, July 6, 1997, at E3.

[43] *See id.*

[44] David A. Hyman, *Medicine in the New Millenium: A Self-Help Guide for the Perplexed*, 26 AM. J. L. & MED. 143, 148–49 (2000).

Section 3

COMPARISONS WITH OTHER TECHNOLOGIES AND OTHER LEGAL REGIMES

15 Saving the Patent Law from Itself: Informal Remarks Concerning the Systemic Problems Afflicting Developed Intellectual Property Regimes

J. H. Reichman
Bunyan S. Womble Professor of Law
Duke University School of Law
Durham, North Carolina 27708

I. INTRODUCTION

Other contributors to this volume have explored the extent to which specific problems encountered in the field of genetic engineering reflect shortcomings and weaknesses of the patent system.[1] I propose to cast these problems into a

[1] For pessimistic views, *see, e.g.*, Rebecca S. Eisenberg, "Reaching through the Genome" (Chapter 10); *see also* Arti Kaur Rai, *Regulating Scientific Research: Intellectual Property Rights and the Norms of Science*, 94 NW. U. L. REV. 77, 92–94, 109–115 (1999); Rebecca S. Eisenberg, *Bargaining Over the Transfer of Proprietary Research Tools: Is this Market Failing or Emerging?* in EXPANDING THE BOUNDARIES OF INTELLECTUAL PROPERTY: INNOVATION POLICY FOR THE KNOWLEDGE SOCIETY 223–49 (Rochelle Dreyfuss *et al.*, eds. 2001) [hereinafter BOUNDARIES OF INTELLECTUAL PROPERTY]; Rai and Eisenberg, *Bayh-Dole Reform and the Progress of Biomedicine*, 66 LAW &

broader perspective by briefly describing the deeper systemic crisis that has overtaken developed intellectual property regimes in the past 50 years.[2]

In 1989, I wrote that biogenetic engineering and computer software engineering industries were both posing, with more dramatic economic consequences, the same fundamental problems that the worldwide intellectual property system had failed to solve with respect to industrial designs from the nineteenth century on.[3] While these unsolved problems grew more acute with the number and importance of the technologies at issue, the underlying question was not technology specific. It remains today what it was 50 or even 100 years ago—namely, how should we protect investments in cumulative and sequential applications of know-how to industry without impeding follow-on innovation and without impoverishing the public domain?[4]

Paradoxically or not, this conviction makes me rather skeptical of the burgeoning proposals to reform domestic patent laws in order to accommodate technology-specific phenomena. Like Richard Epstein, F. Scott Kieff, and their forerunner, Edmund Kitch,[5] I continue to believe that the patent system deals relatively well with truly nonobvious or discontinuous inventions that require complex organizational efforts to transfer research results from laboratories to commercial endeavors. Tinkering with the mature patent paradigm without addressing the deeper problems that have unhinged the worldwide intellectual property regime could produce a wasted effort, and might well yield the worst of both worlds—that is, a poor patent system operating in a progressively more dysfunctional intellectual property regime as a whole.

Here Professor Epstein's illuminating meditation seems particularly helpful and revealing. He sees a growing tension between property rights and

---

CONTEMP. PROBS. 289 (2003). For more optimistic views, *see* Richard A. Epstein, "Steady the Course: Property Rights in Genetic Material" (Chapter 8); F. Scott Kieff, "Perusing Property Rights in DNA" (Chapter 7).

[2] *See* J. H. Reichman, *Legal Hybrids Between the Patent and Copyright Paradigms*, 94 COLUM. L. REV. 2432 (1994) [hereinafter, *Legal Hybrids*]; J. H. Reichman, *Charting the Collapse of the Patent-Copyright Dichotomy*, 13 CARDOZO ARTS & ENT. L. J. 475 (1995) [hereinafter *Collapse of the Patent-Copyright Dichotomy*].

[3] *See* J. H. Reichman, *Computer Programs as Applied Scientific Know-How*, 42 VAND. L. REV. 639, 656–69 (1989). *See also* Samuelson, Davis, Kapor & Reichman, *A Manifesto Concerning the Legal Protection of Computer Programs*, 94 COLUM. L. REV.2208, 2332–65 (1994).

[4] *See most recently* J. H. Reichman, *Of Green Tulips and Legal Kudzu: Repackaging Rights in Subpatentable Innovation*, 53 VAND. L. REV. 1753 (2000) [hereinafter *Green Tulips*]. *See also* WILLIAM KINGSTON, DIRECT PROTECTION OF INNOVATION 1–124 (W. Kingston ed., 1989); James E. Bessen & Eric S. Maskin, SEQUENTIAL INNOVATION, PATENTS, AND IMITATION (MIT Dep't of Econ. Working Paper No. 00–01, Jan. 2000); Rochelle C. Dreyfuss, *Information Products: A Challenge to Intellectual Property Theory*, 20 N.Y.U. J. INT'L L. & POL. 897 (1988).

[5] Edmund W. Kitch, *The Nature and Function of the Patent System*, 20 J.L. & ECON. 265 (1977). *See also* A. Samuel Oddi, *Un-Unified Economic Theories of Patents: The Not-Quite-Holy Grail*, 7 NOTRE DAME L. REV. 267 (1996).

the public domain, and he wants "All or Nothing"—one or the other, but not both.[6] Confronted with a choice between the demands of those who depend on access to the commons and the needs of private property rights holders, he concedes that both institutions play important roles,[7] but insists that we should avoid a mixed system.[8]

II. FALLACY OF THE "ALL-OR-NOTHING" APPROACH

The notion that there exists a clear antinomy between the public domain and a realm of exclusive intellectual property rights has never been true. Free market economies have always operated under a mixed system. Indeed, our whole competitive system, as handed down from the nineteenth century, rests on three tiers or spheres of industrial activity in which different types of rules interact, as indicated in Fig. 15.1.

Historically, the tensions between regimes of exclusive intellectual property rights and the public domain were moderated by ancillary sets of liability rules[9] protecting entrepreneurs against unfair competition in general and the misappropriation of trade secrets or confidential information in particular. Under property rules, one cannot take the entitlement in question without prior permission of the owner. In this sense, property rules are "absolute permission rules."[10] In contrast, liability rules operate on the principle of "take now, pay later," which means that nonowners can "take the entitlement without permission of the owner, so long as they adequately compensate the owner later."[11]

Trade secret and other unfair competition laws protect against certain market-destructive forms of conduct that enable second-comers to free-ride on a first-comer's investments in innovative applications of know-how to industry. Know-how consists of information about how to achieve some technical or

[6] Epstein, *supra* note 1.

[7] For recent discussion of the role of the public domain, *see, e.g.*, James Boyle, *The Second Enclosure Movement and the Construction of the Public Domain*, 66 LAW & CONTEMP. PROBS. 33 (2003). *See generally*, SYMPOSIUM: THE PUBLIC DOMAIN, 66 LAW & CONTEMP. PROBS. 1–483 (Winter/Spring 2003).

[8] Epstein, *supra* note 1.

[9] "Whenever someone may destroy the initial entitlement if he is willing to pay an objectively determined value for it, an entitlement is protected by a liability rule." Guido Calabresi and A. Douglas Melamed, *Property Rules, Liability Rules, and Inalienability: One View of the Cathedral*, 85 HARV. L. REV. 1092 (1972).

[10] Robert P. Merges, *Institutions for Intellectual Property Transactions: The Case of Patent Pools*, in EXPANDING THE BOUNDARIES OF INTELLECTUAL PROPERTY: INNOVATION POLICY FOR THE KNOWLEDGE SOCIETY 123, 131 (Rochelle Dreyfuss *et al.*, eds., 2001).

[11] *Id.*

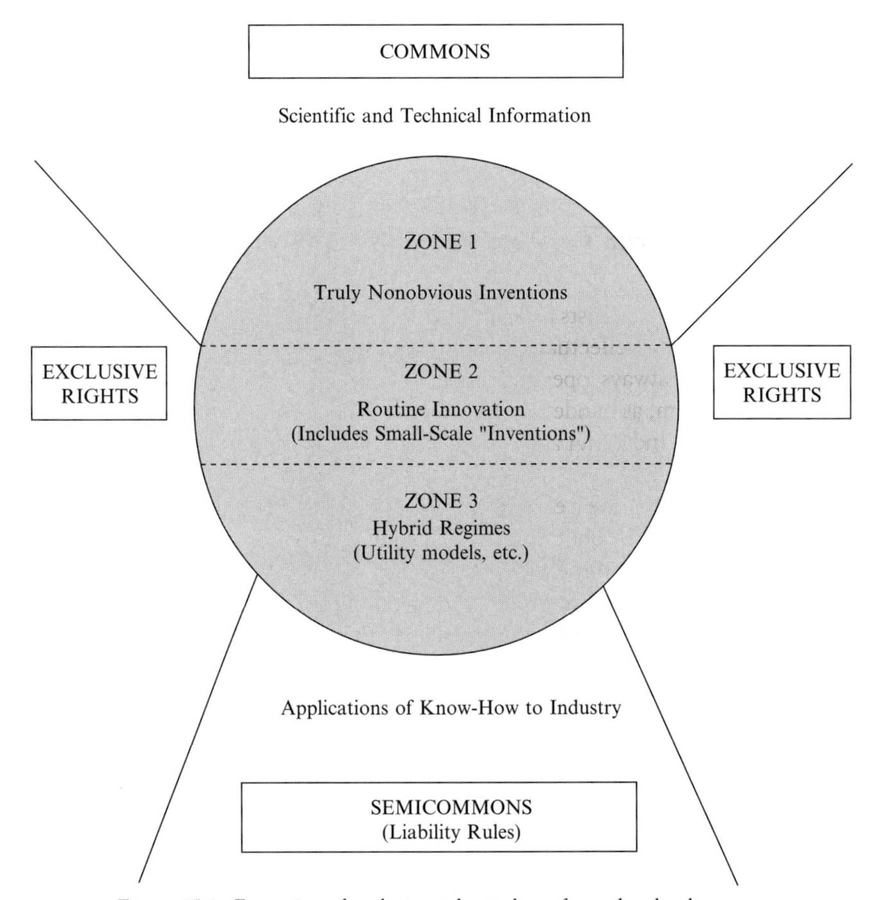

Figure 15.1. Expansion of exclusive rights in lieu of actual or legal secrecy.

commercial advantage over competitors, typically by means of novel methods or processes of production.[12] Such information may or may not be secret. If secret, it may be held only under actual, but not legal, secrecy, which in turn affects the degree of protection the law affords. In common-law countries, laws protecting novel applications of know-how to industry as developed by routine engineers normally partake of liability rules, not exclusive property rights. Any entitlements these laws confer are, as Professor John C. Stedman observed

[12] Stephen P. Ladas, PATENTS, TRADEMARKS, AND RELATED RIGHTS: NATIONAL AND INTERNATIONAL PROTECTION 1617 (1975).

decades ago, merely "disappearing rights," which independent discovery by others may extinguish at any time.[13]

The salient characteristic of laws protecting incremental applications of know-how to industry is that they do not prohibit reverse engineering by honest means. This feature promotes improvements and follow-on innovation while contributing indirectly to the technical community's overall costs of research and development.[14] In effect, these laws endow first-comers with a modicum of natural lead time in which to recoup their investments and establish distinguishing trademarks.

When second-comers appropriate secret know-how by improper means, the law obliges the violators to compensate investors only for the quantum of natural lead time they forfeited. In all cases, liability rules govern in the sense that, without permission, second-comers may extract and improve preexisting industrial applications of know-how as long as, in the absence of any contractual agreement to the contrary, they either defray the costs of reverse engineering or pay the equivalent costs of having usurped lead-time advantages by improper means.

Because most innovation consists of cumulative and sequential applications of know-how to industry by routine engineers at work on common technical trajectories, free market economies after the industrial revolution depended primarily on the liability rules of unfair competition law and only tangentially on the exclusive property rights of patent law, which protected a relatively small set of nonobvious inventions beyond the reach of routine engineers. This traditional state of affairs was encapsulated in the notion that intellectual property law provides only "islands of protection in a sea of free competition." It accounts for the three tiers or spheres of activity in which different types of rules interact, as depicted in Fig. 1.

III. THE SEMICOMMONS AS A NATURAL, OPEN-SOURCE COMMUNITY

Figure 1 subdivides the realm of industrial property law into three spheres: a commons, a semicommons, and a totally private domain in which exclusive property rights prevail. At the top (upstream), there exists a true commons

[13] John C. Stedman, *Trade Secrets*, 23 OHIO STATE L. J. 4, 21 (1962). *See generally* RESTATEMENT (THIRD) OF UNFAIR COMPETITION §§1, 38 (rejecting the recognition of exclusive rights in intangible trade values).

[14] *See most recently* Pamela Samuelson & Susan Scotchmer, *The Law and Economics of Reverse Engineering*, 111 YALE L. J. 1575 (2002); *see generally* Reichman, *Legal Hybrids*, *supra* note 2, at 2434–36, 2520–28.

in which scientific and technical information flow freely through the universe of discourse as an unregulated input, much of it government-generated or government-funded. This is a true commons, as Epstein defines it,[15] and it has played a fundamental role in our national system of innovation, as Paul Uhlir and I recently demonstrated.[16]

Underlying the worldwide intellectual property system (which classically rests on the patent and copyright paradigms[17]), there lies the vast domain of routine engineers who develop applications of know-how to industry without securing patent protection. The routine engineers working on common technical trajectories basically constitute an interrelated group that operates under a de facto sharing ethos. They form a natural, open-source community built around the practice of reverse engineering and the availability of adequate lead time under the liability rules governing trade secrets and confidential information.[18]

In this endeavor, routine engineers depend on the reciprocal insights and contributions that the relevant technical communities derive from the public domain—the shared body of knowledge that underlies the common technical trajectories—and on their inability to remove novel insights and cumulative contributions to know-how from the semicommons. The collective knowledge available from the public domain advances by dint of the small-scale contributions of single innovators. These contributions are statistically predictable in the sense that they inhere in what was already known about existing technical paradigms. Improvements result from trial and error, from new compilations of existing data, from insight, effort, and investment, but they are by definition the kinds of things routine engineers, following their commercial and technical instincts, are expected to produce.

The progressive development of know-how is thus a community project that benefits from the countless small-scale contributions to the prior art by individuals who draw from the public domain to make improvements, and who thereby enrich the public domain by generating new information that others in the technical community may exploit to their own advantage. These cumulative but predictable additions to the prior art partake of "skilled efforts," and they are undoubtedly individual achievements (*leistung*); but they do not attract exclusive property rights precisely because they normally do not surpass the ability of the routine engineers who comprise the relevant technical communities.

[15] *See* Epstein, *supra*, note 1.

[16] *See* J. H. Reichman and Paul F. Uhlir, *A Contractually Reconstructed Research Commons for Scientific Data in a Highly Protectionist Intellectual Property Environment*, 66 LAW & CONTEMP. PROBS. 315, 325–60 (Winter/Spring 2003).

[17] *See Collapse of the Patent-Copyright Dichotomy*, *supra*. note 2.

[18] As previously explained, this realm traditionally operated as a semicommons (in Epstein's terms) because it was regulated by liability rules, especially those governing trade secrets and confidential information. *See supra* notes 13–14 and accompanying text.

Historically, the legal protection of know-how—at least in common law countries—reflects the community-based character of innovation and its dependence on shared information and access to the information commons that is the heritage of the larger technical community. The legal protection of know-how was accordingly organized around liability rules that discouraged certain forms of conduct harmful to that community as a whole. This regime did not allow single innovators to remove their contributions from the semicommons by means of exclusive property rights. On the contrary, the traditional role of these liability rules was to overcome the entrepreneur's risk of market failure due to improper conduct without creating barriers to entry or impeding qualified participants from either accessing the contents of the semicommons or from using those contents to make follow-on applications of others' routine innovations.

IV. FROM SEMICOMMONS TO ANTICOMMONS: THE UNBRIDLED PROLIFERATION OF EXCLUSIVE RIGHTS

Under the classical system of industrial property law that I have described so far, the strong exclusive property rights of the domestic patent laws were confined, at least in principle, to a relatively small zone occupied by truly nonobvious inventions, as depicted in Zone 1 of Fig. 1. At the same time, the relatively large zone of industrial activity governed by liability rules depended in practice, if not in theory, on the continued availability of actual or legal secrecy to provide investors with natural lead time. What happened, beginning with the advent of industrial design in the nineteenth century and continuing with the rise of software and biogenetic engineering in the twentieth century, was that *secrecy as a trigger for liability protection became increasingly scarce or irrelevant*. The routine engineer's applications of commercially valuable know-how to industry no longer remained behind lock and key in the investor's business office. Rather, they were embodied on or near the face of products distributed in the open market, which any second-comer could duplicate without need of incurring the time and costs of reverse-engineering.

Under these conditions, which I have elsewhere tried to encapsulate in the phrase "incremental innovation bearing know-how on its face,"[19] the liability rules built around actual or legal secrecy break down, and there is a chronic shortage of natural lead time. Would-be investors in applications of know-how to industry understand that their contributions can be instantly duplicated without the second-comer having to defray the costs of reverse engineering ("zero lead time" problems).

[19] *See Legal Hybrids*, *supra* note 2, at 2511–19.

To address the perceived risk of market failure that this chronic shortage of natural lead time logically generates, courts and legislatures everywhere in the developed world have sought to fill the space heretofore occupied by liability rules rooted in actual or legal secrecy with a proliferation of exclusive property rights. To this end, they have combined two strategies that undermine the patent system as classically conceived and that largely account for the "thickets of rights" and other related predicaments that occupy the forefront of attention in many areas of intellectual property law today, including that of genetic engineering.[20]

The first strategy was to keep lowering the nonobviousness standard (and to expand the copyright paradigm previously intended for artistic works) so as to rescue commercially valuable slivers of innovation that had nowhere else to go. Sound familiar? The domestic patent and copyright laws—sources of powerful legal monopolies—are thus used as roving unfair-competition laws to overcome market failure by means that create heavy social costs. This approach entails using a cannon where a pistol would do.

The second strategy was to enact *sui generis* regimes of exclusive property rights to deal with technology-specific applications of know-how to industry that appear to elude the ever more distorted patent and copyright paradigms. Over time, those who promote a proliferation of hybrid or deviant *sui generis* exclusive property rights to deal with slivers of innovation invariably become obsessed with protecting investors against unauthorized follow-on applications. The hybrid regimes thus tend to evolve into miniature patent laws—patent-like hybrids—that choke competition, impede access to the semicommons, and disrupt the technical community's normal cycle of building on each other's innovations.[21] This phenomenon becomes especially pernicious in countries where the enactment of hybrid regimes of exclusive property rights encounters no constitutional obstacles.

In effect, a mindless proliferation of exclusive property rights has thus been filling the systemic space previously occupied by liability rules, as indicated in Zones 2 and 3 of Fig. 15.1. This expansion of exclusive rights into the realm of routine applications of know-how to industry converts the collective knowledge available to the technical community as a whole into artificial private preserves, which have to be negotiated and combined in order to support investment in research and development. It generates the anticommons effects—or at least the perceived threat of them—currently experienced in many quarters, and it fosters

[20] *See esp.* Michael A. Heller & Rebecca S. Eisenberg, *Can Patents Deter Innovation? The Anticommons in Biomedical Research*, SCIENCE, May 1 1998, at 698. *See also* J. Walsh *et al*, *Research Tool Patenting and Licensing and Biomedical Innovation*, in W. COHEN & S. MERRILL, PATENTS AND THE KNOWLEDGE-BASED ECONOMY (forthcoming 2003) (finding uneasy reliance on ad hoc accommodations to avoid risk of anti-commons effects).

[21] *See esp. Green Tulips*, *supra* note 4, at 1749–53, 1766–71.

proposals for reform, some of them quite radical, which supporters of the patent system view with alarm.

V. DATABASE PROTECTION OR HOW TO ELEVATE THE COSTS OF INNOVATION ACROSS THE ENTIRE ECONOMY

Before outlining my own views of how to address the problems identified above, I want to stress that new developments could soon make the economic consequences of these problems far more acute than in the recent past. The unresolved problem of how to avoid a *kudzu*-like proliferation of strong exclusive rights controlling slivers of innovation has recently given rise to a cancerous deviant regime that threatens to attack the integrity of the upstream commons through which all scientific and technical information had previously been filtered. I refer to the European Commission's Directive on the Legal Protection of Databases,[22] which will eventually be implemented in some 50 countries seeking entry to or affiliation with the European Union (EU).

The *sui generis* regimes of database protection mandated by this Directive confer a strong and potentially perpetual exclusive property right in collections of data and information as such. These regimes (and pending US proposals that would institute a similar regime here) convert data and information—the previously unprotectable raw materials or basic inputs of the modern information economy—into the subject matter of a new exclusive property right that is paradoxically more powerful in key respects than either the patent or the copyright paradigms.

A detailed explanation and analysis of the EU's database protection regime lies beyond the scope of these remarks.[23] Nevertheless, to see why this regime is a great magnitude worse than other hybrid exclusive rights, one should understand how radical a change it would make in the traditional economic role of intellectual property rights (IPRs) generally. Until now, the economic function of IPRs was to *make markets possible where previously there existed a risk of market failure due to the public good nature of intangible creations*. Exclusive rights make embodiments of intangible public goods artificially appropriable, create markets for those embodiments, and make it possible to exchange payment for access to these creations.

[22] Directive 96/9 of the European Parliament and the Council of 11 March 1996 on the legal protection of databases, 1996 O.J. (L. 77), 20 [hereinafter E.C. Directive on the Legal Protection of Databases].

[23] *See, e.g.*, J. H. Reichman, *Database Protection in a Global Economy*, 2002 REVUE INTERNATIONAL DE DROIT ECONOMIQUE 455, 463–78; Reichman & Uhlir, *supra* note 16, at 385–95.

In contrast, an exclusive intellectual property right in the contents of databases *breaks* existing markets for downstream aggregates of information, which were formed around inputs of information largely available from the public domain. It conditions the very existence of all traditional markets for intellectual goods on:

- the willingness of information suppliers to supply at all (they can hold out or refuse to deal)
- their willingness not to charge excessive or monopoly prices (*i.e.*, more than downstream aggregators can afford to pay in view of their own risk management assessment)
- the willingness and ability of information suppliers to pool their respective chunks of information in contractually constructed cooperative ventures.[24]

This last constraint is perhaps the most telling of all. In effect, the *sui generis* database regimes create new and potentially serious barriers to entry to all existing markets for intellectual goods owing to the multiplicity of new owners of upstream information in whom these regimes invest exclusive rights, any one of whom can hold out and all of whom can impose onerous transaction costs (analogous to the problem of multimedia transactions under copyright law). This tangle of rights is known as the "anticommons" effect,[25] and database protection laws appear to be ideal generators of such a phenomenon.

To see how the consequences of this regime could affect the field of genetic engineering, consider the following scenarios.

- Genomic scientist A publishes research results in a reputable journal and makes the data available for verification purposes. But he retains all rights in the data (and the database) that are needed to make the discovery operational. As a result, the data no longer enter the public domain; they remain in databases controlled by universities, scientific groups, or private companies. All the problems we now experience with regard to Material Transfer Agreements—including the prospect of reach-through clauses—may now surface in regard to the underlying data and databases themselves.[26]

- Genomic scientist B (and her university or firm) obtain a patent for genetic material and they disclose whatever is needed to meet the requirements of patent law. After 20 years, the patent expires, with the benefits we anticipate.

[24] The analysis here was based on J. H. Reichman, *Database Protection in a Global Economy*, *supra* note 23, at 482–84.

[25] *See* Michael A. Heller & Rebecca S. Eisenberg, *Can Patents Deter Innovation? The Anticommons in Biomedical Research*, SCIENCE, May 1, 1998.

[26] For details, *see* Reichman & Uhlir, *supra* note 16, at 398–404.

But B and her firm or university may still retain all rights to the data and technical information that support the disclosure, including all the updates and cumulative additions, and these resources never enter the public domain in the EU (and maybe not here either). Gradually, firms or entities controlling more and more of these primary data collections could exercise a considerable degree of control over the future of genetic and genomic research, regardless of what the patent law says or does.[27]

From a broader perspective, there is a built-in risk that, under the new *sui generis* database regimes, too many owners of information inputs will impose too many costs and conditions on all the information processes we now take for granted in the information economy. At best, the costs of research and development activities seem likely to rise across the entire economy, well in excess of benefits, owing to the potential stranglehold of data suppliers on raw materials. This stranglehold will increase with market power if most databases are owned by sole-source providers and if the comparative advantages from owning large, complex databases tend progressively to elevate barriers to entry over time.

VI. SAVING THE SYSTEM FROM ITSELF

The deeper problem we started tracking at the outset of this paper was that, as "secrecy" increasingly failed to provide a meaningful trigger for the liability rules that traditionally protected investors in routine innovation, the developed economies aggressively expanded exclusive property rights to fill the gap. The social costs of alleviating risk aversion and market failure by such means become prohibitively high, however, as concerns about the future of genetic research suggest, and these dubious solutions could ultimately destabilize national systems of innovation that rely on them to excess. When legislatures institute exclusive rights to stimulate investment in commercially valuable applications of know-how to industry—in place of traditional liability regimes—they disrupt the natural sharing mechanisms that benefit routine engineers, impede follow-on innovation, foster thickets of rights and other barriers to entry that slow the pace of innovation, and generally impoverish the public domain on whose inputs future innovators depend.

The solution is to devise a new and more rational set of liability rules, not rooted in actual or legal secrecy, that could restore the historical buffer zone

[27] For details, *see id.* at 361–413. *Cf. also* Walter W. Powell, *Networks of Learning in Biotechnology: Opportunities and Constraints Associated with Relational Contracting in a Knowledge-Intensive Field*, in EXPANDING THE BOUNDARIES OF INTELLECTUAL PROPERTY, *supra* note 1, 223–49 (2001); Pamela Samuelson, *Mapping the Digital Public Domain: Threats and Opportunities*, 66 LAW & CONTEMP. PROBS. 147, 158–159, 163–165 (Winter/Spring 2003).

between exclusive intellectual property rights and free competition. In other words, for the patent system to promote the commercialization of truly nonobvious inventions at acceptable social costs (the function that Epstein, Kieff, and Kitch rightly esteem), we must take the pressure off its nonobviousness standard by formulating an alternative set of liability rules to protect investment in cumulative and sequential innovation as such.[28]

The object is to provide qualifying innovators with some functional equivalent of the lead-time laws previously rooted in secrecy that would require second-comers making follow-on applications to compensate first-comers for the costs of developing novel applications of know-how to industry during a relatively short period of time. As I have previously demonstrated, assessing the quantum of compensation presents few hard problems with respect to small-scale innovations because the bulk of their value derives from public-domain inputs by definition.[29] Even when the scale of innovation at issue attains a greater magnitude, the social costs of valuation by imperfect means will usually be lower, in most cases, than those attendant on the enforcement of strong exclusive property rights.[30]

I have recently tried to show that the proposed "compensatory liability regime" could stimulate more investment in subpatentable innovation at lower social costs than either abusive extensions of patent rights or the hybrid regimes of exclusive property rights that currently abound.[31] My latest work in progress, coauthored with the regulatory economist, Tracy Lewis, will try to show ways in which liability rules could also provide healthy alternatives to patent protection when innovative entrepreneurs needing direct protection of investment are willing to trade the hassles and costs of the patent system for an effective compensatory alternative.[32]

I cannot, in these informal remarks, conduct a more detailed discussion of specific proposals for a compensatory liability regime that would operate in

[28] Providing a compensatory liability rule for at least subpatentable innovations should allow the nonobviousness standard to rise to the point where it covers truly discontinuous inventions, and not slivers of incremental innovation, because courts no longer need to use patent law to combat free-riding duplication. *See, e.g.*, *Green Tulips*, *supra* note 4, at 1797. Here I note parenthetically that William Kingston would go farther and introduce a "multiples of investment" test even for nonobvious inventions, which could enable second-comers to trade money for time under a compulsory license. *See* William Kingston, *Intellectual Property Needs Help From Accounting*, 2002 E.I.P.R. 508. This paper takes no position on those proposals.

[29] *See Green Tulips*, *supra* note 4, at 1783–86 (proposing and justifying a range from 3% to 9% of gross returns, that would normally average out at 6%).

[30] *See* Tracy R. Lewis and J. H. Reichman, *Potential Role of Liability Rules in Stimulating Local Innovation in Developing Countries: A Law and Economics Primer*, paper presented to the Conference on International Public Goods and Transfer of Technology under a Globalized Intellectual Property Regime, Duke Law School, April 4–6, 2003.

[31] *See generally*, *Green Tulips*, *supra* note 4.

[32] *See* Lewis & Reichman, *supra* note 30.

tandem with the patent system. Let me emphasize here, however, that we are not proposing "compulsory licenses" or other nonvoluntary mechanisms that cut back on a patentee's exclusive rights once these have been properly granted.[33] A compensatory liability regime provides a clear-cut entitlement to relief, unlike the vague prescriptions of traditional unfaircompetition laws;[34] but the relief in question constitutes *an entitlement to compensation only*, and it entails no further power to interdict follow-on applications of a protected innovation.[35] In this sense, the proposed regime provides an *automatic license* that both defines and limits the basic entitlement, without cutting back on any grant of exclusive property rights.

A compensatory liability regime performs at least two functions that Professor Epstein says he likes. One is to give nominal compensation for "small amounts contributed from members of a large cohort."[36] The other is to allow use of routine innovation by institutions under a preset formula or arbitration.[37] In short, my proposal envisions a compensatory default rule with relatively small royalties that would help entrepreneurs who invested in routine innovation to recuperate the costs of research and development from value-adding second-comers without creating barriers to entry or otherwise diminishing the technical inputs available from the public domain.

VII. A COMPREHENSIVE SOLUTION

A comprehensive solution to the problems at hand would then require us to adapt these compensatory liability principles to the needs of today's database industry, with a goal of stimulating greater investment in the development of electronic databases without stifling research and without creating information cartels. Under such a regime, a database producer could earn the right to compensation for data borrowed by a third party for value-adding follow-on applications during a specified period of time. However, the database originator should normally lack the power to prevent follow-on applications requiring independent investment so long as a reasonable royalty were paid for the data borrowed from the first-comer's

[33] For this topic, *see* Jerome H. Reichman with Cathy Hasenzahl, *Non-voluntary Licensing of Patented Inventions Part I—Historical Perspective, Legal Framework Under TRIPS and an Overview of the Practice in Canada and the United States of America, Part II-The Canadian Experience*, UNCTAD/ICSTD Capacity Building Project on Intellectual Property and Sustainable Development, Geneva, Switz., Sept.-Oct. 2002.

[34] *But see* Dennis Karjala, *Misappropriation as a Third Intellectual Property Paradigm*, 94 COLUM. L. REV. 2594 (1994); Wendy J. Gordon, *On Owning Information: Intellectual Property and the Restitutionary Impulse*, 78 VA. L. REV. 149 (1992).

[35] *See generally Green Tulips*, *supra* note 4, 1793–96.

[36] *See* Epstein, *supra* note 1.

[37] *See id.*

own collection.[38] Needless to say, the first-comer should remain equally free to borrow back data from the second-comer's own follow-on applications for similar competitive purposes in return for similar compensatory payments.

Even this solution, however, could fail to resolve the difficulties that public science may encounter in accessing data for nonprofit research purposes in a highly protectionist intellectual property environment. A codified database protection law might actually provide some worthwhile exceptions for scientific research, although none have so far been put on the table.[39] Yet, more and more data formerly available to public science can now be privatized in the online environment by means of a combination of electronic fencing devices and one-sided electronic contracts of adhesion, and these practices would not necessarily depend on database protection rights as such.[40] Even databases that consist primarily of government-generated or government-funded data could be privatized by such means, in which case access for research purposes may depend on the contractual dispositions to preserve it that funding agencies or academic institutions were moved to impose.[41]

Without delving further into these matters here, my point is that the nonprofit research institutions that have heretofore played such a critical role in our national system of innovation will face new and serious threats under these conditions. On the one hand, they can, of course, join the enclosure movement and profit from it. Thus, universities that now transfer publicly funded technology to the private sector can also profit from the licensing of databases.[42] On the other hand, the ability of researchers to access and aggregate

[38] *See, e.g.*, J. H. Reichman & Pamela Samuelson, *Intellectual Property Rights in Data?*, 50 VAND. L. REV. 51, 145–151 (1997). For less ambitious proposals sounding in misappropriation theory, *see id.* 139–145 Neither a minimalist misappropriation regime nor a compensatory liability regime would do much harm, but the latter is more procompetitive because it stimulates follow-on applications while rewarding first-comers with potential lottery effects. This helps to overcome the sole-source problem that plagues markets for complex databases. *See generally Database Protection in a Global Economy*, *supra* note 23, at 478–84.

[39] *See, e.g.*, *Database Protection in a Global Economy*, *supra* note 23, at 465, 472–73.

[40] *See, e.g.*, Jane C. Ginsberg, *U.S. Initiatives to Protect Works of Low Authorship*, *in* EXPANDING THE BOUNDARIES OF INTELLECTUAL PROPERTY, *supra* note 1, 55, 68–72 (2001); Samuelson; *supra* note 27, at 163.

[41] *See, e.g.*, Reichman & Uhlir, *supra* note 16 at 417–56. In this respect, cyberspace distribution of information restores the "power of the two-party deal" to the point where intellectual property rights are often less needed to avoid market failure than they were, say, when the printing press was invented. Under these conditions, indeed, a primary effect of any new intellectual property right may be to strengthen the de facto monopoly power of providers by superimposing a legal monopoly that makes it ever harder for users, consumers, and researchers to elicit public interest safeguards at either the judicial or the legislative level. *See, e.g.*, J. H. Reichman & Jonathan A. Franklin, *Privately Legislated Intellectual Property Rights: Reconciling Freedom of Contract with Public Good Uses of Information*, 147 U. PA. L. REV. 875, 897–99, 929–50 (1999).

[42] *See* Reichman & Uhlir, *supra* note 16 at 396–409.

the information they need to produce upstream discoveries and innovations may be compromised both by the shrinking dimensions of the public domain and by the demise of the sharing ethos in the nonprofit community, as these same universities and laboratories view each other as competitors rather than partners in a common venture.[43]

Any long-term solution must accordingly look to the problems of the research communities and of nonprofit users of data generally, in an increasingly commoditized information environment. As commoditization proceeds and IPRs multiply, the functions of the public domain that are now taken for granted will increasingly need to be reconstructed contractually by the nonprofit actors engaged on specific projects. If the research communities decide to address these challenges frontally, they may seek, of their own initiative, to recreate by consensus and agreement a dynamic research commons that could ensure a continuous flow of raw materials through the national innovation system, notwithstanding the pressures for commodization in the private sector. In other words, universities and laboratories that depend on sharing access to data may have to stipulate their own "treaties" and arrangements to ensure unimpeded access to commonly needed raw materials in a public or quasi-public dimension, even though each institution separately engages in transfers of information to the private sector for economic gain.[44]

Proposals to construct a kind of nature conservancy[45] or voluntary "E-commons" for public access to scientific and technical information are currently under investigation in the U.S. As applied to scientific research, this project envisions a series of measures to strengthen exchanges of data and information for nonprofit purposes without impeding data providers from further commercialization in the private sector. Ideally, both funding agencies and universities would participate in constructing contractual templates and supplementary norms to govern this E-commons and the means of administering it.[46] If these initiatives proved successful, then similar efforts would have to be undertaken at the international level as well in order to extend the benefits of a dynamic E-commons to scientists and other research communities around the world.

[43] *See id.* at 398–404.

[44] For details, *see* Reichman & Uhlir, *supra* note 16, at 416–56 (Part IV).

[45] *See, e.g.*, James Boyle, *A Politics of Intellectual Property: Environmentalism for the Net?* 47 DUKE L. J. 87 (1997). *See also* James Boyle, *The Second Enclosure Movement and the Construction of the Public Domain*, 66 LAW AND CONTEMP. PROBS. 33, 53–74 (2003).

[46] *See generally* Reichman & Uhlir, *supra* note 16, at 425–56. This project was first discussed at a major conference on the public domain held at Duke University Law School in November, 2001, and the National Academies sponsored a follow-up workshop on this same topic in September, 2002. *See Proceedings of the Symposium on the Role of Scientific and Technical Data and Information in the Public Domain* (forthcoming, National Academies Press 2003).

16

Biotechnology's Uncertainty Principle

Dan L. Burk* and Mark A. Lemley†

*Oppenheimer, Wolff & Donnelly Professor of Law
University of Minnesota Law School
Minneapolis, Minnesota 55455

†Professor of Law
Boalt Hall School of Law
University of California at Berkeley
Berkeley, California 94720-7200
Of Counsel
Keker & Van Nest LLP
San Francisco, California 94111

ABSTRACT

Patents have proven to be important to the growth and financing of the American biotechnology industry, but it remains unclear whether current patent standards are suited to the needs of this industry. Patent law has a general set of legal rules to govern the validity and infringement of patents in a wide variety of technologies. With very few exceptions, the statute does not distinguish between different technologies in setting and applying legal standards. Rather, those standards are designed to adapt flexibly to new technologies, encompassing "anything under the sun made by man."[1] In theory, then, we have a unified patent system that provides technology-neutral protection to all kinds of technologies.

However, we have recently noticed an increasing divergence between the rules actually applied to different industries. Biotechnology provides one of the best examples. In biotechnology cases, the Federal Circuit has repeatedly held that uncertainty in predicting the structural features of biotechnological inventions renders them nonobvious, even if the prior art demonstrates a clear plan for producing the invention. At the same time, the court claims that the uncertain nature of the technology requires the imposition of stringent patent enablement and written description requirements that are not applied to patents in other disciplines. Thus, as a practical matter, it appears that although patent law is technology-neutral in theory, it is technology-specific in application. We provide evidence for this claim in Part I. While this chapter focuses on biotechnology, which presents an extreme example, our findings have implications for other industries as well, notably small-molecule chemistry.

Part II explores how the application of the same legal standards can lead to such different results in diverse industries. Much of the variance in patent standards is attributable to the use of a legal construct, the "person having ordinary skill in the art" (PHOSITA), to determine obviousness and enablement. The more skill those in the art have, the less information a patentee has to disclose, but the harder it is to find an invention nonobvious. The level of skill in the art affects not only patent validity, but also patent scope. Because both claim construction and the doctrine of equivalents turn on the understanding of the PHOSITA in certain circumstances, judgments the court makes about those industries affect the scope of those patents that do issue.

One reading of the biotechnology cases is that the Federal Circuit believes that biotechnology experts know very little about their art—at least, this seems the clear implication of the court's holdings and accompanying analysis. We do not challenge the idea that the standards in each industry should vary, nor the idea that that variation depends in part on the level of

[1] *Diamond v. Chakrabarty*, 447 U.S. 303, 309 (1980) (citing S.Rep.No.1979, 82d Cong., 2d Sess., 5 (1952); H.R.Rep.No.1923, 82d Cong., 2d Sess., 6 (1952)).

skill in that industry. As we have explained elsewhere, patent law should be technology-specific because the industries it affects are not homogenous.[2] We think the use of the PHOSITA provides needed flexibility for patent law, permitting it to adapt to new technologies without losing its essential character. We fear, however, that the Federal Circuit has not applied that standard properly in biotechnology. The court has a static perception of the field that was set in its initial analyses of biotechnology inventions, but does not reflect the realities of the current industry.

In Part III, we offer a very preliminary policy assessment of these industry-specific patent cases. We suggest that the special rules the Federal Circuit has constructed for biotech cases are rather poorly matched to the specific needs of the industry. Indeed, in some ways, the Federal Circuit cases have it exactly backwards. We offer a few suggestions as to what a consciously designed biotechnology patent policy may look like. In doing so, we hope to lay the groundwork for broader exposition of those ideas and suggestions for implementing them.

I. HETEROGENEITY IN THE PATENT LAW

Intellectual property law generally aims to solve the "public goods" problem that arises in regards to creative activity. Legal rights in inventions allow creators to control and profit from goods that are costly to produce but virtually costless to reproduce or to appropriate once they have been created. A variety of intellectual property systems have been promulgated to deal with this problem for different, if occasionally overlapping, areas of subject matter. These various degrees and modes of legal protection carry different scopes and lengths of protection, hopefully roughly appropriate to their subject areas. Copyright is generally addressed to artistic or aesthetic works, although it now includes software in its ambit; patent law generally addresses industrial or technological inventions; trade secrecy covers a wide range of valuable business assets. Each of these modes of protection covers a wide swath of subject matter; specialized statutes, sometimes called *sui generis* laws, are relatively rare.[3] As a practical matter,

[2] Dan L. Burk and Mark A. Lemley, *Policy Levels in Patent Law.* 89 VA. L. REV. (forthcoming 2003).

[3] At various times commentators have called for *sui generis* protection of specific subject matter. *See, e.g.*, Peter S. Menell, *Tailoring Legal Protection for Computer Software*, 39 STAN. L. REV. 1329, 1364–65 (1987) (computer software); Pamela Samuelson, *Creating a New Kind of Intellectual Property: Applying the Lessons of the Chip Law to Computer Programs*, 70 MINN. L. REV. 471 (1985) (same); Dan L. Burk, *Copyrightability of Recombinant DNA Sequences*, 29 JURIMETRICS J. 469, 530–31 (1989) (biotechnology); Kenneth D. Crews, *Looking Ahead and Shaping the Future: Provoking Change in Copyright Law*, 49 J. COPYRIGHT SOC'Y USA 549, 564 (2001) (" 'One-size-fits-all' ultimately fits

Congress cannot enact a new form of intellectual property statute each time a new technology arises.[4] Nevertheless, there are drawbacks to encompassing many types of subject matter within one broad system.

A. The History of the Uniform Patent System

A patent statute, one of the first laws made by Congress, passed in 1790. Since that time, a patent statute has been a constant feature of the U.S. legal landscape.[5] While the nature of the patent system went through some rather dramatic changes in the first 50 years of the Republic—beginning with a requirement that two cabinet officials must personally review and sign off on any patent[6] and swinging to the other extreme with an automatic registration system subject to caveats[7]—by 1836, the essential features of modern patent law were in place.[8] Despite periodic revisions, most recently in 1952, the basic structure of the patent system has remained unchanged for 165 years.

Technology, of course, has changed dramatically during that time. The "useful arts" envisioned by the framers were mechanical inventions useful in a primarily agrarian economy. Since that time, the country has gone through several periods of dramatic innovation in a wide variety of fields. As late as 1950, though, most inventions were still mechanical in nature. It is only in the last

---

few."). *Cf.* Nancy Gallini & Suzanne Scotchmer, *Intellectual Property: When Is It the Best Incentive System?* in INNOVATION POLICY AND THE ECONOMY 51, 53 (Adam B. Jaffe *et al.*, eds., 2001) ("[I]ntellectual property regimes should be designed so that the subject matter of each one has relatively homogenous needs for protection."). The Semiconductor Chip Protection Act of 1984 is one of the few examples where Congress heeded such encouragement. It created a unique form of intellectual property in the "mask works" embodying semiconductor chip circuit designs. *See* Semiconductor Chip Protection Act of 1984, 17 U.S.C. §§ 901–914 (2000)).

[4] *See* Louis Kaplow, *The Patent-Antitrust Intersection: A Reappraisal*, 97 HARV. L. REV. 1813, 1819–20 (1984); Richard Stern, *The Bundle of Rights Suited to New Technology*, 47 U. PITT. L. REV. 1229, 1261 (1986).

[5] *See, e.g.*, Bruce Bugbee, THE GENESIS OF AMERICAN PATENT AND COPYRIGHT LAW 126, 143 (1967); Edward C. Walterscheid, *To Promote the Progress of Useful Arts: American Patent Law and Administration*, 1787–1836, pts. 1 & 2, 79 J. PAT. & TRADEMARK OFF. SOC'Y 61 (1997); 80 J. PAT. & TRADEMARK OFF. SOC'Y 11 (1998). Even before that time, the U.S. colonies granted patent rights. *See* Robert P. Merges *et al.*, INTELLECTUAL PROPERTY IN THE NEW TECHNOLOGICAL AGE 109 (3d ed. 2003).

[6] This was a feature of the short-lived Patent Act of 1790. *See* Walterscheid, *supra* note 8; Edward C. Walterscheid, *Charting a Novel Course: The Creation of the Patent Act of 1790*, 25 AIPLA Q. J. 445, 519–520 (1997).

[7] The 1793 Act replaced the cumbersome cabinet-level review with a registration system. Under this system, patents were granted without examination unless a competitor or other interested party filed a "caveat"—essentially a request to be notified and given a chance to object if someone patented in a particular field. *See* Walterscheid, *supra* note 5, at 73.

[8] *See* Merges *et al.*, *supra* note 5, at 109–110.

half-century—and to a large extent in the last 25 years, as John R. Allison and Mark A. Lemley show[9]—that patent law has lost its primarily mechanical character, branching out into biotechnology, semiconductors, computer hardware and software, electronics, and telecommunications.

What is notable about this history is that the fundamental rules of patent law were set in a world where inventions were mechanical. Because inventions in the past were far more homogenous than they are today,[10] it made sense to have a unified set of rules for dealing with those inventions. The application of those old rules to new technologies has not been free from controversy. Some have suggested that the unified rules suitable for the old, homogenous world are no longer appropriate in today's increasingly complex innovative landscape.[11] But without changing the rules themselves, in the last dozen years, the Federal Circuit has applied those rules in a way that effectively creates different standards for different industries.[12] In the sections that follow, we examine the legal treatment of one such industry—biotechnology—in detail.

B. Biotechnology Patent Cases[13]

In stark contrast to the Federal Circuit decisions in other technologies,[14] recent decisions involving genetic material have imposed a stringent disclosure standard for patenting macromolecules.[15] The court has placed particular emphasis

[9] *See* John R. Allison & Mark A. Lemley, *The Growing Complexity of the United States Patent System*, 82 B.U.L. REV. 77, 87–90 (2002).

[10] *Id.* at 79–80.

[11] *See, e.g.*, S. Benjamin Pleune, *Trouble With the Guidelines: On Urging the PTO to Properly Evolve With Novel Technologies*, 2001 J. L., TECH. & POL'Y 365 (arguing for DNA-specific legislation). For a critical analysis of such proposals, see Dan L. Burk, *Biotechnology and Patent Law: Fitting Innovation to the Procrustean Bed*, 17 RUTGERS COMP. & TECH. L. J. 1 (1991).

[12] Hodges observes that computers and biotechnology are treated differently in the written description cases, though he limits his focus primarily to biotechnology. Robert A. Hodges, Note, *Black Box Biotech Inventions: When a "Mere Wish or Plan" Should be Considered an Adequate Description of the Invention*, 17 GA. ST. U. L. REV. 831, 833 (2001). Others have complained that even within industries the standard may not be applied consistently. *See, e.g.*, Glynn S. Lunney Jr., *E-Obviousness*, 7 MICH. TELECOMM. & TECH. L. REV. 363, 365 & n.13 (2001).

[13] For background on the science of biotechnology, see generally ROBERT P. MERGES, ET AL., INTELLECTUAL PROPERTY IN THE NEW TECHNOLOGICAL AGE: 2002 CASE AND STATUTORY SUPPLEMENT 501–16; Dan L. Burk, A *Biotechnology Primer*, 55 U. PITT. L. REV. 611 (1994).

[14] For example, the Federal Circuit has articulated very loose, almost trivial standards for disclosure of computer software. *See, e.g.*, *Northern Telecom, Inc., v. Datapoint Corp.*, 908 F.2d 931 (Fed. Cir. 1990) (holding that only minimal disclosure is needed for enablement, because implementation is "a mere clerical function to a skilled programmer"). For elaboration of how software cases differ from biotech cases, see Dan L. Burk & Mark A. Lemley, *Is Patent Law Technology-Specific?*, 17 BERKELEY TECH. L.J. 1155 (2002).

[15] We acknowledge Lawrence Sung's contrary view that the Federal Circuit's biotechnology cases are simply decided on their individual facts and do not reflect any patterns. *See* Lawrence M.

on the written description requirement of section 112, which forces the patentee to specifically describe the claimed invention as part of the disclosure. The justification for such a detailed description is to demonstrate to others of ordinary skill that the inventor in fact has the invention in her possession, the assumption being that a sufficiently detailed description would not be possible if the inventor were speculating or guessing about its features.[16] This requirement is separate from (and potentially more stringent than) the enablement requirement. Although the two are closely connected, satisfying one requirement does not necessarily satisfy the other. The classic example offered by one court is the situation in which the description of a particular chemical compound enables one of ordinary skill to make other related compounds, yet those other compounds are not described in the patent disclosure. The first compound is both enabled and described; the others are only enabled.[17]

This venerable chemical patenting hypothetical has been brought to life by the Federal Circuit's biotechnology opinions. For example, in *Fiers v. Revel*,[18] the court considered the decision of the Patent Office in a three-way interference over patent applications claiming the human DNA sequence that produces the protein fibroblast beta-interferon (β-IF).[19] One of the applicants, Revel, relied for priority upon his Israeli patent application, which disclosed methods for isolating a fragment of the DNA sequence coding for β-IF and for isolating messenger RNA coding for β-IF. The court considered whether the disclosure in Revel's Israeli application satisfied the written description

---

Sung, *On Treating Past as Prologue*, 2001 U. ILL. J.L. TECH. & POL'Y 75, 107. *See also* John W. Schlicher, *Biotechnology and the Patent System: Patent Law and Procedures for Biotechnology, Health Care and Other Industries*, 4 U.BALT. INTELL. PROP. L.J. 121, 127 (1996) ("I do not understand the Court of Appeals for the Federal Circuit to have created a subset of patent law doctrines for biotechnology."). But other commentators appear to recognize that something unusual is happening in the case of biotechnology. *See, e.g.*, Hodges, *supra* note 12, at 832; Janice M. Mueller, *The Evolving Application of the Written Description Requirement to Biotechnological Inventions*, 13 BERKELEY TECH. L.J. 615 (1998); Harris A. Pitlick, *The Mutation on the Description Requirement Gene*, 78 J. PAT. & TRADEMARK OFF. SOC'Y 209 (1998); Margaret Sampson, *The Evolution of the Enablement and Written Description Requirements Under 35 U.S.C. § 112 in the Area of Biotechnology*, 15 BERKELEY TECH. L.J. 1233 (2000). It seems readily apparent to us, as to the majority of other commentators, that the biotechnology cases consistently depart from the standards applied in other industries.

[16] Of course, in the case of constructive reduction to practice, or filling a "paper patent" without having actually made the invention, the inventor is in some sense speculating or guessing about the features of an invention not yet built. But even in that instance, the underlying assumption in patent law is that the inventor "has" the invention mentally, and so can give a sufficiently detailed description of that inventive conception—physically creating the invention is straightforward.

[17] *In re DiLeone*, 436 F.2d 1404, 1405 n.1 (C.C.P.A. 1971).

[18] *Fiers v. Revel*, 984 F.2d 1164 (Fed. Cir. 1993).

[19] In biotechnology terms, we say that the DNA sequence in question "codes for" the protein.

requirement and could therefore support a U.S. application. The Federal Circuit upheld a determination by the Board of Patent Appeals and Interferences that Revel's disclosure was not an adequate description, largely because it failed to disclose the actual sequence of the DNA molecule at issue. According to the court's reasoning, disclosing a method for obtaining the molecule was not the same as disclosing the molecule itself:

> An adequate written description of a DNA requires more than a mere statement that it is part of the invention and reference to a potential method for isolating it; what is required is a description of the DNA itself.... A bare reference to a DNA with a statement that it can be obtained by reverse transcription is not a description; it does not indicate that Revel was in possession of the DNA.[20]

Since the Revel application did not disclose the sequence for the molecule claimed, the court characterized it as disclosing merely "a wish, or arguably a plan, for obtaining the DNA."[21] Under *Fiers*, an inventor does not conceive of a DNA invention until she actually creates it.[22]

A similar conclusion was reached in a subsequent case, *Regents of the University of California v. Eli Lilly*.[23] The patent at issue covered a microorganism carrying the DNA sequence coding for human insulin. The patentee supported this claim by disclosing a method for obtaining the human cDNA,[24] as well as the amino acid sequences for the insulin protein and the corresponding insulin DNA sequence in rats. Relying on the *Fiers* opinion, the court concluded that the written description requirement again was not met: "Describing a method of preparing a cDNA or even describing the protein that the cDNA encodes, as the example does, does not necessarily describe the DNA itself."[25]

[20] *Id.* at 1170–71.

[21] *Id.*

[22] *See also Adang v. Fischhoff*, 286 F.3d 1346 (Fed. Cir. 2002) (disclosure of genetically altered tobacco plant did not enable claim to genetically altered tomato plant); *Hitzeman v. Rutter*, 243 F.3d 1345 (Fed. Cir. 2001) (conception of biotechnology invention simultaneous with reduction to practice). To be sure, the court stopped short of creating an absolute rule, noting that "[t]here may be situations where an organism's performance of certain intracellular processes might be reasonably predictable, and evidence of such predictability might be sufficient to support a finding of conception prior to reduction to practice." *Id.* at 1357. But even here the court's language focuses on organic processes, not DNA sequences.

[23] 119 F.3d 1559 (Fed. Cir. 1997).

[24] cDNA, or complementary DNA, is produced by reverse transcribing the messenger RNA transcript of genomic DNA. David Freifelder & George M. Malacinski, ESSENTIALS OF MOLECULAR BIOLOGY 278 (2d ed. 1993). This process reverses the usual flow of genetic information from DNA to RNA, but the cDNA transcript is not necessarily identical to the genomic DNA template, as mRNA sequence may have been edited after translation. *Id.*

[25] *Fiers v. Revel*, 984 F.2d at 1170–71.

In reaching these results, the Federal Circuit has been adamant that the degree of specificity required for an adequate description of nucleic acids requires description of "structure, formula, chemical name, or physical properties."[26] In *Eli Lilly*, because "[n]o sequence information indicating which nucleotides constitute human cDNA appears in the patent . . . the specification does not provide a written description of the invention."[27] The court seems particularly incensed by applicants who designate a macromolecule by generic or functional terms, such as "vertebrate insulin cDNA":

> A definition by function . . . is only an indication of what the gene does, rather than what it is. It is only a definition of a useful result rather than a definition of what achieves that result. Many such genes may achieve that result. The description requirement of the patent statute requires a description of an invention, not an indication of a result that one might achieve if one made that invention. Accordingly, naming a type of material generally known to exist, in the absence of knowledge as to what that material consists of, is not a description of that material.[28]

The failure to describe more than one or two nucleotides is a particular problem where the patent claims are drawn to a broad class of nucleotides. For example, Revel's claim covered all DNA molecules that code for β-IF, but "[c]laiming all DNAs that achieve a result without defining what means will do so is not in compliance with the description requirement; it is an attempt to preempt the future before it has arrived."[29]

The Federal Circuit's construction of the written description requirement as requiring precise sequence data gains particular significance whenever claims are drawn to an entire genus, or family, of molecules. The patent discussed in the *Eli Lilly* written description analysis claimed a broad family of DNA molecules coding for insulin in different mammalian species, but it disclosed only one species of DNA, that coding for rat insulin. The court held this to be insufficient to describe the broad class of cDNAs coding for mammalian or vertebrate insulin.[30] Although declining to specify exactly what would be needed to support a broad claim, the court cited previous chemical cases dealing with related groups of small molecules. Based on these cases, the court declared that macromolecules should be treated in the same fashion: The patentee need not show every member of a claimed genus, but is required to

[26] *Id.* at 1171.
[27] *Eli Lilly*, 119 F.3d at 1567.
[28] *Id.* at 1568 (citations omitted).
[29] *Fiers*, 984 F.2d at 1171.
[30] *Eli Lilly*, 119 F.3d at 1567.

show a "representative" number of cDNAs illustrating or defining the common structural features of a "substantial" portion of the genus.[31]

A similarly broad claim was rejected in the *Amgen* case as failing the standard for enablement rather than written description.[32] There, the patentee claimed nucleic acid sequences coding for the protein erythropoetin, or for other proteins with the same biological function. The trial judge concluded that because Amgen was unable to specify which analogs might have the biological properties claimed, the claims were not enabled.[33] The Federal Circuit panel, however, held that the district court had reached the right conclusion for the wrong reason. While the district court focused on the thousands of EPO analogs that could be created by substituting amino acid residues in the polypeptide chain, the appellate court focused on the patentee's failure to disclose the DNA molecules that would code for those analogs.[34] Since the claims were directed to DNA sequences, the issue was not the enablement of the EPO analogs but rather the enablement of the myriad DNA sequences, which the court held could not be made and used on the basis of a few examples.[35]

In an important recent decision, the Federal Circuit backed off somewhat from its categorical insistence on structure in biotechnology disclosure cases. In *Enzo Biochem v. Gen-Probe*,[36] the court adopted the PTO's Guidelines on Written Description.[37] Those guidelines provide that biotechnology inventions normally must be described by structure, but may also be described by "functional characteristics when coupled with a known or disclosed correlation between function and structure."[38] The court specifically identified antibody claims as ones that might be described by function—*i.e.*, by describing the antigen to which they bind.[39] Its holding was more limited, however. It held that the deposit of three actual DNA sequences created a factual question as to whether the deposited sequences could satisfy the written description requirement for claims covering those sequences and a broader genus. Because the deposited sequences inherently included the structure of the gene, the court in *Enzo* had no opportunity to endorse claims based entirely on proof of function

[31] *Id.* at 1569.

[32] *Amgen, Inc. v. Chugai Pharm. Co.*, 927 F.2d 1200, 1212 (Fed. Cir. 1991).

[33] *Id.* at 1205.

[34] *Id.* at 1212–14.

[35] *Id.*

[36] 296 F.3d 1316 (Fed. Cir. 2002).

[37] *Id.* at 1324–25.

[38] U.S. PTO, Guidelines for the Examination of Patent Applications Under the Written Description Requirement, 66 FED. REG. 1099, 1106 (Jan. 5, 2001).

[39] *Enzo*, 296 F.3d at 1324. *See also Amgen Inc. v. Hoechst Marion Roussel, Inc.*, 314 F.3d 1313, 1332 (Fed. Cir. 2003) (adopting *Lilly*, but holding that it did not apply to a patent that identified cells producing a human protein that was itself already well-known).

rather than structure. The court did not repudiate, and indeed relied upon, the *Eli Lilly* baseline rule that disclosure of structure was required.

The same concerns that characterize the Federal Circuit's jurisprudence of biotechnology disclosure—the inadequacy of methodological disclosure, the requirement to specify sequence or structure, and the uncertainty of selection within large classes of homologous molecules—have shaped the Federal Circuit's biotechnology obviousness cases. However, in the case of obviousness, the issue has been the presence of such factors in the prior art rather than in the inventor's disclosure. Thus, the Federal Circuit held in *In re Bell* that a claim to DNA coding for human insulin-like growth factor (hIGF) was not obvious even though the prior art disclosed the amino acid sequence for the hIGF proteins and a method for using that information to obtain the corresponding DNA molecule.[40] Under similar facts in *In re Deuel*, the court found claims directed to DNA coding for heparin-binding growth factors (HBGFs) were not obvious in light of prior art disclosure of a partial amino acid sequence and a method for using that information to obtain the corresponding DNA molecule.[41]

Each decision rested largely upon the court's perception that the actual sequence of the claimed DNA molecules was uncertain or unpredictable from the prior art. In both cases, the court dismissed as irrelevant the biological relationship between the molecules disclosed in the prior art and those claimed by the patent. The amino acid sequences of the proteins disclosed in the prior art are ultimately determined by the sequence of RNA nucleotides coding for the protein, which is, in turn, determinative of the cDNA claimed in the patent.[42] The correspondence of nucleotide sequences to amino acid sequences is well known as key to the "central dogma" of molecular biology: the transfer of genetic information from DNA to RNA to protein chains. However, particular amino acids can correspond to more than one nucleotide sequence, introducing uncertainty into the inverse relationship: that of amino acid sequence to nucleotide sequence. Because of this redundancy or "degeneracy" in the genetic code, the court noted in *Bell* that a vast number of possible sequences—about 10^{36}—might code for the protein sequences disclosed in the prior art. The plaintiff claimed only one of these, in essence having searched among a large number of possibilities to select the particular cDNA sequence coding for hIGF.

[40] *See In re Bell*, 991 F.2d 781, 784 (Fed. Cir. 1993).

[41] *See In re Deuel*, 51 F.3d 1552 (Fed. Cir. 1995).

[42] Neither *In re Bell* nor *In re Deuel* dealt with genomic DNA (gDNA) sequences, which are transcribed by cellular proteins to produce a messenger RNA molecule. *See* Freifelder & Malacinski, *supra* note 24 (describing the transcription process). Both cases considered nonnaturally occurring cDNA sequences, which are reverse transcribed from messenger RNAs. The correspondence between gDNA and RNA may be very different than that of cDNA to RNA, especially in eukaryotic organisms where the processing of RNA transcripts may be extensive. *Id.*

Numerous commentators have pointed out that such a search is relatively routine using tried and true techniques of molecular biology.[43] But prior art disclosure of a method, even an admittedly obvious method, was held insufficient to cure such uncertainty of structure. In rejecting the DNA claims in *Bell* and *Deuel*, the court rejected "the [US Patent and Trademark Office (PTO)] focus on known methods for potentially isolating the claimed DNA molecules" as "misplaced because the claims at issue define compounds, not methods."[44] Prior to *Bell*, the opinion in *Amgen* had stressed the uncertainty of the methods for gene location available at the time of invention: While "it might have been feasible, perhaps obvious to try, to successfully probe a human gDNA library with a monkey cDNA probe, it does not indicate that the gene could have been identified and isolated with a reasonable likelihood of success. . . . [T]here was no reasonable expectation of success in obtaining the EPO gene by the method that Lin eventually used."[45] The court arguably just got the science wrong; by the time of the research at issue in *Bell*, such methods for searching a large universe of molecules were perhaps painstaking and time-consuming, but had an established likelihood of success.

Yet the court defined the issue in *Bell* and *Deuel* not as a matter of the uncertainty of obtaining a particular sequence, but of the uncertainty of predicting or visualizing from the prior art what sequence would be found. Even in the *Amgen* opinion, the court hinted that the key to macromolecular obviousness lay in the prediction of an exact sequence, as "[n]either the DNA nucleotide sequence . . . nor its exact degree of homology with the [prior art] monkey EPO gene was known at the time."[46] And in *Deuel*, the court explicitly held that "until the claimed molecules were actually isolated and purified, it would have been highly unlikely for one of ordinary skill in the art to contemplate what was ultimately obtained. *What cannot be contemplated or conceived cannot be obvious*."[47] Thus a likelihood—or even a certainty—of finding a DNA molecule with particular properties was deemed essentially irrelevant to whether structural claims to that molecule are obvious.[48]

[43] *See, e.g.*, Anita Varma & David Abraham, *DNA is Different: Legal Obviousness and the Balance Between Biotech Inventors and the Market*, 9 HARV. J.L. & TECH. 53 (1996); Philippe G. Ducor, PATENTING THE RECOMBINANT PRODUCTS OF BIOTECHNOLOGY (1998); Arti K. Rai, *Intellectual Property Rights in Biotechnology: Addressing New Technology*, 34 WAKE FOREST L. REV. 827 (1999); Arti K. Rai, *Addressing the Patent Gold Rush: The Role of Deference to PTO Patent Denials*, 2 WASH. U.J.L. & POL'Y 199 (2000).

[44] *In re Deuel*, 51 F.3d at 1558; *see also In re* Bell, 991 F.2d at 785.

[45] *Amgen, Inc. v. Chugai Pharm. Co.* 927 F.2d at 1208–1209.

[46] *Id.*

[47] 51 F.3d at 1558 (emphasis added).

[48] *Cf. Rhone-Poulenc Agro v. DeKalb Genetics Corp.*, 272 F.3d 1335, 1357–1358 (Fed. Cir. 2001) (holding that addition of a second transit peptide to a string of amino acids with a transit peptide and a fragment of a second transit peptide was not obvious because the amino acids were structurally different).

The corollary to this holding is that a molecule will be obvious if the sequence is discernible in the prior art, even if its function is not. Prior art description of the "general idea of the claimed molecules, their function, and their general chemical nature"[49] is insufficient to render a molecule obvious. Some commentators have suggested that this formulation of obviousness stands some danger of collapsing into the standard for anticipation[50]; under section 102 of the Patent Act, an invention lacks patentable novelty if its elements are fully described in a prior art reference, and the Federal Circuit's obviousness requirement could be read to require such a prior art anticipation as the effective standard for obviousness.[51] But unlike the requirements for anticipation, the Federal Circuit's biotechnology obviousness standard appears to require that the sequence of the DNA be predictable from the prior art, and not necessarily explicitly described. For example, the court in *Deuel* suggests that for "a protein of sufficiently small size and simplicity, . . . lacking redundancy, each possible DNA would be obvious over the protein."[52] Although the Federal Circuit has not explicitly held so, one would also suspect that disclosure in the prior art of a substantial number of homologous sequences would render a new homologue predictable, and so render it obvious—just as the court has held that disclosure of a substantial number of homologues is enough to satisfy the written description requirement for a genus of homologues.

The Federal Circuit's biotechnology obviousness cases are consistent with the court's earlier holdings, such as the rejection on disclosure grounds of Revel's claim to all DNA sequences coding for β-IF.[53] Due to degeneracy in the genetic code, Revel could not adequately describe the claimed invention as DNA coding for β-IF; an astronomically large number of possible sequences might do so. And if a functional or narrative description in a patent is insufficient to properly describe a DNA molecule coding for β-IF, the presence of a functional or narrative description of β-IF protein in the prior art would be insufficient to render the molecule obvious. According to the court, one cannot describe what one has not conceived, and what cannot be contemplated or conceived cannot be obvious. Just as disclosure in a patent of a method for

[49] 51 F.3d at 1558.

[50] *See* Rebecca Eisenberg & Robert P. Merges, *Opinion Letter as to the Patentability of Certain Inventions Associated with the Identification of Partial cDNA Sequences*, 23 AIPLA Q. J. 1, 32 (1995).

[51] Indeed, the Federal Circuit has several times suggested that the two patent standards are closely linked, characterizing obviousness as a sort of continuum with anticipation as the "epitome" or "ultimate" endpoint of obviousness. *See, e.g., In re Baxter Travenol Labs*, 952 F.2d 388 (Fed. Cir. 1991); *Jones v. Hardy*, 727 F.2d 1524 (Fed. Cir. 1984); *Connell v. Sears, Roebuck & Co.*, 722 F.2d 1542 (Fed. Cir. 1983).

[52] 51 F.3d at 1559.

[53] *See* discussion *supra* notes 18–21.

obtaining a particular cDNA is inadequate to properly describe the invention, so disclosure in the prior art of a method for obtaining a particular cDNA cannot render the claimed invention obvious.

The conceptual linkage of obviousness and enablement to the depiction of macromolecular sequences in, respectively, the prior art or the patent disclosure dictates a particular and predictable result for the availability and scope of such biotechnology patents. The expected outcome is that DNA patents will be numerous but extremely narrow. Under the Federal Circuit's precedent, a researcher will be able to claim only sequences disclosed under the stringent written description rules—the actual sequence in hand, so to speak. And as Judge Learned Hand observed long ago, a claim that covers only the thing invented is a weak claim indeed.[54] At the same time, the inventor is shielded from obviousness by the lack of such explicit and detailed disclosure in the prior art. This lack of effective prior art seems to dictate that anyone who has isolated and characterized a novel DNA molecule is certain to receive a patent on it. But the inventor is certain to receive a patent only on that molecule, as the Federal Circuit appears to regard other related molecules as inadequately described until their sequence is disclosed.

The set of axioms underlying this set of results forms a logical framework that may be extended to some—but not all—other biotechnology inventions. For example, one would conclude from the Federal Circuit's analysis in these cases that a cDNA should be obvious in light of its corresponding mRNA,[55] since the former is reverse transcribed from the latter, and there is no redundancy or degeneracy in the correspondence between the nucleotides in the two molecules.[56] However, an mRNA or corresponding cDNA need not render obvious the genomic DNA (gDNA) from which it is derived because in many organisms the gDNA will include intervening sequences, or introns, that are not predictable from the mRNA sequence.

Perhaps more important than the extension of the Federal Circuit's logic to other classes of molecules is the extension of its logic to other patent doctrines. For example, as we have indicated with regard to software, patent scope is a function of the obviousness and written description requirements. Under the court's decisions, the literal scope of biotechnology patents will be quite narrow: Patent claims are confined to the DNA sequences actually

[54] *See Philip A. Hunt Co. v. Mallinckrodt Chem. Works*, 177 F.2d 583, 585–586 (2d Cir. 1949) (noting that it may be impossible to write claims of appropriate scope without using functional language to describe variants).

[55] mRNA, or messenger RNA, is the complementary molecule produced from transcription of genomic DNA. *See* Freifelder & Malacinski, *supra* note 24, at 159.

[56] Lubert Stryer, BIOCHEMISTRY 132 (3d ed. 1988); James D. Watson *et al.*, MOLECULAR BIOLOGY OF THE GENE 610–11 (4th ed. 1987).

generated and disclosed rather than those enabled by the patentee. While that scope may be broadened by the doctrine of equivalents,[57] the recent trend to limit the scope of the doctrine of equivalents[58] may mean that the biotechnology industry will be characterized by large numbers of narrow patents.

C. The Divergent Standards

Patent practitioners often focus on a single technology area, and so may tend to take the court's rules in that area for granted. Even a casual juxtaposition of the biotechnology and software cases, however, shows dramatic differences in applying what are nominally the same legal rules.[59] District courts have recognized the difference, applying the Federal Circuit rules in different ways depending on the technology at issue, and one Federal Circuit judge has forthrightly acknowledged that the biotechnology written description cases apply a different legal

[57] The very parsimonious reading that the Federal Circuit gives to obviousness in biotechnology cases seems to leave wide latitude for findings of equivalence in nucleotide infringement cases. *See Wilson Sporting Goods Co. v. David Geoffery & Assoc.*, 904 F.2d 677 (Fed. Cir. 1990) (testing equivalence by inquiring whether a hypothetical claim encompassing the accused product would have been obvious at the time of invention).

[58] The courts have recently strengthened other limits on the doctrine of equivalents, notably prosecution history estoppel and the doctrine of dedication to the public domain. *See, e.g.*, *Festo Corp. v. Shoketsu*, 122 S. Ct. 1831 (2002) (prosecution history estoppel applies broadly, though not absolutely); *Johnson & Johnston Assoc. v. R.E. Serv. Co.*, 238 F.3d 1347 (Fed. Cir. 2001) (en banc) (equivalents disclosed in the patent but not claimed are dedicated to the public domain). *See also* Matthew J. Conigliaro, *et al.*, *Foreseeability in Patent Law*, 16 BERKELEY TECH. L.J. 1045 (2001) (discussing both doctrines). Those limitations may prevent any patent from being read too broadly under the doctrine of equivalents.

[59] Commentators have observed that the Federal Circuit's biotechnology written description cases apply a standard quite different from the written description precedent in other areas. *See, e.g.*, Mueller, *supra* note 15; Sampson, *supra* note 15; Limin Zheng, Note, *Purdue Pharma L.P. v. Faulding Inc.*, 17 BERKELEY TECH. L.J. 95, 95 (2002). While there are a number of recent written description cases outside the biotechnology context, all of them involve patentees who changed their claims during prosecution to cover a competitor's product. *See, e.g.*, *Turbocore Div. of Demag Delaval Turbomachinery Corp. v. Gen. Elec. Co.*, 264 F.3d 1111 (Fed. Cir. 2001); *Hyatt v. Boone*, 146 F.3d 1348 (Fed. Cir. 1998); *Gentry Gallery, Inc. v. Berkline Corp.*, 134 F.3d 1473 (Fed. Cir. 1998). *See also* Janice M. Mueller, *Patent Misuse Through the Capture of Industry Standards*, 17 BERKELEY TECH. L.J. 623, 639–40 (2002) (distinguishing the biotechnology cases from written description decisions in other areas, especially *Union Oil Co. of Cal. v. Atl. Richfield Co.*, 208 F.3d 989 (Fed. Cir. 2000)). *Cf.* Matthew L. Goska, *Of Omitted Elements and Overreaching Inventions: The Principle of Gentry Gallery Should Not Be Discarded*, 29 AIPLA Q.J. 471, 484 (2001) (arguing that the written description requirement makes sense, but that it should not be applied to original claims as it has been in the biotechnology cases). Other commentators have pointed out that the nonobviousness standard in biotechnology is lower than in other industries. *See, e.g.*, Sara Dastgheib-Vinarov, *A Higher Nonobviousness Standard for Gene Patents: Protecting Biomedical Research from the Big Chill*, 4 MARQ. INTELL. PROP. L. REV. 143, 154 (2000); John Murray, Note, *Owning Genes: Disputes Involving DNA Sequence Patents*, 75 CHI.-KENT L. REV. 231, 247 (1999).

standard from the court's other written description opinions.[60] The easiest way to see this difference may be to imagine the court's language from one discipline applied to another. In *Fonar*, for instance, the court said:

> As a general rule, where software constitutes part of a best mode of carrying out an invention, description of such a best mode is satisfied by a disclosure of the functions of the software. This is because, normally, writing code for such software is within the skill of the art, not requiring undue experimentation, once its functions have been disclosed.[61]

Replace software with DNA, though, and the following would result:

> As a general rule, where [DNA] constitutes part of a best mode of carrying out an invention, description of such [DNA] is satisfied by a disclosure of the functions of the [DNA]. This is because, normally, [identifying such DNA] is within the skill of the art, not requiring undue experimentation, once its functions have been disclosed.

This is *exactly* antithetical to the actual rule in biotechnology cases, as stated by *Eli Lilly*:

> A definition by function . . . is only an indication of what a gene does, rather than what it is. It is only a definition of a useful result rather than a definition of what achieves that result. Many such genes may achieve that result. The description requirement of the patent statute requires a description of an invention, not an indication of a result that one might achieve if one made that invention. Accordingly, naming a type of material generally known to exist, in the absence of knowledge as to what that material consists of, is not a description of that material.[62]

Conversely, of course, application of the biotechnology rule to software would radically change the law. The legal rules are the same, but the application of those rules to different industries produces results that bear no resemblance to each other.[63]

[60] *See, e.g.*, *Moba, B. V. v. Diamond Automation*, 325 F.3d 1306 (Fed. Cir. 2003) (Rader, J., concurring) (noting the industry-specific nature of the written description doctrine); *Gummow v. Snap-On Tools*, 58 U.S.P.Q.2d 1414 (N.D. III. 2001) (holding that mechanical patents require less disclosure than biotechnology patents due to the uncertainty in biotechnology).

[61] *Fonar Corp. v. Gen. Elec. Co.*, 107 F.3d 1543, 1549 (Fed. Cir. 1997).

[62] *Regents of the Univ. of Calif. v. Eli Lilly & Co.*, 119 F.3d 1559, 1568 (Fed. Cir. 1997).

[63] Nor are obviousness, disclosure, and patent scope the only doctrines which show such an industry-specific variation. The requirement that an invention have general utility, which has been all

Professor R. Polk Wagner has argued that these differences need not concern us greatly, because they are merely case-specific differences rather than systematic variations by industry.[64] We simply disagree with that reading of the cases. The court's systematic conclusions in different cases, its reliance on industry-specific precedent from case to case, its focus on uncertainty in the biotechnological arts, and its emphasis in biotechnology cases on proof of structure—a discussion totally absent from the software cases—all point in the direction of industry-specific rather than fact-specific differences in legal rules.

II. MODULATING TECHNOLOGY-SPECIFICITY

Besides divergent results, our survey of the biotechnology patent cases also highlights an important reciprocal relationship between obviousness and disclosure. In biotechnology, where highly detailed disclosure is required to satisfy the enablement and written description standards, similarly detailed disclosure in the prior art is required to render the invention obvious. The Federal Circuit takes the patentability requirements of nonobviousness and disclosure as firmly tied to a common standard. The use and misuse of that common standard, then, is central to the development of technologically tailored patent rules.

A. The Role of the PHOSITA

The common standard connecting the requirements of obviousness and disclosure is the requirement in each statutory section that obviousness and the sufficiency of disclosure must be considered from the perspective of the "person having ordinary skill in the art," sometimes known by the acronym of

but eliminated in most fields of technology, *see Juicy Whip, Inc. v. Orange Bang, Inc.*, 185 F.3d 1364 (Fed. Cir. 1999) (holding that a patented device is useful if there is a demand for it), is alive and well in the life sciences. The Supreme Court imposed a stringent requirement on pharmaceutical inventions in *Brenner v. Manson*, 383 U.S. 519 (1966). The Federal Circuit has relaxed that requirement, *see In re Brana*, 51 F.3d 1560, 1567 (Fed. Cir. 1995), but the court still requires more proof of experimentation in order to satisfy the utility requirement in biotechnology and pharmaceuticals than elsewhere. *See* U.S. Patent & Trademark Office, Utility Examination Guidelines, 60 FED. REG. 36263 (July 14, 1995) (describing the law as setting different standards for the life sciences); Timothy J. Balts, *Substantial Utility, Technology Transfer, and Research Utility: It's Time For a Change*, 52 SYRACUSE L. REV. 105 (2002) (describing and criticizing the higher utility standard applied to life sciences); Philippe Ducor, *New Drug Discovery Technologies and Patents*, 22 RUTGERS COMPUTER & TECH. L.J. 369, 431–33 (1996); *cf.* Rebecca S. Eisenberg & Robert P. Merges, *Opinion Letter As To the Patentability of Certain Inventions Associated With the Identification of Partial cDNA Sequences*, 23 AIPLA Q. J. 1 (1995) (arguing that the utility doctrine may bar the patenting of "expressed sequence tags" that can be used to identify human gene sequences).

[64] R. Polk Wagner, "(Mostly) Against Exceptionalism" (Chapter 19). *See also* Sung, *supra* note 15 (making the same argument for the biotechnology cases).

PHOSITA.[65] Much of the case law concerning the PHOSITA arises out of the consideration of the obviousness standard found in section 103 of the patent statute.[66] Although originally developed as a common-law doctrine, the non-obviousness criterion was codified in the 1952 Patent Act as a requirement that the claimed invention taken as a whole not be obvious to one of ordinary skill in the art at the time the invention was made.[67]

The PHOSITA is equally central to calibrating the legal standard for patent disclosure. As the *quid pro quo* for her period of exclusive rights over an invention, the inventor must fully disclose the invention to the public. The first paragraph of section 112 requires that this disclosure enable "any person skilled in the art" to make and use the claimed invention.[68] The parallel language suggests that the inventor's compliance with the requirement of enablement should be measured with reference to a standard similar or identical to that in section 103; indeed, the language appears to tie the enablement requirement to nonobviousness via this shared metric.[69]

This same language sets the metric for several related disclosure doctrines as well. First, the definition of enablement affects the patentability requirement of specific utility, as the invention must operate as described in the specification if the inventor is to enable one of ordinary skill to use it.[70] Additionally, compliance with the independent requirements of adequate written description and best mode disclosure is measured with reference to the understanding of a "person skilled in the art."[71] And finally, the definiteness of patent claims, which must be written so as to warn members of the public just what is and is not covered by the patent, has traditionally been assessed with regard to the knowledge of one having ordinary skill in the art. If the terms of the claims are not comprehensible to such a person, then they fail the requirements of section 112.[72]

[65] John O. Tresansky, *PHOSITA—The Ubiquitous and Enigmatic Person in Patent Law*, 73 J. PAT. & TRADEMARK OFF. SOC'Y 37 (1991); *see also* ROBERT L. HARMON, PATENTS AND THE FEDERAL CIRCUIT § 4.3 (5th ed. 2001); Joseph P. Meara, Note, *Just Who is the Person Having Ordinary Skill in the Art? Patent Law's Mysterious Personage*, 77 WASH. L. REV. 267 (2002). The first known use of the term PHOSITA appears to be in Cyril A. Soans, *Some Absurd Presumptions in Patent Cases*, 10 IDEA 433, 438 (1966).

[66] 35 U.S.C. § 103 (2000).

[67] *Id.*

[68] 35 U.S.C. § 112 ¶ 1 (2000).

[69] The language of the two statutes is not identical, however, and one might draw a distinction between one of ordinary skill and "any person skilled," on the theory that the latter standard includes those with less than ordinary skill. More on this *infra*.

[70] *See, e.g., Newman v. Quigg*, 877 F.2d 1575, 1581-82 (Fed. Cir. 1989).

[71] *See, e.g., In re Wands*, 858 F.2d 731 (Fed. Cir. 1988).

[72] The Federal Circuit's recent decision in *Exxon Res. & Eng. Co. v. United States*, 265 F.3d 1371, 1376 (Fed. Cir. 2001), however, holds that indefiniteness is a pure question of law. How

The PHOSITA is nothing if not versatile, and may also show up as a convenient metric in other unexpected areas, including judicially created patent doctrines. Claim construction requires reference to how the PHOSITA would understand terms in the patent claims.[73] The PHOSITA reappears in some formulations of the standard for infringement by equivalents. In its germinal opinion on the doctrine of equivalents, *Graver Tank*, the Supreme Court indicated that the equivalence of elements of an allegedly infringing device and those of a claimed invention might be tested by determining whether the elements were known in the art to be substitutes for one another.[74] The Federal Circuit strengthened this use of the PHOSITA by making the "reasonable interchangeability" of elements—judged from the perspective of one of ordinary skill in the art—a fundamental test for equivalence.[75] A great deal of patent doctrine therefore rests on the measurement of some legal parameter against the skill and knowledge of the PHOSITA.

This is not to say the PHOSITA has any actual skill or knowledge. Like her cousin, the reasonably prudent person in tort law,[76] the PHOSITA is something of a juridical doppelgänger,[77] embodying a legal standard for patentability rather than the actual capability of any individual or group of individuals.[78] Courts have, on occasion, equated the knowledge of a given individual, such as a patent examiner, with that of the PHOSITA.[79] But courts walk a fine line between taking the skill of an examiner or other artisan as probative evidence of the level of skill in the art and equating the skill of such persons with the characteristics of the hypothetical PHOSITA.[80] Further, unlike any

---

the court will resolve the understanding of the PHOSITA as a legal matter is not entirely clear, though it undertakes a similar burden in construing patent claims. *See Cybor Corp. v. FAS Techs.*, Inc., 138 F.3d 1448, 1454-1455 (Fed. Cir. 1998) (en banc).

[73] *See* Craig Allen Nard, *A Theory of Claim Interpretation*, 14 HARV. J. L. & TECH. 1, 6 (2000).

[74] *See Graver Tank Mfg. Co. v. Linde Air Prods. Co.*, 339 U.S. 605, 609 (1950).

[75] *See Hilton Davis Corp. v. Warner-Jenkinson*, 62 F.3d 1512, 1519 (Fed. Cir. 1995) (en banc), *aff'd in part & rev'd in part on other grounds*, 520 U.S. 17 (1997).

[76] *See, e.g., Panduit Corp. v. Dennison Mfg. Co.* 810 F.2d 1561, 1566 (Fed. Cir. 1987) (comparing the PHOSITA to the "reasonable man" in tort law).

[77] *Id.* (characterizing the PHOSITA as a "ghost").

[78] *See, e.g., Stewart-Warner Corp. v. City of Pontiac*, 767 F.2d 1563, 1570 (Fed. Cir. 1985); Michael H. Davis, *Patent Politics*, (visited Nov. 15, 2002) http://papers.ssrn.com/sol3/papers.cfm?abstract_id=282056 (observing that the PHOSITA standard is "undeniably fictional"); David E. Wigley, *Evolution of the Concept of Non-Obviousness of the Novel Invention: From a Flash of Genius to the Trilogy*, 42 ARIZ. L. REV. 581, 598-599 (2000).

[79] *See* Tresansky, *supra* note 65 at 58 (collecting cases).

[80] *See, e.g., In re Mahurkar Double Lumen Hemodialysis Catheter Patent Litigation*, 831 F. Supp. 1354, 1361-1362 (N.D. Ill. 1993) (Easterbrook, J., sitting by designation) (taking the finding of the examiner, as a PHOSITA, to be probative of written description compliance).

actual person of skill in the art, the PHOSITA is endowed with knowledge of all of the relevant prior art references.[81]

This places the standard for patentability on a legally objective, rather than subjective, footing. The PHOSITA standard measures the inventor's achievements against a judicially determined external metric rather than against an expectation based on whatever level of skill the inventor might actually possess. The standard also has the practical effect of avoiding the requirement that judges and other arbiters of patentability be experts in a given field. The PHOSITA standard is thus an ultimate conclusion of law based upon evidence,[82] not dictated by the capabilities or knowledge of the Patent Office examiner, a reviewing judge, or even that of the inventor:

> Realistically, courts never have judged patentability by what the real inventor/applicant/patentee could or would do. Real inventors, as a class, vary in their capacities from ignorant geniuses to Nobel laureates; the courts have always applied a standard based on an imaginary worker of their own devising whom they have equated with the inventor.[83]

The standard is thus objective in the sense that it does not inquire into a particular inventor or artisan's level of skill. But this does not mean that it is static or fixed. Courts consider a number of constituent factors that may be adjusted to modulate the requirements for patentability under different circumstances. The first of these is the definition of the particular "art" in which the PHOSITA is deemed to have ordinary skill. The PHOSITA is generally portrayed as having comprehensive knowledge of the references in the particular art.[84] But the parameters of the art are subject to fluctuation, and thus so is the size and depth of the library of references with which the PHOSITA is presumed to be familiar. For example, in the case of a DNA patent, would the relevant art be biochemistry, molecular biology, cell biology, or biology in general? Courts have attempted to avoid drawing such boundaries by defining the PHOSITA's knowledge as that reasonably pertinent to the problem the inventor was trying

[81] *See In re Winslow*, 365 F.2d 1017 (C.C.P.A. 1966).

[82] *See Panduit Corp. v. Dennison Mfg. Co.*, 810 F.2d 1561 (Fed. Cir. 1987).

[83] *Kimberly-Clark Corp. v. Johnson & Johnson*, 745 F.2d 1437, 1454 (Fed. Cir. 1984). *See also In re Nilssen*, 851 F.2d 1401 (Fed. Cir. 1988) (noting that the Board of Patent Appeals and Interferences was not required to have ordinary skill in the art to apply the standard); *Hodosh v. Block Drug Co.*, 786 F.2d 1136 (Fed. Cir. 1986) (stating that actual inventors cannot be required to have the omniscience of the figurative person of ordinary skill).

[84] Although, as we point out below, this imputed knowledge varies a bit depending upon whether obviousness or disclosure is at issue. *See infra* notes 90-94.

to solve. But this requires that the court engage in the equally mercurial exercise of defining the problem that the inventor had under consideration.[85]

A second PHOSITA variable that may be adjusted to different circumstances is the level of skill that would be considered "ordinary." Unlike the inventor, who almost by definition is presumed to be one of extraordinary skill,[86] the PHOSITA standard contemplates some median or common level of skill. In assessing that common level, courts may take into account a long list of factors, including the approaches found in the prior art, the sophistication of the technology involved, the rapidity of innovation in that field, and the level of education typical of those in the field.[87] The courts have also endowed the PHOSITA with mediocre personality traits; she is conceived of as someone who adopts conventional approaches to problem solving, and is not inclined to innovate, either via exceptional insight or painstaking labor.[88]

Some care must be exercised in characterizing the PHOSITA, as it is tempting to do so on the basis of an unfounded presumption that the PHOSITA remains constant from section to section of the patent statute. On the contrary, some commentators have recognized the possibility that the imaginary artisan found in these different statutory sections, though bearing the same denomination, might display different and even inconsistent characteristics as between the different sections.[89] The PHOSITA for purposes of obviousness may not necessarily be the PHOSITA for purposes of enablement, written description, definiteness, or equivalence. Because she is a legal construct designated to embody certain legal standards, the PHOSITA could change depending on the purpose she is serving at the time. Understanding this difference is critical, because the Federal Circuit's linkage of obviousness and enablement depends on the easy equation of the PHOSITAs.

Some disparity of this sort does, in fact, appear in the judicial characterization of the PHOSITA in the contexts of obviousness and of enablement. The section 103 PHOSITA appears to be something of a problem solver who the courts set to work hypothetically tackling the problem solved by the inventor.[90] The obviousness PHOSITA is admittedly not an especially inspired

[85] *See e.g.*, *George J. Meyer Mfg. Co. v. San Marino Elec. Corp.*, 422 F.2d 1285, 1288 (9th Cir. 1970) (deeming the PHOSITA in optical bottle inspection art to be aware of prior art in optical missile tracking field).

[86] *Standard Oil Co. v. Am. Cyanamid Co.*, 774 F.2d 448, 454 (Fed. Cir. 1995).

[87] *See, e.g.*, *Bausch & Lomb Inc. v. Barnes-Hind, Inc.*, 796 F.2d 443 (Fed. Cir. 1986) (listing pertinent factors); *see also Helifix Ltd. v. Blok-Lok Ltd.*, 208 F.3d 1339 (Fed. Cir. 2000) (stating that district court erred by failing to consider these factors).

[88] *Standard Oil Co. v. Am. Cyanamid Co.*, 774 F.2d 448, 454 (Fed. Cir. 1985).

[89] *See* Tresansky, *supra* note 65 at 52-53.

[90] *See Orthopedic Equip. Co. v. United States*, 702 F.2d 1005 (Fed. Cir. 1983); *In re Grout*, 377 F.2d 1019 (C.C.P.A. 1967).

problem solver, as she is imagined to remain stuck in the rut of conventional thinking.[91] But the obviousness PHOSITA is still someone who is trying to solve new problems. By contrast, the PHOSITA of the first paragraph of section 112 shows no such innovative tendency; she is simply a user of the technology. If the enablement PHOSITA shows any problem-solving ability, it is in tapping the prior art to fill in gaps left by the inventor's disclosure—a rather different skill than that of the obviousness PHOSITA.[92]

The two PHOSITAs also differ in the date at which knowledge is imputed to them. The knowledge of the obviousness PHOSITA is assessed at the time of invention, while the enablement PHOSITA is aware of information available at the time a patent is filed. Due to the passage of time, the latter universe of references is likely to be larger. The temporal disparity is even stronger when the doctrine of equivalents PHOSITA is employed; this entity knows of all developments up to the date of infringement.[93] But conversely, knowledge of hidden or nonpublic references that may serve as prior art under section 103 is not necessarily imputed to the PHOSITAs who make or use the invention under section 112, as such references are not readily available to the public.[94]

B. Misapplication of the PHOSITA Standard

The PHOSITA approach, in general, represents the proper standard for patent law. Basing the required proof on the level of skill in the art makes logical sense.[95] At the simplest level, this approach is intended to benefit the public; people who work in a given technology must understand the patent as it relates to the prior art, so it makes sense to take into account what that person knows in order to decide whether a patent is obvious or has been enabled. From a policy standpoint, the practicality of working in different technologies requires a flexible approach to determining disclosure or obviousness, and the PHOSITA approach gives a court that flexibility. In this sense, patent law is inherently technology-specific, in essence offering different and fact-sensitive standards of disclosure and obviousness for different technologies.

[91] *Standard Oil Co. v. Am. Cyanamid Co.* 774 F.2d 448 (Fed. Cir. 1985).

[92] *See* Tresansky, *supra* note 65 at 54.

[93] *See Warner-Jenkinson Co., Inc. v. Hilton Davis Chem. Co.*, 520 U.S. 17, 37 (1997) (holding that equivalence is tested at the time of infringement). Indeed, were it otherwise, the doctrine of equivalents could not feasibly be applied to later-developed technologies.

[94] *See Quaker City Gear Works, Inc. v. Skil Corp.*, 747 F.2d 1446 (Fed. Cir. 1984); *In re Howarth*, 654 F.2d 103 (C.C.P.A. 1981).

[95] *But see* Davis, *supra* note 78, at 32 (use of PHOSITA is "disingenuous and almost foolish").

But even recognizing that the PHOSITA standard dictates that different technologies will be accommodated in different ways, the developments that we have described in biotechnology seem to us extraordinary and difficult to explain solely by reference to the level of skill in these arts. Consider, for example, the extremely stringent disclosure standard developed in the biotechnology cases. If the PHOSITA analysis explains that requirement, it suggests that the Federal Circuit believes that biotechnology researchers need a very high degree of assurance before they are capable of replicating an invention. Computer programmers, on the other hand, apparently require very little assurance—simply an indication of function will do. Similarly, with regard to obviousness, the court appears to believe that computer programmers can fully envision working code from only a suggestion of function, whereas biotechnologists apparently need genetic sequences explicitly spelled out in the prior art to render a molecule obvious. As detailed below, we are not persuaded that the levels of skill in these arts are so different, either for innovators or for users.

In this section, we seek to understand why the Federal Circuit's application of the PHOSITA standard has produced such incongruous results in the industries we studied. In order to identify the source of the anomalies in biotechnology, we look first to the Federal Circuit's application of this standard rather than to the standard itself. One possibility, which has occurred to previous commentators as well as to us, is that the Federal Circuit application of the PHOSITA standard in these technologies is wrong as a matter of science.[96] One reading of these cases is that the Federal Circuit seems to have substituted caricature for a nuanced understanding of the technology. In the biotechnology cases, the court focuses repeatedly on the "uncertainty" inherent in the field, scoffing at claims drawn to molecular function rather than to structure and demanding precise disclosure of any embodiment.[97] The court seems to believe that biotechnology is as much a black art as a science, where the result of experimentation is largely out of the skilled artisan's hands. While the assumption that an art is uncertain may befit a new and undeveloped field, the court has maintained its assumption that biotechnology is an uncertain art long after the industry began to mature. The Federal Circuit has sidestepped the difficulty of determining the level of skill in the art in each case by grounding biotechnology patent standards in a doctrine of structural foreseeability. This solution is attractive to the court, as the requirement of foreseeable structure becomes an axiom from which other patent standards can be neatly

[96] *See, e.g.*, Varma & Abraham, *supra* note 43; Rai, *Addressing New Technology*, *supra* note 43; Rai, *Patent Gold Rush*, *supra* note 43.

[97] *See supra* notes 41-48 and accompanying text (discussing the role of uncertainty in the Federal Circuit's biotechnology jurisprudence).

derived. However, just as we are cautioned by the old maxim that when one has a hammer everything looks like a nail, it would seem that the Federal Circuit, having once crafted a solution based on structural foreseeability, begins to see every DNA patenting problem as a problem of structure. In *Bell* and *Deuel* the court's belief in uncertainty benefits the patentee because it means that knowledge of a protein and a method for deriving the cDNA sequence does not render the cDNA sequence obvious without the disclosure of structure.[98] By contrast, the same assumption about uncertainty hurts patentees in cases like *Enzo v. Calgene*, *Lilly*, and *Amgen* because it precludes them from claiming any DNA sequence they have not actually described in structural terms in the patent specification.[99] All of these holdings are based on the assumption that one ordinarily skilled in biotechnology cannot move conceptually from a protein to a DNA sequence, or from the DNA sequence of one organism to the corresponding DNA sequence of another organism.

Arguably this understanding of the science of biotechnology is simply wrong. Robert Hodges has argued that "the key event is the cloning of the first gene in a family of corresponding genes. Once a researcher accomplishes this very difficult task, the researcher can typically obtain other members of the gene family with much less effort."[100] Indeed, today the process is largely automated. Such research is properly compared to searching a "black box" that contains molecules of known characteristics, if unknown structure; the search is conducted on the basis of what is known—the function—rather than on the basis of what is unknown—the precise structure. The function of the molecule that will be found is predictable, as is the likelihood of finding such a molecule, even if the precise structure of the molecule cannot be predicted.[101]

[98] *See In re Deuel*, 51 F.3d 1552, 1559 (Fed. Cir. 1995); *In re Bell*, 991 F.2d 781 (Fed. Cir. 1993). *Cf.* Fiers v. Revel, 984 F.2d 1164 (Fed. Cir. 1993) (using the same standard in an interference proceeding to benefit one applicant at the expense of another). *But cf. In re Mayne*, 104 F.3d 1339 (Fed. Cir. 1997) (DNA sequence in prior art rendered obvious a claim to an altered version of that sequence that changed only one amino acid).

[99] *See Enzo Biochem v. Calgene, Inc.*, 188 F.3d 1362, 1371 (Fed. Cir. 1999); *Regents of the Univ. of California v. Eli Lilly & Co.*, 119 F.3d 1559 (Fed. Cir. 1997); *In re Goodman*, 11 F.3d 1046, 1052 (Fed. Cir. 1993); *Amgen Inc. v. Chugai Pharmaceutical Co.*, 927 F.2d 1200 (Fed. Cir. 1991).

[100] Robert A Hodges, *Black Box Biotech Inventions: When a Mere "Wish or Plan" Should be Considered an Adequate Description of the Invention*, 17 GA. ST. U. L. REV. 831, 832 (2001). *See also* John M. Lucas, *The Doctrine of Simultaneous Conception and Reduction to Practice in Biotechnology: A Double Standard for the Double Helix*, 26 AIPLA Q. J. 381, 418 (1998) ("Making the inventions of *Amgen*, *Fiers* and *Lilly* today would be routine.").

[101] *See, e.g.*, Alison E. Cantor, *Using the Written Description and Enablement Requirements to Limit Biotechnology Patents*, 14 HARV. J. L. & TECH. 267, 310–11 (2000) (stating that "[t]here is already indication that initial biotechnology techniques are increasingly considered to be more predictable and are more likely to fall into the category of routine experimentation" and citing monoclonal antibodies as an example).

This explanation of the Federal Circuit's jurisprudence in these areas is not altogether satisfactory, as (in biotechnology, at least) it fails to explain the court's indifference to the technology subsequent to *Amgen*. The obviousness decision in *Amgen* clearly rested upon the uncertain likelihood of success in the particular probing methodology used to find the EPO gene. Had the court adhered to this analysis in later cases, carrying forward into subsequent opinions a static impression of biotechnological techniques, the poor fit between patent doctrine and patent policy could be easily explained; indeed some commentators have offered this easy explanation.[102] But in those later cases, the court seems quite indifferent to the certainty or uncertainty of methodological success, fashioning instead a standard based on structural precision and foreseeability that ignores the state of technology, past or present. It seems not so much that the court misunderstood the changes in technology since *Amgen* as that the court simply ignored them.

We should here acknowledge an alternative explanation for the Federal Circuit's biotechnology decisions: The court, rather than stumbling in its application of law to changing technology, is, as a matter of law, deliberately creating a unique enclave of patent doctrine for biotechnology, making patent law indeed technology specific. This alternative explanation seems to us even less satisfactory than the first. If the court is taking the trouble of fashioning individual patentability standards for different areas of subject matter, one would expect that the standards fashioned would be suited to the needs of the different areas addressed. Yet, as we discuss in Part III, the Federal Circuit's biotechnology cases are ill-considered as a policy matter.

C. Obstacles to Applying the PHOSITA Standard Properly

If, as we suggest, the concept of the PHOSITA makes sense, why has the Federal Circuit got it wrong in these industries? There are several reasons. First, we think that there are several structural barriers that make it difficult for courts to accurately assess the level of skill in a complex, technological art. As a practical matter, it is worth emphasizing that judges are at a serious disadvantage in trying to put themselves in the shoes of an ordinarily skilled scientist. Judges generally don't have any scientific background, and at the district court level, at least, most law clerks don't either. Further, district court judges have extremely full dockets with many different types of cases. The average judge may hear no more than one patent case every few years.[103] Few of those will be biotechnology

[102] *See* Wagner, *supra* note 64.

[103] There are roughly 1700 patent cases filed per year. The exact data for the years 1995–1999 can be found in the Derwent Litalert database http://www.derwent.com/intellectualproperty/

cases.[104] A very busy judge must therefore learn not only patent law but also some difficult science in a very short period of time. Expert witnesses can help, but the Federal Circuit has imposed some limits on the extent to which district courts can rely on such evidence.[105] In particular, courts must avoid the temptation to assume that the expert witness *is* a person ordinarily skilled in the art.[106] Even the Federal Circuit, which does not suffer nearly so much from these limitations,[107] is not in a position to fully understand all of the science it encounters.[108] Given these limitations, courts understandably won't get it right every time.[109]

Second, the timing of the PHOSITA analysis complicates the court's task. While the court will determine the level of skill in the art during a pretrial hearing or at trial, the appropriate level of skill in the art is not what people know at the time of trial, but what people knew at the time of the invention (in the case of obviousness) or the filing of a patent application (in the case of

---

litalert.html The data that follow were compiled as of June 1, 2000, and involve cases labeled "patent."

Year	*Number of Cases Filed*
1999	1,652
1998	1,730
1997	1,731
1996	1,514
1995	1,258

Id. Most of these cases settle, however. Kimberly Moore's comprehensive study of all patent cases that went to trial found only 1411 cases in the 17 years from 1983 to 1999, an average of less than 100 cases per year. Kimberly A. Moore, *Judges, Juries, and Patent Cases—An Empirical Peek Inside the Black Box*, 99 MICH. L. REV. 365, 380 (2000). Since there are over 600 district court judges in the United States, it is obvious that most judges get only a few filed patent cases a year, and well less than one patent trial a year. In fact, many judges get even fewer cases than this number would suggest (though others get more), since the concentration of innovation in certain regions and the permissibility of forum shopping in patent cases cause patent cases to be bunched in a few districts. *See* Kimberly A. Moore, *Forum Shopping in Patent Cases: Does Geographic Choice Affect Innovation?* 79 N.C. L. REV. 889 (2001) (analyzing where patent suits are filed).

[104] *See* John R. Allison & Mark A. Lemley, *Empirical Evidence on the Validity of Litigated Patents*, 26 AIPLA Q. J. 185, 217 tbl.5 (1998) (demonstrating that between 1989 and 1996, only 3% of patent cases litigated to judgment involved biotechnology).

[105] *See Vitronics Corp v. Conceptronic, Inc.*, 90 F.3d 1576, 1584 (Fed. Cir. 1996) (stating that courts may rely on expert testimony in construing patent claims only in rare circumstances); *but see Pitney-Bowes v. Hewlett-Packard Corp.*, 182 F.3d 1298, 1308 (Fed. Cir. 1999) (holding that judges may hear expert testimony on the meaning of patent claims, but may not normally *rely* on such testimony). This distinction between admitting testimony to help the judge understand the claims and reliance on such testimony may make conceptual sense, but courts reading this line of cases may be reluctant to hear such evidence at all. Plus, it will not help much in deciding pretrial motions. More recently, the Federal Circuit has taken yet another tack on claim construction, emphasizing the importance of dictionary definitions over either the prosecution history or oral testimony. *Texas Digital Sys. v. Telegenix, Inc.*, 308 F.3d 1193 (Fed. Cir. 2002). In so doing, it seemed to denigrate the role of expert testimony in construing claims, perhaps signaling a return to a variant of the *Vitronics* approach.

enablement).[110] On average, it takes more than 12 years from the time a patent application is filed until final judgment on the merits; it takes even longer from the date of invention.[111] So courts trying to determine the level of skill in the art must learn not just science, but the history of that science. Courts and expert witnesses must shut out of their minds intervening developments in the field. This is notoriously difficult to do. Empirical evidence has demonstrated that people in general, and judges in particular, are subject to a "hindsight" bias: They are likely to reason backwards from what did happen to make assumptions about what was likely to happen *ex ante*.[112] The Federal Circuit has repeatedly recognized the problem of hindsight bias in its obviousness jurisprudence[113] and has built rules designed to cope with it there,[114] but hindsight bias risks infecting the PHOSITA analysis in enablement and claim scope as well. Hindsight bias will normally lead fact finders to overestimate the level of skill in the art, since subsequent advances will suggest that the invention could not have been that difficult to do. This effect is likely to be the most pronounced in technologies that are familiar or readily understood by the trier of fact—that is, in the "predictable" arts. Occasionally, however, hindsight bias may have the opposite

---

[106] *See, e.g., Dayco Prods. v. Total Containment, Inc.*, 258 F.3d 1317, 1324 (Fed. Cir. 2001) ("Our objective is to interpret the claims from the perspective of one of ordinary skill in the art, not from the viewpoint of counsel or expert witnesses."); *Endress & Hauser v. Hawk Measurement Sys.*, 122 F.3d 1040, 1042 (Fed. Cir. 1997) ("The 'person of ordinary skill' in the art is a theoretical construct . . . and is not descriptive of some particular individual"; experts need not themselves be of ordinary skill in the art.).

[107] While relatively few Federal Circuit judges have technology backgrounds, John R. Allison & Mark A. Lemley, *How Federal Circuit Judges Vote in Patent Validity Cases*, 27 FLA. ST. U. L. REV. 745, 751 n.23 (2000), many of their clerks do. Further, the Federal Circuit has more time to consider each case, has the full record before it, and gets many more patent cases, including software and biotechnology cases, than any district court judge would.

[108] Arti Rai argues that the Federal Circuit should defer to the PTO, because the PTO better understands biotechnology. Rai, *supra* note 43. We agree with her that the Federal Circuit makes mistakes in this area. We are not persuaded that the PTO can do any better, however, particularly given the minimal time examiners can spend on any one invention. *See* Mark A. Lemley, *Rational Ignorance at the Patent Office*, 75 NW. U. L. REV. 1495, 1500 (2001) (noting that examiners spend only 18 hours per application on average).

[109] *Cf.* Stephen L. Carter, *Custom, Adjudication, and Petrushevsky's Watch: Some Notes From the Intellectual Property Front*, 78 VA. L. REV. 129, 132 (1992) (worrying that judges may not be particularly good at "judicial anthropology").

[110] *See Arkie Lures, Inc. v. Gene Larew Tackle, Inc.*, 119 F.3d 953, 956 (Fed. Cir. 1997) (holding that PHOSITA analysis must "focus on conditions as they existed when the invention was made" in obviousness cases).

[111] Allison & Lemley, *Empirical Evidence*, *supra* note 104, at 236 tbl.0011 (12.3 years on average). This has been a particular problem in biotechnology cases, particularly because they spend longer in prosecution and because biotechnology patents are often most valuable at the end of their lives. *See, e.g., Enzo Biochem v. Calgene, Inc.*, 188 F.3d 1362, 1371 (Fed. Cir. 1999) (16-year old invention); *Genentech, Inc. v. Novo Nordisk*, 108 F.3d 1361, 1367 (Fed. Cir. 1997) (18-year

effect, notably where certain things known or believed at one time to be feasible turn out later to be more difficult than anticipated.[115]

Finally, the backward-looking nature of the legal system itself creates a problem that is, in some sense, the opposite of the hindsight bias. Legal rules are based on *stare decisis*. The law accumulates nuance over time by respecting and building on the body of existing precedent. Only rarely will courts expressly reject their prior decisions. This system has worked well over time in producing thoughtful legal rules.[116] Judges trained in this process will naturally tend to apply it to factual issues they see repeatedly. Indeed, doing so seems economical as well, since revisiting those factual determinations appears redundant. Thus, once the Federal Circuit has ruled on the level of skill in a particular art, the temptation is strong for both that court and district courts to apply that determination in subsequent cases. This tendency is evident in biotechnology cases. *In re Bell* concluded that the knowledge of an amino acid sequence produced by a gene, coupled with a plan for identifying the DNA sequence of the gene, did not render the DNA sequence itself obvious.[117] *In re Deuel* relied on *Bell*'s conclusion, despite the fact that biotechnology had advanced somewhat between the two inventions.[118] In *Regents of the University of California v. Eli Lilly & Co.*,[119] the court expressly relied on its conclusions about the level of skill in

old invention); Jeffrey S. Dillen, *DNA Patentability—Anything But Obvious*, 1997 WISC. L. REV. 1023, 1038 (noting this time lag).

[112] There is an interesting empirical literature in the behavioral law and economics movement on hindsight bias. The existence of such a bias is well documented. In the behavioral science literature, see, *e.g.*, Baruch Fischhoff, *Hindsight ≠ Foresight: The Effect of Outcome Knowledge on Judgment Under Uncertainty*, 1 J. EXPERIMENTAL PSYCHOL.: HUM. PERCEPTION & PERFORMANCE 288 (1975); Amos Tversky & Daniel Kahneman, *Availability: A Heuristic for Judging Frequency and Probability*, 5 COGNITIVE PSYCH. 207 (1973). In the legal literature, see, *e.g.*, BEHAVIORAL LAW AND ECONOMICS (Cass Sunstein ed. 2000); Russell Korobkin & Thomas S. Ulen, *Law and Behavioral Science: Removing the Rationality Assumption from Law and Economics*, 88 CALIF. L. REV. 1051, 1095 (2000); Eric Talley, *Disclosure Norms*, 149 U. PA. L. REV. 1955, 2000 (2001). There is even empirical evidence that federal judges are subject to hindsight bias. *See* Chris Guthrie *et al.*, *Inside the Judicial Mind*, 86 CORNELL L. REV. 777, 799-805 (2001).

[113] *See, e.g., Al-Site Corp. v. VSI Int'l, Inc.*, 174 F.3d 1308, 1324 (Fed. Cir. 1999); *Monarch Knitting Mach. Corp. v. Fukuhara Indus. & Trading Co.*, 139 F.3d 877 (Fed. Cir. 1998).

[114] *See, e.g., In re Dembiczak*, 175 F.3d 994, 999 (Fed. Cir. 1999) ("Our case law makes clear that the best defense against the subtle but powerful attraction of a hindsight-based obviousness analysis is rigorous application of the requirement for a showing of the teaching or motivation to combine prior art references."). Indeed, the Federal Circuit may have overcompensated, making it very difficult to combine references in order to prove obviousness. *See* Mark A. Lemley & David W. O'Brien, *Encouraging Software Reuse*, 49 STAN. L. REV. 255, 301 (1997) (making this argument). For an extremely strict statement of the legal standard on combining references, see *Winner Int'l Royalty Corp. v. Wang*, 202 F.3d 1340 (Fed. Cir. 2000).

[115] For a detailed discussion of hindsight in biotechnology cases, see Lawrence M. Sung, *On Treating Past as Prologue*, 2001 U. ILL. J.L. TECH. & POL'Y 75.

the art in *Bell* and *Deuel* to determine its conclusions regarding written description.[120] *Fiers* is even more explicit in this regard, creating a firm rule that the conception of a DNA sequence requires a listing of that sequence "irrespective of the complexity or simplicity of the method of isolation."[121]

While apparently logical, the reliance on industry-specific precedent in determining the level of skill in the art is problematic. First, while both obviousness and enablement rely on the PHOSITA construct, the PHOSITA is not necessarily the same for obviousness and enablement even in a single case. Obviousness is tested at the time the invention was made, while enablement is tested at the time the application was filed. Clearly the application cannot be filed until after the date of invention, and in some cases several years elapse between the two.[122] Knowledge in the art can change during this period, sometimes dramatically. Second, and more important, the level of skill in the art will normally change between the dates of different inventions. It is hazardous, therefore, to rely on one court's statement of the level of skill in the art as determinative or even evidentiary of the level of skill in the same art at a different time. The level of skill in the art is a factual question that must be determined anew on the particulars of each case.[123]

---

[116] For arguments suggesting the common law evolves towards efficiency over time, see Richard A. Posner, ECONOMIC ANALYSIS OF LAW 23-27 (1st ed. 1979); George L. Priest, *The Common Law Process and the Selection of Efficient Rules*, 6 J. LEGAL STUD. 65 (1977); Paul H. Rubin, *Why is the Common Law Efficient?* 6 J. LEGAL STUD. 51 (1977). Whether or not this controversial claim is correct, *stare decisis* is clearly entrenched in the legal mindset.

[117] *In re Bell*, 991 F.2d 781, 785 (Fed. Cir. 1993).

[118] *In re Deuel*, 51 F.3d 1552, 1559 (Fed. Cir. 1995). In *Bell*, the prior art disclosed the amino acid sequence for the proteins of interest, and a method for cloning genes. By contrast, the art in *Deuel* disclosed only a partial amino acid sequence. Nonetheless, the passage of several years between the priority dates of the applications (Deuel's application was first filed January 8, 1990, and Bell's application was filed June 16, 1987) was ignored by the court, which did not focus on or even mention when the inventions occurred.

[119] 119 F.3d 1559 (Fed. Cir. 1997).

[120] Example 6 provides the amino acid sequence of the human insulin A and B chains, but that disclosure also fails to describe the cDNA. Recently, we held that a description which renders obvious a claimed invention is not sufficient to satisfy the written description requirement of that invention. *Lockwood*, 107 F.3d at 1572, 41 USPQ2d at 1966. We had previously held that a claim to a specific DNA is not made obvious by the mere knowledge of a desired protein sequence and methods for generating the DNA that encodes that protein. *See, e.g., In re Deuel*, 51 F.3d 1552, 1558, 34 USPQ2d 1210, 1215 (1995) ("A prior art disclosure of the amino acid sequence of a protein does not necessarily render particular DNA molecules encoding the protein obvious because the redundancy of the genetic code permits one to hypothesize an enormous number of DNA sequences coding for the protein."); *In re Bell*, 991 F.2d 781, 785, 26 USPQ2d 1529, 1532 (Fed.Cir.1993). Thus, a fortiori, a description that does not render a claimed invention obvious does not sufficiently describe that invention for purposes of § 112 ¶ 1. Because

A related problem is the equally time-honored tradition of reasoning by analogy. If courts and lawyers can't find precedent directly on point, they will turn to the closest available analog. In the case of biotechnology, the court appears to have taken its understanding of DNA directly from its small-molecule chemistry cases of a generation before. But if reliance on precedent is bad in the case of the PHOSITA, reliance on analogy is worse. Expanding the search for the PHOSITA beyond a narrow definition of the field in question will almost certainly get it wrong, as indeed the court has done in the biotechnology cases. Given the fact-specific nature of the inquiry, the Federal Circuit may need to resist its tendency—well documented in other areas—to substitute its factual conclusions for the district court's.[124] A clear signal by the Federal Circuit that identifying the PHOSITA is a fact-specific question that must be decided anew in each case (perhaps by reference to expert testimony) might go a long way toward solving the problem of substituting precedent and analogy for detailed analysis.[125] Courts should also spend more time and effort fleshing out the PHOSITA, who in many opinions seems to be mentioned only perfunctorily.[126] We offer more ideas for tailoring the treatment of the PHOSITA elsewhere.[127]

> the '525 specification provides only a general method of producing human insulin cDNA and a description of the human insulin A and B chain amino acid sequences that cDNA encodes, it does not provide a written description of human insulin cDNA.

Id. at 1567.

[121] *Fiers v. Revel*, 984 F.2d 1164, 1169 (Fed. Cir. 1993).

[122] The law permits a one-year grace period between any public act and the filing of a patent application. *See* 35 U.S.C. § 102(b) (2000). But many inventors wait even longer between invention and the filing of an application. This is permissible, as long as they do not put the invention on sale or in public use in the interim, and do not abandon it. 35 U.S.C. § 102(c) (2000).

[123] For a detailed discussion, see Dillen, *supra* note 111, at 1039-44. The U.S. Court of Customs and Patent Appeals in *In re Driscoll*, 562 F.2d 1245, 1250 (C.C.P.A. 1977), and the Federal Circuit in *Enzo Biochem. v. Calgene, Inc.*, 188 F.3d 1362, 1374 n.10 (Fed. Cir. 1999), both recognized this. However, it has proven a hard rule to adhere to.

[124] *See, e.g.*, William C. Rooklidge & Matthew F. Weil, *Judicial Hyperactivity: The Federal Circuit's Discomfort With Its Appellate Role*, 15 BERKELEY TECH. L.J. 725 (2000) (noting this problem); Arti K. Rai, *Engaging Facts and Policy: A Multi-Institutional Approach to Patent System Reform*, 103 COLUM. L. REV. 1035 (2003) (arguing that the Federal Circuit's preoccupation with making factual determinations has prevented it from taking a leadership role in guiding legal rules). Empirical data suggests that the Federal Circuit is particularly unlikely to defer to district court rulings construing patent claims. *See* Christian Chu, *Empirical Analysis of the Federal Circuit's Claim Construction Trends*, 16 BERKELEY TECH. L.J. 1045 (2001); Kimberly A. Moore, *Are District Court Judges Equipped to Resolve Patent Cases?* 15 HARV. J. L. & TECH. 1 (2001). Rai also argues that the Federal Circuit has not shown sufficient deference to factfinders in obviousness and disclosure cases. *See* Rai, *Facts*, *supra*.

[125] In this respect we agree with Wagner, who argues that improper determinations of the PHOSITA in one case should not bind courts in a later case. Wagner, *supra* note 64, at 6. But since

III. INNOVATION, INVENTION, AND UNCERTAINTY

The fact that the court has created technology-specific patent rules for biotechnology is not necessarily a bad thing. As we have suggested elsewhere, different industries experience both innovation and the patent system in very different ways.[128] Biotechnology is no different. We don't object, therefore, to the idea that courts treat biotechnology differently. Indeed, we embrace it. Existing law creates a variety of "policy levers" that permit and may even compel the courts to do so. Our concern is instead that courts do not seem to take the actual characteristics of the industry into account. As a result, the specific biotechnology rules the court has created do not work for the biotechnology industry. In this section, we talk briefly about the theories of patent law that fit the economics of biotechnology, and what those theories imply for optimal biotechnology patent policy.

A. Theories of Biotechnology Patents

There are a number of different economic theories of the patent system. These approaches exist in considerable tension. They make different and conflicting predictions about the effect of patents on industries, and dictate different and conflicting prescriptions for the parameters of patent law. We discuss five different theories in detail elsewhere;[129] here we briefly review the two that apply most neatly to biotechnology.

1. Prospect Theory

In 1977, Edmund Kitch offered a new theory of the patent system, one that he said would "reintegrate the patent institution with the general theory of property rights."[130] This property rights or "prospect" theory of intellectual property is rooted in many of the same economic traditions as the classic incentive-to-invent theory, but not on *ex ante* incentives to create as much as on the ability of intellectual property ownership to force the efficient use of inventions and

the Federal Circuit has relied on such prior determinations, we see the current state of affairs as more problematic than he does.

[126] *See* Meara, *supra* note 65 (arguing that the existing factors for determining skill in the art do not work very well, and suggesting ways to refine the PHOSITA inquiry).

[127] Burk & Lemley, *Technology-Specific*, *supra* note 14, at 1202-1205.

[128] Burk & Lemley, *Policy Levers*, *supra* note 2.

[129] *Id.* at Part II.A.

[130] Edmund Kitch, *The Nature and Function of the Patent System*, 20 J. L. & ECON. 265, 265 (1977).

creations through licensing once they are made.[131] The fundamental economic bases of this approach are the "tragedy of the commons" and the hypothetical Coasean world without transactions costs. The tragedy of the commons is a classic economic story: Common property will be overused by people who have access to it, since each individual reaps all of the benefits of his personal use but shares only a small portion of the costs. Thus, lakes open to the public are likely to be overfished, with negative consequences for the public (to say nothing of the fish!) in future years. Common fields will be overgrazed, with similarly unfortunate consequences. Any other exhaustible resource may be misallocated if publicly available.

The conventional economic solution to the tragedy of the commons is to assign resources as private property. If everyone owns a small piece of land (or lake), and can keep others out of it (with real or legal "fences"), then the private and public incentives are aligned. People will not overgraze their own land, because if they do they will suffer the full consequences of their actions.[132] Further, if deal-making between neighbors is costless, as Coase postulated but did not believe,[133] transactions will allow neighbors with large cattle herds to purchase grazing rights from others with smaller herds. Such transactions should occur until each piece of land is put to its best possible use.[134]

In the context of intellectual property, Kitch's article remains one of the most significant efforts to integrate intellectual property with property rights

[131] *See id.* at 276-78; Wendy J. Gordon, *Of Harms and Benefits: Torts, Restitution, and Intellectual Property*, 21 J. LEGAL STUD. 449, 473 (1992); Robert P. Merges, *Of Property Rules, Coase, and Intellectual Property*, 94 COLUM. L. REV. 2655 (1994).

[132] While in theory it is possible for cattle-owners to agree to limit their grazing in the public interest, any such effort at agreement is likely to run into insurmountable problems. Not only will organizing and policing such an agreement take effort that will not be rewarded, but individual grazers have an incentive to free-ride, reaping the benefits of reduced grazing by others while refusing to reduce their own grazing. For more on these problems, see Mancur Olson, THE LOGIC OF COLLECTIVE ACTION (1961). One commentator views this internalization of (positive) externalities as a key function of property. Harold Demsetz, *Toward a Theory of Property Rights*, 57 AM. ECON. REV. 347, 348 (1967). On the other hand, for a rejection of the tragedy of the commons approach in certain contexts, see Carol Rose, *The Comedy of the Commons: Custom, Commerce, and Inherently Public Property*, 53 U. CHI. L. REV. 711 (1986). Rose is surely correct that private division of land is not always efficient. Consider the problematic task walking through your neighborhood would be if every piece of sidewalk were privately owned by a different person, and you were required to obtain permission to take each step. *Cf.* Dan Hunter, *Cyberspace as Place and the Tragedy of the Digital Anticommons*, 91 CALIF. L. REV. 439 (2003); Mark A. Lemley, *Place and Cyberspace*, 91 CALIF. L. REV. 521 (2003) (both criticizing the excessive division of rights currently taking place online).

[133] *See* Ronald H. Coase, *The Problem of Social Cost*, 3 J.L. & ECON. 1 (1960).

[134] *See* Guido Calabresi & A. Douglas Melamed, *Property Rights, Liability Rules, and Inalienability: One View of the Cathedral*, 85 HARV. L. REV. 1089 (1972) (discussing this implication of Coase).

theory.[135] Kitch argues that the patent system operates not (as traditionally thought) as an incentive-by-reward system, giving exclusive rights to successful inventors in order to encourage future inventors, but as a "prospect" system analogous to mineral claims. In this view, the primary point of the patent system is to encourage further commercialization and efficient use of as yet unrealized ideas by patenting them, just as privatizing land will encourage the owner to make efficient use of it.[136] Society as a whole should benefit from this equalization of private with social interests.

Fundamental to this conclusion are three assumptions. First, Kitch argues that:

> a patent prospect increases the efficiency with which investment in innovation can be managed.. . .[T]echnological information is a resource which will not be efficiently used absent exclusive ownership.. . . [T]he patent owner has an incentive to make investments to maximize the value of the patent without fear that the fruits of the investment will produce unpatentable information appropriable by competitors.[137]

This is analogous to the tragedy of the commons argument that only with private ownership do private incentives match social incentives. In the tragedy of the commons, the private incentive to "invest" in a field or lake—for example, by letting it lie fallow or limiting grazing in order to permit it to grow—is less than the social value of such an investment. In the patent context, Kitch makes an analogous argument: The private incentive to improve and market an invention will be less than the social value of such efforts unless the patent owner is given exclusive control over all such improvements and marketing efforts.

Second, Kitch argues that "[n]o one is likely to make significant investments searching for ways to increase the commercial value of a patent unless he has made previous arrangements with the owner of the patent. This puts the patent owner in a position to coordinate the search for technological and market enhancement of the patent's value so that duplicative investments are not made and so that information is exchanged among the searchers."[138] This is the Coase theorem at work. Under that theory, giving one party the power to control and orchestrate all subsequent uses and research relating to the patented

[135] For other property-based views of intellectual property, see, *e.g.*, Kenneth W. Dam, *Some Economic Considerations in the Intellectual Property Protection of Software*, 24 J. LEGAL STUD. 321 (1995); I. Trotter Hardy, *Property (and Copyright) in Cyberspace*, 1996 U. CHI. L.F. 217; Edmund Kitch, *Patents: Monopolies or Property Rights?* 8 RES. L. & ECON. 31 (1986).

[136] Kitch, *Nature*, *supra* note 130, at 270–271, 275 (making the analogy to land explicit).

[137] *Id.* at 276.

[138] *Id.*

technology should result in efficient licensing, both to end users and to potential improvers—assuming, that is, that information is perfect, all parties are rational, and licensing is costless.[139]

Finally, for social benefit to be maximized, the property owner must make the invention (and subsequent improvements) available to the public at a reasonable price—ideally, one that approaches marginal cost as much as is feasible.[140] But a property owner will have no incentive to reduce his prices toward marginal cost unless he faces competition from others. If the property owner is alone in the market, he may be expected to set a higher monopoly price for his goods, to the detriment of consumers (and social welfare). Kitch notes this problem, but does not resolve it. He merely points out that not all patents confer monopoly rights, and that in some cases the creators of intellectual property rights will face competition from the makers of other fungible goods, and therefore that their individual firm demand curves will be horizontal rather than downward-sloping.[141] If one assumes such competition, intellectual property owners may be expected to price competitively, just as producers of wheat do.

Kitch's prospect theory strongly emphasizes the role of a single patentee in coordinating the development, implementation, and improvement of an invention. The analogy to mining is instructive: Kitch's theory is that if we consolidate ownership in a single entity, that entity will have appropriate incentives to invest in commercializing and improving an invention. Indeed, on Kitch's theory, one might think it appropriate to assign rights to prospect for inventions to companies even before they have invented anything, just as we do for the owners of prospecting rights, because doing so will give them the monopoly incentive to coordinate the search.

Kitch's prospect theory draws on economic literature in the Schumpeterian tradition, which in its strong form holds that companies in a competitive marketplace have insufficient incentive to innovate. On this view, only strong

[139] *See* Anastasia P. Winslow, *Rapping on a Revolving Door: An Economic Analysis of Parody and Campbell v. Acuff-Rose Music, Inc.*, 69 S. CAL. L. REV. 767, 780 (1996) (arguing that the Coase theorem suggests that initial assignment of property rights between original creators and improvers is irrelevant). For a discussion of what happens when we relax these unrealistic assumptions, see Mark A. Lemley, *The Economics of Improvement in Intellectual Property Law*, 75 TEXAS L. REV. 989, 1048–72 (1997). On the importance of efficient licensing to the case for intellectual property protection, see Wendy J. Gordon, *Asymmetric Market Failure and Prisoner's Dilemma in Intellectual Property*, 17 U. DAYTON L. REV. 853, 857 (1992).

[140] It is not possible to price intellectual property *at* its marginal cost and still stay in the business of producing new works, since developing those new works requires a fixed investment of resources (time, research money, etc.), one that frequently dwarfs the marginal cost of making and distributing copies of the idea once it has been developed.

[141] Kitch, *supra* note 130, at 274.

rights to preclude competition will effectively encourage innovation.[142] Prospect theory therefore suggests that patents should be granted early in the invention process, and should have a broad scope and few exceptions.

Prospect theory is based on the premise that strong rights should be given into the hands of a single coordinating entrepreneur. Thus, prospect theory necessarily envisions invention as something done by a single firm rather than collectively; as the result of significant expenditure on research rather than the result of serendipitous or inexpensive research; and as only the first step in a long and expensive process of innovation rather than as an activity close to a final product.[143] As a result, prospect theory suggests that patents should stand alone, be broad, and confer almost total control over subsequent uses of the product.[144]

The prospect vision of patents maps most closely to invention in the pharmaceutical industry. Pharmaceutical innovation is notoriously costly and expensive. The pharmaceutical industry reports that it spends as much as $800 million in research and development (R&D) for each new drug produced.[145] While those numbers are almost certainly inflated,[146] there is no doubt that R&D is extremely expensive in the pharmaceutical industry.[147] Further, inventing a new drug is only the beginning of the process. The Food and Drug Administration (FDA) requires a lengthy and rigorous set of tests before drugs

[142] The classic argument cited in favor of monopolists coordinating innovation is Joseph A. Schumpeter, CAPITALISM, SOCIALISM, AND DEMOCRACY 106 (1st ed. 1942). For an application to patent law, see F. Scott Kieff, *Property Rights and Property Rules for Commercializing Inventions*, 85 MINN. L. REV. 697 (2001). *Cf.* Suzanne Scotchmer, *Protecting Early Innovators: Should Second-Generation Products Be Patentable?*, 27 RAND J. ECON. 322 (1996) (suggesting that incentives be weighted towards pioneers). Schumpeter's conclusions have been challenged, both in theoretical and empirical terms. *See* Burk & Lemley, *Policy Levers, supra* note 2, at II.A.2 (discussing this literature).

[143] We follow Joseph Schumpeter in distinguishing between the act of invention, which creates a new product or process, and the broader act of innovation, which includes the work necessary to revise, develop, and bring that new product or process to commercial fruition. *See* Richard R. Nelson & Sidney G. Winter, EVOLUTIONARY THEORY OF ECONOMIC CHANGE 263 (1982) (distinguishing the invention of a product from innovation, a broader process of research, development, testing, and commercialization of that product, and attributing that distinction to Schumpeter); William Kingston, DIRECT PROTECTION OF INNOVATION (1987).

[144] *See supra* notes 130–143 and accompanying text.

[145] *See* Gardiner Harris, *Cost of Developing New Medicine Swelled to $802 Million, Research Study Reports*, WALL ST. J., Dec. 3, 2001.

[146] Among other things, they include substantial marketing expenditures, which should not count as R&D.

[147] Estimates of the average cost of drug development and testing range from $110 million to $500 million; the latter is the industry's figure. *Compare* http://www.phrma.org/publications/publications/profile01/chapter2.pdf with http://www.citizen.org/Press/pr-drugs33.htm

can be released to the market.[148] While the imitation of a drug is reasonably costly in absolute terms, a generic manufacturer who can prove bioequivalency can avoid the R&D cost entirely, and can get FDA approval much more quickly than the first mover. The ratio of inventor cost to imitator cost, therefore, is quite large in the absence of effective patent protection. As a result, it is likely that innovation would drop substantially in the pharmaceutical industry in the absence of effective patent protection.[149] In addition, as a general rule, the scope of patents in the pharmaceutical industry tends to be coextensive with the products actually sold. Patents do not merely cover small components that must be integrated into a marketable product.[150] On the other hand, if patents do not cover a group of related products, imitators can easily design around the patent by employing a close chemical analog to the patented drug.

All of these factors suggest that patents in the pharmaceutical industry should look like those the prospect theory prescribes. There is, in this industry, no serious problem of either cumulative or complementary innovation. Strong patent rights are necessary to encourage drug companies to spend large sums of money on research years before the product can be released to the market. And because much of the work occurs after the drug is first identified, it is important to give patentees the right to coordinate downstream changes to the drug. Prospect theory fits the pharmaceutical industry.

2. Anticommons Theory

While the economic literature on cumulative innovation has generally suggested the grant of divided entitlements as a means of encouraging innovation by both initial inventors and improvers, a more recent body of literature has pointed to the limits of divided entitlements in circumstances in which

[148] PhRMA estimates that the total time spent from the beginning of a research project to the marketing of a successful drug is 14.2 years, 1.8 years of which is due to the FDA approval process. *See* http://www.phrma.org/publications/ publications/profile01/chapter2.pdf

[149] *See, e.g.*, James W. Hughes *et al.*, *Napsterizing Pharmaceuticals: Access, Innovation, and Consumer Welfare*, NBER Working Paper No. 9229 (2002) (finding that eliminating patent protection on pharmaceuticals would cost consumers $3 in lost innovation benefits for every dollar saved in reduced drug prices).

[150] While pharmaceutical companies have tried to find ways to obtain multiple patents on the same basic invention in an effort to extend the life of their patents, these efforts are aberrations that represent a failure of the system, not its normal function. *See* Lara J. Glasgow, *Stretching the Limits of Intellectual Property Rights: Has the Pharmaceutical Industry Gone Too Far?* 41 IDEA 227 (2001) (documenting efforts by pharmaceutical companies to obtain multiple patents on the same basic drug). The patent doctrine of "double patenting" is designed to prevent this sort of abuse. *See, e.g.*, *Eli Lilly & Co. v. Barr Labs.*, 251 F.3d 955 (Fed. Cir. 2001).

transactions costs are positive. Relying on Michael Heller's description of what he calls the "anticommons,"[151] a number of patent scholars have argued that granting too many different patent rights can impede the development and marketing of new products when making the new product requires the use of rights from many different inventions.[152] Underlying this argument are concerns about transactions costs and strategic behavior, which these scholars argue will sometimes prevent the aggregation of the necessary rights.

The anticommons is characterized by fragmented property rights, the aggregation of which is necessary to make effective use of the property.[153] Aggregating such fragmented property rights entails high costs to locate and bargain with the many rights owners whose collective permissions are necessary to complete broader development. This type of licensing environment may quickly become dominated by holdouts who refuse to license their essential sliver of the pie unless bribed.[154] Because a given project will fail without their cooperation, holdouts may be prompted to demand a bribe close to the value of the entire project.[155] And, of course, every property holder needed for the project is subject to this same incentive; if everyone holds out, the cost of the project will rise substantially, and probably prohibitively.

The anticommons problem is really a particular species of a more general problem in economics—the issue of complementarity of products. Complementarity exists where two or more separate components must be combined into an integrated system. Economists have noted that the problem of double (or triple or quadruple) marginalization can occur when different companies own rights to complementary goods.[156] The problem is this: If a product must include components A and B, and A and B are each covered by patents that grant

[151] *See* Michael A. Heller, *The Tragedy of the Anticommons: Property in the Transition from Marx to Markets*, 111 HARV. L. REV. 621 (1998).

[152] *See* Michael A. Heller & Rebecca S. Eisenberg, *Can Patents Deter Innovation? The Anticommons in Biomedical Research*, 280 SCI. 698 (1998). *See also* Arti Kaur Rai, *The Information Revolution Reaches Pharmaceuticals: Balancing Innovation Incentives, Cost and Access in the Post-Genomics Era*, 2001 U. ILL. L. REV. 173.

[153] Heller, *supra* note 151, at 670–672.

[154] On the holdout problem, see generally Mancur Olson, THE LOGIC OF COLLECTIVE ACTION (1961). On its specific application in patent law, see Chapter 9, "Varying the Course in Patenting Genetic Material: A Counter-Proposal to Richard Epstein's Steady Course," by Rochelle Cooper Dreyfuss.

[155] Lloyd Cohen, *Holdouts and Free Riders*, 20 J. LEGAL STUD. 351 (1991).

[156] The double-marginalization theorem shows that it is inefficient to grant two monopolies in complementary goods to two different entities because each entity will price its piece without regard to the efficient pricing of the whole, resulting in an inefficiently high price. For a technical proof of this, see Carl Shapiro, *Setting Compatibility Standards: Cooperation or Collusion?*, *in* EXPANDING THE BOUNDARIES OF INTELLECTUAL PROPERTY 81, 97–101 (Rochelle Cooper Dreyfuss *et al.* eds., 2001) [Hereinafter Shapiro, *Cooperation or Collusion*]. For a description of the problem in practice, see Ken Krechmer,

different companies monopoly control over the components, each company will charge a monopoly price for its component. As a result, the price of the integrated product will be inefficiently high—and output inefficiently low—because it reflects an attempt to charge two different monopoly prices. The anticommons literature builds on this economic work, offering additional reasons to believe that the companies may not come to terms at all.[157]

Complements or anticommons problems can arise either horizontally or vertically in an industry. The problem arises horizontally when two different companies hold rights at the same level of distribution—say, inputs into the finished product. It arises vertically if a product must be passed through a chain of independent companies (such as a monopoly manufacturer who must sell through an independent monopoly distributor), or if patents on research tools or upstream components must be integrated with downstream innovation in order to make a finished product.

The anticommons literature suggests that too many companies have patents on components or inputs into products.[158] The problem is not so much the scope of those patents as it is the number of different rights with different owners that must be aggregated in order to participate in the marketplace. Thus, this literature addresses a dimension of patent rights not considered in any of the theories discussed previously. It is generally at odds with the divided entitlement proposals of cumulative innovation theory. There are two different ways to solve this problem: consolidate ownership of rights among fewer companies or grant fewer patents. Most legal scholars working in the anticommons literature have assumed that the solution is to grant fewer patents, particularly to developers of upstream products like research tools or DNA sequences.[159] Economists, by contrast, tend to assume that the solution to vertical complementarity problems is to vertically integrate—that is, to consolidate rights in a single company.[160]

Communications Standards and Patent Rights: Conflict or Coordination? 3 (2002) (draft working paper, on file with author) (citing examples in which so many different IP owners claim rights in a standard that the total cost to license those rights exceeds the potential profit from the product); Douglas Lichtman, *Property Rights in Emerging Platform Technologies*, 29 J. LEGAL STUD. 615 (2000) (making a double-marginalization argument in favor of vertical integration in computer systems).

[157] There is some evidence casting doubt on whether patents in fact commonly have anticommons characteristics. *See* John P. Walsh, Ashish Arora, & Wesley M. Cohen, *The Patenting and Licensing of Research Tools and Biomedical Innovation* (working paper 2000) (conducting a survey and finding no evidence of anticommons problems in the biotechnology industry). But the theoretical problem certainly exists.

[158] *See*, Matthew Erramouspe, Comment, *Staking Patent Claims on the Human Blueprint: Rewards and Rent-Dissipating Races*, 43 UCLA L. REV. 961 (1996) (making this argument).

[159] *See, e.g.*, Arti K. Rai, *Fostering Cumulative Innovation in the Biopharmaceutical Industry: The Role of Patents and Antitrust*, 16 BERKELEY TECH. L.J. 813 (2001); Philippe Jacobs & Geertrui Van Overwalle, *Gene Patents: A Different Approach*, [2001] EUR. INTELL. PROP. REV. 505, 505 (arguing that patents should not be granted for DNA, but only for downstream medical products).

Obviously, these two different solutions have very different implications for patent policy. As a result, the anticommons literature does not necessarily dictate particular policy results.

In summary, anticommons theory emphasizes the problems of divided entitlements among complements. These problems can occur either horizontally or vertically—horizontally if patents cover different pieces that must be integrated into a product, and vertically if patents cover different steps in a cumulative innovation process. Anticommons theorists point to the risk of bargaining breakdown whenever the development of a product requires permission from the owners of two or more inputs. Different strands of anticommons theory suggest that the solution to this problem is either to consolidate ownership in a single owner—a result reminiscent of prospect theory—or to preclude patent protection altogether for certain types of inputs, particularly upstream research tools.

Anticommons theory maps very well onto the biotechnology industry. The biotechnology industry has some of the characteristics of the pharmaceutical industry, with which it shares certain products.[161] In particular, the long development and testing lead time characteristic of pharmaceuticals is also evident in DNA-related innovation. These delays are due in part to the stringent regulatory oversight exercised over the safety of new drugs, foods, biologics, and over environmental release of new organisms. Another similarity between DNA and pharmaceuticals is that generics who wish to imitate an innovator's drug face substantially lower costs and uncertainty than do innovators in the industry. While the FDA does impose regulatory hurdles even on second-comers, the process is substantially more streamlined than it is for innovators. Indeed, the primary regulatory hurdle a generic company faces is to show that its drug is bioequivalent to the innovator's drug.[162] Assuming bioequivalency, the FDA allows the generic to rely on the innovator's regulatory efforts. The uncertainty associated with developing and testing a new drug is also completely absent for generic competitors; they need only replicate the drug the innovator has identified and tested. Similarly, the hard work involved in producing a cDNA sequence coding for a human protein is in identifying and isolating the

---

[160] Alternatively, anticommons licensing rights can be consolidated into a collective rights organization such as ASCAP or a patent pool, even if the rights themselves remain under separate ownership. For a discussion of collective rights organizations, see Robert P. Merges, *Contracting Into Liability Rules: Intellectual Property Rights and Collective Rights Organizations*, 84 CALIF. L. REV. 1293 (1996).

[161] Biotechnology products appear in a wide variety of economic sectors, from pharmaceuticals to foodstuffs to industrial processes. *See* Dan L. Burk, *A Biotechnology Primer*, 55 U. PITT. L. REV. 611 (1994). Much of our discussion will focus on a subset of biotechnology that includes gene sequences and gene therapy.

[162] For a discussion of this process, see, *e.g.*, *Eli Lilly & Co. v. Medtronic, Inc.*, 496 U.S. 661, 676 (1990).

right sequence; once the sequence is known, a follow-on competitor can easily replicate it. And the existence of numerous functional equivalents to a particular DNA sequence means that patent protection must be broad enough to effectively exclude simple design-arounds, just as pharmaceutical patents must be broad enough to cover chemical analogs.

On the other hand, the total cost of sequencing a particular gene is significantly less than the cost of a more traditional drug design, especially as computers can automate much of the process.[163] And DNA, unlike pharmaceuticals, involves the use of both vertical and horizontal complements. Patentees have acquired thousands of patents on DNA sequences that cover specific genes or, in some cases, fragments of genes.[164] Further, biotechnology companies have patented probes, sequencing methods, and other research tools. Any particular gene therapy requires the simultaneous use of many of these patents, leading to anticommons problems. The problem is exacerbated by "reach-through" licenses in which the owners of upstream research tools seek control of and royalties on the downstream uses of the tool.[165]

Scholars have proposed several different ways of solving these aggregation problems. First, vertical integration of companies may make much of the problem disappear. If biotechnology companies are owned by or allied with pharmaceutical companies, the resulting company may own enough rights to research tools, gene sequences, and implementation methods to go it alone.[166] Alternatively, if the absolute cost of sequencing DNA is sufficiently low, or if the nonproprietary incentives are sufficiently great, the anticommons problem could be solved by refusing to protect certain types of inventions, such as ESTs.[167]

In short, the structure of the biotechnology industry seems likely to run high anticommons risks. Product-development times from creation to market are long and costly, but DNA patents are numerous and narrow. Production of any given product may require bargaining with multiple patent holders. The potential for divided patent entitlements to prevent efficient integration into products is particularly high. Anticommons theory fits DNA.

[163] *See, e.g.*, Robert A Hodges, *Black Box Biotech Inventions: When a Mere "Wish or Plan" Should be Considered an Adequate Description of the Invention*, 17 GA. ST. U. L. REV. 831, 832 (2001) (discussing the increasing automation of gene sequencing).

[164] *See, e.g.*, S.M. Thomas *et al.*, *Ownership of the Human Genome*, 380 NATURE 387, 387 (1996).

[165] *See* Rebecca S. Eisenberg, "Reaching through the Genome" (Chapter 10).

[166] *See* Rai, *Cumulative Innovation*, *supra* note 159. Rai is critical of this form of integration, however.

[167] *See id.*; *cf.* Rebecca S. Eisenberg & Robert P. Merges, *Opinion Letter as to the Patentability of Certain Inventions Associated With the Identification of Partial cDNA Sequences*, 23 AIPLA Q. J. 1 (1995).

In the section that follows, we consider the implications that the prospect and anticommons theories have for biotechnology patent policy. We then talk briefly about some implications of our reasoning for the related issue of pharmaceutical patent policy.

B. Designing Optimal Biotechnology Policy

If any technology fits the criteria of high-cost, high-risk innovation, it is certainly biotechnology. Development of biotechnology products, particularly in the pharmaceutical sector, has been characterized by extremely long development times and high development costs. These delays are due in part to the stringent regulatory oversight exercised over the safety of new drugs, foods, and biologics, and over environmental release of new organisms.[168] Yet the onerous regulatory requirements to which biotechnology is subject may obscure a more fundamental uncertainty that justifies such oversight: Biotechnology products arise out of living systems, and are typically intended to interact with other human or nonhuman living systems. Such interactions, whether physiological or ecological, are enormously complex and the systems involved poorly characterized. As a consequence, the functionality of biotechnology products is always unforeseeable, and always involves a high degree of uncertainty and risk.[169] Thus, while we have argued that the Federal Circuit has been wrong to suggest that identifying and making biotechnological products—the *invention* of those products—is always difficult and uncertain, it is also true that turning those research tools into medicines that can be sold in the market—*innovation*—is time-consuming, complex, and risky.

At the same time, imitators, such as generic manufacturers, who wish to imitate an innovator's drug face substantially lower costs and uncertainty than do innovators in the industry. While the FDA does impose regulatory hurdles on second-comers, the process is substantially more streamlined than it is for innovators. Indeed, the primary regulatory hurdle a generic company faces is to show that its drug is bioequivalent to the innovator's drug.[170] Assuming bioequivalency, the FDA allows the generic to rely on the innovator's regulatory efforts. The uncertainty associated with developing and testing a new drug is completely absent for generic competitors; they need only replicate the drug the innovator has identified and tested. Similarly, the hard work involved in

[168] *See supra* notes 145–48 and accompanying text (discussing the delay and cost associated with pharmaceutical development)

[169] For example, the Centocor sepsis antibody, a highly promising biotechnology treatment, succeeded in passing many years of costly trials but failed in the last phase of FDA approval.

[170] Federal Food, Drug, and Cosmetic Act, § 505(j)(2)(A)(i), 21 U.S.C.A. § 355(j)(2)(A)(i).

producing a cDNA sequence coding for a human protein is in identifying and isolating the right sequence; once the sequence is known, a follow-on competitor can quite easily replicate it.

Consistent with these characteristics and Robert P. Merges' standard economic model, the current Federal Circuit jurisprudence lowers the obviousness barrier for biotechnology.[171] This lower barrier seems at odds with the modern science of biotechnology. The availability of research tools has made routine the isolation and characterization of biological macromolecules. As a result, considerable criticism has been directed against the Federal Circuit's biotechnology obviousness cases.[172] Given such tools, the outcome of a search for a particular nucleotide or protein seems relatively certain, and hence, it is argued, obvious. But if patents are to drive innovation rather than merely invention in biotechnology, courts must take into account the cost and uncertainty of postinvention testing and development.[173] The availability or unavailability of a patent is expected to have little effect on the incentive to engage in preliminary research to, say, use the available tools to secure a macromolecule of interest.[174] But the ready availability of tools for finding a new biotechnology product does not change the high cost and uncertainty entailed in developing a marketable product using that macromolecule. Hence, under Merges' framework, a lowered standard of obviousness might seem to make sense from a policy standpoint not so much to encourage invention but as a way to encourage the development of marketable products.[175]

[171] *See* Burk & Lemley, *Technology-Specific*, *supra* note 14, at 1178–79.

[172] Kenneth J. Burchfiel, BIOTECHNOLOGY AND THE FEDERAL CIRCUIT §6.2, at 84 (1995); Jonathan M. Barnett, *Cultivating the Genetic Commons: Imperfect Patent Protection and the Network Model of Innovation*, 37 SAN DIEGO L. REV. 987 (2000); Philippe Ducor, *New Drug Discovery Technologies and Patents*, 22 RUTGERS COMP. & TECH. L. J. 369 (1996); Arti K. Rai, *Intellectual Property Rights in Biotechnology: Addressing New Technology*, 34 WAKE FOREST L. REV. 827 (1999). *See generally* John M. Golden, *Biotechnology, Technology Policy, and Patentability: Natural Products and Invention in the American System*, 50 EMORY L. J. 101 (2001).

[173] *See* Robert P. Merges, *Uncertainty and the Standard of Patentability*, 7 HIGH TECH. L. J. 1 (1992); Robert P. Merges, *One Hundred Years of Solicitude: Intellectual Property Law, 1900–2000*, 88 CALIF. L. REV. 2187, 2225–27 (2000); Karen I. Boyd, *Nonobviousness and the Biotechnology Industry: A Proposal for a Doctrine of Economic Nonobviousness*, 12 BERKELEY TECH. L. J. 311 (1997).

[174] *See, e.g.*, Robert P. Merges, PATENT LAW AND POLICY 519 (2d ed. 1997).

[175] *See* Robert P. Merges & John Fitzgerald Duffy, PATENT LAW AND POLICY 727–728 (3d ed. 2002) ("section 103 actually has a bigger effect on decisions regarding which technologies to *develop* than regarding which research projects to pursue in the first place."); *see also* Giorgio Sirilli, *Patents and Inventors: An Empirical Study*, 16 RES. POL'Y 157, 164 (1987) (finding that patents give most inventors more incentive to commercialize than incentive to invent). One way to think of this is to conceive of patents as a financing mechanism: By providing definable rights, patents enable companies to obtain the funding they need to turn an invention into a product. *See Picard v. United Aircraft Co.*, 128 F.2d 632, 642–43 (2d Cir. 1942) (patents may serve as a "lure to investors");

Yet in its current jurisprudence, what the Federal Circuit gives biotechnology with one hand, it takes away with the other. Although biotechnology patents are relatively easy to obtain under the obviousness standard, the accompanying enablement and written description standards dramatically narrow the scope of the resulting patents. By requiring disclosure of the particular structure or sequence in order to claim biological macromolecules, the Federal Circuit effectively limits the scope of a patent on those molecules to the structure or sequence disclosed.[176] This standard dictates that the inventor have the molecule "in hand" (so to speak) before being able to claim it. In other words, the inventor can have patent protection for any given molecule only after a substantial investment has already been made in isolating and characterizing the molecule. The result is that everyone who invests in discovering a new molecule will receive a patent, but one that is trivial to avoid infringing, at least literally. Under this standard, no one is likely to receive a patent broad enough to support the further costs of development.[177] Indeed, some promising lines of inquiry, such as the development of drugs custom-tailored to individual DNA, may be foreclosed entirely if a biotechnology patent is not broad enough to cover the small structural variations that inhere in custom drugs.

Golden, *supra* note 172, at 167–172; Mark A. Lemley, *Reconceiving Patents in the Age of Venture Capital*, 4 J. SM. & EMERGING BUS. L. 137 (2000); Fritz Machlup, *Patents*, in 2 ENCYCLOPAEDIA OF THE SOCIAL SCIENCES 461, 467 (1968).

[176] *See, e.g.*, *Regents of the University of California v. Eli Lilly & Co.*, 119 F.3d 1559 (Fed. Cir. 1997) (describing rat insulin DNA did not justify claims to insulin DNA for any other mammals); *Plant Genetic Sys. v. DeKalb Genetics*, 315 F.3d 1335 (Fed. Cir. 2003) (holding a patent claim to a class of genetically engineered plants invalid for lack of enablement because only certain types of plants within the class were described, notwithstanding the pioneer nature of the invention). *But see Amgen v. Hoechst Marion Roussel*, 314 F.3d 1313 (Fed. Cir. 2003) (finding the written description requirement satisfied by a broad claim to cells used to produce EPO, where host cells, unlike DNA, were well known in the art; the written description "requirement may be satisfied if in the knowledge of the art the disclosed function is sufficiently correlated to a particular, known structure."). While *Amgen* certainly reads the written description requirement more laxly than *Lilly*, it appears to have limited its holding to cases in which those of skill in the art already know of a correspondence between function and structure before the invention, something that will not be true in the DNA patent cases.

[177] *See* Kenneth G. Chahine, *Enabling DNA and Protein Composition Claims: Why Claiming Biological Equivalents Encourages Innovation*, 25 AIPLA Q. J. 333 (1997) (arguing for a broader scope of biotechnology patents, extending to proteins with comparable biological activity). Curiously, Merges doesn't see this as a major problem, suggesting that, in general, "the Federal Circuit has overall been quite successful at integrating biotechnology cases into the fabric of patent law." Merges, *Solicitude*, *supra* note 173, at 2228. We think the written description cases and the correspondingly narrow scope afforded biotechnology patents are a more serious problem than Merges acknowledges.

One might question why, if the written description requirement is producing such narrow DNA patents, the biomedical industries consistently cite patent protection as extremely important

Unfortunately, this proliferation of narrow biotechnology patents may be nearly impossible to avoid under the reciprocal structure of obviousness and enablement in current PHOSITA patent doctrine.[178] In order for the invention to avoid obviousness, it must be deemed beyond the skill of the PHOSITA to construct given the level of disclosure in the prior art. Yet this means that in disclosing the invention, the inventor must tell those of ordinary skill a good deal more about how to make and use it, effectively raising the standard for enablement and written description. The Federal Circuit's insistence that the results of biotechnology research are unforeseeable or unpredictable avoids the problem of obviousness but results in an extremely stringent standard for disclosure and description. Once again, the result is not optimal from the perspective of economic policy. We have suggested elsewhere a doctrinal solution to this particular problem, namely, treating the PHOSITA standards in obviousness and disclosure as separate policy-based questions rather than as a common standard.[179]

But even given such doctrinal tools, courts must confront the policy question of the proper scope of patents in the biotechnology industry. The proper focus of biotechnology patent policy is a matter of some dispute. Merges' classic economic framework suggests that the standard of nonobviousness should be low to compensate for the high cost of innovation in the industry.[180] Both the need for effective protection and the anticommons literature suggest that the disclosure requirement should be less strict than it currently is, lest property rights be too disintegrated to permit effective licensing.[181] But if both the nonobviousness and disclosure requirements are lessened, the result will be more patents with broader scopes. This, in turn, will likely produce a large number of

---

to them. *See, e.g.*, Richard C. Levin *et al.*, *Appropriating the Returns from Industrial Research and Development*, 1987 BROOKINGS PAPERS ON ECONOMIC ACTIVITY 783; Wesley M. Cohen, *et al.*, *Protecting Their Intellectual Assets: Appropriability Conditions and Why U.S. Manufacturing Firms Patent (or Not)* (NBER Working Paper No. W7552, Feb. 2000). We think there are two answers. First, the industries that count patents as extremely valuable tend to be chemistry and pharmaceuticals, not biotechnology *per se*, and certainly not those in the business of discovering and using DNA sequences. Second, the biotechnology written description cases are relatively new, and the industry-specific studies are somewhat older, so their understanding of the value of patents may not reflect modern realities either because the survey is old or because those in the industry have not yet internalized the effect of these decisions.

[178] *See, e.g.*, Mark J. Stewart, *The Written Description Requirement of 35 U.S.C. § 112(1): The Standard After Regents of the University of California v. Eli Lilly & Co.*, 32 IND. L. REV. 537, 557–558 (1999) (noting the linkage between the Federal Circuit's view of biotechnology as an uncertain art and the narrowness of the patents that result).

[179] Burk & Lemley, *Technology-Specific*, *supra* note 14, at 1202–05.

[180] Merges, *Uncertainty*, *supra* note 173.

blocking patents.[182] Blocking patents aren't necessarily bad, particularly when they are coupled with mechanisms like the reverse doctrine of equivalents that will relieve bargaining pressures in extreme cases.[183] And they will certainly give biotechnology companies incentives to innovate, at least initially. But they do raise the specter of overlapping first-generation patents choking out innovation, particularly where those first-generation patents are granted on upstream research tools.[184] This is precisely the concern that anticommons theory identifies.

We suggest instead that courts should modify Merges' classic theory. Lowering the obviousness threshold is only one way to encourage investment in uncertain technologies. An alternative is to broaden the scope of the patents that are issued by reducing the disclosure requirement or by strengthening the doctrine of equivalents for a particular industry; doing either will encourage innovation in uncertain industries, not by increasing the percentage chance of getting a patent but by increasing the value of the patent once it is granted. In fact, it seems to us that while Merges is correct to suggest that the standard of patentability should be responsive to the cost and uncertainty of innovation, obviousness is the wrong lever to use in biotechnology.[185] Lowering the obviousness threshold makes it more likely that marginal inventions will be patented but does nothing to encourage inventions that would have met the (already rather modest) obviousness standard anyway. If getting from invention to the market is the costly and uncertain part of the endeavor, it is these more significant inventions that we need to worry about rewarding.[186]

---

[181] Heller & Eisenberg, *supra* note 152; Rebecca Eisenberg & Arti Rai, *The Public and the Private in Biopharmaceutical Research*, http://www.law.duke.edu/pd/papers/raieisen.pdf

[182] For example, suppose a patentee isolates the DNA sequence for human beta-interferon, but because of the lowered disclosure requirement is entitled to claim all mammalian beta-interferon. The lowered obviousness requirement may mean that future inventors can patent rat, bat, and cat beta-interferon, respectively, if they discover those particular sequences; it is well established that a patent on a genus does not necessarily render obvious claims to a previously undisclosed species within that genus. *E.g. In re Baird*, 16 F.3d 380 (Fed. Cir. 1994); *Corning Glass Works v. Sumitomo Elec. U.S.A.*, 868 F.2d 1251 (Fed. Cir. 1989). Those later patents will be subservient to, but block, the original broad patent to mammalian beta-interferon.

[183] For detailed discussions, see Mark A. Lemley, *The Economics of Improvement in Intellectual Property Law*, 75 TEXAS L. REV. 989 (1997); Robert P. Merges, *Intellectual Property Rights and Bargaining Breakdown: The Case of Blocking Patents*, 62 TENN. L. REV. 75 (1994). There is some evidence that the reverse doctrine of equivalents may play a greater role in the biotechnology arena than elsewhere. *See, e.g., Scripps Clinic & Res. Found. v. Genentech*, 927 F.2d 1565 (Fed. Cir. 1991).

[184] *See, e.g.*, Eisenberg & Merges, *supra* note 50; Heller & Eisenberg, *supra* note 152.

[185] *See also* Rebecca S. Eisenberg, "Reaching through the Genome" (Chapter 10) (arguing that the Federal Circuit's low obviousness standard for biotechnology has aggravated the anticommons problem). Merges himself notes that increasing the scope of patents is an alternative to lowering the obviousness threshold. *See* Merges, *Uncertainty*, *supra* note 173, at 47. He doesn't pursue that alternative in his paper, however.

This alternative approach—a fairly high obviousness threshold coupled with a fairly low disclosure requirement—will produce a few very powerful patents in uncertain industries. It will, therefore, solve the anticommons problem often identified with biotechnology, while at the same time boosting incentives to innovate.[187] This calibration of patent frequency and scope seems to us the proper response to the anticommons concern found in much of the biotechnology literature. We worry that the alternate solution proposed by certain commentators—eliminating biotechnology property rights in favor of governmental control over inventions supported by public funds[188]—might unacceptably reduce the incentive for biotechnology companies to move beyond invention to innovation and product development.

Recalibrating patent scope through disclosure would seem to require a much more fundamental rethinking of the Federal Circuit's section 112 jurisprudence. The court currently requires *more* disclosure from patentees in uncertain arts, while our proposal would require *less*. The key to understanding this puzzle is the difference between uncertainty about invention *ex ante* and the uncertainty about innovation (getting the product to the market) *ex post*. The court repeatedly intones the maxim that biotechnology is an "uncertain art."[189] We think, however, that it is not so much invention as product development, production, and regulatory approval that is uncertain in the biotechnology industry. From a policy perspective, the result is the same: Biotechnological inventions need more incentive than other types of inventions if they are actually to make it to the market. But from a disclosure perspective, the difference is quite significant: There is no reason to require heightened disclosure of an invention—and correspondingly narrow its scope—if invention itself isn't uncertain in the art.

Biotechnology, then, is properly described in part by the anticommons theory (too many narrow patents must be aggregated to produce a viable product) and in part by prospect theory (a long and uncertain postinvention development process justifies strong control over inventions). A rational patent policy for DNA would seek to minimize the anticommons problems and give inventors sufficient control to induce them to walk the uncertain path towards commercial development. A variety of policy levers might be employed to this end. The utility and abstract ideas doctrines can restrict the anticommons

---

[186] Indeed, Hunt suggests that lowering the nonobviousness threshold actually creates a tradeoff, increasing the probability of acquiring a patent but reducing the value of any given patent, and therefore possibly weakening the incentive to innovate. Robert M. Hunt, *Nonobviousness and the Incentive to Innovate: An Economic Analysis of Intellectual Property Reform*, working paper 1999.

[187] Heller and Eisenberg, *supra* note 152; Eisenberg and Rai, *supra* note 181.

[188] Eisenberg and Rai discuss this approach. *See id.*

[189] *See, e.g.*, *In re Vaeck*, 947 F.2d 488, 496 (Fed. Cir. 1991) (biotechnology less "predictable" than mechanics or electronics).

problem in a few cases by preventing unnecessary upstream patents (*e.g.*, on ESTs) that threaten to hold up downstream innovation. The written description and enablement doctrines need to be recalibrated to permit broader claiming of inventions. The doctrine of equivalents can play a similar role, perhaps by rejuvenating the doctrine of pioneer patents or by applying the notion of known interchangeability with an eye toward function, not structure. Experimental use may also have a role to play, ensuring that the long development time necessary in the biotechnology industry doesn't interfere with an inventor's ability to patent the ultimate product.[190]

C. Designing Optimal Pharmaceutical Policy

Application of the uncertainty principle courts have used in biotechnology may have pernicious effects in other industries as well. For example, small-molecule chemistry has long had its own discrete set of patentability doctrines, developed in a long line of cases that attempted to accommodate the level of skill in that particular technology.[191] The rules articulated in this line of cases represent something of a compromise between the predictable similarities in the characteristics of molecular families and the difficulty in predicting the effect of structure in three dimensions. As a first approximation, structural relatedness between molecules disclosed in the prior art and a novel molecule claimed in a patent gives rise to a *prima facie* case of obviousness.[192] However, chemical structures depicted two-dimensionally on paper may not accurately reflect the properties of a physical structure that exists in three dimensions. Molecules react with one another in three dimensions, and the three-dimensional configuration dictates the chemical characteristics of the molecule.

Thus, even in small molecules, the three-dimensional complexity arising from what appear on paper to be slight changes in structure may give rise to radically different properties in apparently related molecules. Even with three-dimensional modeling, the effects of such complexity have long been difficult to predict. Such unpredicted characteristics occurred with enough frequency that a rule developed allowing a *prima facie* case of obviousness in small molecules to be rebutted by evidence of unpredictable or unexpected properties in the claimed molecule.[193] The technological assumption built into

[190] Other policy levers may also be relevant to biotechnology. For example, arguments against injunctive relief may be stronger in biomedical cases than with other sorts of inventions. The levers we discuss in the text are the most important for fashioning the incentive to innovate, however.

[191] *See In re Dillon*, 919 F.2d 688 (Fed. Cir. 1990) (recounting history of chemical obviousness cases).

[192] Harold Wegner, CHEMICAL PATENT PRACTICE (1991).

[193] *In re Papesch*, 315 F.2d 381 (C.C.P.A. 1963); *see also* Harold Wegner, *Prima Facie Obviousness of Chemical Compounds*, 6 AM. PAT. L. ASS'N Q. J. 271 (1978).

such a rule appears to be that the PHOSITA in small-molecule chemistry can generally predict the properties of a chemical or group of chemicals, or may occasionally be surprised by their properties, but either outcome is based on the molecules' structural depiction.

The rule in these small molecule cases appears closely related to that announced in Federal Circuit's biotechnology cases. The Federal Circuit has declared that DNA "is a chemical, albeit a complex one,"[194] and has articulated a desire to treat the patenting of macromolecules in the same fashion as the patenting of more traditional organic molecules. In focusing on structural depiction as the linchpin of both obviousness and disclosure, the biotechnology cases rely upon, and appear to extend, the line of chemical cases summarized above. But just as we question the application of these rules to macromolecules, we are similarly uncertain that these special rules for obviousness in small molecule chemical cases are well suited to accommodate current chemical research practice, especially in light of the rules articulated by the Federal Circuit for macromolecules.

In particular, modern techniques of rational drug design and combinatorial chemistry seem to push against this traditional construction of chemical obviousness in much the same way that the routinization of DNA probing pushes against the rules of patentability in the biotechnology cases. For example, small-molecule chemists now search for useful compounds by first specifying the functions that they hope to find.[195] The characteristics of desirable molecules are represented mathematically, in equations depicting functionally equivalent chemical groups and side chains.[196] Based on the predictions of such mathematical models, chemists can then search through large panels of related molecules, selecting those with the closest match to predicted function.[197]

This methodology closely parallels the type of molecular "search" considered in most of the Federal Circuit macromolecule cases, where large libraries of DNA molecules were probed in order to identify those that correspond to an expected functional characteristic—*e.g.*, the propensity to hybridize with probes of a particular nucleotide configuration, and concomitantly the

[194] *In Re Bell*, 991 F.2d 781 (Fed. Cir. 1993); *In Re Deuel*, 51 F.3d 1552 (Fed. Cir. 1995). *But see* Rai, *Addressing New Technology*, *supra* note 43, at 203 (arguing against treating biotechnology cases as analogous to earlier chemical cases).

[195] *See generally* Hugo Kubinyi, *The Quantitative Analysis of Structure-Activity Relationships* in 1 BURGER'S MEDICINAL CHEMISTRY AND DRUG DISCOVERY 497–571 (Manfred E. Wolff ed. 1995).

[196] Richard B. Silverman, THE ORGANIC CHEMISTRY OF DRUG DESIGN AND DRUG ACTION 26–34 (1992) (describing the Hansch equation that correlates biological activity with physicochemical properties of drug candidates).

[197] *See* Dinesh V. Patel *et al.*, *Applications of Small-Molecule Combinatorial Chemistry to Drug Discovery*, 1 DRUG DISCOVERY TODAY 134 (1996); Jan J. Scicinski, *Chemical Libraries in Drug Discovery*, 134 TRENDS IN BIOTECHNOLOGY 246 (1995); Joseph C. Hogan Jr., *Directed Combinatorial Chemistry*, 384 NATURE 175 (1996).

capacity to code for cellular production of particular gene products.[198] Combinatorial chemistry, much like DNA probing, tends to focus upon the function of the end product, removing much of the uncertainty from the outcome of a search for a desired molecule but not necessarily from predicting the precise structure of the molecule that is ultimately found. Indeed, the role of chemical structure is to some extent marginalized, as dissimilar structures with similar functions may be treated as equivalent in narrowing the search. Just as in biotechnology, a focus on structure rather than function may render chemical patent protection ineffective because modern development tools render structure less important to the invention.

Consequently, the industry-specific patent prescriptions for small molecule chemistry increasingly resemble those we have described for biotechnology. To the extent that such research is done in heavily regulated contexts, particularly for pharmaceutical applications, it faces much the same innovation profile as biotechnology. Other stringent regulatory oversight, such as Environmental Protection Agency Toxic Substances Control Act (TSCA) oversight,[199] may affect innovation outlooks similarly. Chemistry and pharmaceuticals, like biotechnology, seem to fit well into prospect theory. Fewer and broader patents, encouraged by relaxing the disclosure doctrines and strengthening the doctrine of equivalents, are most likely to provide the proper encouragement to innovation. A relatively robust utility doctrine can prevent anticommons problems in chemistry by preventing the patenting of numerous analogues to a successful chemical by "inventors" who don't know what the chemical can do.[200]

One policy lever that will likely take on greater importance in the pharmaceutical industry than in biotechnology is patent misuse. Pharmaceutical companies have gone to great lengths to extend the lawful scope of their patents by collusively settling disputes with generic companies,[201] strategically delaying prosecution of patents, and obtaining multiple patents covering the same invention.[202]

[198] *See generally* J. Watson *et al.*, RECOMBINANT DNA 104-07 (2d ed. 1991) (describing techniques for probing libraries of cloned genes).

[199] Toxic Substances Control Act, 15 U.S.C.A. §§ 2601–2671.

[200] Alternatively, Becky Eisenberg has suggested that FDA law can serve to encourage innovation in pharmaceuticals, not just regulate them, by granting industry-specific exclusive rights. The advantage of this industry-specific exclusivity is that it is applied downstream, to products as they enter the marketplace, and not upstream, where anticommons problems are more likely. *See* Rebecca S. Eisenberg, *Reexamining Drug Regulation from the Perspective of Innovation Policy* (working paper 2003).

[201] *See, e.g.*, Herbert Hovenkamp *et al.*, *Anticompetitive Settlement of Intellectual Property Disputes*, 87 MINN. L. REV. 1719 (2003); Maureen A. O'Rourke & Joseph F. Brodley, *Antitrust Implications of Patent Settlement Agreements*, 87 MINN. L. REV. 1767 (2003); Thomas F. Cotter, *Refining the "Presumptive Illegality" Approach to Settlements of Patent Disputes Involving Reverse Payments*, 87 MINN. L. REV. 1789 (2003).

[202] On these latter strategies, see Glasgow, *supra* note 150, at 248–251.

The patent misuse doctrine can play a powerful role in deterring anticompetitive efforts to extend patent rights beyond the scope a rational pharmaceutical patent policy would give.[203]

IV. CONCLUSION

Patent law is becoming technology-specific. The legal rules applied to biotechnology cases bear less and less resemblance to those applied in other industries, and particularly in the software cases. One can debate the wisdom of tailoring patent law to accommodate particular industries.[204] But if the courts *are* to create a new set of legal rules for biotech cases, it only makes sense to try to design those special rules to fit the industry. While there are good policy reasons to treat biotechnology differently than other industries, the current legal rules are not expressly informed by the economics of the industry, but by an *ad hoc* combination of judicial policymaking and *stare decisis*. Not surprisingly, they don't reflect optimal patent policy in biotechnology. We have offered some explanations for this phenomenon, along with a sketch of how an optimal biotechnology patent law might look. Unfortunately, the Federal Circuit's current trends in biotech won't get us there.

Acknowledgments

Thanks to Washington University and to the University of Toronto Faculty of Law Distinguished Visitor program and the Centre for Innovation Law and Policy for their generous support in the preparation of this article. Kristen Dahling, Laura Quilter, and Colleen Chien provided research assistance. John Allison, Rochelle Dreyfuss, Rebecca Eisenberg, Richard Epstein, Dan Farber, Nancy Gallini, Wendy Gordon, Rose Hagan, Bruce Hayden, David Hyman, Brian Kahin, Clarisa Long, David McGowan, Craig Nard, Arti Rai, Herb Schwartz, Polk Wagner, and participants at the Telecommunications Policy Research Conference, the Washington University Conference on the Human Genome, the University of Toronto conference on Competition and Innovation, the University of San Francisco symposium on Defining and Defending Intellectual Property Rights in Biotechnology, and a faculty workshop at the University of Minnesota Law School all provided comments on an earlier draft of this or a related paper.

[203] Alternatively, the problem could be controlled to some extent using policy levers relating to obviousness. Pharmaceutical companies often engage in the practice of "double-patenting": seeking multiple patents on the same or only slightly different technologies in an effort to extend the effective life of their proprietary right. Strengthening the obviousness standard will make it harder to extend patent life through double-patenting, because the doctrine of "obviousness-type double patenting" precludes obtaining two patents that would be obvious in view of one another' unless the patentee disclaims the longer patent term. *See, e.g., Ortho Pharmaceutical Co. v. Smith*, 959 F.2d 936 (Fed. Cir. 1992).

[204] For a detailed discussion of the wisdom of both legislative and judicial industry-specificity, see Burk & Lemley, *Policy Levers, supra* note 2, at Part III.

17

Commentary on the Panel Presentations

Pauline Newman
United States Court of Appeals for the Federal Ciruit
Washington, DC 20439

This topic invites creative thinking about the optimum legal rules for the advancement of genomic and other biological sciences and their practical applications, in the context of the powerful potential for public benefit of these advances. The distinguished scholars on this panel have written provocative papers concerning the laws of intellectual property and their application to the new biology.

Biotechnology's immense promise for human benefit warrants careful attention to the legal structure that will best advance this science and its practical applications. Thus I take concerned notice when a thoughtful scholar such as Professor Reichman finds serious flaws in patent systems, and suggests that the scope of patent protection may be unduly broad. In turn, Professors Burk and Lemley state that the courts have unduly narrowed the scope of patent protection in the field of biotechnology. Both positions conclude that the national interest is ill served.

These divergent positions reflect some of the differences manifest in the burgeoning legal scholarship of intellectual property. In these brief remarks I shall respond particularly to the paper of Professors Burk and Lemley, for they have penetrated deep into the workings of patent systems, and deliver severe criticisms of several judicial rulings, stating that the judges "have it exactly backwards." I welcome this opportunity to comment on the commentators, for it is rarely available to judges to criticize their critics, particularly those who have it exactly backwards.

I shall concentrate on the areas of discussion with which I am least in accord, in the spirit of the discourse. Professors Burk and Lemley argue first that the decisions of the Federal Circuit in the field of biotechnology are specific to that industry, and that this is the cause or perhaps the result of our errors. Indeed, all judicial decisions are specific to the facts presented. The biological sciences have presented the patent law with factual situations not shared with most other fields, for they deal with naturally occurring products, living organisms, extreme complexity, and incomplete scientific understanding. In turn, the patent statute contains general provisions intended for application to the entire breadth of technology. I do not share these authors' thesis that a different law has been applied to biotech inventions, or even that the law has been applied differently. It is more likely that the court, drawing on its broad experience with the patent statute as it has been applied to all technologies, has continued to apply to biological inventions the statutory rules and established policy embodied in the wisdom of precedent.

Precedent has demonstrated that patent law and policy readily accommodate new forms of technology, as illustrated in the bursts of creative and entrepreneurial activity after the Supreme Court's decision in *Diamond v. Chakrabarty*[1] and again after our decision in *In re Alappat*.[2] Only occasionally has statutory change been needed in order to preserve commercial incentive for new technologies, as in the Semiconductor Chip Protection Act.[3] To the extent that Professors Burk and Lemley argue that new law has been disadvantageously created for the biotech industry, I cannot agree. Nor do I think that different law should apply to the new biology, and indeed Professor Reichman in his paper illustrates the uncertainties of shifting to new regimes. Experience shows that the finer tuning of patent law derives from fact-driven application, not change of legal principle.

The fact-driven application of patent principles is conspicuous in fields in which results are "unpredictable," the patent shorthand for the experimental sciences, evolved in response to the statutory requirement of unobviousness. Patentability of chemical and biological inventions has to varying degrees been affected by the unpredictability of aspects of these sciences, as compared with the mechanical and electrical technologies. But if the societal and economic and jurisprudential premises of a patent system are sound, they are universally so, whatever the field in which they are applied.

In the decisions criticized by Professors Burk and Lemley, our court held that the patentee must comply, in the biological arts as in other fields, with the statutory requirement of "written description." A principal focus of these authors' concern is the *Eli Lilly*[4] case, where the court held that the description

[1] 447 U.S. 303 (1980).

[2] 33 F.3d 1526 (Fed. Cir. 1994).

[3] 17 U.S.C. § 901 et seq.

[4] *Regents of the University of California v. Eli Lilly & Co.*, 119 F.3d 1559 (Fed. Cir. 1997).

in the patent of the nucleotide sequence of the rat DNA that encodes rat insulin was inadequate to support claims to the human DNA that encodes human insulin. The inventors did not describe the composition of the DNA for human insulin. The authors here urge that it should have sufficed, to support the broad claims that were rejected by the court, that the patent stated that the DNA for human insulin, whose differences from the rat DNA were not identified, could be obtained using the techniques described for the rat DNA. However, the court applied its precedent which held that in the unpredictable arts a researcher's intention to produce a product, called "a wish, or arguably a plan" in *Fiers v. Revel*,[5] is inadequate to meet the statutory requirement of description of the subject matter for which a patent is sought. This reasoning underlay the court's decision in the interference contest involving Fiers and Revel, a case also disapproved by these authors.

These judicial decisions are founded on the statutory requirement that the inventor must describe that which he seeks to patent. This requirement is fundamental to the purposes of patent systems, for these purposes include not only the grant of exclusivity as a commercial incentive, but the obligation of the inventor to add to the store of knowledge by disclosing—by describing and enabling—that for which exclusive rights are sought. The patent law serves the advancement of knowledge as well as provides commercial incentive.

The requirement of adequate description is not new to the law. However, I recognize that these decisions have engendered controversy, and if in fact they do not comport with the needs of biological science and the biotech industry, if inventions are not being made or are not being developed to practical application and availability because of these decisions, or if injustice results, then correction is required. I take concerned note when committed scholars announce that our decisions are "ill-considered as a policy matter." Yet I search their paper in vain for a practical proposal for change, or an adequate analysis of the consequences of a law that would award broad patents for biological products that have not been described or otherwise made available.

The new biology promises untold benefits for humankind and it is essential that there be not only an effective but also an optimum incentive system for the development of this science and its applications to human benefit. With much of the basic research in the biological sciences government-funded, and with commercial development requiring private investment, resolution of patent rights affects many public and private interests, and involves complex interactions. I am heartened to note that on the present state of the law, entrepreneurial start-ups and innovative advances in the biotech field continue to blossom.

[5] *Fiers v. Revel*, 984 F.2d 1164 (Fed. Cir. 1993).

Based on its increased significance in a world economy based on technology, study of the patent law and its role in technologic advance is important. An incentive system may be based on other than patent principles, and perhaps something better can be devised, as Professor Reichman suggests. The patent system has the advantage of avoiding bureaucratic administration after the patent is granted, and even the examination stage is not taxpayer-supported. The working of the system is based on economic principles, whereby intellectual and entrepreneurial energies are encouraged by the market protection provided by the patent. A patent system well fits technologic advance in a market economy, where the potential of private gain is the incentive to provide new things for public availability. In the United States the diversity of our marketplace and the vigor of our patent activity go hand in hand. It is a matter of serious concern if the law should lapse in service to the policy it implements.

Professor Burk and Lemley find much to criticize in our jurisprudence. I note their remark that "the Federal Circuit believes that biotechnology experts know very little about their art." The authors appear concerned that judges do not take the experts' advice. In actuality we rarely receive one expert's advice—we receive two experts' advice, and it is invariably contradictory. I also note the authors' criticism of what they call the court's "static perception" of the patent law. I am reminded of the obverse of this criticism that was levied at pre-Federal Circuit law, where the absence of static perception by judges left innovators uncertain of how the law would be applied, thus increasing the risk of investment in new technologies. Investment decisions are based on what the law is, and require confidence in its stability. Whatever the virtues of judicial initiative, surely a static perception of the law has a role in the pragmatic value of patent systems.

I have long observed the pragmatic value of patent systems, whereby research that might otherwise not be done or products that might otherwise not be marketed are brought forth with the support of a patent. Yet it is essential to recognize that patent law is itself a balance of policies, for the vigor of our economy depends on its competitive strength as well as its technological leadership. It is essential that the law strike the appropriate balance between an inventor and those who build on the invention, whether by incremental advance, further invention, or competitive pressure. These factors are not absent in judicial administration of the patent law, for their premises underlie not only the questions here discussed, but the current issues of "claim construction" and of "equivalency." These issues are of relevance to biotech inventions, and raise the same overarching policy concerns that have been discussed by this panel.

I also take note of the extensive discussion of "the commons" and the "anticommons." These economist buzz-words start with the premise that the invention that is the subject matter of the commons already exists, whereas often the unavailability of patent protection will simply direct an inventor's efforts in some other direction. The commercial potential is usually evaluated at the threshold, and if the foreseeable profit does not warrant the investment risk, some inventions will not be made and others will not be developed. Professors Burk and Lemley recognize this economic reality; it is important that we avoid over-simplifying the perturbations of innovation in a market economy, which involve many complex and interacting considerations.

The courts, in each case presented for judicial decision, are confronted with a tight synthesis of such complex and interacting considerations, focused on narrow questions of law and fact as applied to new circumstances. While a strength of the common law is its responsiveness to changing conditions, departure from the constraints of precedent should not be easy. The Federal Circuit, since its inception some 20 years ago, has undoubtedly added stability to the patent law, while the nation has seen the most spectacular growth in innovative commerce and technological power in its history. Patent-based entrepreneurial business is particularly evident in the biotech field. I'm heartened by this observation, for I share with my colleagues a deep concern for fulfilling the court's mandate.

I welcome this discussion, in the spirit of the academy and in our shared interest in achieving an optimum system of intellectual property. So I thank all on this panel for their stimulating contributions. I am grateful for the opportunity to join the debate.

18

Commenting on Biotechnology's Uncertainty Principle

Herbert F. Schwartz
Partner
Fish and Neave
New York, New York 10020

ABSTRACT

In "Biotechnology's Uncertainty Principle," Dan L. Burk and Mark A. Lemley state that in biotechnology cases the Federal Circuit has it "exactly backwards" (p. 307), because its application of the written description, enablement, and nonobviousness requirements result in "numerous" patents of "extremely narrow" scope (p. 317). I write to comment on some of the issues raised by the authors.

I. THE WRITTEN DESCRIPTION AND ENABLEMENT REQUIREMENTS

The written description requirement ensures that the inventor had possession of the entire subject matter claimed when the patent application was filed. *Vas-Cath, Inc.* v. *Mahurkar*, 935 F.2d 1555, 1563–64 (Fed. Cir. 1991). The purpose of the written description requirement is to avoid "overreaching" and to prevent the patent applicant from claiming later-developed technology that

0065-2660/03 $35.00

is beyond the scope of the original invention. "Adequate description of the invention guards against the inventor's overreaching by insisting that he recount his invention in such detail that his future claims can be determined to be encompassed within his original creation." *Id.* at 1561 (citation omitted).

Although the purpose of the written description requirement is to ensure that the applicant was in possession of the claimed invention, possession alone is not always sufficient to meet that requirement. *Enzo Biochem, Inc.* v. *Gen-Probe Inc.*, 285 F.3d 1013 (Fed. Cir. 2002). Rather, "adequate written description of genetic material 'requires a precise definition, such as by structure, formula, chemical name, or physical properties.' " *Id.* at 1018 (citations omitted).

The enablement requirement ensures that the specification of a patent teaches those skilled in the art how to make and use the full scope of the claimed invention without undue experimentation. *Genentech, Inc.* v. *Novo Nordisk, A/S*, 108 F.3d 1361, 1365 (Fed. Cir. 1997) (citations omitted). The purpose of the enablement requirement is to ensure that the public receives a disclosure that is equal in breadth to the limited monopoly rights granted to the inventor.

The basic principle that patent claims must not "overreach" beyond the invention echoes the important policies established long ago in landmark cases such as *O'Reilly* v. *Morse*, 56 U.S. (15 How.) 62 (1853) (holding claims invalid as overbroad). In biotechnology cases, the courts have applied the written description and enablement requirements to conform with this long-established policy, as well as to prevent patent applicants from claiming too broadly. *See, e.g.*, *Fiers* v. *Revel*, 984 F.2d 1164, 1171 (Fed. Cir. 1993) ("Claiming all DNA's that achieve a result without defining what means will do so is not in compliance with the description requirement; it is an attempt to preempt the future before it has arrived."); *Genentech*, 108 F.3d at 1366 (citations omitted) ("Patent protection is granted in return for an enabling disclosure of an invention, not for vague intimation of general ideas that may or may not be workable. Tossing out the mere germ of an idea does not constitute enabling disclosure.").

The authors believe that the current written description and enablement standards in biotechnology cases result in patents of "extremely narrow" scope (p. 317). For example, the authors state that "a researcher will be able to claim only sequences disclosed under stringent written description—the actual sequences in hand, so to speak" (p. 317).

However, the scope of the claims in biotechnology cases need not be as narrow as the authors describe. For example, the Federal Circuit has expressed a willingness to uphold broad patent claims when the claims are adequately described and enabled by the specification. In *Amgen, Inc.* v. *Chugai Pharmaceutical Co.*, 927 F.2d 1200 (Fed. Cir. 1991), the claims were directed to the specific DNA encoding EPO that Amgen had cloned, the EPO gene analogs disclosed in the specification, and all other analogs and modifications that would produce a protein that stimulated red blood cell production. The specification only provided details for preparing "a few of the EPO analogs disclosed." *Id.* at 1213. The Federal Circuit stated that although the disclosure "might well justify a generic claim encompassing [the analogs disclosed in the specification] and similar analogs" it provided "inadequate support for Amgen's desire to claim all EPO gene analogs." *Id.* The Federal Circuit clarified that they did "not intend to imply that generic claims to genetic sequences cannot be valid where they are of a scope appropriate to the invention disclosed by an applicant." *Id.* at 1214. *See also, Regents of Univ. of California* v. *Eli Lilly & Co.*, 119 F.3d 1559, 1569 (Fed. Cir. 1997) ("A description of a genus of cDNAs may be achieved by means of a recitation of a representative number of cDNAs, defined by nucleotide sequence, falling within the scope of the genus or of a recitation of structural features common to the members of the genus, which features constitute a substantial portion of the genus."). Thus, the Federal Circuit is willing to uphold generic claims that are commensurate in scope with what is disclosed in the specification.

The authors cite *Amgen, Inc.* v. *Hoechst Marion Roussel, Inc.*, 126 F. Supp. 2d 69 (D. Mass. 2001) as an example of a case where a patent claim was held invalid for lack of written description for failing to describe an adequate number of species within a genus (p. 346, n. 176). The authors misinterpreted this case. In this case, the patent applicant did not attempt to claim a broad genus by disclosing a single species. The claims discussed by the authors were directed to a "non-naturally occurring" EPO "having glycosylation which differ[ed] from" human urinary EPO. *Id.* at 154. The court held the claim invalid for lack of written description because the patent did not describe a non-naturally occurring EPO having glycosylation which differed from human urinary EPO. *Id.* at 155. This finding was based on the fact that different human urinary EPOs had different glycosylations and the patent did not identify which human urinary EPO to use as the standard. *Id.* at 155.

II. THE NONOBVIOUSNESS REQUIREMENT

The authors believe that under the current nonobviousness standard, most claims to DNA would be considered nonobvious, and hence patentable. In *In re Deuel*, 51 F.3d 1552 (Fed. Cir. 1995), the Federal Circuit held that claims directed to DNA molecules coding for heparin-binding growth factors were nonobvious in light of the prior art disclosure of a partial amino acid sequence of heparin-binding growth factors and a method for using that information to obtain the corresponding DNA molecule. *See id.* at 1558 ("A prior art disclosure of the amino acid sequence of a protein does not necessarily render particular DNA molecules encoding the protein obvious because the redundancy of the genetic code permits one to hypothesize an enormous number of DNA sequences coding for the protein.").

However, the Federal Circuit stated that under appropriate circumstances, DNA claims can be rendered obvious by the prior art disclosure of an amino acid sequence and a method of obtaining the corresponding DNA molecule. *See id.* at 1559 (citation omitted) ("A different result might pertain, however, if there were prior art, *e.g.*, a protein of sufficiently small size and simplicity, so that lacking redundancy, each possible DNA would be obvious over the protein."). *See also In re Bell*, 991 F.2d 781, 784 (Fed. Cir. 1993) ("This is not to say that a gene is never rendered obvious when the amino acid sequence of its coded protein is known. Bell concedes that in a case in which a known amino acid sequence is specified exclusively by unique codons, the gene might have been obvious."). Thus, under appropriate circumstances, the Federal Circuit is willing to hold some DNA claims obvious in light of the prior art disclosure of the corresponding amino acid sequence.

III. THE OBVIOUSNESS REQUIREMENT AND THE DOCTRINE OF EQUIVALENTS

The authors argue that the current patentability standard in biotechnology cases could result in narrow patents with a broad scope of equivalence. The authors seem to base this conclusion on the premise that "a patentee is prevented from capturing by equivalence subject matter that would have been obvious at the time he obtained his patent" (pp. 317–318). However, this does not mean that under the doctrine of equivalents a patentee can capture all subject matter that is nonobvious. Under the doctrine of equivalents infringement may be found if the differences between the accused product and a claim limitation are insubstantial. Whether or not a change is insubstantial is determined from the perspective of a person having ordinary skill in the art (PHOSITA). Because the scope of equivalence is determined from the perspective of a PHOSITA, the Federal Circuit is not likely to hold that there is a broad range of equivalence for DNA claims. Further, the Federal Circuit has issued a series of decisions limiting the scope of the doctrine of equivalents.

IV. THE PHOSITA IN BIOTECHNOLOGY

As the authors describe, the written description, enablement, and nonobviousness requirements are determined from the perspective of a PHOSITA at the time the patent application was filed (p. 320). The authors state that "[I]n biotechnology, where highly detailed disclosure is required to satisfy the enablement and written description standards, similarly detailed disclosure is required to render the invention obvious" (p. 320). This is consistent with the fact that all three requirements are measured from the standard of a PHOSITA.

The authors suggest that the PHOSITA "may display different and inconsistent characteristics as between the different statutory subsections" (p. 324). Specifically, the authors suggest "decoupling the treatment of the PHOSITA in obviousness and enablement cases" (pp. 324–325). The authors do not explain why the patent statute would allow the use of different PHOSITAs for different patentability requirements. "Decoupling" of the PHOSITA standard could result in a scenario where it would be relatively easy to obtain patents of broad scope. This is likely to hinder, rather than promote innovation.

The authors argue that the Federal Circuit's application of the PHOSITA standard in biotechnology cases "is wrong as a matter of science" because it assumes that the biotechnology art is unpredictable, when in fact it is predictable (p. 326). While the authors argue that "the Federal Circuit is wrong to suggest that identifying and making biotechnological products is always difficult and uncertain," they admit that biotechnology involves "fundamental uncertainty" due to the fact that "biotechnology products arise of living systems" (p. 344). The fact that biotechnology products arise out of living systems makes biotechnology inherently unpredictable. Thus, it can be argued that even if the scientific methods used in biotechnology become routine, they will still give unpredictable results.

Further, the authors state that "the suggestion that misunderstood science explains the Federal Circuit's jurisprudence in these areas is not altogether satisfactory, as it fails to explain the court's indifference to the technology subsequent to Amgen" (p. 328). The authors state that in biotechnology cases decided subsequent to *Amgen, Inc.* v. *Chugai Pharmaceutical Co.*, 927 F.2d 1200 (Fed. Cir. 1991), the court focused on "structural precision and foreseeability" rather than on the "certainty" or predictability of the biotechnology art (p. 328).

The emphasis on structural precision and foreseeability in the biotechnology cases cited by the authors may be related to the nature of DNA inventions. Typically, a naturally occurring substance (such as a DNA sequence) is not patentable unless and until it is isolated and purified. Thus, in "DNA cases," the Federal Circuit has required the patent applicant to have actually isolated the DNA sequence before patenting it. *See Amgen*, 927 F.2d at 1206 ("We hold that when an inventor is unable to envision the detailed constitution of a gene so as to distinguish it from other materials, as well as a method for obtaining it, conception has not been achieved until reduction to practice has occurred, *i.e.*, until after the gene has been isolated.") *See also In re Deuel*, 51 F.3d at 1558 ("[U]ntil the claimed molecules were actually isolated and purified, it would have been highly unlikely for one of ordinary skill in the art to contemplate what was ultimately obtained. What cannot be contemplated or conceived cannot be obvious.").

Further, in some of the cases decided subsequent to Amgen, the Federal Circuit held certain biotechnology patent claims invalid for failure to satisfy the written description requirement. *See, e.g., Eli Lilly*, 119 F.3d at 1566–1569; *Enzo*, 285 F.3d at 1018–24. The predictability or unpredictability of the art is not a relevant factor in determining written description. Instead, the predictability or unpredictability of the art is a factor typically considered in enablement determinations. *See, e.g., Enzo Biochem, Inc.* v. *Calgene, Inc.*, 188 F.3d 1362, 1372–1374 (Fed. Cir. 1999); *In re Vaeck*, 947 F.2d 488, 496 (Fed. Cir. 1991). The predictability or unpredictability of the art may also be relevant in obviousness determinations. *See, e.g., Amgen*, 927 F.2 at 1208–1209 (holding a claim to an isolated DNA nonobvious because even if it was obvious to try the inventor's probing and screening method to isolate the DNA, there was no reasonable expectation of successfully identifying and isolating the DNA).

V. INNOVATION, INVENTION, AND UNCERTAINTY

The authors conclude that the patentability standards applied in biotechnology cases are not optimal from the perspective of economic policy because "[u]nder this standard, no one is likely to receive a patent broad enough to support the further costs of development" (p. 346). The authors believe that there should be fewer patents of broader scope, and suggest that this could be accomplished by lowering the written description and enablement requirements (p. 350). The authors do not explain why this should be a good thing in the real world. I do not understand why it should be so. Broad patents without clear support are more an invitation to unreasonable litigation than a reasonable solution to the problem.

Broad patents that do not adequately describe and enable their claims may preclude competition and innovation by granting unreasonably broad monopoly rights to the first person to make an invention in a given field, as well as stifle, rather than promote innovation in the development of that field.

The authors' suggestions that narrow patents can be broadened by the doctrine of equivalents (pp. 317–318, 350), or that broad patents can be narrowed by the reverse doctrine of equivalents (p. 348), are also an invitation to litigation. A patent is supposed to put the public on notice as to what is and what is not covered by the patent claims. Expansion of the doctrine of equivalents and reverse doctrine of equivalents would increase the uncertainty as to what is and what is not covered by the patent claims, and thus is likely to promote more, rather than less, litigation.

Acknowledgment

I thank my associate Gloria M. Fuentes for her assistance with this commentary.

19

(Mostly) against Exceptionalism

R. Polk Wagner
Assistant Professor
University of Pennsylvania Law School
Philadelphia, Pennsylvania 19118

I. INTRODUCTION

As befits an expansive regulatory regime concerned with innovation policy, the patent law is inextricably intertwined with the process and details of technological development. As courts and commentators alike have long recognized, both a challenge and a strength of our patent system is the ongoing effort to adapt the legal infrastructure to an ever-changing environment.[1] Like a ship on

[1] *See, e.g., AT&T, Inc. v. Excel Communications, Inc.*, 172 F.3d 1352, 1356 (1999) ("As this brief review suggests, this court (and its predecessor) has struggled to make our understanding of the scope of [35 U.S.C.]§ 101 responsive to the needs of the modern world.")

0065-2660/03 $35.00

the open sea, the direction of the law at any given instant may appear to diverge somewhat from the goal, and small helm corrections will be necessary from time to time.[2] The patent law, by explicit design, is technologically flexible, with significant adjustment points built into the system. That distinctions in treatment will exist between various technologies is as expected (and as unremarkable) as the necessity of steering our metaphorical ship towards its destination.

In "Biotechnology's Uncertainty Principle,"[3] however, Dan Burk and Mark Lemley argue that the patent law is seriously off course, at least with respect to biotechnology. Carefully analyzing the Federal Circuit's jurisprudence, they argue that the court has developed in recent years a distinctly exceptionalist approach—addressing differing technologies under quite distinct doctrinal rules—and that this has yielded a legal infrastructure for biotechnological development that is "exactly backwards."[4] Responding to contemporary academic concerns that a proliferation of patenting in this field will create substantial transaction-cost-related holdups,[5] Burk and Lemley argue for adjustments in the Federal Circuit's doctrinal rules that would have the effect of expanding the power of biotechnological (and especially genomic) patents.

"Biotechnology's Uncertainty Principle" is an important and insightful contribution to the growing literature on the institutional relationships of the patent law.[6] Reasonable people can, however, disagree with the ultimate success of the effort. To that end, in the few pages that follow, I suggest an alternative view of Burk and Lemley's findings—specifically, that their exposition makes a rather compelling case *against* precisely the sort of judicial ventures into technologically specific innovation policy that they recommend. Instead, the example of the struggles to adapt the patent law to the changes in biotechnology illuminates the undesirability of entangling the patent doctrine in broad, policy-driven technological exceptionalism.

The argument is presented in three parts. First, while Burk and Lemley are undoubtedly correct that biotechnological inventions get "treated

[2] For an overview of the art and science of helmsmanship, *see generally* Elbert S. Maloney & Charles Frederic Chapman, CHAPMAN PILOTING SEAMANSHIP & SMALL BOAT HANDLING (63rd ed.), 1999.

[3] See Dan L. Burk and Mark A. Lemley, "Biotechnology's Uncertainty Principle," (Chapter 16) (hereinafter Burk & Lemley, *Uncertainty Principle*). Note that this chapter is part of a group of related works. *See* R. Polk Wagner, *Of Patents and Path-Dependency: A Comment on Burk & Lemley*, 18 BERKELEY TECH. L.J.__ (2003); Burk & Lemley, *Policy Levers in Patent Law.*__VA. L. REV.__(2003); Burk and Lemley, *Is Patent Law Technology-Specific?* 17 BERKELEY TECH. L.J. 1155 (2002) (hereinafter, Burk & Lemley, *Technology-Specific*).

[4] See Burk and Lemley, *Uncertainty Principle*, 3.

[5] See *infra* notes 45–48.

[6] For recent related work, see, *e.g.*, Arti Kaur Rai, *Engaging Facts and Policy: A Multi-Institutional Approach to Patent System Reform*, 103 COLUM. L. REV. 1035 (2003); F. Scott Kieff, *The Case for Registering Patents* (ms. 2003); Burk and Lemley, *Technology-Specific*, supra note 4; R. Polk Wagner, *Reconsidering Estoppel: Patent Administration & the Failure of Festo*, 151 U. PA. L. REV. 159 (2002).

differently" than, say, software or mechanical inventions, this observation alone is certainly no cause for alarm. Submerged in the Burk and Lemley analysis is an important conceptual distinction between two types of technological-specificity: *micro*-specificity, which applies the variable legal rules to specific technological circumstances; and *macro*-specificity, which countenances distinct legal rules across different technologies, and relatively more similar application within related technologies.[7] Determining which of these two forms exists in the biotechnological jurisprudence is critically important, for this explains whether the Federal Circuit has developed (or seeks to develop) an innovation regime especially for biotechnology, or whether the observable distinctions are merely the expected consequence of the patent law's inherent flexibility.

Section III addresses the question of whether the patent law does indeed evince the sort of *macro*-specificity that Burk and Lemley identify. Even accepting their analysis of the relevant case law as correct, I argue that there remain a number of observations and explanations—such as factual misunderstandings, doctrinal confusions between facts and law, or even the unique circumstances surrounding the set of opinions that dominate this area—that point rather strongly in favor of a *micro*-specific framework rather than *macro*-exceptionalism. That is, "Biotechnology's Uncertainty Principle" has identified potentially serious defects in the court's jurisprudence as applied to this technological area. But this, I suggest, does not itself make the case that the doctrine is *macro*-specific, in part because there seems to be little reason to worry that path dependency has emerged.

Indeed, as I argue in Section IV, the several problems with the Federal Circuit's doctrinal development identified by Burk and Lemley seem to strongly support the position that *macro*-exceptionalism is unjustified. Biotechnology's Uncertainty Principle advocates substantial, policy-driven *macro*-specific changes to fundamental standards of patentability, in order to expand the breadth of patents in this area and thus (theoretically) avoid the (theorized) anticommons problem. And while this almost-Kitchian approach dominates alternatives involving the weakening of property rights in biotechnology, there is a third option that seems even better—clarifying and stabilizing the patent law to reduce transaction costs. Indeed, a transaction-cost-focused analysis would suggest that it is Burk and Lemley, rather than the Federal Circuit, that have it "exactly backwards." Given the deep uncertainties underlying the premises of the argument, as well as the promise of increasing transaction costs resulting from a shift from the *micro*-specific to *macro*-specific approach, there are strong reasons to conclude that we should remain (mostly) against exceptionalism.

[7] For example, *micro*-specificity allows that invention A, in a newly developing niche of the software field, would potentially have different application of the disclosure and obviousness standards than invention B, in a mature area of software, or invention C, in a pathbreaking biotechnological area. *See infra* Part II for more explanation.

II. ADAPTATION VERSUS PRESCRIPTION: EXPLORING TECHNOLOGICAL SPECIFICITY AND THE PATENT LAW

That the patent law is significantly "technology-specific" is both easily apparent and fully expected. As noted previously, any law purporting to provide a regulatory foundation for innovation must be able to account for both the broad range of technologies and the rapid pace of changes.[8] To bind the patent law to the technological assumptions of an earlier era or to the maturity of any particular technology would be foolish. And yet there is a limit: Not all technological exceptionalism is necessarily benign. When the jurisprudential approach shifts from adaptation to prescription—from the application of consistent rules to variable facts to the promulgation of distinct rules to implement technology-based innovation policy—courts put at risk the very social progress they seek to enhance.[9]

A. The Two Forms of Technological Specificity: *Macro-* and *Micro-*

Because of the recognition that technological specificity can be either a boon or burden to the patent system, it is critically important at the outset to determine what one means by a "technology specific" patent law. As noted above, there are two distinct conceptual schemas to consider:

Micro-Specificity: The (legal) rules applied to innovations are variable, dependent upon particular technological circumstances.

Macro-Specificity: The (legal) rules are quite distinct across different technologies—even though they are relatively more similar within related technologies.[10]

To illustrate the distinction, consider three inventions, two generally in the software field, and one in biotechnology: The first deals with data management, a relatively stable and mature area of software technology; one relates to machine learning, a relatively immature and undeveloped sub-field; the third considers genomic research, at the high end of biotechnology.

In a regime of *micro*-specificity, each of these inventions will have rather distinct patentability requirements, primarily because of the operation of the patent law's "person having ordinary skill in the art" (PHOSITA) standard. As Burk and Lemley demonstrate, a higher PHOSITA standard (a greater degree of difficulty in the field) implies a lesser standard for obviousness and a

[8] Indeed, as Burk and Lemley note, "different industries experience both innovation and the patent system in very different ways." *Uncertainty Principle*, 40–41.

[9] *See infra* Part IV.

[10] One might call this an "industry-specific" approach, but that implies an economic structure coincident within related technologies, which is perhaps—but not necessarily—the case.

Table 19.1. The *Micro*-Specific Approach

Invention	PHOSITA level	Obviousness standard	Disclosure requirement
Data management	baseline	baseline	baseline
Machine learning	+	−	+
Genomic research	++	−−	++

Table 19.2. The *Macro*-Specific Approach

Invention	PHOSITA level	Obviousness standard	Disclosure requirement[13]
Data management	A	A'	A"
Machine learning	A	A'	A"
Genomic research	B	B'	B"

greater disclosure requirement.[11] Table 19.1 notes these requirements, taking the data management invention as having the lowest (baseline) difficulty level.

In contrast, a regime of *macro*-specificity contemplates that different inventions will be accorded PHOSITA standards on the basis of the invention's technological field, rather than by a more nuanced or particularistic analysis.[12] Thus, (see Table 19.2) the software-related inventions will have roughly the same standards for patentability, but distinct standards from the biotechnological invention. Burk and Lemley do not address this question, but the implication is that the patentability differences in this scheme (*e.g.*, the difference between A' and B') are more pronounced than those in the *micro*-specific context.

As between these two forms of specificity, I take the *micro* form to be both a positive description of the patent law as well as a normatively justifiable

[11] *See* Burk and Lemley, *Uncertainty Principle*, 316–19 (describing biotech jurisprudence); Burk and Lemley, *Technology-Specific*, 1160–73 (describing software jurisprudence). Of course, as Burk and Lemley note, the explicit coupling of the PHOSITA standard in obviousness and disclosure doctrines can obscure some small differences in the way the standard is applied to each requirement. *See* Burk and Lemley, *Uncertainty Principle*, 332–33.

[12] How one might determine the appropriate level of skill for a given field is open to question, of course. Burk and Lemley suggest that software and biotechnology require roughly similar levels of skill—or at least do not deserve the divergent treatment they report in their review of the jurisprudence. *See* Burk and Lemley, *Uncertainty Principle*, 325–26.

[13] Note in this regard that Burk and Lemley suggest that in their ideal *macro*-specific regime, the obviousness and disclosure standards would be "decoupled," or relaxed in the specific case of biotechnological inventions. *See* Burk and Lemley, *Uncertainty Principle*, 350–52. *See also Technology-Specific*, 1202–1205.

position. The chief advantage (and challenge) of the patent law is its ability to provide a set of clear background (*i.e.*, "property") rules upon which private parties can build to invent, invest, and commercialize. Accordingly, the patent law must always retain the flexibility to adapt to new technological developments and economic shifts. In the *micro*-specific context, this flexibility is realized through the use of the PHOSITA standard as the lens through which a number of critical analyses are conducted. As a question of fact that should necessarily vary from particular innovation to particular innovation, the ordinary skill in the art framework grounds the legal abstractions of the patent law to the technological facts in any given case.

The *macro* form of technological exceptionalism, however, is far more problematic. Here, rather than building flexibility and innovation into the stable backdrop of the law, the project is far broader, typically invoking arguments related to the "nature of the technology" or the "structure of the innovation," or perhaps even the normative profile of the participants to support essentially *sui generis* changes in the patent law. *Macro*-specificity shifts consideration of the patent law from a general background principle of property rights to a vehicle for particularistic, technology-specific innovation policy choices. As I note in Part IV later, there are a number of reasons why it is worth at least challenging the efficacy and appropriateness of this development of the patent law.

One important limitation of the Burk and Lemley thesis is that the distinction set forth above remains unaccounted for in their analysis. They do seem to recognize what they describe as "inherent" technological specificity, which might be taken to correspond to what I have described as *micro*-specificity.[14] Yet they also quite clearly perceive (and advocate for) a broader, *macro* version of exceptionalism as a means by which to influence technological development in the biotechnological field.[15] Because, as I argue later, it is by no means clear that the differences they identify result from conscious *macro*-specific behavior on the part of the Federal Circuit, this failure to account for both forms of technological specificity weakens their argument.

B. The Uncertain Effects of the PHOSITA

At various points in their argument, Burk and Lemley seek to connect the PHOSITA standard directly to the scope of the patent grant. This linkage, I suggest, is tenuous at best, for the following reasons.

[14] Burk and Lemley, *Uncertainty Principle*, 325.

[15] Burk and Lemley identify what they suggest are "radical" differences in the legal standards applied to software and biotechnology, respectively. *See* Burk and Lemley, *Uncertainty Principle*, 319–20.

First, both the disclosure requirements and the obviousness requirement are scope-affecting. Obviously, a higher standard of disclosure will force a patentee to claim more closely to what she has described, narrowing the literal scope of the patent.[16] But the obviousness standard will also affect scope: A reduced standard of (non)obviousness will allow a patentee to establish claims "closer" to any relevant prior art.[17] An extremely reduced version of the obviousness requirement—call it "anticipation"—will allow claims that merely avoid the disclosure of the prior art, as well as those that cover more innovative subject matter.[18] Conversely, a higher standard of (non)obviousness will yield claims that are more distinct (in physical terms, more distant) from the prior art, and thus narrower.

Importantly, note the following: (1) the inverse relationship between the obviousness and disclosure standards (at least under current doctrine); and (2) the direct relationship between the scope-effects of the obviousness and disclosure standards. This suggests that, contrary to Burk and Lemley's assumption, the patent scope-effects of changes in the PHOSITA standard will be fundamentally indeterminate, without knowledge of the relative magnitude of the disclosure-related and obviousness-related scope effects.[19]

A second observation concerning the PHOSITA and claim scope is that the standard's effect on the scope and availability of equivalents infringement is also fatally indeterminate. For example, a high degree of skill in the art (a difficult field) implies:

1. a relatively *narrower* doctrine of equivalents, because the possibility of "known interchangeability" between claim elements and their purported equivalents will be reduced;[20]

[16] *See, e.g.*, Burk & Lemley, *Uncertainty Principle*, 317–18.

[17] As well as, of course, enabling the patenting of subject matter that could not otherwise be patented. Indeed, it is this "gatekeeping" function that is perhaps the most important (and apparent) contribution of obviousness. *See generally* Robert P. Merges, *Uncertainty and the Standard of Patentability*, 7 HIGH TECH. L. J. 1 (1992).

[18] *See, e.g.*, 35 U.S.C. § 102.

[19] Perhaps one might assume that any obviousness-based scope effects will be swamped by those related to the disclosure requirement. This seems to me to be a problematic assumption, given the typicality of scope-reducing claim amendments as a means to overcome examiner rejections. Another possibility is to assume that patentees do not care about the scope of their claims *vis-a-vis* the prior art, and are instead claim-aggressive only towards the "outer limit" that the disclosure allows. This seems somewhat more probable but assumes irrational behavior. In any event, neither of these assumptions seems to fit with the Burk and Lemley argument.

[20] *See, e.g.*, Burk and Lemley, *Uncertainty Principle*, 322–23. *See also Graver Tank Mfg. Co. v. Linde Air Prods. Co.*, 339 U.S. 605, 609 (1950); *Hilton Davis Corp. v. Warner-Jenkinson*, 62 F.3d 1512, 1519 (Fed. Cir. 1995) (*en banc*) *aff'd in part & rev'd in part on other grounds*, 520 U.S. 17 (1997).

2. a relatively *broader* doctrine of equivalents, because prior-art based limitations on the doctrine will be less available;[21]
3. a relatively *narrower* doctrine of equivalents, because of the efforts of the Federal Circuit to limit equivalents due to the patent's disclosure;[22]
4. a relatively *broader* doctrine of equivalents, because of the ability to more easily overcome the *Festo* presumption against equivalents in cases where a claim amendment would eliminate infringement of a technology unforeseeable by one of ordinary skill.[23]

Thus, while it is clear that the determination of a PHOSITA standard will *affect* the scope of the doctrine of equivalents, not much more can be safely concluded. Table 3 notes the relative effects of the PHOSITA standard on the scope of the patent.

Again, given the varying results in three columns on the right (scope-effects) for each case, any conclusion concerning the overall scope-effects of a change in the PHOSITA standard would require detailed knowledge concerning the relationship among the various scope-effects. Thus, while it is clear that the PHOSITA standard influences the scope of the patent, the direction and magnitude of that effect is quite indeterminate, and no meaningful conclusions can be drawn concerning the relationship.[24]

Table 19.3. The Scope-Effects of the PHOSITA Standard

PHOSITA level	Obviousness standard	Disclosure standard	Scope-effects (obviousness)	Scope-effects (disclosure)	Scope-effects (equivalents)
Low (easy field)	+	–	–	+	indeterminate
High (hard field)	–	+	+	–	indeterminate

[21] *See, e.g., Wilson Sporting Goods Co. v. David Geoffrey & Assoc.*, 904 F.2d 677, 684 (Fed. Cir. 1990) ("[S]ince prior art always limits what an inventor could have claimed, it limits the range of permissible equivalents of a claim.")

[22] *See, e.g., Sage Prods., Inc. v. Devon Indus.*, 126 F.3d 1420 (Fed. Cir. 1997) (limiting equivalents due to clarity of the patent's disclosure).

[23] *See, e.g., Festo Corp. v. Shoketsu Kinzoku Kogoyo Kabushiki*, 535 U.S. 722, 740 (2002).

[24] Note Burk and Lemley's proposal that the courts "decouple" the linkage between obviousness and disclosure, and—in the biotechnological context—relax the disclosure requirement while maintaining the obviousness standard could in some cases yield the broader patents they seek. *See infra* Part IV.

III. THE HUNT FOR EXCEPTIONALISM: THE TECHNOLOGICAL SPECIFICITY OF PATENT JURISPRUDENCE

As I established in Part II.A, two distinct forms of technological specificity can potentially be applied to the patent law. The first, *micro*-specificity, applies varying standards of patentability according to the specific technological circumstances, meaning that each invention (in theory at least) has a unique, contextual requirement for patentability. The second, *macro*-specificity, applies similar standards of patentability to inventions in the same technological field (or "industry," as Burk and Lemley at times refer to it), while applying distinct standards to different technological fields. This section explores which form(s) of exceptionalism exist in the patent law.

As an initial matter, I note that there is no real question but that *micro* forms of technological-specificity are fundamental to the patent law. That is, whether the patent law is, as Burk and Lemley ask, "technology-specific," strikes me as an easy and rather obvious question.[25] Of course it is: Among other aspects, the ordinary skill in the art standard implements the *micro*-exceptionalism described above. Thus the analysis here considers whether the broader, *macro*, form is descriptive of the modern patent law.

A. Patentability Jurisprudence

As Burk and Lemley note, the Federal Circuit's patentability jurisprudence in the biotechnological area is self-consciously distinct from that in other technological fields, such as software. Indeed, Burk and Lemley juxtapose language from cases considering the disclosure requirement in the software and biotechnological fields, illuminating that in the software context, the Federal Circuit has held that the disclosure of functions is sufficient, while in the biotechnological context, the Federal Circuit has held that the disclosure of genetic function is insufficient.[26] Yet this demonstration, standing alone, does little more than highlight the importance of the PHOSITA standard: In each case, the court was viewing the technology through the prism of a particular level of skill in the art.[27] It is apparent that the court believes that the levels of skill differ in relation to the two technologies at issue; we should expect, not resist, the distinct treatment.

[25] They do acknowledge that the patent law is "inherently technology specific." Burk and Lemley, 325.

[26] *See* Burk and Lemley, *Uncertainty Principle*, 318–19 (quoting *Fonar* and *Regents of the Uni. Of Calif. v. Eli Lilly*).

[27] *Compare Fonar Corp. v. Gen Elec. Co.*, 107 F.3d 1543, 1549 (Fed. Cir. 1997) with *Regents of the Univ. of Calif. v. Eli Lilly*, 119 F.3d 1559, 1568 (Fed. Cir. 1997).

In order to show *macro*-specificity, Burk and Lemley must argue (as they do) that the level of skill in the art is systematically approached differently in the biotechnological cases and that this is an enduring (likely policy-driven) feature of the patent law rather than transient in nature. Yet even when the Burk and Lemley approach to the relevant caselaw is taken as correct, there remain a variety of reasons that the approach to the PHOSITA standard in this area could appear systematic and yet result in a jurisprudence that is far more *micro*-specific than *macro*-specific. For example:

1. *Sample size*. One possibility is that the systematic technological specificity identified by Burk and Lemley is essentially a statistical artifact related to the fairly small number of cases extant in the relevant jurisprudence (*i.e.*, those analyzed by these and other commentators).[28] The implication here is that the purported systemization would fade or disappear as more cases are decided.
2. *Judicial consistency*. Another possibility is that the technological specificity identified by Burk and Lemley is related to the (remarkably) small number of judges who have authored the opinions studied in the article (and others).[29] This suggests that any case-to-case consistency is a reflection of one judge's uniform approach, rather than either a court-wide decision or an enduring feature of the jurisprudence.[30]
3. *Factual error*. Yet another possibility is that the systematic technological specificity identified by Burk and Lemley results from a judicial misunderstanding of the relevant facts.[31] The judges who have addressed these issues

[28] Burk & Lemley cite seven cases as representative of the suggested problematic approach to the issue. They are: *Amgen Inc. v. Chugai Pharma Co.*, 927 F.2d 1200 (Fed. Cir. 1991); *In re Goodman*, 11 F.3d 1046, 1052 (Fed. Cir. 1993); Fiers v. Rivel, 984 F.2d 1164 (Fed. Cir. 1993); *In re Bell*, 991 F.2d 781 (Fed. Cir. 1993): *In re Deuel*, 51 F.3d 1552, 1559 (Fed. Cir. 1995); *Regents of the University of California v. Eli Lilly & Co.*, 119 F.3d 1559 (Fed. Cir. 1997); *Enzo Biochem v. Calgene, Inc.*, 188 F.3d 1362, 1371 (Fed. Cir. 1999); *Enzo Biochem v. Gen-Probe, Inc.*, 296 F.3d 1316 (Fed. Cir. 2002). These cases, or a subset thereof, appear to be the most relevant to the commentators.

[29] Of the opinions cited in footnote 28 as being relevant, every one except for *Goodman* has been authored by Judge Lourie.

[30] There is another important observation to be made here. Given the essentially random assignment of cases to panels of judges, it is extremely unlikely that cases focusing on obviousness or disclosure in the biotechnological area have been uniformly assigned to panels containing Judge Lourie. This fact further suggests that Judge Lourie (perhaps alone among Federal circuit judges) seeks opportunities to use his particular brand of biotechnological PHOSITA analysis. That is, Judge Lourie presumably is far more likely to seize the opportunity to analyze the disclosure (or obviousness) of biotechnological inventions, while his colleagues are more likely to decide these cases on other issues. Absent these sorts of selection-effects, it would be difficult to reconcile the Federal Circuit's professed random case assignment procedure with the pattern of decision-making in this area.

[31] Burk and Lemley initially note this explanation themselves, but seem to suggest that the court has not considered these factual issues in more recent cases. *See* Burk and Lemley, *Uncertainty Principle*, 329–30. Yet the recent cases need not undermine the intuition that the factual

thus far may not fully understand the detailed fact-based distinctions between genomic research and small molecule chemistry, for example, and thus be likely to simply apply a PHOSITA standard from one context in another. Again, this concern is neither indicative of a *macro*-exceptionalist approach nor difficult to remedy going forward.[32]

4. *Fact/Law Confusion*. A fourth possibility is that any systematic technological-specificity identified by Burk and Lemley arises as a result of doctrinal confusion at the Federal Circuit concerning the nature of facts, law, and *stare decisis*. For example, the court may be failing to understand the implications of the distinctly fact-based inquiry into the PHOSITA with respect to appellate review. Or the court may simply be refusing to afford factual findings any deference, in favor of factual analysis of its own.[33]

Any one of these explanations for Burk and Lemley's observed patterns in the patent law is sufficient to undermine their *macro*-specific argument. (The most likely situation, of course, is that a combination of the above exists.)

This is not to suggest, however, that the application of the PHOSITA standard in the biotechnological area appears anywhere near optimal. Indeed, in my view, Burk and Lemley have compellingly identified problems with the Federal Circuit's jurisprudence; at least some of these concerns—especially the Judge-based effects (number 2 above)—suggest that the court would be well-advised to carefully consider the process by which this doctrinal development is occurring.

---

analysis is flawed as a matter of technology: The failure to explicitly revisit technological might suggest that the court continues to believe them to be correct. Note that in the Federal Circuit's most recent effort at biotechnological disclosure standards it remanded with explicit instructions to analyze the relevant technological facts. *See Enzo Biochem Inc. v. Gen-Probe, Inc.*, 296 F.3d 1316, 1324–26 (2002). ("It is not correct, however, that all functional descriptions of genetic material fail to meet the written description requirement.")

[32] Indeed, the court seems to have importantly reaffirmed the basic factual nature of the disclosure inquiry in its most recent precedent. In *Enzo Biochem*, 296 F.3d at 1328, the court stated:

> Although the patent specification lacks description of the location along the bacterial DNA to which the claimed sequences bind, Enzo has at least raised a genuine issue of material fact as to whether a reasonable fact-finder could conclude that the claimed sequences are described by their ability to hybridize to structures that, while not explicitly sequenced, are accessible to the public. Such hybridization to disclosed organisms may meet the PTO's Guidelines stating that functional claiming is permissible when the claimed material hybridizes to a disclosed substrate. That is a fact question. We therefore conclude that the district court erred in granting summary judgment that the claims are invalid for failure to meet the written description requirement. On remand, the court should consider whether one of skill in the art would find the generically claimed sequences described on the basis of Enzo's disclosure of the hybridization function and an accessible structure, consistent with the PTO Guidelines. If so, the written description requirement would be met.

[33] Arti Rai suggests that this is an endemic problem with the Federal Circuit's jurisprudential approach. *See generally, e.g.*, Rai, *supra* note 6.

B. Biotechnological "Jurisprudence"

The other possibilities offered above—that the court is factually mistaken or legally misunderstanding the role and nature of the PHOSITA analysis along a number of dimensions (numbers 3 and 4)—are also worth criticizing. And yet these problems seem quite unlikely to create the sort of path-dependencies that would raise concerns of *macro*-specificity.

In this vein, most commentators appear to assume the future development of this "biotechnological patent doctrine" will continue along the presently observed trajectory.[34] However, the 'mistakes have been made' form of criticism has an easy answer: the use of correct technological facts.[35] The distinctly factual nature of *micro*-exceptionalism provides ample opportunity for future panels of the Federal Circuit to establish their own analysis in any given case. Indeed, an appropriate understanding of the role of the PHOSITA in the patent law would seem to virtually *preclude* the creation and use of categorical rules.[36] The state of the art in such fields is changing rapidly; that one of ordinary skill might have been unable to determine the DNA sequences that would code for EPO from a few examples circa 1984[37] seems nearly irrelevant to the level of knowledge in DNA sequence identification in the late 1990s.[38] Put another way, the explicit references to the "ordinary skill in the art to which [the invention] pertains" might be said to fundamentally *require* the reconsideration of issues of technological fact at each instance[39]—rather than perpetuating imprecise standards, even "decoupl[ed],"[40] as substitutes for technological fact. The correct rule as a matter of doctrine may also be the correct rule as a

[34] *See, e.g.*, Burk and Lemley, *Uncertainty Principle*, 333–4. *See also generally* John M. Lucas, *The Doctrine of Simultaneous Conception and Reduction to Practice in Biotechnology: A Double Standard for the Double Helix*, 26 AIPLA Q.J. 381, 418 (1998) (cited in Burk and Lemley); Arti K. Rai, *Intellectual Property Rights in Biotechnology: Addressing New Technology*, 34, WAKE FOREST L. REV. 827 (1999); Arti K. Rai, *Addressing the Patent Gold Rush: The Role of Deference to PTO Denials*, 2 WASH. U. J. L. & PUB. POL'Y 199 (2000).

[35] As Burk and Lemley seem to acknowledge, at least in part. Burk and Lemley, *Uncertainty Principle*, 329.

[36] To this end, the court's explicit description of aspects of the disclosure requirements in the software context as a "general rule" seems unwise. *See, e.g.*, *Fonar Corp. v. Gen. Elec. Co.*, 107 F.3d 1543, 1549 (Fed. Cir. 1997). And to the extent that the court's failure to explicitly ground PHOSITA analyses in the biotechnological area to factual considerations can be taken to infer a form of the 'general rule' statement noted above, corrective actions should be taken. *See, e.g.*, *Eli Lilly*, 119 F.3, 1568;

[37] *U.S. Patent No.* 4,703,008, entitled *DNA sequences encoding erythropoietin*, was filed November 30, 1984. The '008 patent was at issue in *Amgen Inc. v. Chugai Pharma. Co.*, 927 F.3d 1200 (Fed. Cir. 1991).

[38] *See, e.g.*, Burk and Lemley, *Uncertainty Principle*, 330; Lucas, *supra*, 418.

[39] *See* 35 U.S.C. §§ 103, 112 (2001).

[40] *See* Burk and Lemley, *Uncertainty Principle*, 332.

matter of policy: the courts may not, and should not, "standardize" the person of ordinary skill in the art.[41] *Micro*-specificity will prevail; and unhelpful path-dependencies will be avoided.

C. Reading the Cases: A (Brief) Response to Burk and Lemley

Before moving to a broader critique of exceptionalist schemes, a brief response is in order. That is, in "Biotechnology's Uncertainty Principle" Burk and Lemley seek to brush aside the line of analysis in Part III of this essay, suggesting that it simply results from a different understanding of the relevant case law.[42] But this response is too facile; as I have repeatedly noted above, I have assumed in this part that their view of the relevant doctrine is correct. The point here is that even evaluating the precedent *just as they do* does not resolve the question of whether what we see in the patent law is *macro*-specificity, or if it is merely a version of *micro*-specificity colored by factual error or the unusual circumstances surrounding this line of cases.[43] While it is possible that the Federal Circuit has created a policy-driven, enduring, *macro*-specific doctrine for the field of biotechnology, that issue has yet to be resolved.

IV. (MOSTLY) AGAINST EXCEPTIONALISM

Even if the patent law evinces a technologically-exceptionalist approach—with disparate legal rules applied to different technological fields—there remain significant reasons to believe that the effort to formalize and tailor such exceptionalism, as Burk and Lemley advocate, is seriously misguided.

As several commentators have observed, there is at least some concern that the field of biotechnology in particular has structural and technological features that might make it susceptible to transaction cost and related forms of

[41] An anticipated response to this assertion is that the Federal Circuit seems to consider the prior rulings as having precedential value. *See*, e.g., *Enzo Biochem v. Gen-Probe Inc.*, 2002 WL 487156 *4–5 (Fed. Cir. Apr. 2, 2002) (citing *Eli Lilly* as precedent). This objection is unsatisfactory. First, the court always acknowledges the factual basis of the analysis. Second, notwithstanding the citations, it is difficult to determine the actual weight given to earlier factual determinations in different cases. And third, I previously noted the truly remarkable homogeneity of the relevant Federal Circuit decisions, which suggests that author consistency rather than doctrinal development is at issue. *See supra* note 7 and accompanying text.

[42] Burk and Lemley, *Uncertainty Principle*, 320. They note that Lawrence M. Sung has suggested that the biotechnology cases are little different in application from cases in other technological areas. *See generally* Lawrence M. Sung, *On Treating the Past as Prologue*, 2001 U. Ill. J.L. Tech. & Pol'y 75. This is not, however, the argument I make.

[43] *See supra* Part III.A.

market inefficiencies.[44] Generally referred to by the term "anticommons," the theory suggests that the difficulty in arranging and aggregating the patent rights necessary to actually deliver marketable goods will stymie the participants in this field to a degree that will ultimately reduce the pace of technological development, and thus increase social losses.[45] There are a variety of responses to this perceived problem in the literature, ranging from vertical integration,[46] to the formation of collective rights organizations,[47] to the denial of patenting altogether in some areas of the field.[48] To these, Burk and Lemley add another: the expansion of patent rights in these areas, so as to create conditions more akin to Kitch's "prospect" theory.[49] In Burk and Lemley's view, such expansion, together with modestly increased standards for patentability, will yield fewer and more powerful patents,[50] thus decreasing the characteristics of the biotechnology field that might create an anticommons problem.[51]

[44] *See, e.g.*, Heller and Eisenberg, *Can Patents Deter Innovation? The Anticommons on Biomedical Research*, 280 SCIENCE 698 (1998); Arti Kaur Rai, *The Information Revolution Reaches Pharmaceuticals: Balancing Innovation Incentives, Cost, and Access in the Post-Genomics Era*, 2001 U. Ill L. REV. 173. *But see* John P. Walsh, Asish Arora & Wesley M. Cohen, *The Patenting and Licensing of Research Tools and Biomedical Innovation* (working paper 2000) (empirical survey casts doubt on this view).

[45] *See* Heller and Eisenberg *id.* For a full description of the anticommons theory, see Michael A. Heller, *The Tragedy of the Anticommons: Property in the Transition from Marx to Markets*, 111 HARV. L. REV. 621 (1998).

[46] *See, e.g.*, Arti Kaur Rai, *Fostering Cumulative Innovation in the Biopharmaceutical Industry: The Role of Patents and Antitrust*, 16 BERKELEY TECH. L.J. 813 (2001) (describing and criticizing this trend).

[47] *See, e.g.*, Heller and Eisenberg, *supra.* On collective rights organizations more generally, see Robert Merges, *Contracting into Liability Rules: Intellectual Property Rights and Collective Rights Organizations*, 84 CALIF. L. REV. 1293 (1996).

[48] *See, e.g.*, Heller and Eisenberg; Rai, *Cumulative Innovation, supra.*

[49] *See* Edmund Kitch, *The Nature and Function of the Patent System*, 20 J.L. & ECON. 265 (1977) (broad "prospect" patents allow for better resource allocation to innovation). *See also* Burk and Lemley, *Uncertainty Principle*, 338 (noting the benefits of Kitch's "prospect" theory in the pharmaceutical context).

[50] Note that there are serious problems with Burk and Lemley's assumption concerning the relationship between the frequency of patents and their scope. The suggestion seems to be that strengthening the obviousness requirement while simultaneously weakening the disclosure requirement will yield fewer yet broader patents. *See* Burk and Lemley, *Uncertainty Principle*, 349. An initial problem is the assumption that such changes will affect patent scope in the way that Burk and Lemley suggest: In Part II.B previously, I described the indeterminacy of the relationship between patent scope and simultaneous changes in the obviousness and disclosure requirements—the most that can be said without a series of difficult empirical assumptions is that scope will be affected; the magnitude and direction of the effect is unclear. (The suggestion that those doctrines be "decoupled" might have helped their case, see Burk and Lemley, *Technology-Specific*, at 1205, if they limited their proposal to loosening the disclosure requirement, for example. *Cf.* Burk & Lemley, *Uncertainty Principle*, 349 (positing simultaneous changes).)

While I am generally sympathetic to Burk and Lemley's view that strengthening biotech patents is likely to be a better solution along a number of dimensions than reducing patent scope,[52] a third option appears dominant here, especially given the uncertainties surrounding the premises of the Burk and Lemley argument. That is, the possibility of an anticommons in biotechnology could be ameliorated, perhaps significantly, via relatively straightforward efforts to clarify and stabilize the patent jurisprudence, thereby reducing the transaction costs of combining rights. This point is simply Cosean: While other commentators focus on the appropriate entitlements,[53] the reduction of transaction costs (thus diminishing the importance of those entitlements) may well provide a better payoff.[54] To be sure, we can never eliminate the transaction and related costs inherent in the patent system, and efforts to clarify the law can only achieve part of this goal at best. But all options in this context are "second best" in nature: The goal is to improve the current situation, given the world as we find it.

One important question is how one might seek to clarify patent rights in this context. And while a full treatment of that question is beyond the scope of this essay, one thing we would surely *not* want to do is create the additional jurisprudential and doctrinal confusion that will result from introducing *macro*-exceptionalism to the patent law. In this sense "Biotechnology's Uncertainty Principle" is a virtually perfect indictment of itself: Burk and Lemley go to great

---

Perhaps even more troubling are the assumptions about patent frequency. That there will be fewer patents of course does not logically follow from broader patents. This ignores the *ex ante* incentives of the patent system: Broader (stronger) patents will induce additional incentives to engage in inventive behavior (or at least patenting behavior). This has been empirically verified. *See, e.g.*, Browyn Hall and Rosemarie Ham Zideonis, *The Patent Paradox Revisisted: Determinants of Patenting in the US Semiconductor Industry*, 1980–94, 32 RAND J. ECON 101–125 (2001) (documenting an increase in patenting linked to strengthening the patent law in the 1980s). Thus, broader patents should lead to greater, not fewer, patents. *Cf.* Burk and Lemley, *Uncertainty Principle*, 349. To the extent that Burk and Lemley rely on their proposed increase in the obviousness standard to yield fewer patents, then that simply raises the fatal indeterminacy problem noted above: Will the additional difficulty of obtaining patents (the heightened (non)obviousness requirement) outweigh the effects of broader and stronger patents (due to lower disclosure requirements)? The effects of the Burk and Lemley proposals on the scope and frequency of patents is unclear.

[51] *See*, Burk and Lemley, *Uncertainty Principle*, 349.

[52] *See, e.g.*, R. Polk Wagner, *Information Wants to Be Free: Intellectual Property and the Mythologies of Control*, 103 COLUM. L. REV. 995 (2003) (noting the benefits of broader, as opposed to narrower, property rights in information goods).

[53] See *supra* notes 45–49.

[54] Ronald Coase, *The Problem of Social Cost*, 3 J.L. & ECON. 1 (1960) (noting the importance of transaction costs in entitlement analysis). Note that Heller and Eisenberg suggested in their original, pathbreaking, article that collective rights organizations—another transaction-cost-reducing mechanism—might be a solution to any anticommons problem. *See* Heller & Eisenberg, *supra* note, 45.

lengths to demonstrate the troublesome aspects of *macro*-specificity in the patent law, reserving special criticism for the Federal Circuit's inability to adequately understand the innovation policy needs of the modern biotechnology industry. Having demonstrated (they suggest) the error of the court's technology-specific ways, one might expect that the next step is to suggest doctrinal adjustments to eliminate the inter-industry differences they suggest are harming technological development. Instead, they suggest that the court apply an entirely new jurisprudential framework, based on underlying principles of innovation policy. (Not to mention that the new doctrine would "decouple" the relatively uniform PHOSITA standard, thus requiring *at least* two detailed factual analyses to replace one.[55]) Implicit in their proposal is the idea that this new jurisprudential framework would (a) require reconsideration of obviousness and disclosure standards in all technological fields served by the patent law,[56] and (b) be subject to (and under obligation of) revision by the court anytime the background conditions for innovation in a particular field change.[57] And all this policy-driven, unconstrained decision-making with incredible importance for the future of technological development is being placed into the hands of the Federal Circuit—the very body that Burk and Lemley suggest has done such a poor job to this point.

In sum, this does not seem to be a proposal that is likely to increase the certainty and stability of the patent law in either biotechnological or other areas. Indeed, a transaction-cost-focused analysis would suggest that it is Burk and Lemley, rather than the Federal Circuit, that have it "exactly backwards."[58] Given the deep uncertainties underlying the premises of the argument,[59] as well as the promise of increasing transaction costs resulting from a shift from the *micro*-specific to *macro*-specific approach, there seem to be strong reasons to conclude that we should remain (mostly) against exceptionalism.

Acknowledgments

Thanks to Jeff Leftsin, Lee Petherbridge, Arti Rai, Reed Shuldiner, and participants at the Washington University Human Genome Conference for helpful comments on earlier drafts.

[55] *See*, Burk and Lemley, *Technology-Specific*, *supra* note 4, 1202.

[56] That is to say *all* technological fields.

[57] One conjures up troubling images of the Federal Circuit reviewing evidence concerning the availability of early-stage capital investment vehicles, interest rates, market conditions, and the like in a particular field prior to establishing the levels of (non)obviousness and disclosure required.

[58] *Cf.* Burk & Lemley, *Uncertainty Principle*, 307.

[59] *See supra* Parts II and III.

Section 4

TRANSACTIONS OVER GENETICS IN ACADEMIA AND BUSINESS

20 O Brave New Industry, That Has Such Patents in It! Reflections on the Economic Consequences of Patenting DNA

Iain M. Cockburn

Professor, Finance and Economics Department
Boston University School of Management
Boston, Massachusetts 02215

I. INTRODUCTION

The issuance of increasing numbers of patents on DNA sequences raises a variety of interesting and difficult questions, which span many fields of inquiry and attract extraordinarily wide attention. For economists, perhaps the most significant of these revolve around the efficiency of the industries and institutions that create and use this technology. What has been, or will be, the impact of gene patents on the pace of innovation in pharmaceuticals, agriculture, and other industrial applications of biotechnology? What will happen to public and private investment in relevant kinds of research and development (R&D) and human capital? What will happen to prices of products that directly or indirectly embody these inventions? What will happen to industry structure in the DNA-using sectors, and how will related upstream and downstream markets and institutions respond?

There are at present few concrete analytical or empirical answers to these questions, and this essay is confined to speculating about some possible trends and outcomes. I focus principally on one of the most visible and most

0065-2660/03 $35.00

important consequences of issuing patents on genomic information: the shifting of the boundary between noncommercial and for-profit research in biomedical science, and the appearance at this interface of a venture funded, for-profit "tool sector" in the biotechnology industry—of which genomics companies are some of the most visible and controversial exemplars. The long-term effects of this vertical disintegration of the pharmaceutical industry are difficult to fathom. With good reason, optimists anticipate that entrepreneurial upstream entry supported by patent rights in genes and DNA sequences will raise the overall productivity of the biomedical research sector. Also with good reason, pessimists fear the opposite: that exclusion-based property rights in genomic information will result in misallocation of research effort, thus lowering system productivity, and that the norms and institutions of open science that fostered the postwar revolution in molecular biology will be significantly weakened.

In the remainder of this essay, I begin by reviewing the role played by patents in the business model of companies formed to create and exploit genetic information and gene-based discoveries. I go on to evaluate the societal role of these companies in the context of an interpretation of the recent history of the pharmaceutical industry as the evolution of a previously highly successful configuration of institutions into a new and more complex industry structure whose long–term efficiency is far from clear. On the one hand, the emergence of a vibrant, entrepreneurial group of "tool" companies at the interface between publicly funded science and the traditional "Big Pharma" firms offers potentially substantial economic gains from factors such as the creation of markets that explicitly price upstream technology, higher productivity through increased specialization and tailored incentives, and efficient unbundling and reallocation of risk among investors. On the other, I argue that a reconsideration of the basic economics of vertical integration and of the nature of competition between the new entrants and the incumbent firms that develop, manufacture, and distribute pharmaceuticals, should give some pause for thought. There is at least the possibility that gene patents might destroy as well as create value. The prospect of "locking up" significant portions of molecular biology could induce inefficiently high levels of R&D spending, and draw biotechnology companies, pharmaceutical firms, universities, government entities, purchasers of health care, and the legal system into a costly struggle for associated economic benefits that consumes scarce resources, undermines critical institutions, and results in poor investment decisions.

Absent concrete evidence on which of these benefits and costs associated with strong patent rights over DNA-related inventions is likely to prove to be a significant determinant of the pace of innovation in biomedical technologies, it is difficult to arrive at concrete policy recommendations. I conclude with the unsurprising observation that the need for empirical research on these issues is clearly indicated.

II. TOOL COMPANIES AND THEIR BUSINESS

In the late 1990s it became fashionable to refer to biotechnology firms as falling into two categories: "tool companies" and "product companies." Product companies, typically founded in the 1980s, focus their efforts on creating marketable products. Classic product companies from the early years of the biotechnology industry are Amgen, Biogen, Chiron, and Genentech. Companies such as these overcame very substantial risks and were able to finance product development over very long periods of time to bring therapeutic proteins to market, and investors were rewarded with commensurate returns from high-margin proprietary products. Many other product-oriented companies could not overcome technological, financial, and regulatory hurdles, and either went out of business, were acquired, or made significant changes in strategy. Tool companies, by contrast, seek returns from selling discovery technologies to downstream product developers. The tool companies that have sought proprietary rights in genomic information, such as Incyte, Human Genome Sciences, Millennium, Celera Genomics, Sequenom, Orchid, deCODE, Gemini Genomics and others, are of particular interest here, though it should be noted that "tools" also includes instrumentation, drug delivery systems, tissue/protein/antibody libraries, DNA/protein chips, combinatorial chemistry techniques, microfluidics and other technologies, as well as services such as bioinformatics software. Tool companies were thought to face less risk of product failure and to be able to reach profitability more quickly, and were therefore able to raise substantial amounts of capital. Though there have been few outright failures among tool companies, returns have generally been disappointing, and in recent years the tide appears to have turned—venture capitalists are showing renewed interest in product companies, and pure tool companies are now integrating into product development.

For both types of companies, patent protection has been crucially important. For the product companies, patents appear largely to have been used in the classic Big Pharma mode to limit imitation and exclude direct competitors. Tool companies, on the other hand, have tried to obtain broad patents on what are arguably data and methods, and to use complex contractual arrangements to capture downstream rents accruing to companies that use them to develop and sell end products. While tool companies have occasionally attempted to obtain such "reach-through" rights in the form of explicit patent claims to diagnostic and therapeutic products derived from their technology, typically these contracts are based indirectly on the blocking power of claims to novel nucleotide sequences, proteins, assays, libraries, or screening methods.

III. VERTICAL DISINTEGRATION OF THE PHARMACEUTICAL INDUSTRY

In large part, the growth of the biotechnology industry can be understood as a process of vertical disintegration of the pharmaceutical sector. Biotechnology "tool" companies have inserted themselves into the industry value chain at the interface between two very different sets of organizations: the upstream non-profit academic research sector, and the downstream for-profit pharmaceutical firms. Using patents on, among other things, DNA sequences, they have succeeded in taking over a certain amount of research activity in both sectors and have forced some interesting institutional adjustments in university-industry relations. Any resulting redistribution of rents is of intense interest to industry participants, but from an economic perspective the only thing that matters is whether this vertical restructuring of the industry has increased or decreased the present value of total future value creation.

It may be helpful to place this vertical restructuring in historical context by reviewing the post-war evolution of the industry.[1] This can be characterized in terms of two distinct eras: the "Good Old Days," which ended somewhere in the early 1980s, followed by the "Modern Era" from the mid 1980s to the present.

In the Good Old Days, there was a clear division of effort between upstream nonprofit institutions that did curiosity-driven basic research, and downstream for-profit companies that did market-oriented applied research. In the for-profit sector, almost all firms were large and fully integrated from drug discovery, through clinical development, regulatory affairs, manufacturing, and marketing. Most commercial drug discovery activity was conducted in-house, and at least in the early part of this period was dominated by large-scale "random screening" programs with limited requirements for deep knowledge about fundamental physiological processes at the molecular level. Licensing activity was driven largely by downstream concerns: rights to sell drugs that were already approved (or were in the late stages of clinical development) would be acquired in order to maintain efficient levels of utilization of manufacturing or marketing assets, or, in the international context, to take advantage of local knowledge and access to regulators and distribution channels. Upstream technology was largely acquired either "for free" by reading journals and attending conferences, or by purchasing tangible inputs and services, such as instruments or highly skilled graduates.

Pharmaceutical firms appropriated returns from R&D through a combination of extensive patenting of production processes and end products,

[1] *See, e.g.*, Cockburn, I., R. Henderson, L. Orsenigo, and G. Pisano "Pharmaceuticals and Biotechnology." Chapter in D. Mowery (ed.) *U.S. Industry in 2000: Studies in Competitive Performance*, National Research Council, Washington DC, 1999, 363–398.

proprietary know-how, brands, regulatory barriers to entry, and favorable product market conditions. Most of these firms were mature organizations, tracing their roots back many decades, even to the nineteenth century chemical industry. Their large and sustained investments in R&D, marketing assets, and human and organizational capital were largely financed from internal cash flow. Competitive advantage was driven by firms' ability to effectively manage product market interactions with regulators and end-users, and to "fill the pipeline" with a steady succession of internally developed blockbuster drugs. In turn, the productivity of internal R&D appears to have been driven by economies of scale and scope in conducting research, efficient allocation of resources in internal capital markets, and the ability to capture internally and externally generated knowledge spillovers.

In the upstream nonprofit sector, taxpayers (and, to some extent, philanthropists) supported curiosity-driven research conducted at cottage industry scale inside government labs, universities, research institutions, and teaching hospitals. Legal constraints and a strong set of social norms limited commercial or contractual contacts between the world of open science and pharmaceutical firms in important ways. Resource allocation in the nonprofit sector was driven by peer-reviewed competition for grants on the basis of scientific merit and the reputation of individual researchers. The importance of establishing priority and reputation drove early and extensive publication of results, and social norms (and requirements of granting agencies) promoted routine sharing of research materials. Nonprofit researchers concentrated largely on fundamental science and filed very few patents.

The aforementioned is, of course, a gross oversimplification. Many drug companies invested significant resources in "blue sky" basic research. Specialist research boutiques generated and sold technology to large firms. Public sector institutions conducted screening programs for drug candidates, and many academic researchers had close financial and contractual links with drug companies through individual consulting arrangements and institutional research grants and contracts. Funding priorities reflected political pressure, intellectual fashions, and the dynamics of the Matthew Effect, as well as pure scientific merit. Importantly, the "waterfall" model of vertical knowledge spillovers, with a one–way flow of ideas and information down a gradient running from upstream basic science to downstream applied research and clinical practice, is only partially accurate. Nobel-winning work in basic science was done in for-profit labs, and nonprofit institutions were an important source of data, techniques, and expertise in late-stage drug development, epidemiology, and postmarketing follow-up. There was significant movement of ideas, candidate molecules, research materials, research results and individuals back and forth across the for-profit/nonprofit divide.

Notwithstanding these caveats, it is still possible to summarize the vertical structure of the industry in the Good Old Days as being essentially binary, with a clear distinction drawn between upstream open science and a downstream commercial sector dominated by large, highly integrated firms. By contrast, in the Modern Era the industry structure became much more complicated. After decades of stability and consolidation, in the late 1970s the for-profit side of the industry experienced significant entry as an intermediate sector emerged between academic research institutions and Big Pharma. By the mid-1990s several thousand biotechnology ventures had been launched, and several hundred had survived to reach sufficient scale to be an important force in the industry. Existing vertical relationships were disrupted and reformed, with consequences that are still far from clear. The new companies straddled and blurred the divide between for-profit and nonprofit research. Though they were, for the most part, overtly profit-oriented, they also had much closer and more explicit links to nonprofit research institutions, with close personal, geographical, cultural, and contractual ties to universities, research institutes, and government labs. Academic scientists played a particularly significant role in the founding of these companies, either moving out of academic employment or participating actively in both worlds.[2] Though some of the new companies saw themselves as horizontal competitors with Big Pharma, most assumed the role of specialist suppliers of leading edge technology to a less science-intensive clinical development, manufacturing, and marketing sector.

A variety of interlinked economic and legal forces appear to have brought about this change. Perhaps the most salient of these are the developments in law and administrative practice that have brought much of molecular biology and the life sciences within the arena of the patent system. Very clearly, without patent rights for inventions in areas such as isolation and purification of proteins, DNA sequences, monoclonal antibodies, knockout and transgenic organisms, gene expression systems, etc. (or at least the prospect of obtaining and enforcing them), many biotechnology companies could not exist as currently conceived. It seems very clear that without strong patent protection the product companies could not have entered as horizontal competitors to Big Pharma. The situation for tool companies would likely be somewhat different—without patents they might well be able to compete to some extent on the basis of cost advantages, execution speed, know-how, or information kept as trade secrets—thought it is difficult to imagine them being able to raise financing or establish themselves as going concerns without patent protection. But patents are not the whole story, and it is worth reflecting on other factors that contributed to the rise of the biotechnology sector.

[2] Zucker, L., Darby, M. and Brewer, M. *Intellectual Human Capital and the Birth of U.S. Biotechnology Enterprises*, AMERICAN ECONOMIC REVIEW, March 1998; 88(1): 290–306.

Among the most frequently cited of these are the passage of the Bayh-Dole Act, and the rise of a venture capital industry (and ultimately a stock market) that was willing and able to support inexperienced companies entering a market with a 7- to 10-year product development cycle. The Bayh-Dole Act relaxed barriers to the licensing of the products of government-sponsored research, and though the casual impact of the Bayh-Dole Act is debatable, it certainly facilitated licensing of biomedical research by universities.[3] Venture capital has been a vital aspect of the biotechnology sector from its very beginning, and in many ways its growth has been contingent on the expansion of venture and private equity funding. Other developments in the capital markets may also have contributed to the rise of the biotechnology sector. At least in the United States, there appears to have been a significant increase in investors' tolerance for risk, as evidenced by the falling equity premium imputable from stock market returns. New financial technologies were also developed for pricing and managing risk. After a few well-hyped early successes, investors became comfortable with the idea of "high science for profit," developed a shared language and conceptual framework for valuing these new ventures, and (periodically) have been willing to support the new sector with substantial injections of capital.

Equally significant, however, are the organizational and managerial effects of the changes in the technology of pharmaceutical research that arose from the revolution in life sciences. One important factor was the rapid increase in the cost and scale of basic research projects. Another was that drug discovery became progressively more science-intensive, with increased emphasis on understanding and exploiting "deep" understanding of physiology at the molecular level. As "rational drug design" took center stage, changes in the nature of research activity were accompanied by complementary changes in the internal structure and incentives of commercial R&D organizations. Drug companies began to look more like universities and behave more like universities, with increasing emphasis on collaboration, publication, and exchange of (precompetitive) information.[4] This was accompanied by increased willingness to consider external sources of technology, whether in the form of research projects conducted as joint ventures, or through strategic partnerships, or simply outsourced contract research. This created an environment in which specialist research firms could expect if not to prosper, at least to survive. At the same time the growing costs and complexity of academic research projects forced

[3] Mowery, D., Nelson, B., Sampat, B., and Ziedonis, A., *The Growth of Patenting and Licensing by U.S. Universities: An Assessment of the Effects of the Bayh-Dole Act*, RESEARCH POLICY, 2001, (30): 99–119.

[4] Cockburn. I. and R. Henderson, *Absorptive Capacity, Coauthoring Behavior, and the Organization of Research in Drug Discovery*, JOURNAL OF INDUSTRIAL ECONOMICS, 1998, 46(2): 157–182.

successful scientists to acquire managerial and organizational skills—making them better equipped and more favorably disposed toward business ventures, and looking much more like entrepreneurs and managers to outside investors or business partners. As ever-increasing resource requirements, and growing societal pressure to justify their budgets, pushed universities and other government–funded institutions to become more tolerant of "just-off-campus" commercial activity, or even to actively encourage it, this rising cadre of scientist-entrepreneurs were well positioned to take advantage of the opportunities created.

By 1990, it was clear that biotechnology was here to stay. Though investors' interest waxed and waned, fresh waves of entrants were able to take advantage of periodic openings of the financing window. Technology fashions came and went (antibodies good, antibodies bad, antibodies good after all) but the industry adjusted to what appears to be a permanent new tripartite vertical structure. The waterfall model has been replaced by a whirlpool, with rapid and chaotic circulation of information and materials between nonprofits, Big Pharma, and the biotechs, supported by a complex set of contractual agreements.

Recent years have seen a wave of consolidation among the Big Pharma companies, which may be a competitive response to increasing competitive pressure from upstream, but more likely reflects rising costs of late-stage drug development (Pfizer is now reported to spend over $5 billion per year on R&D) and pressures to support high fixed costs of marketing and distribution with a broad product line.

IV. IS THE NEW INDUSTRY STRUCTURE EFFICIENT?

Most economists maintain the presumption that profit-seeking market outcomes tend to be socially optimal. But is it obvious that this restructuring of the industry was socially desirable? One place to begin to address this question is to ask, "What was wrong with the Good Old Days?" By most measures, the productivity of the pre-1980 biomedical research sector was outstanding. Academic output in the life sciences in the decades following World War II was remarkably high, with major advances made in understanding physiology and the molecular basis of disease. On the for-profit side of the fence, the pharmaceutical industry used these advances to create drugs that dramatically reduced mortality, human suffering, the economic burden of illness, and treatment costs for many diseases and conditions: (H2-antagonists, infectious diseases, depression, etc.) Substantial benefits to patients were accompanied by high returns to investors. So what was broken that needed fixing? To what problems was disintegration of the industry the solution?

In attempting to answer these questions, it may be worth revisiting the elementary economics of vertical integration. In a world with perfect information, competitive markets, and no transactions costs, there is no need for vertical integration. Stepping away from this benchmark, it has long been clear that large, vertically integrated firms are an efficient response to a number of real-world problems. These include the inability to diversify risk where capital markets are incomplete or imperfect, the inability to minimize transactions costs when complete contracts cannot be written, the inability to capture spillovers or other externalities, and a variety of familiar difficulties that arise from flaws in markets for information. In fact, there is a strong presumption that vertical integration is the first, best solution to economic problems such as financing and management of multiple projects that are long-term, risky, complex, involve activities that are costly to monitor, require substantial project-specific unrecoverable investments, and have shared costs and vertically complementary outcomes—*i.e.*, pharmaceutical R&D!

If vertical integration is not possible, then the efficiency of the unintegrated alternative will depend on the ability of the upstream and downstream parties to contract with each other. Vertical contracts are the subject of a large and rich literature in economics, to which justice cannot be done here. The following considerations are likely to be important in the present context.

Consider the stylized case of an upstream tool company that holds a valid and enforceable patent on a gene coding for a target, whose claims will be infringed by any attempt by a downstream pharmaceutical company to develop a marketable drug. The pharmaceutical company, in turn, blocks access to the end-user with its own product or use patents. The familiar logic of the double-marginalization problem dictates that instead of the tool company charging a monopoly price (reach-through royalty rate) to the pharmaceutical company, which then independently determines its monopoly price as a function of a (now higher) marginal product cost and final demand, the two parties should agree to charge a price to the end-user that maximizes joint surplus and then divide it between them. The classic question is whether the two parties can agree on a division of surplus, and whether bargaining costs will eat up any efficiency gains.

Bargaining is likely to be easy and efficient when both participants can agree on the payoff, neither has an informational advantage, and both are equally risk averse. However, in this context these assumptions are violated, and it is quite likely that the two firms will find it hard to agree. The tool company will tend to have overinflated expectations of the value it brings to the table, and the pharmaceutical company will be in stronger bargaining position given its greater size, wider range of other opportunities, and potentially a credible threat to invent around the tool company's patent (or litigate it to death). Both sides will likely have plenty of private information (the pharmaceutical company will be better informed about market prospects and product development risks, and the tool company will be better informed about its technology) and incentives to act opportunistically on that information, raising the costs of drawing up a contract, or inducing the parties to make defensive investments.

To cap it all, imperfect capital markets mean that the tool company will occasionally be facing a very real threat of bankruptcy. When the funding window is closed, cash-poor companies are easily pressured into entering an agreement on adverse terms: a low fixed fee rather than a high reach-through royalty rate, plus exclusivity provisions which limit its ability to sell its technology elsewhere or exploit it through internal development. Add a little more realism to this picture by introducing the costs of coordinating contracts with multiple tool vendors, potential anticommons problems created by overlapping rights, and uncertainty about the ultimate validity and enforceability of broadly written patents, and it becomes increasingly difficult to be optimistic about efficient outcomes being reached in licensing negotiations.

Pushing pessimism even further, one can argue that the tool companies may even have reduced current and future total value creation. One way in which value creation is reduced is when resources may have been wasted on bargaining and other transactions costs. Another, perhaps more subtle problem, is that prices in the market for these technologies may have been significantly distorted by informational asymmetries, thin markets, bad (or good) bargaining outcomes, and other problems. Using market prices as signals for resource allocation works well from a social perspective when prices reflect the marginal opportunity cost of the resources employed. When market failure drives a wedge between prices and marginal opportunity costs, markets send the wrong signals, and poor decisions result.

Furthermore, pushing the boundary between open science and for-profit research further upstream may have weakened and undermined a fragile component of the overall biomedical innovation system. Historically the academic research enterprise has been driven by social norms and resource allocation procedures that ignored market signals and commercial concerns. To the extent that "creeping propertization" results in decreased information sharing and an increased emphasis on product market potential over scientific merit in funding decisions, there may be serious long-term consequences for the future vitality and productivity of academic research.[5]

[5] *See* Eisenberg, R., *Property Rights and the Norms of Science in Biotechnology Research*, YALE L. J. 1987, 97:177–223, and Rai, A. *Regulating Scientific Research: Intellectual Property Rights and the Norms of Science*, NW. U. L. REV. 1999, 77:94–129. Though see the interesting counter-argument of Kieff, S. *Facilitating Scientific Research: Intellectual Property Rights and The Norms Of Science—A Response to Rai And Eisenberg*, 2000, 95:691.

There are, of course, a number of arguments in favor of supporting a tool sector through issuing strong, broad patents on upstream basic technology. First, basic technology tends to have broad applicability, often in ways that are very difficult to anticipate. Disclosure of early-stage technologies in patent applications may therefore promote spillovers and raise social returns to a much greater degree than narrower, follow-on patents. Development of tool technology in secret is undoubtedly socially costly.

Second, relying on incumbent firms to develop tools may result in delayed development. Incumbents may have incentives to slow down technology development in order to avoid cannibalizing existing products. They may also shelve or abandon new technology that threatens other sources of quasi-rents. Interestingly, limiting proprietary rights in early-stage technologies can reinforce the competitive position of incumbents. Indeed incumbents sometimes take actions on their own to limit proprietary rights in early-stage technologies. The "Strategy of the Commons" argument suggests, for example, that incumbent firms can deter entry into their markets by putting new technology in the public domain.[6] By denying entrants the ability to establish patent rights, their ability to raise capital and establish a proprietary market position is sharply limited. The SNPs consortium has been suggested as an example of this dynamic in action. (An interesting variant of this strategy is to sponsor university research, but only on condition that it be licensed nonexclusively.)

Third, while large, integrated firms minimize some costs, they may also raise others. Gains from integration come at the cost of creating internal bureaucracies to coordinate and control activity. These systems are costly to maintain and may cause rigidity, organizational "slack" and a bias towards conservative decisions—limiting the ability of these firms to respond to new technological opportunities. It is widely believed that new enterprises are faster and more cost effective at developing new technologies. It may well be the case that there are gains from specialization and "focus" and that tool companies therefore have significant cost advantages in doing research.

Fourth, the prospect of obtaining broad patent rights in early-stage technologies may stimulate socially valuable investment in R&D—and further rapid innovation as second movers invent around the first round of patents on a new technology. Models of sequential innovation highlight the importance of balancing the division of rents between first movers and second movers for equilibrium levels of R&D. Reluctance to grant strong rights to early innovators may therefore have deleterious effects.

Last, though the "gold rush" and "land grab" metaphors commonly employed to describe genome patenting raise the specter of socially wasteful rent

[6] Agrawal, A. and Garlappi, L., *Public Sector Science and the Strategy of the Commons (Abridged)*, Best Paper Proceedings, Academy of Management, 2002.

dissipation, such "racing" behavior may also have beneficial effects. Competitive races finish faster. Falling behind in a protracted race may cause weak competitors to drop out, weeding out bad ideas or poorly conceived enterprises. Indeed, recent developments in game theoretic modeling of technology races suggest that in some circumstances social surplus can be raised by awarding patents early rather than late in the development of a technology.[7]

V. CONCLUDING THOUGHTS

Vertically disaggregated industries are not necessarily inefficient, and specialized research firms can play an important role in the right circumstances.[8] One can be optimistic about efficiency being raised by increased vertical specialization in industries where horizontal intrasegment competition is high, where specialization reduces costs, where vertical coordination is relatively unimportant, where prices reached in the market for the upstream technology accurately reflect marginal opportunity costs, and where bargaining and contracting are easy and effective.

Is this the case in early-stage pharmaceutical research? Several aspects of the economic relationship between tool companies and Big Pharma suggest otherwise. Muted price signals from end-users, high levels of uncertainty, imperfect information, high transactions costs and serious contracting problems, and limited competition in specific areas of technology all make finding an efficient vertically disintegrated solution less likely. If this is the case, then vertical restructuring induced by regulatory or technological change may have adverse effects on social welfare. "More and stronger patents" could make things worse if they induce excess entry upstream or exacerbate contracting problems.

Anecdotal evidence and the relatively low stock market returns from tool companies support this pessimistic view. "Reach-throughs" may not be, as one attorney has claimed, "over," but the apparently broad claims of patents on DNA sequences have not yet translated into the ability to extract a significant share of the rents accruing to downstream incumbents. In part this reflects the superior bargaining position of the downstream firms, which have largely been able to dictate contractual terms to tool companies. But it also reflects what

[7] Empirical evidence for this phenomenon is thin. For the case of pharmaceutical R&D see Cockburn, I. and Henderson, R., *Racing to Invest? The Dynamics of Competition in Ethical Drug Discovery*, JOURNAL OF ECONOMICS AND MANAGEMENT STRATEGY, 1994, 2(3), pp. 481–519.

[8] See, for example, the case of specialist engineering firms in the chemicals industry, as documented bin Arora, A. and Gambardella, A. in *Evolution of Industry Structure in the Chemical Industry*, in Arora, A., Landau, R. and Rosenberg, N. (eds) CHEMICALS AND LONG TERM ECONOMIC GROWTH, New York: Wiley, 1998.

Richard Nelson called "the simple economics of basic scientific research"—patents or no patents, capturing the value that ultimately derives from fundamental early-stage research is extraordinarily difficult for profit-oriented organizations. Those firms that succeeded in doing this have, historically, been large, stable, highly integrated firms, sufficiently diversified in product markets to capture spillovers and financially strong enough to be able to effectively manage risk internally.

Stand-alone, pure play tool companies do not therefore seem likely to replicate the success of product winners like Amgen. Falling stock market valuations may reflect a realization by investors that large portfolios of gene patents are unlikely to confer significant access to blockbuster downstream revenues. In fact, licensing revenues may for the most part be confined to one-time payments or periodic user fees, with any royalties eventually realized from sales of downstream products shared with other tool providers. Many tool companies have been forced to change their strategies. Some have switched to emphasizing product development, others have moved toward much closer relationships with downstream firms, emphasizing long-term mutual interests, proprietary nondisclosed information, and close coordination, *i.e.*, a "quasi-integration solution." Tools (and associated patents) that the passage of time reveals to be truly valuable are likely to be acquired by downstream firms–potentially raising fresh issues in the antitrust area about vertical foreclosure.

One thing that gene patents seem to do effectively is create powerful incentives for new entrepreneurial companies to enter the pharmaceutical industry as vertical competitors against the established firms. But it is far from clear that these new entrants have, on net, increased value creation in the industry. The new entrants do appear to have dramatically reduced the costs of finding (then using) biologically significant sequence information compared to alternatives. Competitive pressure appears to have rapidly pushed down the cost of gene sequencing, and brought the global effort to sequence the human genome to completion much faster. The effort induced by incentives to searching for patentable DNA sequences may also have had the benefit of generating spillovers in other technologies. But these achievements must be set against the costs of racing behavior, whether they be socially wasteful duplicative effort, or simply the opportunity cost of employing extra resources to finish faster.

Other than inducing potentially inefficient levels of entry and investment into the tool sector, the impact of gene patents, at least in the medium term, may be quite small. On the positive side, they prompted voluminous disclosure of fundamentally important information, though, to some extent, this information was being created and published elsewhere. On the negative side, in some highly publicized cases gene patents appear to be being used in ways that limit nonprofit research activity or otherwise raise the costs of doing research.[9] Relatively low marginal costs of generating some types of applications for gene patents also appear to have had adverse consequences: early in the gene patent "gold rush" the Patent Office was flooded with ultimately fruitless applications on ESTs, straining its resources and likely lowering the quality of examination. Anecdotal reports circulating in 2001 suggesting that some genomics companies had more than 60,000 applications pending do nothing to assuage these concerns.

The longer-term impact on the productivity of the "biomedical-industrial complex" of extending the domain of patents deeper into the sphere of pure research is potentially very serious but difficult to assess. On this issue, as on many others raised in this essay, hard data are difficult to obtain and evaluate. Property rights in DNA sequences continue to generate controversy, and the stakes for industry participants, and ultimately for patients, are very high. In this arena, legal decisions on patent rights perforce become industrial policy: By shaping incentives of individuals, corporations and universities, and altering the balance of power between upstream and downstream players, decisions reached on the particular merits of a case, or highly technical points of law may have unintended consequences for the structure and performance of the entire industry.

The need for more research is clearly indicated—though given delays of several decades between performing basic science and measurable impacts on human health, ambiguous results on the impact of recent developments (such as the entry of the genomics companies) may take a very long time to accumulate. Thirty years after the dawn of the biotechnology industry, when called upon to assess the impact of changing industry structure on research productivity, the prudent response is almost certainly to quote Chou En-lai's response, when asked by Henry Kissinger to comment on the impact of the French Revolution on western civilization: "Too early to tell."

[9] Myriad Genetics' exclusive licensing of the BRCA1 gene is often claimed to restrict academic research. *See* Schissel, A, Merz, J, Cho, M., *Survey confirms fears about licensing of genetic tests*, NATURE 402:118, 1999.

21

Pharmacogenomics, Genetic Tests, and Patent-Based Incentives

Michael J. Meurer
Associate Professor of Law
Boston University School of Law
Boston, Massachusetts 02215

ABSTRACT

Pharmacogenomics promises to revolutionize medicine by using genetic information to guide drug therapy. Genetic tests should help doctors improve drug safety and efficacy by better matching patients and drugs. This chapter evaluates the effectiveness of patent-based incentives to create genetic tests, and

0065-2660/03 $35.00

considers the optimal mix of public and private sector pharmacogenomic research and development (R&D). Drug patent owners have a strong incentive to develop genetic tests that predict adverse drug reactions and allow them to market drugs that otherwise would be shelved. Incentives are also strong for genetic tests that are created as part of the drug development process. Incentives tend to be weaker for genetic tests that are used in conjunction with existing drugs. Drug patent owners might gain or lose profit from the introduction of genetic tests into existing drug markets. Profits may fall because of lost sales, or profits may rise because drugs are more valuable to appropriate patients, and because drugs become more differentiated.

Public sector R&D should target genetic tests that are likely to be underprovided by the private sector because private returns are low relative to social returns, or private costs are high relative to social costs. Private returns are relatively low when the rate of adoption of a genetic test is apt to be low, when test results increase consumer heterogeneity and consumer bargaining power, and when a test reveals information relevant to the use of more than one drug. Private costs are relatively high when test innovators need to obtain costly patent and trade secret licenses.

I. INTRODUCTION

The patent system, academic medical centers, and federal research funding come together to provide a complicated set of incentives to promote medical R&D. Pharmaceutical and medical device innovations are provided mostly by the private sector, and depend heavily on patent protection. Medical and surgical procedure innovations are provided mostly by the public sector and largely outside the domain of the patent system. This division of labor is not rigid. Private sector R&D often builds on basic research funded by the government and conducted in universities. Universities obtain patents, use private sector funds, and enter ventures with private companies for medical R&D. However, not much is known about the optimal mix of public and private incentives for encouraging medical R&D. This chapter takes a step in that direction by analyzing the effect of these incentives on medical R&D in the field of pharmacogenomics.

Pharmacogenomics applies genomics to drug therapy.[1] Medical researchers hail the progress of genomics research and proclaim its potential for

[1] Pharmacogenomics also plays a role in drug research and development. This chapter focuses mainly on the drug therapy application of pharmacogenomics. Some authors distinguish between pharmacogenetic and pharmacogenomic approaches to drug therapy; the distinction is not important for my purposes. *See* Allen Buchanan, Andrea Califano, Jeffrey Kahn, Elizabeth McPherson, John Robertson, & Baruch Brody, *Pharmacogenetics: Ethical Issues and Policy Options*, 12 KENNEDY INST. OF ETHICS J. 1, 1–2 (2002).

revolutionizing drug therapy. With luck, we will soon enjoy the benefits of customized drugs prescribed in optimal dosage with minimized side effects. This revolution rests on the belief that much of the variation in patient response to drugs is determined by measurable genetic variation in patients. New genetic tests and knowledge of the links between genes and drug response will make better drug therapy possible.

Thus far, public sector funding and R&D has been critical to the creation and deployment of pharmacogenomic innovations.[2] As the field matures, much innovative activity will probably shift to the private sector. In particular, drug patent owners are likely to generate most future innovation in this field. One reason is that pharmacogenomic research is likely to become a routine part of the drug discovery process, and the development of genetic tests will be ancillary to drug development. Genetic tests may also be able to resuscitate patented drugs that were abandoned because of problems with adverse reactions. Nevertheless, it may be desirable to continue public sector support of pharmacogenomics.

Patents do not always provide adequate private incentive to develop genetic tests. Genetic tests designed to improve therapy with existing drugs might be underprovided by the private sector. Private incentives may be too low because private returns are low relative to social returns, and because private costs are high relative to social costs. The main contribution of this chapter is an analysis of the private and social returns to pharmacogenomic innovation in markets for existing patented drugs. Introducing genetic tests could increase or decrease the profit from patented drugs used in conjunction with the tests. Similarly, social value could rise or fall after tests are introduced. I will try to identify factors that cause private returns to fall below social returns.

Social value generated in the market for a patented drug tends to rise with the introduction of a genetic test because of the social benefit from the better matching of patients and drugs. Social value is also influenced by the change in sales of the patented drug that is caused by testing. The introduction of testing can decrease sales because of a price increase that pushes untested patients out of the market. But testing can also increase sales, for example, when the test avoids harmful side effects and makes a drug newly available.

The introduction of a genetic test unambiguously increases profit from the sale of an associated patented drug in cases in which the drug cannot be marketed absent the test. This might occur because the drug has severe side effects for a subset of patients that cannot be identified without the test. This might also occur if the drug is efficacious for only a small subset of patients. Testing also increases profit in a market in which duopolists offer drugs that are imperfect substitutes. Testing raises profit to each firm through two different

[2] *Cf.* Antonio Regalado, *Mining the Genome*, 102 TECHNOLOGY REV. 56, 63 (1999) (the NIH recommends significantly increased support for computational biology in American universities).

effects. First, testing allows doctors to better match patients with one of the two drugs; this leads to higher demand for each of the drugs.[3] Second, testing differentiates the two drugs and gives more market power to each of the two firms.

The introduction of a genetic test might also decrease monopoly profit from the sale of a patented drug that is marketed even without the test. Profit might fall because certain patients learn they are not well matched with the drug. This effect may be offset if the drug maker is able to increase the price enough. Genetic tests can also hurt profits from the sale of drugs because the tests give private information to patients about their valuation for the drugs. Private information can increase consumer bargaining power and reduce seller profit.

In addition to direct threats to drug profit, the private sector incentive to invent genetic tests might be too weak for two other reasons. First, the use of more than one patented drug might be influenced by a single genetic test. In this case, a test that has a favorable effect on the profitability of each drug is a public good that is apt to be underprovided.[4] Second, the private costs of genetic test innovation are likely to be higher than the social costs because the private costs of licensing database access and rights to patented research tools are mere transfers that do not count as social costs.

The existence of independent diagnostic firms that compete with drug patent owners to develop genetic tests is a factor weighing against public sector R&D. Competition to get a patent is a potent stimulant of innovation that can more than offset problems appropriating the full social value of a genetic test. Independent inventors have an incentive to invent genetic tests based on the prospective profit from the sale or licensing of the tests, or from the assignment of the patent on the test to a drug patent owner.[5] Drug patent owners are likely to acquire such patents because drugs and genetic tests are complements marketed most efficiently by a single firm. Normally, drug patent owners have a strong and socially optimal incentive to push for broad adoption of genetic tests. But in some cases profit maximization requires suppression of a genetic test. Surprisingly, such suppression can be socially desirable.[6]

[3] This statement is true if the two drugs serve a roughly equal market share after testing. It is possible that the profits from one drug will fall if the genetic tests reveal that most patients are better matched with the other drug.

[4] This is true, even though the test and the affected drugs are all patented. Some tests will inherently reveal information relevant to more than one drug, and the test patent owner will not be able to selectively disclose that information.

[5] Independent inventors are likely to face obstacles when competing with drug patent owners when they control access to drugs and clinical data in a way that discourages independent inventors from developing a genetic test.

[6] Part III discusses the diffusion of genetic tests. There are a variety of factors outside the scope of my model that will affect the profit and social value from the use of genetic tests.

II. PHARMACOGENOMICS AND THE PROMISE OF CUSTOMIZED DRUG THERAPY

Genomics is the study of the function and structure of genes and gene products.[7] It has become a vital new research tool of life-science industries, one that promises more efficient drug research and development. This efficiency arises because genomics improves the understanding of the disease process,[8] helps identify drug targets,[9] guides drug design,[10] and ultimately reduces the cost and increases the success rate of R&D.[11] Genomics also has great potential as applied to drug therapy. This branch of genomics, called pharmacogenomics, is less developed than the application to drug discovery, but it may be just as significant.

A. Genomics and the Practice of Medicine

Pharmacogenomics promises tailored medicine. Future doctors are likely to routinely collect genetic information from patients who need medication so doctors can avoid causing adverse drug reactions and select the optimal medication and dosage.[12] Genetic differences between patients apparently explain why some patients, but not others, suffer from harmful side effects

[7] *See* US Dept. of Energy Human Genome Program, *Genomics and Its Impact on Science and Society: The Human Genome Project and Beyond*, March 2003, *available at* http://www.ornl.gov/TechResources/Human_Genome/glossary/glossary_p.html (last visited April 18, 2003).

[8] *See id.* at http://www.ornl.gov/TechResources/Human_Genome/publicat/primer2001/6.html

[9] *See* Allen D. Roses, *Pharmacogenetics and the Practice of Medicine*, 405 NATURE 857 (2000); Comment: *Panning For Biotechnology Gold: Reach-Through Royalty Damage Awards For Infringing Uses Of Patented Molecular Sieves*, 39 IDEA: J. L. & TECH. 553, 554 (1999) (describing high-throughput screening of drug candidates); Regalado, *supra* note 2, at 63 (gene expression profiles help predict toxic effects from a drug candidate, which allows researchers to exclude losers at an early stage).

[10] *See* US Dept. of Energy Human Genome Program, *supra* note 7, available at http://www.ornl.gov/TechResources/Human_Genome/publicat/primer2001/6.html

[11] *See, e.g.*, Gary A. Pulsinelli, *The Orphan Drug Act: What's Right With It*, 15 SANTA CLARA COMPUTER & HIGH TECH. L.J. 299, 304, 339 (1999) (Phamacogenomics will reduce the cost of research and development of the drugs because of the ability to pinpoint the population that would be affected by the drug.); James Kling, *Opportunities Abound in Pharmacogenomics*, THE SCIENTIST, (May 10, 1999) available at http://www.the-scientist.com/yr1999/may/prof_990510.html (last visited Mar. 27, 2002) (New hardware devices will cut drug development costs by providing genetic information at a lower cost and in less time.); Andrew Pollack, *DNA Chip May Help Usher in a New Era of Product Testing*, N.Y. TIMES, Nov. 28, 2000, available at www.nytimes.com/2000/11/ 28/science/28TOXI.html (Toxicogenomics aims to judge the toxicity of food additives, drugs, and cosmetics by using gene chips to measure the pattern of gene activity in response to exposure to a chemical. If this method is successful, it should reduce the cost of testing.); Arti Kaur Rai, *The Information Revolution Reaches Pharmaceuticals: Balancing Innovation Incentives, Cost, and Access in the Post-Genomics Era*, 2001 U. ILL. L. REV. 173, 189–90 (2001) (Genomics promises faster and cheaper preclinical drug research).

caused by certain drugs.[13] Research is under way to devise genetic tests that will, for example, help doctors avoid fatal side effects from an AIDS drug,[14] severe diarrhea from a chemotherapy drug,[15] and allergic reactions to penicillin.[16] Genetic testing could also soon be used to exclude certain patients from receiving a drug because the patients' metabolism renders the drug ineffective.[17] Someday, doctors are likely to test the genetic makeup of tumors, viruses, bacteria, and other pathogens[18] and to use the results to select appropriate

---

[12] *See* Roses, *supra* note 9; US Dept. of Energy Human Genome Program, *supra* note 7, available at www.ornl.gov/hgmis/medicine/pharma.html ("Pharmacogenomics is the study of how an individual's genetic heritance affects the body's response to drugs. Pharmacogenomics holds the promise that drugs might one day be tailor-made for individuals and adapted to each person's own genetic makeup. Environment, diet, age, lifestyle, and state of health all can influence a person's response to medicines, but understanding an individual's genetic makeup is thought to be the key to creating personalized drugs with greater efficacy and safety."). *But see* John Robertson, Baruch Brody, Allen Buchanan, Jeffrey Kahn, & Elizabeth McPherson, *Pharmacogenetic Challenges for the Health Care System*, 21 HEALTH AFFAIRS 155, 157 (2002) (not yet clear that genetic effects are significant and predictable enough to significantly change drug therapy).

[13] *See* Laviero Mancinelli, Maureen Cronin, & Wolfgang Sadee, *Pharmacogenomics: The Promise of Personalized Medicine*, AAPS PHARMACUETICA, (March 7, 2000) available at www.pharmsci.org/scientificjournals/ pharmsci/journal/4.html (last visited Mar. 27, 2002); *but see* Lars Noah, *The Coming Pharmacogenomics Revolution: Tailoring Drugs to Fit Patients' Genetic Profiles*, 43 JURIMETRICS J. 1, 10 (2002) (cautioning that not all adverse drug reactions will be prevented by pharmacogenomics).

[14] *See* Mark Schoofs, *AIDS-Drug Side Effect Linked to Genes*, WALL ST. J., Feb. 28, 2002, B2. (Scientists have shown that people with certain genetic patterns are vulnerable to potentially fatal side effects from a leading AIDS drug.)

[15] *See id.* (A company is developing a genetic test that will screen out patients who suffer severe diarrhea from a chemotherapy drug called irinotecan.)

[16] *See* Pollack, *supra* note 11 (Scientists have found 260 genes that are differentially activated in people allergic to penicillin compared to people who are not.)

[17] *See* John Weinstein, *Pharmacogenomics—Teaching Old Drugs New Tricks*, 343 NEW ENGL. J. MED. 1408 (2000) (SNPs affecting drug metabolism explain why certain drugs are not effective in treating a subset of patients); David Stipp, *Blessing From the Book of Life*, 141 FORTUNE F-21, 24, Mar. 6, 2000 (a SNP chip will be able to determine that a medication will not work on a particular patient whose liver enzymes break down the medication too quickly).

[18] *See* Regalado, *supra* note 2, at 62 (scientists have distinguished two types of leukemia solely by comparing the gene expression profiles for cells affected by the two types of leukemia); Stipp, *supra* note 16, at F-21, 24 (gene chips will be able to distinguish different types of T-cell cancers); Andrew Pollack, *Telling the Threatening Tumors from the Harmless Ones*, N.Y. TIMES, Apr. 9, 2002, available at http://www.nytimes.com/2002/04/09/science/09GENE.html (last visited Apr. 9, 2002) (the National Cancer Institute and Merck are developing a test based on a genetic fingerprint that distinguishes breast cancer that requires chemotherapy from breast cancer that does not). "Children with acute lymphoblastic leukemia, for instance, already undergo several separate tests costing about $1,000 to help subclassify the cancer to help determine therapy, he said. But in a newly published study involving 327 patient samples, one of the largest DNA chip studies to date, Dr. Downing and colleagues showed that the genetic patterns could classify the cancers as accurately as all the other tests combined, if not more so." *Id.* "Millennium Pharmaceuticals . . . has found that if a particular gene in a melanoma skin tumor is inactive, the tumor is more likely to spread throughout the body." *Id.*

medications.[19] Finally, future doctors are likely to use genetic information about patients' metabolism to tailor the appropriate dosage and schedule for administering drugs.[20]

B. Genetic Testing Technology

Genetic testing is likely to be deployed in three different ways to facilitate the practice of pharmacogenomics. First, preventive screening will identify genes that cause or have an association with a disease.[21] Ideally, preventive screening will be coupled with preventive treatment of patients facing elevated risk of disease.[22] Second, general purpose testing will produce information relating to drug metabolism and potential adverse drug reactions.[23] Third, specific genetic testing of diseased cells and pathogens will improve the precision of tailored drug therapy.[24]

[19] *See* Andrew Pollack, *When Gene Sequencing Becomes a Fact of Life*, N.Y. TIMES, Jan. 17, 2001, *available at* www.nytimes.com/2001/01/17/business/17AIDS.html (sequencing HIV can help a doctor choose from among 15 drugs available to fight AIDS).

[20] *See* U.S. Dept. of Energy Human Genome Program, *supra* note 7.

[21] Patients can be screened for particular mutations like BRCA1 and BRCA2, and for markers called SNPs that are linked with genes that are implicated in some disease. *See* Allen C. Nunnally, *Commercialized Genetic Testing: The Role of Corporate Biotechnology in the New Genetic Age*, 8 J. SCI. & TECH. L. 1, 12 (2001); US Dept. of Energy Human Genome Program, *supra* note 7, available at http://www.ornl.gov/TechResources/Human_Genome/publicat/primer2001/6.html (several hundred genetic tests are currently in use mainly to detect rare genetic disorders).

[22] *But see* Gina Kolata, *Test Proves Fruitless, Fueling New Debate on Cancer Screening*, N.Y. TIMES, April 9, 2002, available at http://www.nytimes.com/2002/04/09/science/09CANC.html (last visited Apr. 9, 2002) (describing the debate among oncologists about the value of cancer screening tests and the problem of false positives).

[23] *See* Lori B. Andrews, Maxwell J. Mehlman, & Mark A. Rothstein, GENETICS: ETHICS, LAW AND POLICY 423 (2002) (genetic test that helps determine how patients metabolize a variety of drugs); National Institute of General Medical Sciences, *Your Genes and Your Medicines*, available at http://www.nigms.nih.gov/funding/htm/yrgenes2.html (last visited Mar. 26, 2002). ("In most cases, research to find normal variations in the genes for the proteins that handle medicines in the body will involve a simple test. In many cases, researchers will rub the inside of a volunteer's cheek with a cotton swab and then examine the DNA in those cheek cells."); Arti K. Rai, *Fostering Cumulative Innovation in the Biopharmaceutical Industry: The Role of Patents and Antitrust*, 16 BERKELEY TECH. L.J. 813, 842 (2001) (SNP gene chips will be used to predict adverse drug reactions). Kathleen Giacomini, PhD, professor, and chair of biopharmaceutical sciences in UCSF's School of Pharmacy predicts: "[I]n the future, you may go in for a doctor's visit and have your blood drawn for a genotype to be done which would indicate what genes you have for drug transporters, drug targets, or drug elimination enzymes. Then after you are diagnosed, a pharmacist could interpret the panels of genetic results and advise on which drugs would be 'best' for your particular genes." *What Is Pharmacogenomics?* IMPACT: THE UCSF FOUNDATION'S ONLINE MAGAZINE, *available at* http://www.ucsf.edu/foundation/impact/archives/1999/17 pharmacog.html (last visited Mar. 26, 2002).

[24] *See supra* note 18.

Genetic testing technology captures two different kinds of data: gene sequence data and gene expression data. Gene sequence testing reveals selective information about the DNA of a tested patient or the DNA of a pathogen. For example, Myriad Genetics makes a test of human DNA that detects the presence of mutated genes that cause a form of inherited breast cancer.[25] Visible Genetics developed a sequencing test of the HIV virus that helps tailor prescription of AIDS drugs.[26] Companies are working to develop a low-cost method capable of sequencing a patient's entire genome.[27] Such tests are necessary before pharmacogenomics can achieve its full potential.

Besides gene sequence data, doctors also need to know which genes are active in particular cells. This information is obtained by observing the messenger RNA (mRNA) produced in a cell. The mRNA is an essential intermediary in the production of a protein in a cell. Biologists say that a gene is expressed when the cell produces the protein that the gene encodes. A newly developed device that marries microchip technology with molecular biology produces gene chips that monitor gene expression. These gene chips work by detecting all of the mRNA present in a cell. This information gives a snapshot of which genes are being expressed and in what quantity.[28] The data from the gene chip is called a gene expression profile. This profile has therapeutic value when it is correlated with optimum drug therapy.[29]

Before routine pharmacogenomic therapy becomes a reality, many factors must fall into place.[30] Doctors and patients must have access to kits

[25] *See* Rita Rubin, *To Test, or Not to Test, for Breast Cancer Genes*, USA TODAY.COM (Jan. 10, 2002) *available at* http://www.usatoday.com/news/health/cancer/2002-01-09-usat-breastcancer-gene.htm (last visited Apr. 17, 2003).

[26] *See* Chael Needle, *Gene Genie: Visible Genetics Makes Good on the Wish for Better Resistance Testing Technology*, A&U AMERICA'S AIDS MAGAZINE (Apr. 17, 2003) *available at* http://www.aumag.org/www_aumag_org/archives/archives_contents.cfm?a_id=3730&c_id=86 (last visited April 17, 2003).

[27] *See* David Orenstein, *A Genetic Hole in One: David Deamer's "Nanopore" Device Might Eventually Allow Doctors to Decode Your DNA — While You Wait*, Business2.0 (October 2001) *available at* http://www.business2.com/articles/mag/0,1640,16900,FF.html (last visited April 17, 2003).

[28] *See* Regalado, *supra* note 2, at 59–62 (DNA chips identify a gene expression profile for a cell, the gene chips recognize the type and quantity of mRNA in a cell to get a snapshot of what proteins are being expressed in a cell).

[29] Predicting that in the future "each drug [will] be bundled with a specific set of diagnostic tests for those positions in the human genome which alter drug response." Charles R. Cantor, 1 GENELETTER 3 April 2000 *available at* http://www.geneletter.org/04-01-00/features/pharmacogenomics.html (last visited Mar. 26, 2002).

[30] *See Personal Pills: Genetic Differences May Dictate How Drugs Are Prescribed*, SCIENTIFIC AMERICAN, available at http://www.sciam.com/1998/1098issue/1098infocus.html (last visited Mar. 26, 2002) (William Haseltine, the CEO of Human Genome Sciences, warns that diagnostic tests can be unreliable in part because environmental factors have a significant influence on drug behavior).

and equipment to sequence genes or produce gene expression profiles.[31] Several companies have shown an interest in producing kits and equipment,[32] but it is not yet clear how willing health care providers are to pay for these items.[33] Equally important, doctors and diagnostics manufacturers must have access to theories and data that yield improved drug therapies.[34] The public and private sectors have both started producing the data and compiling the databases that are required for genomic research.[35] As genomics plays a bigger role in drug research, theories about disease mechanisms and the links between genes and diseases will frequently emerge during drug R&D.[36] Theories linking genetic data to drug therapy will also be produced by academic researchers and diagnostic companies interested in making and selling genetic tests.[37]

III. THE IMPACT OF GENETIC TESTS ON PROFIT AND SOCIAL WELFARE IN THE MARKET FOR EXISTING DRUGS

A. The Basic Model

The market for prescription drugs is difficult to model because of the complicated relationship between the patient who demands the drug and the pharmaceutical company that sells the drug. Typical patients have relatively little to say about the variety or quantity of drug they will purchase. Instead, that choice is made by a doctor acting on behalf of the patient facing possible constraints imposed by the party who pays for the drug, the pharmacist, and government

[31] *See* Andrew Pollack, *When Gene Sequencing Becomes a Fact of Life*, N.Y. TIMES, Jan. 17, 2001, available at www.nytimes.com/2001/01/17/business/17AIDS.html (the company Visible Genetics plans to give away small sequencing machines to clinical laboratories, hospitals, and doctors' offices, and make a profit from the sale of kits required to do a test).

[32] *See* Andrews, *et al.*, *supra* note 23, at 423 (listing three gene tests that will soon be available).

[33] *See infra* notes 41 and Part IV.A.

[34] Some observers are skeptical about rapid progress in the application of genomics to medicine. *See, e.g.*, Leslie Roberts, *SNP Mappers Confront Reality and Find It Daunting*, 287 SCIENCE 1898 (2000); Neil A. Holtzman & Theresa M. Marteau, *Will Genetics Revolutionize Medicine?* 343 NEW ENG. J. MED. 141 (2000).

[35] *See* Arti Kaur Rai, *Regulating Scientific Research: Intellectual Property Rights and the Norms of Science*, 94 NW. U. L. REV. 77, 146 (1999) (American universities have reached a consensus that SNPs data should be dedicated to the public domain through a public database); Scientific American, *supra* note 30 (The firm Genset has created a private database of 60,000 SNPs that mark genes that cause disease or differing drug reactions).

[36] The drug company Abbot will use the Genset SNP database during clinical trials to identify patients who do not respond to tested drugs. They will then turn these results into a diagnostic test to screen out patients who do not respond. *See* Scientific American, *supra* note 30. Drug companies monitor the effect of approved drugs on patients; in the future, this monitoring will produce information that aids tailored drug therapy. *See* Roses *supra* note 9.

regulations.[38] Despite these complications, economists have had some success analyzing this market in a traditional framework.[39] To keep my model simple, I assume that a doctor will prescribe drugs in a way that maximizes the payoff to the patient.

In the basic model, a patented drug is given to a patient population; it is effective for half of the population and has no effect on the other half. Suppose that a genetic test is available that can identify whether a particular patient will gain a benefit from use of the drug. Suppose that a fraction α of the population is given the test. One-half of those tested will be a "good match" and gain a benefit V from use of the drug. The other half will be a "bad match" and gain no benefit from the drug.[40] The remaining fraction $1-\alpha$ of the population will be uninformed about how well they match with the drug and gain an expected benefit of $V/2$ from use of the drug. The following table summarizes the information about patient types, probabilities, and valuations.

Suppose that the drug-maker sets a uniform price for the drug. The seller knows the information in the Table 21.1, but does not know whether a particular patient is uninformed, a good match, or a bad match. The seller incurs a constant marginal cost of C from the production and distribution of the drug where $C < V/2$. The profit maximizing price depends on the value of α. If no patients are informed, then all patients have a valuation of $V/2$ and the monopoly price is $V/2$. If all the patients are informed, then only good matches will purchase, and the monopoly price is V.[41] If there is a mix of informed and uninformed patients, then the monopoly price is $V/2$ if the fraction informed is sufficiently small, and the monopoly price rises to V when the fraction informed is sufficiently large.

---

[37] *See* Rob James, *Differentiating Genomics Companies*, 18 NATURE BIOTECHNOLOGY 153, 155 (2000) (The company Gene Logic has developed a database of gene expression profiles for normal cells that can be compared to the gene expression profiles for diseased cells, and cells exposed to toxic substances.); ASSOCIATED PRESS, MIAMI HERALD, Mar. 25, 2002, available at http://www.miami.com/mld/miami/business/2932460.htm (last visited Mar. 27, 2002) (IBM and the Mayo Clinic plan to create a medical database including 4 million patient records and genetic data from the patients.)

[38] *See* Sarah Ellison, Iain Cockburn, Zvi Griliches, and Jerry Hausman, *Characteristics of Demand for Pharmaceutical Products: An Examination of Four Cephalosporins*, 28 RAND J. ECON. 426 (1997).

[39] *See, e.g., id.* (showing the demand for drugs is sensitive to price by presenting evidence of high cross-price elasticity in demand for brand-name drugs with generic substitutes, and significant elasticities for therapeutic substitutes).

[40] After the introduction pharmacogenomics, drugs that are currently prescribed to almost all patients will be prescribed only to those patients with a genetic profile indicating drug effectiveness. *See* Pulsinelli, *supra* note 11, at 339.

[41] *See* Noah, *supra* note 13, at 18 (making similar observation). *But see* Patricia Danzon & Adrian Towse, *The Economics of Gene Therapy and of Pharmacogenetics*, 5 VALUE IN HEALTH 5, 9 (2002) (payers are likely to bargain hard to contain drug price increases linked to genetic tests).

Table 21.1.

	Uninformed	Good Match	Bad Match
Probability	$1-\alpha$	$\alpha/2$	$\alpha/2$
Valuation	V/2	V	0

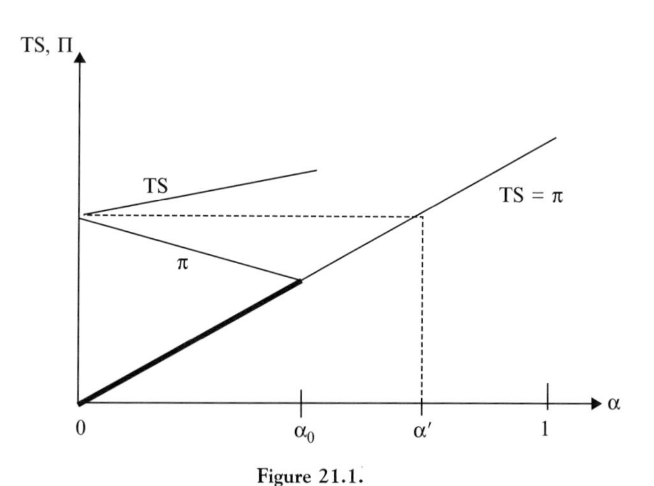

Figure 21.1.

The profit (π) to the seller from setting a price of *V/2* is:

$$\pi = (1 - \alpha/2)\,(V/2 - C),$$

and the profit to the seller from setting a price of *V* is:

$$\pi = \alpha/2(V - C).$$

The optimal price is *V/2* when $\alpha \leq \alpha_0$, and *V* otherwise, where:

$$\alpha_0 = \frac{2V - 4C}{3V - 4C}.$$

Figure 21.1 displays profit and total surplus (TS) using solid lines (the bold line is relevant to the version of the model analyzed in the next section). The figure shows that the profit to the seller falls as the fraction of the informed grows as long as the fraction informed, α, is less than α_0. Profit grows with the fraction that is informed when α is sufficiently large (*i.e.*, greater than α_0).

The effect of the information from genetic testing on profits is easy to understand. Better information has a negative effect on profit because the bad matches drop out of the market.[42] Better information also has a positive effect on profit; the good matches have a higher valuation than the uninformed patients do, and when there are enough good matches, the seller raises the price. Figure 21.1 shows that profit falls for $\alpha \leq \alpha_0$ because the price is fixed at V/2, and the only effect of genetic testing is that sales are lost to bad matches. For $\alpha \geq \alpha_0$, profit rises with α because only good matches buy the drug at the price of V, and increasing genetic testing increases the fraction of patients who know they are good matches. The fraction $\alpha' \equiv (V - 2C)/(V - C)$ shown in Fig. 21.1 indicates the positive fraction of informed patients that yields the same level of profit achieved when no patients are informed. When $\alpha > \alpha'$, genetic testing increases profit because the positive influence of the price increase is greater than the negative influence of lost sales.

Economists measure the social value derived from the drug market by adding consumer surplus to profit; this sum is called total surplus. Figure 1 displays the total surplus as a function of the fraction of patients who get genetic testing. The total surplus in the drug market is easy to calculate. Consumer surplus is zero when the price equals V, because only good matches buy the drug, and the price equals their valuations. Therefore, total surplus equals profit when $\alpha \geq \alpha_0$. For smaller values of α, there is positive consumer surplus. Uninformed patients do not get any consumer surplus at a price of V/2 because that price equals their valuation. Bad matches do not get any surplus because they do not buy the drug. But good matches get consumer surplus equal to $V - V/2$. Thus, the expected consumer surplus when $\alpha \leq \alpha_0$ is $\alpha/2\ (V/2) = \alpha V/4$. Total surplus when $\alpha \leq \alpha_0$ is:

$$TS = V/2 - (1 - \alpha/2)\,C.$$

Total surplus increases with the fraction of informed patients, except for a discontinuous drop at $\alpha = \alpha_0$. Total surplus drops at the point where the seller switches from a relatively low price of V/2 to the higher price of V. The drop is large enough that total surplus in the interval (α_0, α') is lower than total surplus when $\alpha = 0$. The availability of genetic testing has an obvious positive effect on total surplus; testing leads to better matching of patients with drug treatment. The social gain from testing is offset by the seller's response to testing—a price increase. The price increase excludes the uninformed patients from the market and causes a social loss. Either the positive or negative effect may predominate so total surplus may either rise or fall after genetic testing is introduced.

The surprising results from the previous analysis are summarized in the following proposition.

[42] The negative effect of lost sales on profit is mitigated because the seller avoids the cost of making the drug for bad matches.

Table 21.2.

	Uniformed	Good Match	Bad Match
Probability	$1-\alpha$	$\alpha\theta$	$\alpha(1-\theta)$
Valuation	θV	V	0

Proposition 1. The introduction of genetic testing reduces profit on the interval $(0, \alpha')$, and it reduces total surplus on the interval (α_0, α').

Proposition 1 is helpful in thinking about the incentive effects of pharmacogenomics. A drug-maker in a marker similar to the one in the model would oppose the introduction of genetic testing unless the testing reaches a relatively large fraction of the market. Intuitively, genetic testing causes profit to fall because of lost sales. Profit only recovers to the level without testing when there are enough patients who have been informed that they are a good match and so have a high valuation that leads to a price increase. A social planner might also oppose the introduction of genetic testing when the fraction of informed patients falls in the interval (α_0, α'). Total surplus in this interval is low because the output restriction caused by monopoly pricing is substantial and more than offsets the benefit from better matching.[43]

B. The Information Content of the Test

The basic model assumed that the drug worked in exactly half of the patient population. This subsection generalizes the basic model by assuming that some fraction θ benefits from the drug and the remaining $1-\theta$ get no benefit. Table 21.2 displays the probabilities and valuations associated with the three types of patients. Informed patients who are a good match get a benefit of V, uninformed patients get an expected benefit of θV, and bad matches get a benefit of 0. The fraction of uninformed patients is $1-\alpha$, the fraction of good matches is $\alpha\theta$, and the fraction of bad matches is $\alpha(1-\theta)$.

If $\theta V \geq C$, then the basic model of the previous section applies, and the solid lines in Fig. 1 show profit and total surplus.[44] Profit initially falls in α, and grows when $\alpha \geq \alpha_0$. Total surplus equals profit when $\alpha \geq \alpha_0$, but diverges from

[43] The socially optimal policy is difficult to evaluate when α takes on other values. When $\alpha \leq \alpha_0$, introducing the genetic test raises *ex post* total surplus, but reduces drug profit and might inefficiently diminish the incentive to invent the patented drug. When $\alpha > \alpha'$, the social planner might oppose the genetic test even though it raises total surplus because consumer surplus falls to zero, or because the incentive to innovate is excessive.

[44] Price equals θV and profit equals $(1 - \alpha + \alpha\theta)(\theta V - C)$ if $\alpha \leq \alpha_\theta$, and price equals V and profit equals $\alpha\theta (V - C)$ if $\alpha \geq \alpha_\theta$, where α_θ is defined to be $[\theta V - C]/[(2 - \theta)\theta V - C]$.

profit when the fraction informed is relatively small, *i.e.*, $\alpha \leq \alpha_0$. Now, suppose that a relatively small fraction of patients gain a benefit from the drug, specifically, $\theta V < C < V$. In this case, the drug-maker will not sell the drug if there is no genetic testing because the expected value of the drug to an uninformed patient is less than the marginal cost of the drug.[45] Thus, testing is crucial to the sale of the drug. Given testing, the drug-maker will set the price at V and sell to the good matches. The seller captures all consumer surplus so profit and total surplus are equal. The expression for profit is:

$$\pi = \alpha\theta(V - C).$$

The bold line in Fig. 21.1 and the solid continuation of that line represents both profit and total surplus in this case.

An interesting policy issue arises from the relationship between the informational value of a test and the magnitude of θ, the fraction of patients benefiting from the test. The policy issue is captured by the question: How should the magnitude of θ affect public sector support for genetic test R&D? One plausible response is that public support should target drugs that can be used with a genetic test and cannot be used without a test, *i.e.*, drugs such that $\theta V < C < V$. I think that response is mistaken. A better response considers two factors: Whether the private sector has adequate incentive to develop the test; and the informational value of the test. Intuitively, a genetic test has little informational value if θ is close to zero, or if θ is close to one. If a drug is almost never effective for anyone, then there is not much information revealed when a test is performed that indicates the drug will not work for a specific patient. Similarly, not much information is revealed when the drug is almost always effective, and a test shows that it is effective for a specific patient. Rough intuition might suggest that the test is most informative when $\theta = 1/2$. Actually,

[45] A similar problem arises when side effects rather than efficacy are the issue. Some drugs are effective but cause severe side effects in a small subset of the population. Such drugs cannot be used unless doctors have a way to screen out patients who might suffer the side effects. Work is under way on a genetic test that can screen out patients that could suffer fatal side effects from an AIDS drug. *See* Schoofs, *supra* note 14; Jochen Duelli and Ashish Singh, *Tailoring Drugs to Patient Gains Ground*, BOSTON GLOBE, G5, June 30, 2002 (the drug Lotronex was linked to severe intestinal problems and some deaths, but the FDA has allowed Glaxo-SmithKline to reintroduce the drug with patient restrictions based on pharmacogenomic information) (the cancer drug Herceptin is designed exclusively for patients with multiple copies of the HER-2/neu gene) (Glaxo is developing a genetic test that will identify the 5% of patients who could suffer potentially fatal side effects from its AIDS drug Ziagen); Melody Petersen, *Whistle-Blower Says Marketers Broke the Rules to Push a Drug*, N.Y. TIMES, March 14, 2002, available at http://www.nytimes.com/2002/03/14/business/14DRUG.html (last visited March 14, 2002) (a whistle-blower claims that Warner-Lambert marketed the drug Neurontin to doctors for more than a dozen unapproved conditions) (doctors are allowed to prescribe medicines for uses not approved by the FDA, but drug-makers are not allowed to promote unapproved uses) (in 2000 78% of the prescriptions for Neurontin were for off-label uses).

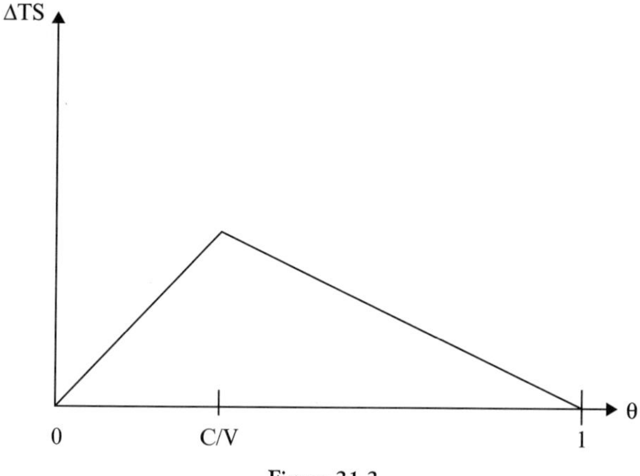

Figure 21.2.

it turns out the social value of a test is highest when $\theta = C/V$, a fraction that is likely to be much smaller than one-half.

One measure of the social value of the test is the difference between total surplus when every patient is tested and the total surplus when no patient is tested.[46] That difference is graphed in Fig. 21.2 as a function of θ. The expression for the social value is $\theta(V - C)$ when $\theta < C/V$. This is simply the expected value of the drug when it is administered to every patient who can benefit from it. The expression for social value is $(1 - \theta)\,C$ when $\theta \geq C/V$. This is simply the expected cost saving gained when the drug is not given to patients who cannot benefit from it.[47] The two expressions differ because in the first case, when there is no testing, the drug is not marketed, but in the second case, when there is no testing, the drug is sold to everyone. The social value grows with the probability that the drug is effective for small values of θ, and it falls in relation to that probability for large values of θ.

The result displayed in Fig. 21.2 supports a policy that targets genetic tests designed for drugs that work for a fraction of the patient population $\theta \approx C/V$, not a policy that targets tests with $\theta < C/V$.[48] Thus, policymakers should not target drugs simply because they can only be used with a test. Instead, they

[46] A more complicated measure is required when I relax the assumptions that the genetic test is costless and will be used by the entire patient population.

[47] Generally, a social benefit occurs if the marginal cost of drug therapy is greater than the marginal cost of the genetic test.

[48] Implementing this policy depends on the government's ability to measure θ and the other relevant variables. I suspect a good estimate of θ is possible given data on the fraction of patients that respond to a drug.

Table 21.3.

	Uninformed	Good Match	Bad Match
Probability	$1 - \alpha$	$\alpha/2$	$\alpha/2$
Valuation	$(V + W)/2$	V	W

should target drugs such that the genetic test yields the greatest social value. Figure 21.2 is helpful in identifying such tests, but it does not capture all the relevant factors. Obviously, the size of the patient population matters as much as the fraction of the population that can be helped. Further, the analysis should incorporate distributional concerns about providing treatment for small patient groups. The distributional concerns are well articulated in the analogous context of orphan drug policy.[49] Finally, government intervention is not warranted within the framework of this model unless the private incentive to develop genetic tests falls short of the socially optimal incentive. I will defer discussion of that issue until Section IV.B.

C. Tests with No Medical Value

In my second refinement of the basic model, I suppose that every patient gains a benefit from the use of the patented drug, but some patients are a good match and they get a large benefit V, and other patients are a bad match and they get a smaller benefit $W > C$. The valuations and probabilities of each type of patient are displayed in Table 21.3.

A test that informs patients whether they are a good or bad match has no "medical value" in this case, in the sense that it is socially desirable for every patient to take the drug.[50] But a test has economic value to certain patients because it gives them private information that allows them to extract surplus

[49] Commentators have raised the related question of whether the orphan drug statute should be used to promote drug discovery to serve "orphan genotypes." *See* Mark A. Rothstein & Phyllis Griffin Epps, *Ethical and Legal Implications of Pharmacogenomics*, 2 NATURE REVIEWS GENETICS 228 (2001) (questioning whether orphan drug statutes should cover drugs intended to treat uncommon genotypes); Arti K. Rai, *Pharmacogenetic Interventions, Distributive Justice, and Orphan Drugs: The Role of Cost-Benefit Analysis*, 19 SOCIAL PHILOSOPHY & POLICY (2002) (discussing the role of genomic data in segregating patient populations in smaller disease categories, thereby increasing the problem of sustaining research directed toward diseases affecting rare genotypes); Noah, *supra* note 13, at 17 (discussing orphan genotypes).

[50] Andrew Pollack, *supra*, note 18 ("In some cases, there are no treatments available. So genetic fingerprinting may merely tell patients how quickly they can expect to die, without allowing doctors to do anything about it.")

from the seller. The drug patent owner has an incentive to block the development of this sort of test.[51]

There are three different profit-maximizing prices in this case, depending on what fraction of patients are informed. If the fraction of informed patients is relatively small, *i.e.*, $\alpha \leq \alpha_1$, then the seller maximizes profit by choosing a price of $(V + W)/2$ and selling to good matches and uninformed patients. Profit is:

$$\pi = (1 - \alpha/2)\,[(V + W)/2 - C].$$

If the fraction of informed patients is in an intermediate range so that $\alpha_1 \leq \alpha \leq \alpha_2$, then the seller maximizes profit by choosing a price of W and selling to all patients. Profit is:

$$\pi = W - C.$$

If the fraction of informed patients is relatively large, *i.e.*, $\alpha_2 \leq \alpha \leq 1$, then the seller maximizes profit by choosing a price of V and selling only to good matches. Profit is:

$$\pi = \alpha/2(V - C).$$

The critical values of α are given by the following expressions:

$$\alpha_1 = \frac{2(V - W)}{V + W - 2C}, \quad \alpha_2 = \frac{2(W - C)}{V - C}.$$

I assume that $\alpha_2 < 1$, *i.e.*, $V - W < W - C$, otherwise setting a price of V is never optimal.

Figure 21.3 displays profit and total surplus as a function of the fraction of patients who are informed. Notice that profit is maximized when none of the patients is informed. Profit falls for small values of α because bad matches drop out of the market and the price is fixed at $(V + W)/2$. For intermediate values of α, profit is not sensitive to the fraction of patients who are informed because every patient purchases regardless of test results when the price is W. For large values of α, profit rises as more patients are informed because only good matches purchase at the price of V.

Total surplus maximization, in this case, simply requires that all patients buy and use the drug. Total surplus is maximized when no patients are informed and the price is set at a level such that they all purchase.[52] Total surplus is also maximized when an intermediate fraction of the patients are informed, $\alpha_1 \leq \alpha \leq \alpha_2$, and the seller chooses a price of W and sells to all patients. Consumer surplus is maximized for intermediate values of α, because

[51] *Cf.* Robertson *et al.*, *supra* note 12, at 75 (drug patent owner may block development of genetic test).

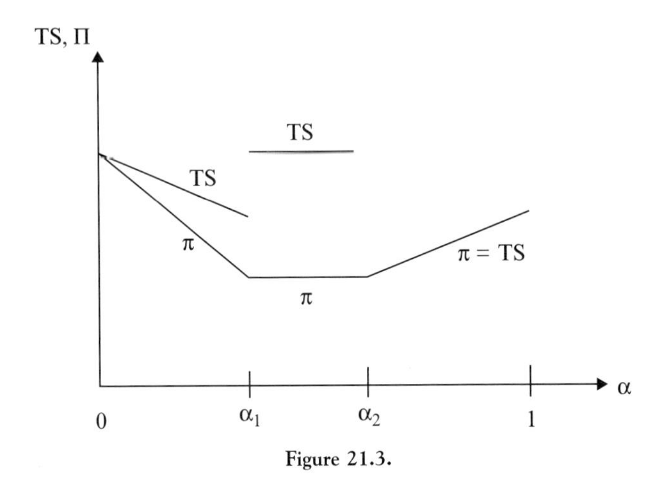

Figure 21.3.

all patients buy the drug and the price is lower than for other values of α. In contrast, consumer surplus is zero when no patients are informed. These observations reveal the antagonistic interests of patients and the drug patent owner in this case. The drug patent owner wants to maximize the size of the pie and take all of that pie by suppressing the genetic test. Patients might prefer to absorb the cost of developing and deploying the therapeutically unnecessary test to increasing their bargaining power against the drug seller.

D. Differentiated Drugs in a Duopoly Market

The model in this section features two drugs available to treat a certain population of patients. One drug, labeled A, is more effective or safer for half of the population, and the other drug B is better for the other half of the population. A test is available that will inform patients whether they are better matched with drug A or B.[53] Suppose that a good match bestows a benefit V,

[52] When price $P = (V + W)/2$, consumer surplus is derived by good matches. Each good match gets utility of $V - P = (V - W)/2$, so the expected consumer surplus is $(\alpha/2)(V - W)/2$, and total surplus is just the sum of consumer surplus and profit. When $P = W$, consumer surplus is derived by good matches and uninformed patients. Each good match gets utility of $V - P = V - W$. Each uninformed patient gets utility of $(V + W)/2 - P = (V - W)/2$. Expected consumer surplus is $(\alpha/2)(V - W) + (1 - \alpha)(V - W)/2 = (V - W)/2$. Since profit $\pi = W - C$, total surplus equals $(V + W)/2 - C$.

[53] For a possible example *see* Alison Davis, *News Release, First Awards Made in NIH Effort to Understand. How Genes Affect People's Responses to Medicines*, April 4, 2000, *available at* http://www.nigms.nih.gov/news/releases/pharmacogenetics.html (last visited Mar. 26, 2002) (an NIH grant is funding research to discover which genes affect widely variable responses to the three main types of asthma drugs).

Table 21.4.

	Uninformed	Good Match A	Good Match B
Probability	$1-\alpha$	$\alpha/2$	$\alpha/2$
Valuation of A	V/2	0	V
Valuation of B	V/2	0	V

and a bad match gives no benefit. The valuations and probabilities of each type of patient are displayed in Table 21.4.

I will analyze the effect of a genetic test on prices, profits, and total surplus in a duopoly market for drugs A and B. Suppose that one firm has a patent on drug A and another firm has a patent on drug B. Suppose the firms simultaneously set the price for the two drugs given that some fraction α of the patients will take the genetic test. Uninformed patients get equal expected value from drugs A and B. Informed patients strongly prefer the drug that is a good match for them. An article by Michael Meurer and Dale O. Stahl characterizes the Nash equilibrium for this problem.[54]

Proposition 2. If the fraction of informed patients is large enough, *i.e.*, $\alpha \geq \alpha_0$, then both sellers choose a price $P_A = P_B = V$, and profit for each firm is:

$$\pi = (\alpha/2)(V - C).$$

If the fraction of informed patients is lower, $\alpha \leq \alpha_0$, then the sellers choose prices from a symmetric mixed strategy equilibrium, and the same expression represents expected profit.[55]

Intuitively, when a large fraction of patients is informed, each of the firms sells only to patients who are a good match at the price of V. Neither firm is interested in cutting its price to $V/2$ to sell to the uninformed patients. Such a price cut would increase sales, but not enough to offset the lost revenue from the good matches. In contrast, when the fraction of informed patients is lower, the two firms waver between targeting just good matches and trying to capture the uninformed patients. The mixed strategy equilibrium requires that each firm sets the price at V with the same positive probability, and otherwise sets a price somewhere in an interval less than or equal to $V/2$. Each firm introduces a random element into its pricing decision to avoid being easily undercut by its rival.

Figure 21.4 displays the profit of a duopolist and the total surplus as a function of the fraction of patients tested. Figures 21.1 and 21.4 both show that

[54] *See* Michael Meurer & Dale O. Stahl, *Informative Advertising and Product Match*, 12 INT'L J. INDUSTRIAL ORG. 1 (1994).

[55] Proof of results stated in this section and Fig. 21.3 are contained in Meurer & Stahl. *See id.*

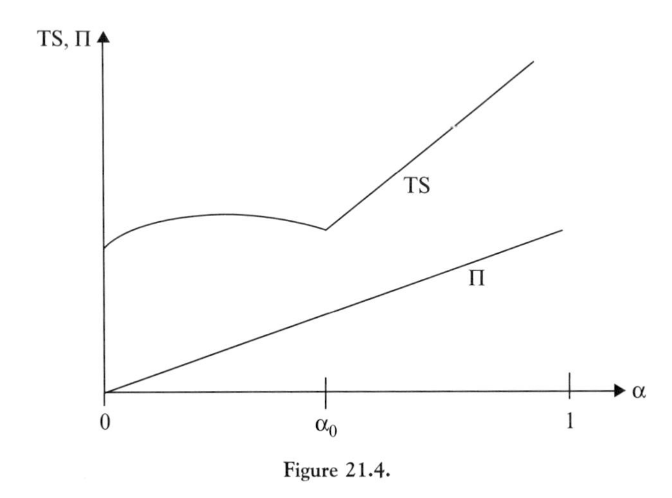

Figure 21.4.

total surplus declines over an intermediate range of values of α. The same explanation applies to the monopoly and duopoly markets. As the fraction of untested patients declines, prices rise and the untested patients are forced out of the market. For other values of α, total surplus grows because increased testing leads to better matching of patients and drugs.

The private incentive to develop genetic tests is probably weaker in the duopoly market depicted in this section than the monopoly market in Section III.A.[56] Testing has two favorable effects on the profit of the duopolists. First, it increases sales by increasing the number of good matches when the fraction tested is high. Second, it decreases competition for untested patients when the fraction tested is low. But the private incentive to develop tests is diminished by a classic free-riding problem. The value of the test flows equally to the two firms in this model, and either firm would be tempted to delay test development in hopes that its rival will provide the public good. A patent on the genetic test is not likely to solve the free-riding problem. To see why, suppose firm A develops and patents a test. Firm A cannot credibly commit to a policy that prohibits the use of the test in conjunction with drug B, and at the same time, authorize use of the test in conjunction with drug A. A physician that performs the test and finds out that a patient is a good match with B and a bad match with A cannot deliberately forget the information, and cannot ethically ignore it. Thus, the

[56] An interesting topic for future research is the incentive to develop genetic tests for use in markets with unpatented drugs. I conjecture that the test developer could capture a larger share of the surplus generated by the test when the drug market is relatively competitive.

free-riding problem is likely to cause underinvestment in genetic tests used to choose among substitute patented drugs.[57]

IV. INCENTIVES TO CONDUCT AND DEVELOP GENETIC TESTS

A. The Incentive to Conduct Genetic Testing

Section III reveals that the rate of adoption of genetic tests has subtle and significant effects on profit in complementary drug markets. In some markets, the rate of adoption will be 100%,[58] but in other markets, adoption will be slowed by many factors. Tort law, cost-containment concerns of health care purchasers, and norms of good medical practice will encourage testing.[59] Optimists expect that physicians will rapidly embrace pharmacogenomics and it will soon be incorporated into the standard diagnostic repertoire. Laggards would be encouraged to embrace genetic testing for fear of a malpractice claim.[60] In a promising example, Patricia Danzon and Adrian Towse document significant savings to health care payers from the adoption of a genetic test used in conjunction with drug treatment of a particular form of breast cancer.[61] But skeptics warn that the pharmacogenomic revolution will be slowed by limited insurance coverage,[62] privacy concerns,[63] and physician resistance or inertia.[64]

[57] The firms might overcome the free-riding problem by jointly developing the test, or if allowed by antitrust law, by jointly marketing the test and their drugs.

[58] Certainly, testing will always be done when it screens out patients who would suffer lethal side effects.

[59] *See* Buchanan, *et al.*, *supra* note 1, at 4 (payers are likely to embrace genetic tests that reduce expenditures on drugs); Danzon & Towse, *supra* note 41, at 8–9 (willingness to pay for testing depends on health benefit from treatment and avoiding adverse reactions and the cost of testing); *See* Rothstein & Epps, *supra* note 49 ("[E]thical concerns, economic considerations and the threat of malpractice liability are likely to encourage physicians to begin testing for and prescribing medications designed for use by specific, smaller groups of individuals ... [but] budgetary constraints imposed by insurers could slow acceptance of drugs developed through pharmacogenomics ....").

[60] Possible tort liability for failure to warn about drug risks will push pharmacists and drug companies to encourage pharmacogenomic medicine. *See* Rothstein & Epps, *supra* note 49.

[61] *See* Danzon & Towse, *supra* note 41, at 11 (The FDA has approved three diagnostic tests that indicate whether a patient will benefit from the chemotherapy drug trastuzumab. The cost is less than $100 per test.).

[62] *See* Rothstein & Epps, *supra* note 49 (questioning the extent of insurance coverage related to pharmacogenomic drug therapy); Rai, *supra* note 11, at 202 (many Medicare beneficiaries do not have prescription drug coverage).

[63] *See* Andrews, *et al.*, *supra* note 23, at 430–431 (discussing privacy risks of genetic testing associated with pharmacogenetics and pharmacogenomics).

[64] *See* Buchanan, *supra* note 1, at 16; Noah, *supra* note 13, at 22–23 (physicians may be slow to accept pharmacogenomics).

Further impediments will be created by the high cost of developing and implementing the new tests, and by the proliferation of costly new tailored drugs.[65] Optimists respond that the cost of testing is likely to fall as the industry gains experience in making and performing tests, and there are likely to be economies of scale and scope in testing.[66]

Industry participants will surely influence the rate of adoption to advance their goals. Generally, health care payers and drug companies have divergent interests in promoting genetic tests. In some markets, both groups will favor tests; in some markets, drug companies will resist tests; and in other markets, health care payers will resist tests. In existing markets with a single patented drug, drug companies will push for complete adoption of tests that accomplish useful sorting at a marginal cost less than the marginal cost of the drug. Looking back to Fig. 21.1, we see that drug company profit is maximized when all patients are tested.[67] Drug companies are apt to promote the test by marketing to physicians and perhaps bundling tests with drugs.[68] Figure 21.1 also shows that consumer surplus is maximized when the fraction of patients tested is $\alpha_0 < 1$.[69] Thus, a managed care organization or other party acting to maximize consumer surplus might prefer limited adoption of genetic testing.[70] In contrast, when the genetic test is not medically necessary because all patients will take

[65] One factor that might contribute to high testing cost is the need to obtain upstream gene patent licenses to perform tests. High cost of license to perform test for breast cancer related genes, and gene related to Canavan's disease. The analysis in Section III.A shows that the price of currently available drugs might also rise. Noah, *supra* note 13, at 27 (payers may not be enthusiastic about pharmacogenomics).

[66] *See* SCIENTIFIC AMERICAN, *supra* note 30 ("Affymetrix expects to be able to mill through 100,000 SNPs dispersed through a patient's genome in several hours, for as little as a few hundred dollars."); Pollack, *supra* note 18 ("Some scientists say the gene chips may be too expensive and difficult to use in the average clinic. But others say the tests will come down to a few hundred dollars apiece. While powerful computers may be needed to find the patterns initially, once they are found a sample can be analyzed on a laptop. And the number of genes that need be tested may also shrink. Merck started out analyzing 25,000 genes but found that only 70 were needed to predict breast cancer outcome.")

[67] Similarly, drug companies will promote widespread adoption of tests that bring a drug to the market that has been shelved because of adverse reactions or low rates of efficacy.

[68] *See* Robertson *et al.*, *supra* note 12, at 159. *Cf.* Iain Cockburn, *Comments on: The Proper Scope of IP Rights in the Post-Genomics Era* (by Arti K. Rai), B. U. J. SCI. & TECH. L. (2001) (genomics may disrupt current drug marketing practices).

[69] That value of α maximizes the difference between total surplus and profit. Notice that α_0 also minimizes profit. The interest of the drug company and consumers are strongly opposed.

[70] Managed care organizations are likely to weigh the costs and benefits of using any diagnostic test. *Cf.* Muin J. Khoury & Jill Morris, *Pharmacogenomics and Public Health: The Promise of Targeted Disease Prevention*, *available at* http://www.cdc.gov/genomics/info/factshts/pharmacofs.htm (last visited Mar. 26, 2002) (discussing a cost-benefit analysis of testing for the factor V Leiden allele, which is associated with an elevated risk of venous thrombosis among women who take oral contraceptives).

the drug regardless of the test outcome, then drug companies will oppose tests because the tests give health care payers information that augments their bargaining power against drug companies in the market for drugs. Genetic tests might not diffuse as widely in markets with patented substitutes or markets with unpatented drugs because the test seller does not (fully) internalize the increased profit from drug sales. If the test seller cannot bundle the test with the complementary drugs, then it will set a price above marginal cost that will slow test adoption.

B. The Incentive to Develop Genetic Tests

The genetic tests required for customized drug therapy will be costly to develop, thus, I explore whether the private sector has an adequate development incentive. Drug innovators have a strong incentive to develop genetic tests because the tests are likely to reduce the cost of clinical trials for *new drugs*.[71] If pharmacogenomics becomes an essential step in drug discovery, then drug patent owners will routinely develop complementary genetic tests.[72] It is less clear whether incentives are adequate for tests that are designed for use with *existing drugs*. Typically, patents on tests provide adequate incentives, but patents fail to provide adequate incentives under certain market conditions, and public support should supplement private sector R&D. I evaluate the adequacy of private incentives by comparing the private costs and benefits to the social costs and benefits of test development. Private costs diverge from social costs mostly because of licensing fees. Private benefits diverge from social benefits for reasons related to the structure of the market for drugs, the structure of the market for genetic tests, and the rate of adoption of tests.

The private cost of developing a genetic test includes both the transfer payments required to license relevant patents and gain access to genomic databases and the real costs associated with research personnel, instruments, and material.[73] The social cost of developing a test excludes transfer payments

[71] *See* Danzon & Towse, *supra* note 41, at 10.

[72] But Sections III.A and III.C show that the innovator may suppress the tests if they are not medically useful or if they will not be widely adopted.

[73] A likely regulatory cost arises from FDA oversight of genetic tests. *See* Michael J. Malinowski & Robin J. R. Blatt, *Commercialization of Genetic Testing Services: The FDA, Market Forces; and Biological Tarot Cards*, 71 TUL. L. REV. 1211, 1229–1232 (1997) (describing FDA regulation of predictive genetic test kits and services); Buchanan, *supra* note 1, at 6. In the future, the economically relevant cost to develop a genetic test may be quite low because the information required for the genetic test will be produced during the normal course of drug development. In other words, since pharmacogenomic information will be gathered as part of an efficient drug development project, then that same information can be used at no cost during the development of genetic testing. *See* Rai, *supra* note 10, at 191.

related to patent licenses and database access. The firm views a wealth transfer as a cost, but it is not a social cost because it does not consume resources. Thus, private and social costs are equal only when transfer payments are zero. Patents on research tools could add significantly to the cost of pharmacogenomic research.[74] For example, gene patents owners are likely to collect royalty payments from companies who make gene chips used to predict adverse drug reactions.[75] The research exemption in patent law is very limited and will not diminish these costs,[76] and pharmacogenomics is outside the reach of the limitations imposed on enforcement of medical process patents.[77] It is possible, though, that these patent licensing costs are overstated.[78] Many existing gene patents may not survive validity challenges.[79] And many genes are unpatentable because of gene sequence data placed in the public domain.[80] License fees

[74] *See* Janice M. Mueller, *No "Dilettante Affair": Rethinking the Experimental Use Exception to Patent Infringement for Biomedical Research Tools*, 76 WASH. L. REV. 1, 8 (2001) (commenting on high transaction costs associated with obtaining patent rights for gene chip); Rebecca S. Eisenberg, *Re-Examining the Role Of Patents in Appropriating the Value of DNA Sequences*, 49 EMORY L. J. 783, 6–9 (2000) (genomics companies aim to protect information rather than molecules with their gene patents); Andrew Pollack, *3-D RNA Folds and Molds Like a Key for a Specialized Work*, N.Y. TIMES (Jan. 21, 2002) *available at* http://www.nytimes.com/2003/01/21/science/21SHAP.html (discover of aptamers (short strands of RNA that interact with proteins) believes basic research on aptamers has been slowed because of his reluctance to license his patent); Philippe Ducor, *New Drug Discovery Technologies and Patents*, 22 RUTGERS COMPUTER & TECH. L. J. 369, 388 (1996) ("A cursory search of U.S. patent titles pertaining to such screening methods identified over forty-two issued patents in the past three years alone, a tally that does not account for patents which may claim such methods without stating so in their titles."); Rebecca S. Eisenberg, *The Shifting Balance of Patents and Drug Regulation*, 19 HEALTH AFFAIRS 119, 127 (2001) (Merck supports the SNPs consortium to reduce the number of upstream patents and avoid licensing costs); Rebecca S. Eisenberg, "Reaching through the Genome," (Chapter 10) (reach through licenses and claims are used by upstream gene patent owners to gain downstream revenue); *See* Rai, *supra* note 23, at 816–817 (drug developers need to get licenses to SNP patents to develop certain targeted drugs).

[75] *See id.* at 842.

[76] *See Roche Prods., Inc. v. Bolar Pharmaceuticals, Inc.*, 733 F. 2d 858 (Fed. Cir.), *cert. denied*, 469 U.S. 856 (1984). The 271(e) research exemption is limited to research designed to gather data related to FDA regulation. *See* Mueller, *supra* note 74, at 25–27. For a full discussion of the research exemption, see Rebecca S. Eisenberg, *Patents and the Progress of Science: Exclusive Rights and Experimental Use*, 56 U. CHI. L. REV. 1017 (1989).

[77] *See* Nunnally, *supra* note 21, at 20 (2001).

[78] *See* F. Scott Kieff, *Facilitating Scientific Research: Intellectual Property Rights and the Norms of Science—A Response to Rai and Eisenberg*, 95 NW. U. L. REV. 691, 701–703 (2001).

[79] The utility requirement is likely to be a problem for many gene patents. *See Brenner v. Manson*, 383 U.S. 519 (1966). Gene patent claims that are valid may be read narrowly so that downstream genetic test developers are not infringers. *See* Kieff, *supra* note 78, at 700 (EST patents could be read narrowly).

[80] *See* Rai, *supra* note 23, at 832–833 (pharmaceutical companies are building public domain genomic databases to ward off potential patents); Nunnally, *supra* note 21, at 12–13 (the SNP consortium will not patent its discoveries and hopes to thwart SNP patents by others).

required for access to genomic databases could also add significantly to development cost.[81] The high fixed cost of producing a database means that the market is not competitive,[82] but prices are constrained by the existence of nonprofit databases,[83] relatively weak intellectual property protection,[84] and the possibility of entry.

The private benefit from developing a genetic test depends on the structure of the market for the drug or drugs that are used in conjunction with the test. In many markets, only a drug patent owner would pursue development of a genetic test designed for use with a patented drug.[85] The patent owner could deter competing innovators by threatening a patent infringement suit if developing the genetic test requires use of the patented drug. Furthermore, the patent owner enjoys cost advantages in developing a genetic test because of access to the drug and patients' genetic information acquired during clinical testing.[86]

The benefit that a drug patent owner receives from developing a genetic test is the sum of the profit from selling the test plus the increase in profit from sale of the drug that is used in conjunction with the test. If the drug is not currently marketed, then the patent owner has a strong incentive to develop a genetic test that would get the drug on the market.[87] For example, a drug might be kept off the market because of adverse reactions. A test might allow doctors to screen out patients who are likely to suffer an adverse reaction, making the drug marketable. If the drug is already marketed, then the drug patent owner

[81] *See supra* notes 35–37.

[82] *See* J. H. Reichman & Pamela Samuelson, *Intellectual Property Rights in Data?* 50 VAND. L. REV. 51 (1997) (database markets tend toward natural monopoly).

[83] *See* Nunnally, *supra* note 21, at 12–13 (the SNP consortium hopes to achieve economies of scale by avoiding redundant research).

[84] Trade secret law is the main source of intellectual property protection of genomic databases. Databases get fairly little protection under copyright law. *See Feist Publications v. Rural Telephone Service*, 499 U.S. 340 (1991). Jane C. Ginsburg, *No Sweat? Copyright and Other Protection of Works of Information After Feist v. Rural Telephone*, 92 COLUM. L. REV. 338 (1992). But Congress has considered establishing new intellectual property rights that apply to databases. Reichman & Samuelson, *supra* note 58, at 95–112 (1997) (explaining forces that push the United States toward adoption of a database statute).

[85] Drug patent owners may develop a test directly or indirectly. Big pharmaceutical firms have started to collaborate with universities and biotech start-ups to make genetic tests used for drug therapy. *Id.*

[86] *See* Allen C. Nunnally, Scott A. Brown, & Gary A. Cohen, *Intellectual Property and Commercial Aspects of Pharmacogenomics*, in PHARMACOGENOMICS: SOCIAL, ETHICAL, AND CLINICAL DIMENSIONS, ed. MARK A. ROTHSTEIN (2003).

[87] The model in Section III.B exaggerates the strength of the private incentive to develop a test for a drug that is not currently marketed. The model assumes that all patients who should use the drug derive the same benefit of V. In a more realistic model with heterogeneous valuations, the drug company would not capture all the consumer surplus from introducing the drug, and therefore, profit would be less than total surplus.

may gain or lose from the introduction of the genetic test. Section III shows that the profit from drug sales depends on the fraction of the patient population that is tested. For example, if the test is not medically useful, then testing always hurts profit. If testing is medically useful, then testing every patient maximizes monopoly profit.[88] In contrast, if the fraction tested is less than α' as displayed in Fig. 21.1, then even medically useful genetic testing cuts profit for drugs that are already marketed. To achieve a high rate of testing, the firm should set a low price on the genetic test, and capture its profit through the price of the drug.[89] In fact, the firm should offer the test free of charge to promote widespread adoption. The test and the drug are complements and patients care about the total expected price of drug therapy, not the separate prices of testing and drugs. Essentially, genetic testing plays a role similar to advertising. Just as a firm offers advertising information at no charge, it should offer the testing information at no charge in order to gain a favorable change in the demand for the drug.[90]

The private benefit from test development equals the social benefit when the test developer captures the increase in total surplus caused by adoption of the test. Sections III.A and B suggest drug patent owners will effectively capture the social benefit from test development when medically useful tests are widely adopted. In contrast, when the fraction of patients tested is relatively low, there can be wide divergence between the private and social benefit. For example, Fig. 21.1 shows profit declines while total surplus grows from test introduction when the fraction of patients tested is less than α_0. Section III.C shows that there is no social benefit from development and introduction of tests that are not medically necessary. For such tests, private and social benefits are equal and maximized when no patients are tested and no test is developed. Section III.D shows that the private benefit from test introduction falls short of the social benefit in duopoly markets. The free-rider problem discourages firms from developing a test used in conjunction with substitute patented drugs owned by different firms.

Finally, consider test development incentives when multiple firms race to get a patent on a test.[91] This problem is outside the scope of the model in

[88] If the genetic test is used by all potential patients then in the basic model the profit from sale of the drug is $\pi = 1/2\,(V - C)$. In contrast, the profit if the genetic test is not developed and all patients are uninformed is $\pi = V/2 - C$. The difference between these two profit levels is $C/2$. Thus, the change in profit to the firm from developing the genetic test is $C/2$. Intuitively, testing increases profit because it allows the firm to avoid the cost of making and selling drugs to patients who will not benefit from the drugs.

[89] Assuming the marginal cost of testing is low compared to the marginal cost of drug therapy.

[90] Drug companies might be constrained by antitrust concerns. *See* Noah, *supra* note 13, at 21–22 (discussing antitrust issues presented by bundling drugs and genetic tests).

[91] Biotech firms are starting to enter this market. *See* SCIENTIFIC AMERICAN, *supra* note 30.

Section III, but two points deserve comment. First, the expected profit from the test is lower because the profit is discounted by the probability of winning the patent race, and because a genetic test inventor who does not own the patent on the associated drug gets a smaller profit than the drug patent owner could.[92] Second, the expected profit from a patent on a genetic test is more potent in stimulating R&D by firms in a patent race. It is possible the private incentive to develop the genetic test exceeds the social incentive. This can be true even though the expected profit from patenting a test is lower. The difference from the single inventor case arises because firms are driven to win the winner-take-all patent race. A social planner does not care which firm wins the race, but, of course, the firms care. Therefore, even if the winner earns far less than the expected social value from competing in the race, firms might invest more than the social optimum.

V. CONCLUSION

Federal subsidies and public sector research currently play significant roles stimulating the development of the first generation of genetic tests designed to customize drug therapy. If pharmacogenomics fulfills its promise, then we should expect that, as this sector of the pharmaceutical industry matures, most of the R&D on genetic tests will shift to the private sector. Drug manufacturers have a strong incentive to develop tests during the drug development process. They also have a strong incentive to develop tests that allow them to market drugs that have been shelved because of adverse reactions or low efficacy rates. Drug manufacturers and independent diagnostic manufacturers are apt to play the leading role in developing genetic tests used in conjunction with drugs that are currently marketed, but the presence of certain factors may dampen private incentives to develop tests for existing drugs and require continued public sector participation in genetic test R&D. The case for public subsidy is strongest when genetic tests are not widely adopted, when tests are designed for use with

[92] The test inventor can profit either by selling the test to patients or by assigning the patent to the drug patent owner. Because the test and the drug used in conjunction with the test are complementary products, assigning the genetic test patent to the drug patent owner is likely to be more profitable. The maximum profit that the drug patent owner can derive from introducing a genetic test is $C/2$. That amount is the upper bound on the profit available to an outside inventor who patents the genetic test. It is likely that the drug patent owner and the inventor of the genetic test would share that amount, thus, the private gain to the inventor is lower than in the case when the drug patent owner is also the inventor of the genetic test. The drug patent owner would market the test when profitable, but would still be willing to acquire the patent and suppress the test if usage was unprofitable.

substitute drugs manufactured by different companies, or when the private cost of test development is high because of licensing costs.

Public sector support for development of genetic tests can take many different forms. Presently, federal grants directly support public sector pharmacogenomic research,[93] and indirectly support pharmacogenomics through subsidies encouraging the production of research inputs (like gene data) that are used in the development of genetic tests.[94] The government can encourage adoption of genetic tests through drug law and health insurance regulation.[95] Finally, the drug laws may be used to subsidize the development of drugs designed to treat "orphan genotypes."[96] An interesting question for future research is the optimal mix of these subsidies.

Acknowledgments

I thank Fran Miller and participants at the Washington University Conference on the Human Genome Project and IP Law (April 2002) for their helpful comments. I also thank Tariq Ashrati for able research assistance.

[93] *See* Davis, *supra* note 53 (the NIH made nine grants in the year 2000, totaling $ 12.8 million for research in pharamacogenetics).

[94] *See* Pollack, *supra* note 18 ("[The] International Genomics Consortium [is] a group of companies and academic institutions that wants to analyze at least 10,000 tumors to develop standard methods [of genetic analysis] and to generate genetic signatures that would be public, not patented.")

[95] *See* Rai, *supra* note 11, at 203–205 (Pharmacogenomics will put pressure on the federal government to subsidize drug purchases by people lacking prescription drug insurance.)

[96] Pharmacogenomics will identify genotypes that should not be treated with existing drugs. *See* Roses, *supra* note 9 (pharmacogenetics will encourage parallel drug development to treat the heterogenous forms of some diseases). Drug companies may not be willing to develop new drugs to treat small patient groups who have a rare genotype. *See* Danzon & Towse, *supra* note 41, at 12 (socially valuable genetic tests may impair incentives to develop new drugs to serve small niches). The Orphan Drug Act spurs development of orphan drugs by providing marketing exclusivity, tax credits, and grants. The act may apply to drugs that treat rare genotypes. *See* Pulsinelli, *supra* note 11, at 310; Buchanan, *supra* note 1, at 12 (suggesting federal subsidies for research to develop drugs for orphan genotypes).

22

The Effect of Intellectual Property on the Biotechnology Industry

James H. Davis* and Michele M. Wales†

*Senior Vice President and General Counsel
Human Genome Sciences, Inc.
Rockville, Maryland 20850-3338

†Assistant General Counsel, Intellectual Property
Human Genome Sciences, Inc.
Rockville, Maryland 20850-3338

I. BRINGING NEW DRUGS TO PATIENTS

Biotechnology today holds the key to tomorrow's medicine. Through biotechnology, doctors of tomorrow will have far more weapons in their fight against disease, injury, and age. But biotechnology is a research-intensive industry, requiring highly trained and dedicated scientists, expensive laboratories and manufacturing facilities, and sufficient money and time to see a project through from its initial discovery to its final implementation. The 1990s saw an amazing explosion of new biotechnology companies, new products, and the promise of more to come. But continued growth after this decade depends on the constant infusion of new capital to continue to fund critical research. The ability to

0065-2660/03 $35.00

raise new capital is directly linked to the proper functioning of intellectual property laws.

Unlike companies in many other industries, biotechnology companies do not gain competitive advantages based on their ability to manufacture a product more quickly, easily, or cheaply than their competitors. Instead, biotechnology companies exert a competitive advantage by virtue of exclusive intellectual property rights granted to them by the United States and foreign governments for their discoveries. The key intellectual property right relied on by biotech companies is the utility patent, which grants its owner the right to exclude others from practicing its patented invention for a limited period of time. Owing to their reliance on patents, biotechnology companies have a greater sensitivity to changes and developments in patent law than do many other industries.

Biotechnology is a rapidly expanding sector of the economy. Between 1993 and 2001, sales and revenues of biotechnology companies more than tripled, and market capitalization[1] grew by more than sevenfold.[2] Products produced by the biotechnology industry have improved the lives of millions of patients around the world. As of 2001, the U.S. Food and Drug Administration (FDA) had approved more than 117 biotech drugs and vaccines, and those products have been used to treat more than 250 million people (almost as many people as lived in the United States last year).[3] More than 75% of those drug and vaccine approvals occurred in the period from 1995 to 2000. As of 2001, more than 350 drug candidates, with the potential to treat more than 200 diseases and disorders, were in clinical trials.[4]

Biotechnology products enhance economic productivity by reducing the duration of disease and disability, providing care to patients for whom no therapy previously was available, or replacing older therapies that are more costly, more invasive, or less efficacious than the newer biotechnology modalities. Biotechnology products often can reduce the length of hospitalization or disability associated with older technologies, saving the patient and his or her insurance company thousands of dollars.[5] In short, biotechnology products are

[1] The total amount of money invested in the US biotechnology industry.

[2] Biotechnology Industry Organization, Biotechnology Industry Statistics 1993–2001, available at *http://bio.org/er/statistics.asp*

[3] *http://workmall com/wfb2001/united_states/united_states_people.html*

[4] *Supra* note 2.

[5] For example, a NIH-sponsored study found that a new stroke treatment used promptly produced a net savings of more than $4 million for every 1000 patients treated. PhRMA. PHARMACEUTICAL INDUSTRY PROFILE 2000 available at http://www.pharma.org/publications/industry/profile00.html Another study found that the use of biotechnology-derived granulocyte colony stimulating factors (naturally occurring proteins that were manufactured recombinantly) reduced the incidence of infections associated with bone marrow transplants and resulted in cost savings of approximately $14,000 per transplant. BIO. *Biotechnology's Impact on Diseases of the Elderly: A White Paper*, available at http://www.bio.org/news/white_paper.html

of immense value both to individual patients and to society. Today, these products have the ability to decrease suffering, increase lifespan, and enhance productivity, and there is every expectation that tomorrow these benefits will only be more fully realized.

Such enormous benefits, however, do not come without cost. The cost of finding and developing new biotechnology products is very high. Currently, it takes 10 years or more of research, whether in a pharmaceutical company or a biotechnology company, and costs close to $800 million to develop a promising compound to the point where it can be made into a commercially available drug. Even the initial research to identify a potential drug candidate often takes 3 to 5 years. It has been estimated roughly that only one in 5000 small-molecule candidate compounds turns out to be an efficacious pharmaceutical agent.[6] While it is hoped that the use of biotechnology will substantially increase the likelihood that a candidate compound ultimately will prove efficacious, it remains a considerable effort, even in biotechnology, just to separate the potentially efficacious compounds from the merely interesting.

Yet, the research to identify a promising candidate is only the first step. That drug candidate then must be tested in humans to prove that it is both safe and effective. These human clinical trials can last 5 to 7 years. Even if trials generate sufficient data to demonstrate to the inventor that a drug is safe and effective, the FDA still must review that data prior to approving the compound for a single intended use. This review process itself can take 1 to 2 years. Even then, there is no guarantee that the approved drug product will prove to be commercially successful. Development of a biotechnology product is an expensive, time-consuming process that is fraught with the danger that the product will not work, will not be approved, or will not be commercially successful.

Fortunately, the biotechnology industry has enjoyed the support of the private capital markets to fund this long and arduous research and development process, and this private investment is critical to the survival of the biotechnology industry. Investment provides biotech companies with the resources necessary to fund research and development for new products. Fewer than 25% of the 1300 biotech companies existing in 2001 were publicly held,[7] and many biotech companies have limited cash remaining. Obviously, both public and private biotechnology companies will need additional infusions of capital over the next few years to fulfill the promise of biotechnology. Yet, as discussed above, biotechnology companies and their investors face significant risks in the research and development process that most, or all, of their investment will be lost. This risk would be unacceptable to investors if there was not a significant possibility that a successful

[6] Clemente, C.L., Remarks at the MERCOSUR Conference, New York, NY, U.N. Plaza Hotel, (April 27, 1998).

[7] *Supra* note 2.

product would be highly profitable. The key to that future profitability is the intellectual property protection of patents, which gives biotechnology companies a sufficient term of exclusivity to recoup their investments.

II. THE U.S. PATENT SYSTEM: PROMOTING INNOVATION IN THE BIOTECH INDUSTRY

By recognizing that patents stimulate innovation in all technologies, the framers of the United States Constitution provided patent rights to inventors to "promote the progress of Science and the Useful Arts."[8] A patent is a right granted to an inventor by a government. Patent rights permit the owner to exclude others, for a limited period of time, from practicing the technology that he or she invented. Ownership of patent rights permits a biotechnology company to assure its investors that a product developed by the company cannot simply be copied by another. In this way, investors can have confidence that a competitor will not be permitted to steal the fruits of the research funded by their investment.

In the United States, the patent system has often been likened to a contract between the federal government and an inventor.[9] In essence, there is an implicit contract whereby the inventor discloses his or her invention to the public and, in return, the government gives the inventor the right to prevent others from practicing the invention for a limited period of time. Thereafter, the public has the right to freely practice the invention.

Patent applications must describe an invention with sufficient thoroughness that a skilled worker in the field of technology corresponding to the invention would be able to make and use the invention for its intended purpose.[10] Implicit in this requirement to describe how to use the claimed invention is a requirement that the invention have a practical use. In addition, a patent application must disclose the most preferred embodiment of the invention.[11] A patent application must include claims that clearly describe the subject matter for which the applicant seeks to obtain exclusive rights. Those claims must be sufficiently clear that the skilled artisan is able to discern the boundaries of the subject matter that is encompassed by the claims.[12] When the U.S. Patent & Trademark Office receives a patent application, a patent examiner determines

[8] Article I, Section 8.

[9] Often designated the contract theory of patent law. *See, e.g.*, Rosenberg, PATENT LAW FUNDAMENTALS (2001) West Group.

[10] 35 U.S.C. § 112 ¶1.

[11] 35 U.S.C. § 112 ¶1.

[12] 35 U.S.C. § 112 ¶2.

whether the application complies with these requirements. The examiner also compares the subject matter claimed in the application with defined types of previous disclosures (the "prior art") and determines whether the claimed subject matter is sufficiently "inventive" with regards to the prior art to merit the grant of a U.S. patent.

Comparing the claimed invention with the prior art, the patent examiner investigates whether the same invention is described in the prior art. If the invention, with every characteristic recited in a claim of a patent application, is described in the prior art (*i.e.*, the prior art "anticipates" the invention), then the applicant is not entitled to a patent for the claimed invention.[13] Even if the patent examiner determines that the prior art does not describe the same invention,[14] the examiner will refuse to allow a patent to issue if the claimed invention is an obvious modification of technology disclosed in one or more prior art references.[15] Only if the prior art does not anticipate the claimed invention nor renders it obvious is the patent applicant entitled to a patent. Once the patent examiner is satisfied that the patent application, including the claims, meets all of the conditions of patentability, the application is allowed to issue as a patent.

The claims in an issued patent define the scope of the inventor's ability to exclude others. One who practices technology within the scope of a valid patent claim is said to "infringe" the patent. A patent claim can be infringed literally[16] or, under certain circumstances, by practicing technology that is not within the literal boundaries of the claims but differs in only insubstantial ways.[17]

Only after a patent issues is the patentee entitled to prevent others from practicing the technology defined by the patent claims.

Issued patents have a defined term,[18] after which the patent claims can no longer be enforced.[19] Patents normally expire 20 years after the filing date of the patent application. The average effective patent term for products other

[13] 35 U.S.C. § 102.

[14] For example, if the prior art describes a similar technology, but not one that has every characteristic recited in the claim of the patent application.

[15] 35 U.S.C. § 103(a). Methods of determining legal obviousness of a patent claim in view of one or more prior art references are defined by a large body of court decisions. The basic method is described in *Graham v. John Deere Co.*, 383 U.S. 1 (1966).

[16] *I.e.*, by practicing technology that falls within the literal boundaries of the subject matter defined in the claims.

[17] Such infringement is commonly called infringement "by equivalents" or infringement "by application of the Doctrine of Equivalents." *Warner-Jenkins Co. v. Hilton Davis Chemical Co.*, 520 U.S. 17 (1997).

[18] Generally, U.S. and foreign patent terms are now 20 years from the earliest claimed filing date for the patent application or for a priority patent application. US patents that issued from an application filed before June 8, 1995 have a term of 17 years from the patent issuance date.

[19] Under certain circumstances, the term of a patent can be extended to account for delays by the U.S. Patent & Trademark Office, regulatory review, or other delays.

than drug products is about 18.5 years[20] because these nondrug products can be immediately sold to the public. However, in the United States and elsewhere, drugs cannot be sold to the public until they have been carefully researched and tested and approved by the appropriate government regulatory body (*e.g.*, the FDA in the U.S.). As previously noted, the research and testing necessary to satisfy government regulators is both expensive and time-consuming.

Unfortunately for investors, the patent term for a drug product does not begin to run when the drug is approved. Rather, as soon as a new patent application is filed, the 20-year patent term begins. Since regulatory approval often is not obtained until many years after the patent term has begun, the average effective patent term[21] for a drug patent is only 11 to 12 years.[22]

Thus, a patent system that rewards biotechnology companies with enforceable claims is critical for realizing the profits necessary to encourage the original risky investment. In contrast, an overly restrictive view of the patent rights would lead investors to fear that they might not be able to reap the rewards of the technical advances that their investment created. Thus, in a world of inappropriately narrowly drawn patent rights, investment in biotechnology would lag behind the value that could otherwise have been delivered by the biotechnology industry.

Patents stimulate innovation in biotechnology companies by encouraging the dedication of people, capital, and time to the search for a new drug. If the search is successful, those who have invested will obtain a substantial reward for taking the risk.

But patents do not only stimulate innovation in the company receiving investor money; patents also stimulate innovation in third parties. Patent applications are published on an established schedule. After November 2001, most U.S. patent applications are published whether or not they issue as patents, similar to the practice for international and foreign patent applications. Publication of patent applications provides the public with access to the information described in the application prior to the issuance of the patent. Regardless of whether a patent ever issues from the application, the scientific contribution has been placed in the public knowledge. Patents and patent applications stimulate innovation, regardless of whether commercially useful patent rights are granted to the patent applicant.

[20] http://www.pharma.org/publications/publications/profile00/chap8.phtml#patent

[21] The effective patent term is the period of time extending from the date on which regulatory approval is received until the expiration date of the patent.

[22] In the U.S., only a limited amount of the time consumed obtaining regulatory approval can be recovered, and in many other countries no patent term extension can be had on account of regulatory review. Even with patent term restoration that may be available in some instances in the U.S. the maximum patent term for these drug patents is capped at 14 years. 35 U.S.C. § 155 *et seq.*

Even where an issued patent precludes others from practicing technology claimed in the patent, a patent can spur innovation. Rather than obtaining a license to a patent, one can instead design an alternative way of practicing the technology—a way that is not within the scope of the patent claims. This "design around" often leads to new, patentable inventions. In such cases, the need to avoid infringing the patent rights of others actually spurs innovation.

Thus, patents are the catalyst that brings together the creativity of modern science with the capital of the American economy to produce the products of tomorrow. Patents reassure investors that their risky contribution of large sums of money to a biotech company can be recouped if a product is eventually approved. Without this investment, companies cannot afford research and development costs associated with discovering new drugs and bringing these drugs to patients. Patents also stimulate innovation in third parties by disclosing technology and by encouraging "design around" products or methods. These many benefits, however, do not flow inexorably from patents. Patents are creatures of law and therefore subject to interpretation, which can either be restrictive or expansive. Thus, the ways in which patent law is interpreted and applied are critical to ensuring that the level of investment in the biotechnology industry continues to match the promise of the products that the biotechnology industry can provide.

III. COMMON CRITICISMS OF BIOTECHNOLOGY PATENTS

Despite the clear benefits offered by the patent system, there are many critics who complain that patents do significant harm, and that the harm may not outweigh the good done by the system. However, these criticisms are unjustified when one examines all of the facts.

Critics of biotechnology often complain that patents for biotechnology-related inventions increase the cost of medicine, citing the ability of a manufacturer to preclude competition in the marketplace for recently developed products. These fears are misplaced.

If other companies were permitted to copy biotechnology or pharmaceutical products as soon as they were approved, no rational drug company would expend the cost and effort of developing new drugs. It is the patent system's exclusive rights that give biotechnology and pharmaceutical companies confidence that their investment in discovering and bringing to market new predictive, diagnostic, therapeutic, and preventive agents can be recaptured. In the absence of an effectively functioning patent system, follow-on manufacturers could "cherry-pick" favorable research results without incurring the costs of either the fruitful research or the many failed research programs that are necessarily conducted in the search for efficacious agents. Even in the absence of "generic biologicals," follow-on manufacturers could conduct narrowly focused clinical trials on recently approved biological products and reap the rewards of years of research conducted by the drug's inventor.

The limited period of exclusivity granted to a drug's inventor by the patent system permits the inventor and investors to recoup the very significant research and development expenses made for that drug. The overwhelming majority of drug candidates fail to be safe, effective, and approved drugs, and the research and development expenses laid out for these failed drug candidates cannot be recovered through sales. Thus, the cost of a new drug must reflect not only the cost of discovering and developing that drug but also the costs of all the efforts that did not lead to a successful drug. In view of the unpredictable nature of drug development, the patent system must reward not only the successful development of a safe, effective, and accepted drug, but also reward those bold enough to search for and develop new drugs.

Critics of biotechnology-related patents also complain that patents directed to "research tools" inhibit biological research. As an initial matter, one must be conscious that the label "research tool" prejudges the usefulness of an invention. That is, one party's research tool is another party's final product. For example, a microscope might be useful as little other than a research tool to many. Nonetheless, enormous effort and ingenuity has gone into making and refining microscopes, and a new microscope invention is not a "research tool" to its inventor, but a final product to be sold and used by others. The same degree of effort and ingenuity would likely not have been expended were microscope developers not able to obtain financial benefit for their work. Likewise in biotechnology, isolated "Taq polymerase" (the DNA polymerase found in the hot-water dwelling bacterium Thermus aquaticus) is a patented research tool that has revolutionized molecular biology. Indeed its inventor, Kary Mullis, shared the 1993 Nobel Prize in Chemistry for this revolutionary work. Similarly, developers of unique cell lines, polynucleotides, biochemical reagents, and biomolecule libraries expend significant effort and expense to provide useful research tools. These efforts should not go unrewarded because they fall into the category of "research tools." Many scientific breakthroughs could not be made without the development of new or improved research tools that prove to be as important as the breakthrough product itself.

Critics also assert that biotechnology patents unnecessarily deter research that uses the patented technology. To remedy this perceived problem, they often suggest that a "research use" exception from patent infringement should be codified or announced by a court, beyond the limited exemptions that currently exist.[23] However, such an exemption is not warranted.

The biotech industry recognizes that academic research is invaluable in furthering scientific progress. Furthermore, use of patented technology in academic research often can lead to improved or previously unrecognized uses for the patented technology. Indeed, these uses might well be to the patentee's benefit. Thus, biotechnology companies often are motivated to provide their technology to academic and nonprofit researchers under appropriate limitations.[24] Moreover, biotechnology companies have little or no incentive to enforce their patents against academic researchers. It must be kept in mind that the cost of litigating a patent infringement suit is many hundreds of thousands, if not millions, of dollars. In the absence of serious injury, a biotechnology company (or any other rational actor) is unlikely to initiate litigation against a nonprofit researcher unless the researcher is aiding the company's competitors or otherwise significantly harming the company.

Another criticism of biotechnology patents is that overlapping patents and patents directed to similar technology create a web of conflicting rights that can be difficult to negotiate. Michael A. Heller and Rebecca S. Eisenberg[25] call this web a "patent thicket," and suggest that the conflicting rights can lead to an "anticommons" effect. An analogy in real estate is that a "commons"[26] tends to be overused and depleted, since the users do not need to bear the cost of overuse. In contrast, an anticommons[27] leads to underuse of the property because of the difficulty obtaining user rights from all of the stake holders.

Commercial entities have negotiated patent thickets in the past, and can likely continue to do so. Even though the transactional costs of sorting through conflicting patent rights may add to the cost of commercial biotechnology ventures, they are not onerous, certainly not to the extent that patentability standards should be altered to avoid creation of a perceived "thicket." Indeed those transactional costs are minimal compared to the enormous research and development costs of a new biotechnology product. If companies see an opportunity to develop and market a novel biotechnology product, they will find economic solutions to ensure that the product has the opportunity to prosper. Rather than considering the abundance of biotech patents as a "thicket," it is better to think of this abundance as a "forest" of patents, through which many different paths run. A company perceiving such a forest will look at the different paths to find the one that best suits its passage through the forest. Some paths

---

[23] *See, e.g.*, Feit, *Biotechnology Research and the Experimental Use Exception to Patent Infringement*, J. PAT. OFF. SOC'Y. 71:819 (1989) for a review of court opinions regarding the common law exception and 35 U.S.C. § 144 *et seq.* for the statutory exceptions.

[24] *See Testimony of the Biotechnology Industry Organization on Competition and Intellectual Property Law and Policy in the Knowledge-based Economy*, before the Federal Trade Commission and the Department of Justice, pp. 10–12 (February 26, 2002).

[25] Heller, Michael A., Eisenberg, Rebecca S. *Can Patents Deter Innovation? The Anticommons in Biomedical Research*, SCIENCE 280:698–701 (May, 1998).

[26] A property to which many have access at little or no cost.

[27] A property from which many have the right to exclude others, and which none has an absolute right to use.

will require a company to design around existing patents. Other paths will require licenses from patent holders, while others will lead to collaboration efforts between various patent holders. Other industries such as the semiconductor, computer software, aircraft, and Internet industries have faced these issues in the past and found solutions through creativity, licenses, and in some cases, industrywide cross-licenses and patent pooling arrangements.[28] There is no reason why these and other creative solutions cannot be used in the biotech industry.

Consider the present situation for pharmaceuticals. Numerous patents exist, each claiming one or more pharmaceutical compositions containing an active agent. Apart from the active agent, many pharmaceutical compositions have similar components (*e.g.*, adjuvants, fillers, lubricants, dispersing agents, timed-release agents, preservatives, and coatings), are manufactured by similar processes, and are stored and shipped in similar containers. Patents owned by many different companies may protect all the various components that go into a final pharmaceutical product. However, pharmaceutical companies have successfully negotiated this forest either through licensing arrangements or through creative solutions that solve the same problem in a different way. Like their pharmaceutical counterparts, biotechnology companies have the ability to negotiate through the forest and find the path that leads to the entry of a new product.

Finally, some critics of biotechnology patents complain that biotechnology companies are "patenting nature." With the growth and notoriety of biotechnology, concerns have arisen regarding what exactly is being patented. Regardless of the efforts of organizations such as the Biotechnology Industry Organization[29] to educate the public about biotechnology issues, including biotechnology patents, the public remains concerned about the patenting of "nature," because the public is not well-informed and often misled. A common misunderstanding is that patents are being issued directed to human beings. The Supreme Court's 1980 decision in *Diamond v. Chakrabarty*[30] indicated that patentable subject matter includes "anything under the sun" made by the hand of man, whether living or not. However, products that exist in nature cannot, as such, be patented. In order for a naturally occurring product to be rendered patentable it must be isolated, modified, or otherwise altered from its naturally occurring condition.

[28] *See, e.g.*, Shapiro, *Navigating the Patent Thicket: Cross Licenses, Patent Pools, and Standard-Setting*, available at http://haas.berkeley.edu/~shapiro/thicket.pdf

[29] *See, e.g.*, Biotechnology Industry Organization, *Primer: Genome and Genetic Research, Patent Protection and 21st Century Medicine*, available at http://www.bio.org/genomics/primer.html (July 2000).

[30] 447 U.S. 303 (1980).

For example, an enzyme called a DNA polymerase occurs in nature, in the bacterium *Thermus aquaticus*. Presumably, the enzyme has existed in that organism for many thousands, or even millions, of years. The enzyme, Taq polymerase, cannot be, and has not been, patented in its naturally occurring form. However, Taq polymerase has been isolated from the organism in which it occurs in nature and used in a biochemical reaction that replicates DNA sequences in a laboratory setting. Today, Taq polymerase and its derivatives are among the most common biotechnology reagents in use, it being a component of the polymerase chain reaction (PCR) mixture. PCR has been used in innumerable ways, in many thousands of variations, and is one of the building blocks upon which the biotechnology industry has been built. In nature, Taq polymerase is a tiny component of a bacterium growing in hot springs—of no effective use to anyone. Isolated, Taq polymerase has been derivatized in many ways to yield useful biochemical reagents. These many derivatives exist today because patent protection was available to protect the results of research that brought Taq polymerase and its derivatives to the public.

IV. PATENT POLICIES ENDANGERING THE BIOTECHNOLOGY INDUSTRY

Critics have advocated patent policies specific for biotechnology in the U.S. and elsewhere that would distort the innovation-enhancing characteristics of the patent system for biotechnology inventions. For example, some have suggested[31] that the standards for assessing obviousness be altered for biotechnology inventions. Any attempt to change such standards and the resulting uncertainty regarding how the currently developed legal tests will be applied to biotechnology inventions could lead to investor insecurity in this industry. This could have a devastating effect on the availability of research capital, which, in turn, could stifle the innovation and development of new biotech products. There is nothing about biotechnology patent claims that warrants treatment different than claims in other technical areas.

In fact, court decisions[32] have recognized that patent laws developed in the context of patents and applications for technologies other than biotechnology (especially including patent law precedent relating to patents and applications relating to chemicals) are equally applicable to patents and patent applications relating to biotechnology. The written description requirement of 35 U.S.C. § 112, first paragraph, and the utility requirement of 35 U.S.C.

[31] *See* comments of Rebecca S. Eisenberg in Steinberg, *Will Genomics Spoil Gene Ownership?* THE SCIENTIST 14(17):1 (4 September 2000).

[32] *E.g., Univ. California. v. Eli Lilly and Co.*, 43 USPQ2d 1398 (Fed. Cir. 1997).

§ 101 did not need to be interpreted or applied in different ways to accommodate biotechnology inventions.[33] Rather, the existing laws and interpretations developed in the context of patent disputes involving the pharmaceutical industry were found applicable to biotechnology inventions.

The application of the pharmaceutical industry's precedents to the biotechnology industry is appropriate. The compositions protected by biotechnology patents are similar to compositions protected by pharmaceutical patents. Both biotechnology compositions and pharmaceutical compositions are chemicals, having a particular structure that can be written down and claimed in a precise manner. Because the structural building blocks of both are so well understood, patent claims in both technologies are capable of being designed around. And thus, just like the pharmaceutical industry, the biotechnology industry needs and relies on an appropriately broad claim scope to prevent competitors from making minor modifications in the structure of the molecule that falls outside the scope of the patent claims.

It is critical for courts and patent offices to recognize that the protection afforded biotechnology companies by patents having claims narrowly limited to the precise sequences shown in the patent specification can fail to provide biotechnology companies with the protection they need to promote innovation and prevent copyists from robbing them of their inventions. If biotechnology patent claims are forced to be so narrow that a competing company could make minor alterations in designing around a new biotechnology drug, investors will not be able to recoup their investments, and new investment in biotechnology (and the corresponding advances in medical technology and standards of living) will grind to a halt. And given the high cost of biotechnology research, government funding of research could not practically replace private investment.[34]

Another existing problem is the lack of consistent recognition of biotechnology rights internationally. In European countries, for example, methods of treating human beings are excluded from patentability. Restrictions regarding patentable subject matter and other requirements for patentability of biotechnology inventions vary significantly from country to country. In addition to patentability limitations, other internationally variable legal issues that restrict the innovation-promoting effects of national patent systems include restrictions on the ability of patentees to enforce their patents, price controls by national governments, compulsory licensing, and acceptance of parallel imports.

The international nature of the problem makes harmonization difficult outside the context of international agreements. The Agreement on Trade-Related

[33] Compare the interim Written Description and Utility Guidelines with the final versions of those guidelines, including the discussions of the case law published with the final versions of the guidelines.

[34] Market capitalization in the biotechnology industry increased by approximately $175 billion between 1999 and 2000. *See* note 2.

Aspects of Intellectual Property Rights (TRIPS) came into effect on January 1, 1995, and attempted to address some of the international issues of interest to biotechnology companies. TRIPS is expected to reduce difficulties regarding discrimination among various technologies before national patent offices and enhance the seriousness with which nations approach enforcement of intellectual property rights. If biotechnology products were guaranteed fair assessment for patentability worldwide, biotechnology companies would have even greater incentive to innovate and develop new products. Furthermore, the economic benefits that biotechnology companies bring to the countries in which they operate might be more widely enjoyed if nations would more fully recognize strong protection of patents and other intellectual property rights.

In two areas, patent policy has improved significantly of late. The American Inventor's Protection Act (AIPA) of 1999 addressed the loss of patent term that could have resulted from the switch to the 20-year-from-priority-date patent term in the U.S., coupled with ongoing delays of biotechnology patent applications before the U.S. Patent & Trademark Office. Although biotechnology companies are not yet able to regain all of the patent term lost through no fault of their own, the AIPA significantly improved the situation. Another area of promise relates to the timeliness and quality of examination of biotechnology patent applications before the U.S. Patent & Trademark Office. The U.S. Patent & Trademark Office should be encouraged to continue its reforms in those areas.

V. CONCLUSION

Biotechnology has provided our society with benefits that were barely imaginable 30 years ago. Biotechnology products have the potential to predict, diagnose, treat, and prevent innumerable diseases and disorders in patients around the world, and to reduce the overall costs of health care. Strong patent protection and fair claim scope for biotechnology inventions provide investors with the necessary incentive to invest in the biotechnology companies, an investment crucial for product development. The biotechnology industry is an invaluable resource for the betterment of mankind, and the patent system must be carefully monitored to ensure that it does not unduly restrict the benefits that biotechnology companies can provide. Loss of strong intellectual property protection for biotechnology inventions would not lead to greater innovation, and would not lead to cheaper pharmaceuticals. Instead, it would lead to a stifling of innovation and the absence of new products to treat the world's unmet medical needs. The United States today has the most creative and best-funded biotechnology industry in the world. A fundamental cornerstone of that success has been, and will need to continue to be, a strong intellectual property system that provides broad and effective protection for biotechnology inventions.

23

Are Real Business People So Easily Thwarted?

Edward T. Lentz
Patent Attorney
New Lisbon, New York 13415

During the course of this conference, questions have been raised by presenters and members of the audience on multiple occasions as to whether or not the patent law, as it is presently being applied by the Court of Appeals for the Federal Circuit to biotechnology inventions, is encouraging innovation or stifling it, and, following therefrom, whether or not a legislative solution to the "problem" should be sought. The questions seem primarily to fall into two categories. One might be called the Patentability category (or, perhaps, the Goldilocks category):

- Is it too easy to obtain patents on "small inventions?" Is it too hard to get patents on "big inventions?" Or, does the court have it just about right?[1]

The other, I'll call the Research category:

- Should patents be granted on research tools and diagnostic markers? Should basic and clinical research be exempt from patent infringement? Should compulsory licenses for research use be made available?

Patenting of biotechnology inventions has always been controversial. The recent clamor is not new, as was suggested by at least one presenter. It is only more clamorous now because there are more of us involved now than there

[1] The U.S., of course, is not the only jurisdiction wrestling with this issue. A European Patent Office Board of Appeals put the issue this way: "[A] balance must be found between the actual technical contribution to the state of the art and the terms in which the invention is claimed so that the patent scope is fair." Mycogen, T 694/92 (1997).

0065-2660/03 $35.00

used to be. The point is that there have been alarmists shouting "The sky is falling" since the first natural products patents were issued decades ago. Each time a new biotechnology has been invented and has become the subject of patent claims, the shouting has reached a new crescendo. We experienced this phenomenon, for example, with the patenting of bacteria, and then of animals, plants, genes, ESTs, SNPs, stem cells, embryo cloning, etc. Each time, the concerns expressed are the same: biological research will be stifled; clinical research will be hampered; companies will go out of business; farmers will lose control of their crops; respect for life will be diminished.

But what has our actual experience been? First, although (as many before me have acknowledged) we are sorely lacking in empirical data on which sound argument should be footed, we must nonetheless acknowledge that biotechnology research has advanced at an astonishing pace and that the biotechnology industry, particularly in the United States, has been enormously successful. So, if the patent law as it is applied to biotechnology is not optimum, it at least has not been so bad as to outweigh the factors that have contributed to these successes.

Second, we (researchers, physicians, businesspersons, and lawyers) have learned to cope with obstacles presented by the patents of others and to use the system to obtain our own patents to our own benefit. In part, we do this by entering into patent licenses and other technology transfer transactions. Patent and technology licenses, material transfer agreements, research use licenses, research collaborations, and other technology transfer transactions have become standard tools for conducting research.

Third, (and this observation is perhaps most vulnerable to the lack of supporting empirical data) patents have not blocked research. Instead, they have fostered research by providing a niche of exclusivity by forcing would-be copyists to invent around or to pursue alternative avenues of research, and by providing negotiating leverage to some who would otherwise lack any leverage at all. The so-called "patent thicket" that researchers must find their way through, while often described as a bad thing, can in fact be a good thing in that it is technology-forcing and because it facilitates technology exchanges.

Fourth, as one of their intended effects, patents encourage dissemination of information and data over secrecy. While it is certainly true that peer pressure and the "publish or perish" environment would require some disclosure of scientific advances, I doubt that the extent of disclosures would be nearly so great were it not for the comfort provided by patent protection. Indeed, many corporate policies and technology transfer agreements prohibit public disclosure until after a patent application has been filed.

Finally, the law continues to self-correct through the ongoing evolution of jurisprudence. While we may fairly complain about the slow pace of this process, we should recognize that the law does evolve, albeit incrementally, and that it usually does so in a way that is consistent with good public policy.

Recent examples of this self-correcting process as it relates to the patentability of biotechnology inventions include the decisions by the Court of Appeals for the Federal Circuit in *Amgen v. Chugai,*[2] *Fiers v. Revel,*[3] *Regents of the University of California v. Eli Lilly,*[4] and *Schering and Biogen v. Amgen.*[5] These, along with other cases make clear that a patent cannot be obtained with broader claims than the invention that was actually made. In other words, a patent cannot claim an invention that has not yet been made, notwithstanding that it may be within the ordinary skill in the art to achieve such ambition. Put yet another way, a "big" patent cannot be granted for a "small" invention.

On the other side of the coin, cases such as *In Re Bell*[6] and *In re Deuel*[7] make clear that prior art disclosures of genes that are similar to a newly discovered gene do not render obvious the particular sequence of the newly discovered gene, even if the methods of discovering the sequence of the newly discovered gene are obvious. In other words, it is fairly easy to obtain a "small" patent but difficult to obtain a "big" patent.

Collectively, these and other cases are calibrating the balance between the level of inventiveness required to obtain a patent and the scope of that patent. So, while it may be relatively easy to obtain a patent on yet another cDNA molecule or protein sequence, the scope of the patent given in reward for that contribution, absent unusual circumstances, will be commensurately narrow. Thus, patent claims to a whole gene or cDNA will not normally be available to a scientist whose contribution to the art is the discovery of a partial gene sequence, *e.g.*, an EST.[8] Patents are awarded for inventions or discoveries,

[2] 927 F.2d 1200, 18 U.S.P.Q.2d 1016 (Fed. Cir. 1991).
[3] 984 F.2d 1164, 25 U.S.P.Q.2d 1601 (Fed. Cir. 1993).
[4] 119 F.3d 1559, 43 U.S.P.Q.2d 1398 (Fed. Cir. 1997).
[5] Civ. No. 01-01230 (Fed. Cir. 2002).
[6] 991 F.2d 781, 785, 26 USPQ2d 1529, 1532 (Fed. Cir. 1993).
[7] 51 F.3d 1552, 1558, 34 USPQ2d 1210, 1215 (1995).
[8] It is interesting to see how legal tribunals in other countries are approaching these same issues. A comparison to a case decided by a Technical Board of Appeal in the European Patent Office may be instructive. In the case styled, Monokine/FARBER, T0111/00–3.3.4, decided February 14, 2002, the Board voiced no objection to the scope of a claim directed to a specified human gene sequence and any other sequence having at least 90% identity, but, instead, rejected the claim for lack of inventive step, *i.e.*, obviousness, over prior art that taught a mouse sequence that is 78% identical. The Board was not persuaded by argument that the specific human cDNA sequence could not have been predicted and therefore was not obvious, because, after all, "a specific DNA sequence must be composed of a succession of defined deoxyribonucleotides, whichever this is and that, therefore, it cannot be considered inventive for that sole reason." Similarly, in a case in which an applicant attempted to claim a use of genes encoding certain xylanases despite not having cloned the xylanases, the Board rejected the claims for having an inadequate disclosure but indicated that it would not require disclosure of specific sequences. Instead, what was needed was information specific to the genes to be cloned such as the starting strain, a partial sequence, the host strain, screening methods, etc. T. reesei/VALTION, T 0222–3.3.8

not for wishes, plans, or goals. Therefore, fears of "large patents" being awarded for "small inventions" are largely unfounded.

Thus, I submit that the Court of Appeals for the Federal Circuit is getting the balance between patentability and scope just about right. Of course, we will not know this until further case law development occurs, and until we gain more experience. For example, we do not yet know how high the Federal Circuit will raise the bar on claim breadth.

Still needed is case law that will permit *some* scope beyond a specifically disclosed sequence of a newly discovered gene or protein, because limiting claim scope to only a specifically disclosed sequence will allow the patent to be too easily circumvented.[9] Perhaps the Enzo case will present the court with an opportunity to do so.[10] I expect that the court will move slightly towards the EPO position by permitting limited additional scope, depending upon the extent to which the patentee provides guidance as to the kinds of sequence changes that can be made without adversely affecting the desired function.

With respect to the Research category, meaningful case law is beginning to develop concerning at least the following issues:

- "reach-through claims," *i.e.*, claims to undefined agents that act on a newly discovered target;[11]
- how far upstream in the research and discovery pipeline the safe harbor of 35 U.S.C. 271(e)(1) extends;[12]

[9] The doctrine of equivalents permits patentees to enforce their patents against insubstantially different sequences. *See, e.g., Boehringer Ingelheim Vetmedica, Inc. v. Schering-Plough Corporation and Schering Corporation*, Civ. No. 02-1026, -1027 (Fed. Cir. 2003), in which the Court found that a claim to growing and isolating a specified porcine virus strain was infringed by the propagation of a different strain of the same virus. The Court was not dissuaded by evidence that the accused strain was attenuated rather than virulent like the claimed strain, did not react with a monoclonal antibody that recognizes the claimed strain, grows poorly in pig lung macrophages unlike the claimed strain, and has 73 nucleotides that are different than in the claimed strain, because these differences are not substantial in the context of claimed method of growing and isolating the virus.

[10] In *Enzo v. GenProbe*, cited above, the Court remanded the case back to the District Court for consideration of the extent to which a patentee was entitled to claim variants of specific DNA sequences.

[11] In *University of Rochester v. G.D. Searle*, a District Court recently considered the validity of claims directed to a "method for selectively inhibiting PGHS-2 activity in a human host, comprising administering a non-steroidal compound that selectively inhibits activity of the PGHS-2 gene product to a human host in need of such treatment." The Court concluded that the patent lacks an adequate written description, stating, "It is . . . necessary that the patent set forth enough detail to allow a person of ordinary skill in the art to understand what is claimed and to recognize that the inventor invented what is claimed. This the . . . patent fails to do." (W.D.N.Y., 00-CV-6161L)

[12] Three District Courts have considered this question. Two of these decisions are noteworthy. In *Integra Lifesciences v. Merck KGaA* (S. D. Calif.; 96CV-1307 JMF), the Court held that upstream drug discovery activities using patented assay methods do not fall under the § 271(e)(1) exemption. In *Bristol-Myers Squibb Co. v. Rhone-Poulenc Rorer, Inc.* 2001 U.S. Dist. LEXIS 19361 (SDNY), the Court held that testing of patented compounds for pharmaceutical activity is exempt.

- whether or not holders of patents directed to research tools are entitled to seek compensation based on "reach through royalties," *i.e.*, royalties on sales of products discovered through use of a patented research tool.[13]

I submit that there is not sufficient evidence to support making any change to the law affecting research uses of patented inventions at this time. As noted above, in general, it appears that technologies are not being left undeveloped, owing to the existence of blocking patents. Furthermore, during this conference, I have heard only one anecdotal report of a case in which a patent is apparently inhibiting basic clinical research. Even if these reports are confirmed, and even if there are not good reasons (from a public policy standpoint) for such inhibition, a single case (or even a small number of cases) is not a reason to change an otherwise well-functioning system.[14]

In any event, compulsory licensing cannot be the answer. Compulsory licensing eliminates or at least diminishes market forces and thereby results in artificial valuation. Furthermore, there is no evidence that technologies will be better exploited to the public benefit by forcing patentees to allow others to work in the same area as the patentee or its chosen licensees. Indeed, this fact was one of the underlying reasons for passage of the Bayh-Dole Act, allowing inventors or their institutions to retain exclusive ownership of their government-funded inventions. And, there is no doubt that the Bayh-Dole Act, by providing the incentive of patents and profits, has been enormously successful.

In conclusion, I submit that any legislative fix is at best premature because in spite of criticisms, the system appears to be working, there is no indication that it will not continue to function, there is no clearly better model, and, last but not least, it is not clear that Congress is capable of effecting rational legislative reform on a subject that is as complicated and contentious as the patent law as it relates to biotechnology.

[13] In *Bayer v. Housey Pharmaceuticals, Inc.*, 169 F. Supp. 2d 328; 2001 U.S. Dist. LEXIS 17905; 61 U.S.P.Q.2D 1051 (D. Del. 2001), the Court refused to dismiss an allegation that imposition of reach through royalties in a license agreement was a misuse of a patent to a drug discovery screen. About a year later, the same Court held that the reach through royalties were not a patent misuse because the grant of the license was not conditioned on the payment of reach through royalties. *Bayer v. Housey Pharmaceuticals, Inc.*, 228 F. Supp. 2d 467; 2002 U.S. Dist. LEXIS 21061 (D. Del. 2002).

[14] Since this conference, a few papers have appeared which present data on the effect of patents on the use of genetic diagnostics. For example, Cho *et al.*, *Effects of patents and licenses on the provision of clinical genetic testing services*, JOURNAL OF MOLECULAR DIAGNOSTICS, Vol 5, Number 1, Pages 3–8, February 2003, in which the authors conclude that "patents and licenses have a significant negative effect on the ability of clinical laboratories to continue to perform already developed genetic tests." The authors also note, however, that their study "does not address the issue of whether patents provided a major incentive for the initial research that led to the patent and development of the genetic tests that the laboratories subsequently stopped providing."

Section 5

DISPUTES OVER GENETICS IN ACADEMIA AND BUSINESS

24

One Size Fits All?

Robin Jacob
Judge of the Chancery Division of the High Court of England and Wales
Judge in Charge of the Patents List
Strand, London, WC2A 2LL
UK

Mark Twain's Connecticut Yank went to Warwick Castle, England. There he was magically transported back to the time of King Arthur. He did well. He was made First Minister. He was called Sir Boss. What was the first thing he did in power?

> "The very first official thing I did, in my administration—and it was on the very first day of it, too—was to start a patent office; for I knew that a country without a patent office and good patent laws was just a crab, and couldn't travel anyway but sideways or backwards."[1]

Was Sir Boss right? Or, more specifically, was he right for modern day conditions, for all technologies and for all countries?

I do not propose to answer that question in detail; it is a very big one. All that I can say is that it does seem inherently unlikely that "One size fits all" is a correct analysis of the benefits of patents (or other IP rights, for that matter) for all peoples everywhere. Certainly that was not the view of many presently advanced industrial countries at earlier stages of their development. Here are two examples:

[1] I have always wanted to use this quote in a lecture. How fortunate to be able to do so on the banks of the Mississippi—it was from the call of the men who measured the depth of the river for the pilots that Mr Clemens took his nom-de-plume. "Mark Twain" meant that the depth was 2 fathoms—dangerously close to the bottom.

1. In England, one of the first countries to have a patent system, you could get a patent for stealing ideas from foreigners and bringing them into the country for the first time. The first recorded British patent for an invention was indeed a stolen idea—it was granted in 1449 to John of Uthnam, a Dutchman who had special techniques for making stained glass windows. He was given a monopoly, plus tax breaks, in return for which he had to do the stained glass windows for Eton College and a Cambridge College. He also had to train apprentices in the new techniques.

> "... and because the said art has never been used in England and John intends to instruct divers lieges of the king in many other arts never used in the realm beside the said art of making glass, the king retains him therefor for life at his wages and fees and grants that no liege of the King learned in such arts shall use them for a term of twenty years against the will and assent of John, under a penalty of £200, whereof two parts shall be rendered to the king and one part to John, any liege who cannot levy that sum to suffer imprisonment without delivery save by the king's special command."

2. In America, copyrights to foreign authors were refused unless their works were first published in the USA. As a result of this process spanning the whole of the nineteenth century and much of the twentieth, America was the world center of piracy. An English judge put it this way in 1880:

> "Persons in the position of the defendants, that is of agents for an American publisher, must be taken to know that Americans are in the habit of printing and exporting piratical works and they must therefore know that they import books from America at the risk of their containing what is piratical and are thus committing an unlawful act and of being liable to be sued without notice.[2]

In those early days national interests, particularly the interest in encouraging national industry, prevailed over international equality. The growth of an international system, commencing with the great Treaty of Paris in 1883, began to change things. But the change was slow. National interests still prevailed. For instance, in many countries failure to exploit a patented invention by local manufacture was a ground for an exclusive licence. Trade-Related Aspects of Intellectual Property Rights (TRIPS) has largely changed that. But no one should underestimate the resentment it causes in underdeveloped and developing

[2] *Per* Sir George Jessel MR (1880) 15 Ch. D 501.

countries. I have had the benefit of attending two World Intellectual Property Organization (WIPO) seminars, one in Egypt for Arab nations in 1996 and most recently, this year in India for judges from Asia. The common complaints were two-fold: that TRIPS would destroy local pharmaceutical industries and that they would keep the price of medicines beyond the reach of the people. Make no mistake, these are complaints which not only have force but also cannot be ignored. They cannot be answered by words like "property." Nor can they be answered by vague promises of economic benefit. If you ask how Westerners would react if the boot were on the other foot, you may understand better. Look how fast people in the U.S. reacted during the anthrax scare—Cipro was patented by a foreign company and there was instantly much talk of compulsory licensing. That was just a scare—think how people in poor countries weigh respect for patents against a shortage or impossibly high prices of drugs for large-scale diseases.

In no field is the complaint more acute than in patents on or related to human genes. It becomes coupled not only with a pressing need for cheap medicines but also with emotional appeals to natural law—how can anyone "own" genes? The complaint is worsened by the fact that no one can design around gene patents. One of the driving forces used to justify the patent system, the incentive to design around, is gone. Another man's gene patent is an absolute bar—a complete block for the other person—in relation to whatever that gene does. This is a very important fact about such patents, a fact that has important effects and creates important conceptions in first world countries.

Gene patents are apt to have a very long reach. Consider the breast cancer gene patents on BRCA1 and BRCA2. These appear to affect all scientists researching those diseases and thus were greeted with much resentment. Can it really be right that only one commercial concern should have control over any use of these genes, either by regulating testing for breast cancer or treatment of the disease and other cancers affected by those genes? Can the patent system justify a monopoly over all testing for these cancers? I am told that the Myriad, who own the patents, are trying to insist not merely on a monopoly on testing kits, but on conducting the tests to the exclusion of hospitals and doctors as well. To outsiders that sort of reach may bring the patent system as it stands into question.

This brings me to the first subject I wish to discuss, the partly allied topics of compulsory licences and parallel imports: first compulsory licences. I believe that most considerations of these have been and are superficial and ill-informed. In my experience as a practicing lawyer there were two periods when first compulsory licences were available in the UK. When I came into practice in the late 1960s, importers of generic drugs from non-patent countries (mainly Italy at the time) were beginning to make use of a provision of our patent law which had been introduced as a result of experience during the First World War.

The pre-war success of the great German dye and chemical industry was seen to have been partly based on the successful use of patents to confine industrial know-how to Germany, to the disadvantage of home manufacture. The "solution" was to provide that anyone could, on application to the Patent Office, have a licence "on such terms as seems fit."

In the early 1960s, aspiring importers sought to use this provision[3] to obtain compulsory licences for already-successful drugs. The earliest targets were the early psychotropic drugs,[4] shortly followed by the benzodiazapines.[5] The battles over these drugs largely concerned the royalty rate—should it be enough to compensate the companies for ongoing research and promotion costs as well as profit? Should it be fixed or should it fall with prices? Not only did patentees pursue fair royalties (which I do not think they received) but also they used the procedural system to delay. Under the law as it stood the licensee had no license until all the terms were settled—so the more you could argue, the greater the delay. Looking back I think it is fair to conclude that the entire business was essentially a failure from society's point of view. Some middle men made money, and the State (the UK pays for most Patented pharmaceuticals), saved a little but not much. Research and risk money were adversely affected.

In 1977 this type of licence was abolished, but another emerged as a result of the extension of the term of a patent from 16 to 20 years and the removal of any possibility of extension of patent term.[6] By way of transition, "new existing patents" (those who had more than eleven years to run) were granted an automatic extension of term to 20 years but were subject to "licences of right" for the last four of these. Again the middlemen emerged, and again the battles were about rates of royalty. But this time the drug companies did better legally,[7] though in practice I think the middlemen made a lot more money because their sales were so much more than under the prior regime.

What is to be concluded about compulsory licences? First and foremost, unless they are well thought-out, they are not likely to be effective in the way intended. In particular they are a very poor way of dealing with drug prices, especially if they are available to middlemen rather than being confined

[3] By this time this had been re-enacted as s.41 of the Patents Act 1949.

[4] E.g., Smith Kline and French's "Stelazine" (trifluoperazine)

[5] First Hoffmann-La Roche's Librium (chlordiazepoxide) and Valium (diazepam) and then a number of follow-ups.

[6] Until 1977 a patentee who could show that he had had "inadequate remuneration" could obtain an extension of his patent for up to 10 years. Drug companies could often do this – demonstrating many "lost years" and only late recompense for early, mainly abortive, research.

[7] SmithKline got 46% of their pre-licence price for Tagamet (cimetidine). Contrast that with 26% which was about the highest achieved under the s.41 regime. Even at 46%, the generic companies, within a few months, took the entire "generic" market (*i.e.*, that met by doctors prescribing generically rather than by the trademark).

to promoting local manufacturers. Middlemen are not interested in low prices—what they want are higher prices so they can sell at that price minus a little bit. Compulsory licences are, on the other hand, a very good way for lawyers and accountants to make money. No doubt better systems can be devised than we had in the UK. But I also doubt that rewarding unproductive middlemen is the best way to get cheaper drugs, even for markets which by any practical or moral standard should have them.

Compulsory licences for local manufacturers, on the other hand, may have a proper purpose. Whether such licences should be available is an important question, one allied to the subject of parallel imports.

What about compulsory licences for research tools? Not patents on scanners, stethoscopes, and the like, all of which as considered artifacts. It is patents on probes, tools employing a genetic code, that may cause problems. On the one hand, some legal systems provide an exception to the scope of protection for research. This may be wide or narrow, but it is an all or nothing solution. Surely better is an approach which recognizes that the patentee should have some reward but also controls the size of that reward: a compulsory licence. There are many who say that a complete monopoly over a gene is inappropriate, just because you were first to isolate it, or even if you are first to isolate it and know what it is for. It may be that well-designed compulsory license provisions would be helpful in this area. It is not as though compulsory licence provisions cannot be made to work; in fact they work very well, for instance, in relation to musical copyrights.[8] There is, I think, too great a danger in calling patents "property" and arguing on that basis that they must be inviolate. A lot depends on the nature of the invention—one that is not only high risk but involves billions in research and years in time to bring to market stands in a different economic category than an isolated gene, especially one isolated by more or less standard techniques.

I turn to parallel imports. The old English rule was that the patentee has consented to his product being on the market anywhere in the world if he sold it without any fetter. Only if he labeled the product in such a way as to bring home to those who dealt in it (including subsequent purchasers) the product's license for sale in limited territories could he use his patents outside those territories to prevent parallel imports. Other countries have different rules, and there may be different rules about other intellectual property rights. The European Union, for instance, has settled on a rule for trademarks stating that unless there is express consent or conduct unequivocally indicating such consent, no trademarked

[8] Cases about these before the Copyright Tribunal in the UK show that they are, or can be, just as complex as compulsory licences for patents. It would not be right to dismiss the compulsory licences in copyright as an inappropriate analogy because they are just a question of a "penny a play."

goods can be imported into the EU without the trademark owner's consent.[9] So Kodak film made in Rochester, New York by Kodak and sold in the US will, if imported into the EU, infringe upon Kodak's EU trademarks.

My suggestion is that rational world, or at least regional, rules for parallel imports are seriously needed, particularly in the case of pharmaceuticals. Drug companies cannot expect to make serious money in poor countries. No one in such places can afford Western prices. So either the companies sell at a much lower price or not at all. But they cannot afford to sell at lower prices if the drugs immediately leave the poor country and reappear as parallel imports in high-price developed countries, which is exactly what middlemen will do if they can. Strong barriers to parallel imports are needed in the name of humanity in this kind of case. If they are not available then once again, either the drugs will not be sold at all in the poor countries or will only be sold at a high price. Either situation is a strong incentive for local manufacture via compulsory licensing. Putting it bluntly, if the drugs are not made available cheaply, the case for compulsory licensing becomes unanswerable. And they can only be sold cheaply if they are confined to the markets where they are sold.

That makes it important to put in place two systems to ensure such confinement. First, one can reasonably ask that some sort of system in the country or region of import to prevent export be put in place. No one can expect the system to work on its own—that would be expecting too much of local court and administrative systems. So the first world countries need to amend their own laws to ensure that patent enforcement works well against "grey" drugs from third world countries. The old British rule for patents is not likely to be enough. Something like the EU's rule for trademarks is optional—proof positive that consent was given for re-sale in the first world when drugs were sold in the third world. Otherwise there is infringement.

I turn to my next main topic, again one alluded to by Sir Boss. He wanted "good patent laws." If one stands back and takes a view of the world, the differences in substantive patent law are, in principle, not that great. This is because the main objections to validity are lack of novelty and obviousness. Things like first to invent and first to file or grace periods and restrictions on the sort of thing that can be patented (computer programs, methods of medical treatment, etc.), important though they are, are second order compared with the great duo of novelty and obviousness.

I said "in principle" because what matters in practice is how the rules are enforced. In her biography of Jeremy Bentham, Mary Mack said two things of penetrating wisdom:

[9] A purist would say this is bizarre because is it is contrary to the central function of a trademark—to denote trade origin. If a mark does that, it fulfills its task—in a global market it is irrelevant where the goods are actually made or sold.

"Substantive laws are not self-enforcing. They pre-suppose an efficient adjectival law."

"The law is everything and in it procedure and evidence come first."

I venture to suggest that the various systems devised for judging whether patents are valid or infringed are, frankly, not very good. No one outside America, for instance, regards the use of jury trial as remotely sensible. William Penn and his collaborator, Mead, were put on trial in London for sedition. Having heard the case, the jury retired to consider their verdict. They decided to grant an acquittal. The Judge refused to accept the verdict and directed them to convict. They refused. The Judge had them locked up without food or water. They held out. Eventually the impasse was overcome by an application to the Court of King's Bench which overruled the Judge, thus establishing the right of a jury to come to its own verdict if it chooses in the face of the evidence. That lesson was brought to America, and it remains an important bastion of freedom. But Americans have taken it to an illogical conclusion, that the right of man is to have his case tried by a body of people who cannot possibly understand.[10]

To right the potential for this kind of injustice, Americans created the Federal Court of Appeal. But this works only after an incorrect verdict and all the damage it can do. Even that court is likely to be severely tested in high technology cases. After all, its judges (and law clerks) are themselves not experts. Some things which seem to outsiders as clever and thus inventive are actually pretty mundane to those who actually understand the art. Some might say, for instance, that it is wrong to assert that it is "inventive" to ascertain a gene sequence by standard techniques, assigning patentability to that which is really no more than the predictable result of routine research.

Conversly, the English use specialist patent judges. They have to give reasons. They can be assisted in understanding the case by a scientific advisor, not a member of the court. In high technology cases I think we are at the limit of what is doable. It is not easy even for an experienced patent lawyer-judge to understand unaided new and very complex technologies loaded with jargon.

On both sides of the Atlantic in the common law courts, the English use party-appointed experts, the expensive process of discovery and extended cross-examination. The costs of this are not merely financial. Large cases take

[10] I am aware that many U.S. trial lawyers express some confidence in trial by jury even for patent cases. I have to say that that confidence is not shared by any foreign lawyer I have ever met, whether from the common law or inquisitorial systems, whether in Europe or Asia or Australasia. Nor have I met a scientist who is happy to have high technology questions judged by laymen.

scientists off their proper job. I had one case, for instance, where the clients reckoned that fighting it cost them three months research time.

There is also the problem of biased party experts—some scientists will say anything for money. We in England are trying to keep them under control by saying that their first duty is to the court (among other things). On the other hand, Americans are often looking for performance rather than a solid expert view, as I would do if I were aiming for a jury trial or even a trial before a non-specialist Judge.

But even if you do have experts who are trying to be fair and non-partisan there is a basic problem in the adversarial system. It is an old problem. The young Learned Hand said in 1901:

> "The trouble with all this is that it is setting the jury to decide where doctors disagree"[11]

And it is not much better if it is "setting" a judge alone to decide really complex matters.

That great jurist James Fitzjames Stephens articulated the problem in a different way even earlier:

> "Few spectacles, it might be said, can be more absurd and incongruous that that of a jury composed of twelve persons who, without any previous scientific knowledge or training, are suddenly called upon to adjudicate in controversies in which the most eminent scientific men flatly contradict each other's assertions."[12]

There is also the problem of expert selection, a problem which is very old and again gives one an uneasy feeling that something is wrong. An English Judge stated the point fair and square as long ago as 1873:[13]

> "Now, in the present instance, I have the evidence of experts on the one side, and on the other, and, as usual, the experts do not agree in their opinion. There is no reason why they should. As I have often explained, since I have the honour of a seat on this bench, the opinion of an expert may be honestly obtained, and it may be quite different from the opinion of another expert, also honestly obtained. But the mode in which evidence is obtained is such as not to give the fair result of scientific opinion to the court. A man may go, and does, sometimes,

[11] [1901] HARVARD LAW REVIEW.

[12] Address to the Juridical Society, 1859.

[13] Sir George Jessel MR in *Thorne v. Worthington Skating Rink Co.*, LR 6 Ch.D n. at p.415.

> to half a dozen experts. He takes their honest opinion: he finds three in his favour and three against him; he says to the three in his favour: "Will you be kind enough to give evidence?" He pays the ones against him their fees and leaves them alone; the other side does the same. It may not be three out of six, it may be three out of fifty. . . . I am sorry to say the result is that the court does not get the assistance from the experts, which, if they were unbiased and fairly chosen, it would have a right to expect."

On the continent of Europe the tendency is to use a comparatively inexpert tribunal which itself consults an "expert." Whether that individual is really an expert is often questionable. And he may often answer the "wrong" question or take ages to come up with any answer at all. He is effectively the judge, which has lots of attendant dangers, especially when there is no opportunity to test his views.

It is further unsatisfactory to have different tribunals judging what is in substance the same case in different countries. You can get a world patchwork of decisions at great expense and delay. There are two sorts of problems: different national jurisdictions and getting an adequate tribunal. I think the problems can be linked and ultimately *should* be linked for high technology cases such as gene patents.

What then is the answer? So far as the different jurisdiction problem goes there are three in principle. We carry on as we are, we let the courts of one country decide for other countries, or we have an international tribunal. I set aside "carry on as we are" for the moment to look at the other two solutions. The second answer confers extra territorial jurisdiction on national courts. The recent drafts for a new Hague treaty go down that line. I think it is wrong and dangerous. You cannot have German or English or Estonian or Greek courts deciding that an American patent is being infringed in the US and ordering an American defendant to stop manufacture in America. Especially unacceptable would be foreign enforcement of U.S. jury verdicts.

What about the third solution? This would involve a treaty, and it would crucially require that a proper tribunal be established. That is why there is link between the two problems of a modern judicial system for patents and the cross-border difficulties. Is it possible to do better than any of the existing models? I think so. I think a court with experienced patent lawyer-judges, assisted (as members of the court or possibly as a specialist jury) by scientists can be made to work. I do not think, however, that scientists should be permanent members of the court. In a field such as genetics you really need people who are or who have recently been active. Otherwise they are not really experts themselves and could be dangerous—witness the sort of problem to be had in the European Patent Office where the Office relies too much on its own expertise and not enough on properly prepared impartial party-provided expert evidence.

Is such a solution likely? I think so, in the long run. Europeans are gingerly going down that line, possibly in the next few years. If not this time, then next time round—say 10 years—language is an enormous complication. Gradually English as the language of business and technology will be sufficiently widespread to make this possible. But the European solution is likely to be modeled on the sort of court I have mentioned—it is a hybrid of the common law and continental approaches.

Looking at the problem more widely, the needs of international trade are more and more requiring regulation at a supra-national level. The WTO tribunals deal with international trade, disputes between states. I suspect that one day there will be international patent courts, that the WTO tribunals are just precursors of things to come. The days of the nation state for big patent disputes are numbered. The number of days is big but finite.

Postscript: The above is for the most part a talk I gave in St. Louis, Missouri in April 2002. Not much has changed since then, but readers might like to note that some of the problems I mention are addressed in a discussion paper published by the Nuffield Council on Bioethics (www.nuffieldbioethics.org) and the Report of the Commission on Intellectual Property Rights (www.iprcommission.org). Both these documents (and the resulting discussions) are worth reading. There has been some progress along the lines I predicted in regards to international patent litigation in that the EU is aiming to bring in a full-fledged EU patents court by the year 2010.

25 Some Empirical Evidence on How Technologically Complex Issues Are Decided in Patent Cases in U.S. District Courts

Roderick R. McKelvie
Partner
Fish & Neave
Washington, DC 20006

Former U.S. District Court Judge, District of Delaware

If you asked Scott Kieff about our communications leading up to this conference, he would probably tell you I was a little anxious about what I could contribute to this discussion. Graciously, Sir Robin Jacob has thrown me a bone that I can gnaw on. That is, I would like to take a few minutes to talk about how we in the United States resolve patent disputes. I will try not to be too defensive as I describe our system of civil justice and comment on a few of Sir Robin's suggestions for improving how we resolve intellectual property disputes.

It is correct that, under our Constitution, we try patent infringement cases, including cases relating to biotechnology disputes, in the United States District Courts, and more often than not, we try them to a jury of citizens selected from the community. (In Delaware, for example, we select our juries from a list of registered voters and licensed drivers.) It is not quite correct that we are trying cases to people who cannot possibly understand the matters at issue. Nor is it correct that the jury has the final word in these cases, and that we judges defer to a jury's decision if it is not supported by evidence.

Let me start with some information on litigation in the United States District Courts, and specifically on the increase in the number of patent infringement cases. I have gathered the numbers in Table 25.1 from an annual publication put out by the Administrative Office of the United States Courts.

As you can see, patent infringement litigation is booming. In the 8 years from 1994 to 2001, the number of patent infringement cases increased from 1617 to 2520. While the number of patent cases filed has been increasing, the number of cases that go to trial peaked in 1997 and 1998 at 103, and

0065-2660/03 $35.00

Table 25.1. Certain Cases Filed in U.S. District Courts 1994–2001[1]

Year	Criminal cases filed	Civil cases filed	Patent cases filed	Patent cases tried to jury	Patent cases tried non-jury	% of patent cases reaching trial	% of all cases reaching trial
1994	45,473	236,391	1617	64	26	5.9	3.5
1995	45,788	248,335	1723	47	42	5.9	3.2
1996	47,889	269,132	1840	54	47	6.0	3.0
1997	50,363	272,027	2112	54	49	5.6	3.0
1998	57,691	256,787	2218	62	41	5.1	2.6
1999	58,889	260,271	2318	61	37	4.5	2.3
2000	61,832	259,517	2484	56	29	3.9	2.2
2001	61,898	250,907	2520	55	28	3.3	2.2

[1]Criminal Cases filed, Judicial Business of the United States Courts, Table D-1
Civil Cases filed, Judicial Business of the United States Courts, Table C-3
Patent Cases Filed, Judicial Business of the United States Courts, Table C-2A
Patent Cases Tried to Jury, Judicial Business of the United States Courts, Table C-4
Patent Cases Tried Non-Jury, Judicial Business of the United States Courts, Table C-4
Percent Patent Cases Reaching Trial, Judicial Business of the United States Courts, Table C-4
Percent All Cases Reaching Trial, Judicial Business of the United States Courts, Table C-4.

declined to a low of 83 in 2001. I expect that you can probably find a correlation between the decline in the number of trials and two significant court decisions: the Federal Circuit and Supreme Court's decisions on claim construction in *Markman* (*Markman v. Westview Instruments, Inc.*, 517 U.S. 370 (1996), where the Supreme Court found claim construction disputes should be resolved by the court rather than the jury) and the Federal Circuit's decision in *Festo* (*Festo Corp. v. Shoketsu Kinzoku Kogyo Kabushiki* Co., 234 F.3d 558 (Fed. Cir. 2000) (en banc), *cert. granted*, 533 U.S. 915 (U.S. June 18, 2001), where the Federal Circuit found prosecution history estoppel bars a claim for infringement under the doctrine of equivalents where, during the prosecution of the patent, the applicant amended a claim limitation for reasons relating to patentability).

The last two columns on this table show two other interesting trends. The decline in the percentage of all cases reaching trial is continuing, dropping from 3.5 to 2.2%. And while the percentage of patent cases reaching trial is still higher than the percentage for all cases, it too is declining and at a more rapid rate, from 5.9 to 3.3%. The numbers in Table 25.2 show these percentages have dropped substantially over the 28 years from 1974 to 2001. They also show that the absolute number of trials in our federal courts has also declined substantially, from a high of 12,500 or so in 1985 to approximately 5,400 in 2001.

Under U.S. rules, in cases such as patent infringement where a party has a right to a jury trial, either party can elect to have the case tried to a jury. In 1994, when approximately 90 patent cases went to trial, 64 cases were tried to a jury and 26 cases were tried to a judge. In 2001, when 83 patent cases went to trial, 55 were tried to a jury and 28 were tried to a judge.

It is not correct that the jury has the final word in these cases. Nor is it correct that the trial judges must defer to a jury's decision without regard to whether it is supported by the evidence. There are numerous points during the trial process where the trial judge has the opportunity to ensure that correct results are reached. Actually, the most significant stage in this process for a trial judge is after the trial, when the trial judge reviews the verdict and trial record to ensure that the verdict is supported by the evidence. The trial judge gives his or her analysis in a post trial opinion, which is then reviewed by the judges of the Court of Appeals for the Federal Circuit. More than once I have set aside a jury verdict in a patent case that was not supported by the evidence.

Sir Robin has suggested that we consider looking to specialist courts to resolve these disputes, courts with specialist patent judges and, perhaps, an international tribunal made up of both judges and scientists. In the U.S., we do not try our cases in specialist courts or to specialist judges. Patent infringement cases can be filed in any one of the 94 United States District Courts, where the judges may hear criminal cases and a variety of civil matters including securities, bankruptcy, and civil rights cases. In order to ensure some degree of uniformity, however, all appeals in patent infringement cases do go to a

Table 25.2. U.S. District Court Cases Reaching Trial 1974–2001[1]

	Total cases				Patent cases				Diversity cases			
Year	Non-Jury	Jury	% of Jury Trials	% Reaching Trial	Non-Jury	Jury	% of Jury Trials	% Reaching Trial	Non-Jury	Jury	% of Jury Trials	% Reaching Trial
1974	5070	3312	39.5	8.7	83	6	6.7	10.5	1120	2321	67.4	13.2
1975	5210	3512	40.3	8.4	111	15	11.9	14.5	1268	2375	65.2	13.1
1976	5245	3544	40.3	8.1	69	15	17.9	10.2	1341	2303	63.2	12.9
1977	5531	3516	38.9	7.8	85	9	9.6	10.8	1291	2277	63.8	12.0
1978	5842	3531	37.7	7.6	77	7	8.3	11.0	1374	2275	62.3	12.0
1979	6027	3603	37.4	6.9	60	13	17.8	9.1	1321	2178	62.2	10.9
1980	6171	3920	38.8	6.5	82	17	17.2	12.0	1398	2339	62.6	10.8
1981	6714	4702	41.2	6.6	73	19	20.7	11.1	1614	2825	63.6	11.3
1982	6538	4788	42.3	6.1	76	30	28.3	11.5	1582	2830	64.1	10.1
1983	6561	5064	43.6	5.4	88	24	21.4	11.9	1558	2976	65.6	9.1
1984	6525	5555	45.9	5.0	67	23	25.6	9.0	1619	3144	66	8.8
1985	6292	6278	49.9	4.7	65	20	23.5	8.6	1626	3531	68.5	9.0
1986	6063	5645	48.2	4.4	63	26	29.2	8.2	1626	3080	65.4	8.0
1987	5614	6299	52.9	5.0	52	37	41.6	8.6	1516	3562	70.1	8.0
1988	5698	5920	50.9	4.9	54	54	50	9.6	1508	3051	66.9	6.8
1989	5486	5676	50.9	4.8	67	38	36.2	8.4	1486	2932	66.4	6.8
1990	4480	4783	51.6	4.3	62	34	35.4	8.5	1162	2549	68.7	6.4
1991	4133	4294	50.9	4.0	47	39	45.3	7.8	980	2241	69.6	5.9
1992	3751	4287	53.3	3.5	38	52	57.8	6.8	852	2134	71.5	5.2
1993	3622	4118	53.2	3.4	47	47	50	6.4	811	1965	70.8	5.5
1994	3460	4450	56.3	3.5	26	64	71	5.9	704	1910	73	4.9
1995	3317	4126	55.4	3.2	42	47	52.8	5.9	635	1718	73	5.1
1996	3206	4359	57.6	3.0	47	54	53.5	6.0	645	1652	71.9	4.8
1997	2802	4557	61.9	3.0	49	54	52.4	5.6	554	1767	76.1	4.3
1998	2452	4331	63.9	2.6	41	62	60.2	5.1	527	1562	74.8	3.9
1999	2226	4002	64.3	2.3	37	61	62.2	4.5	487	1429	74.6	2.8
2000	2001	3779	65.4	2.2	29	56	65.9	3.9	479	1286	72.9	3.4
2001	1768	3633	76.3	2.2	28	55	66.3	3.3	420	1395	76.9	3.6

[1]Civil Cases Terminated, by Nature of Suit and Action Taken. Judicial Business of the United States Courts, Table C-4.

Table 25.3. Patent Cases Filed by District 1997–2001[1]

	FY97	FY98	FY99	FY00	FY01
Central District of California	165	196	227	275	254
Northern District of California	163	161	147	175	154
Northern District of Illinois	125	117	140	152	152
Delaware	80	99	95	103	141
Southern District of New York	80	80	87	112	156
Eastern District of Virginia		80			

[1]U.S. District Courts Civil Cases Commenced by Nature of Suit and District During the Twelve Month Period Ended Sept. 30, 2001, Table C-3, Special, Judicial Business of the United States Courts.

court of limited jurisdiction, or a court of specialists, the Court of Appeals for the Federal Circuit. While we don't have specialized trial courts for patent infringement cases, what has happened is the "market" for patent litigation has moved toward certain districts as the courts favored by parties filing patent infringement actions.

Table 25.3 identifies the districts with the most patent infringement cases and shows the number of those cases filed in each of those districts for the past 5 fiscal years.

Each year, the highest number of patent infringement cases are filed in the Central District of California, which includes Los Angeles. Until last year, that district was followed by the Northern District of California, which includes San Francisco and the Silicon Valley; the Northern District of Illinois, which includes Chicago; and Delaware, the state where a large number of businesses are incorporated. As Herbert F. Schwartz has noted in conversations during this conference, the interesting trend shown in this table is in the filing of cases in the Southern District of New York, which includes Manhattan. In 5 years from 1997 to 2001, the number of patent cases filed in the Southern District has gone from 80 to 156. During this same period, the number of cases filed in the Northern District of California has stagnated and actually dropped. If you ask a trial judge or a litigator why this is so, he or she will probably tell you this drop in the number of cases filed in the Northern District is related to local rules recently adopted in that court. Certain rules may be slowing up the disposition of patent infringement litigation and discouraging lawyers from filing cases in that court.

Table 25.4 includes the information that is more significant to me. We have approximately 684 judges sitting in 94 districts. Some district courts have only a few judges. In Delaware, for example, we have 4 district court judges. Other districts have many more judges. For example, the Central District of California has 27 district court judgeships. With 2520 patent cases filed a year,

Table 25.4. Number of Patent Cases Filed Per Judgeship in 2001[1]

	Patent filings	Judgeships	Patent cases per judge
Central District of California	254	27	9.41
Northern District of California	154	14	11.00
Northern District of Illinois	152	22	6.91
Delaware	141	4	35.25
Southern District of New York	156	28	5.57

[1]U.S. District Courts Civil Cases Commenced by Nature of Suit and District During the Twelve Month Period Ended Sept. 30, 2001, Table C-3, Special, and U.S. District Court—Judicial Caseload Profiles for Twelve Month Period Ending September 30, 2001, Judicial Business of the United States Courts.

the typical district court judge (if there is such a thing) should have 3 or 4 patent cases on his or her docket at any given time. In Delaware, the number is 35; actually I have 50 patent infringement cases assigned to me. New York is now up to about 5.7 patent cases per judge. Each judge in the Northern District of California has 11 patent cases.

Lawyers will tell you that there may be a number of reasons why a relatively higher number of patent infringement cases are filed in these districts, including how quickly the judges move the cases on their dockets, how familiar the judges may be with these types of cases, and the relative sophistication of the jury pool.

In the typical civil case, the judge will not have much information on the potential jurors beyond their name, age, marital status, education, occupation, and employer. I have prepared Table 25.5 from information I gathered during 13 patent infringement trials. In the typical case, we will try the case to a jury of 9. To select the 9, we call forward 15 prospective jurors and allow each side to exercise peremptory challenges to strike 3. I began tracking this information to see if I could pick up a pattern of lawyers striking by gender or education.

As you can see from the summary at the bottom of the table, in Delaware, approximately 28% of the jurors called forward to sit have college degrees. On average, the lawyers in the patent infringement cases tend not to use peremptory challenges to strike the more or less educated prospective jurors. That is, if I have a jury of nine in a patent infringement case, six are are high school graduates and three are college graduates.

The *CellPro* case has been mentioned a couple of times during the program. I was the trial judge in that case. (*Johns Hopkins University v. CellPro, Inc.*, 931 F. Supp. 303 (D. Del. 1996) *affirmed in part, vacated in part*, 152 F.3d 1342 (Fed. Cir. 1998)) We had two trials. I set aside the verdict after the first

Table 25.5. Use of Peremptory Challenges in Certain Patent Cases Tried in the Distric of Delaware

Caption	C.A.No.	Striking	Jurors struck by Plaintiff	Jurors struck by Defendant	Jurors chosen
Johns Hopkins v. Cellpro 7/24/95	94–105	panel 14 (H.S.+4=4) strikés (H.S.+4=2) jury 8 (H.S.+4=2)	23 yr. female, H.S.+4 36 yr. male, H.S.+2 43 yr. female, H.S.+1	43 yr. male, H.S.+2 52 yr. male, H.S.+1 47 yr. male, H.S.+6	45 yr. female, H.S. 45 yr. female, H.S.+4 40 yr. female, education not listed 49 yr. female, H.S.+4 68 yr. female, H.S.+1.5 48 yr. female, H.S.+2 53 yr. female, education not listed 31 yr. female, H.S.+3
Berg v. Molex 9/21/95	94–470	panel 15 (H.S.+4=7) strikes (H.S.+4=3) jury 9 (H.S.+4=4)	56 yr. male, H.S.+4 53 yr. female, H.S. 60 yr. male, H.S.+4	30 yr. female, H.S.+2 55 yr. male, H.S.+6 35 yr. male, H.S.+2	57 yr. female, H.S. 66 yr. female, H.S.+4 31 yr. male, H.S.+4 33 yr. male, H.S.+4 27 yr. male, H.S.+5 43 yr. female, education not listed 47 yr. male, education not listed 49 yr. female, H.S.+1 22 yr. female, H.S.+2.5
Georgia Pacific v. USG 1/29/96	94–489	panel 17 (H.S.+4=1) strikes (H.S.+4=1) jury 9 (H.S.+4=0)	26 yr. female, H.S.+4 43 yr. female, H.S. 27 yr. female, H.S.+2	60 yr. female, not listed 42 yr. female, H.S. 63 yr. female, H.S. 22 yr. male, H.S.	70 yr. male, H.S.+2 49 yr. female, H.S. 50 yr. female, education not listed 64 yr. female, H.S.

(*Continues*)

Table 25.5. *(Continued)*

Caption	C.A.No.	Striking	Jurors struck by Plaintiff	Jurors struck by Defendant	Jurors chosen
			50 yr. male, 11th grade		65 yr. male, 10th grade 65 yr. female, H.S.
Johns Hopkins v. Cellpro 3/4/97	94–105	panel 15 (H.S.+4=4) strikes (H.S.+4=3) jury 9 (H.S.+4=1)	69 yr. male, H.S. 35 yr. female, H.S.+2 67 yr. female, 8th grade	44 yr. male, H.S.+8 37 yr. male, H.S.+4 49 yr. male, H.S.+4	44 yr. female, H.S. 55 yr. male, H.S. 58 yr. female, H.S. 54 yr. female, H.S. 58 yr. male, H.S. 54 yr. male, H.S. 58 yr. female, H.S.+5 55 yr. female, H.S.+1.5 58 yr. male, education not listed
Ampex v. Mitsubishi 3/31/97	95–582	panel 17 (H.S.+4=8) strikes (H.S.+4=0) jury 9 (H.S.+4=8)	66 yr. female, H.S.+1 59 yr. male, H.S. 69 yr. female, H.S.+1 43 yr. male, H.S.	51 yr. male, H.S. 49 yr. male, H.S. 45 yr. male, H.S.+1 66 yr. male, H.S.+1	51 yr. female, H.S. 63 yr. male, H.S.+4 47 yr. male, H.S.+6 46 yr. male, H.S.+9 62 yr. male, H.S.+4 55 yr. female, H.S.+4 38 yr. female, H.S.+4 39 yr. male, H.S.+4 +68 yr. male, H.S.+6
Manchak v. Sevenson Environmental 6/9/97	95–709	panel 15 (H.S.+4=3) strikes (H.S. + 4=2) jury 9 (H.S.+ 4=1)	42 yr. female, H.S. 31 yr. male, H.S. 40 yr. female, H.S.+2.5	26 yr. female, H.S.+4 52 yr. male, H.S. 39 yr. female, H.S.+4.5	38 yr. female, H.S.+2 80 yr. female, H.S.+2 43 yr. female, H.S.+1 27 yr. male, H.S. 57 yr. male, H.S.

					48 yr. male, H.S. 28 yr. female, H.S.+4 26 yr. male, H.S. 37 yr. female, H.S.+3
Quantel v. Adobe 9/8/97	96–18	panel 17 (H.S.+4=3) strikes (H.S.+4=3) jury 9 (H.S.+4=0)	32 yr. female, not listed 35 yr. male, H.S.+4 43 yr. male, H.S.+4 35 yr. female, H.S.	23 yr. female, H.S.+4 39 yr. male, H.S.+2 44 yr. male, H.S.+2 57 yr. male, not listed	70 yr. male, 9th grade 46 yr. female, H.S. 65 yr. female, H.S. 59 yr. male, 6th grade 66 yr. female, H.S. 26 yr. male, H.S. 45 yr. female, H.S.+2 68 yr. female, H.S. 66 yr. male, 10th grade
Allied Signal v. Cooper Automotive 6/1/98	96–540	panel 15 (H.S.+4=4) strikes (H.S.+4=3) jury 9 (H.S.+4=1)	52 yr. female, H.S.+4 67 yr. male, H.S.+7 37 yr. male, H.S.+4	21 yr. female, not listed 36 yr. female, H.S.+2 42 yr. female, H.S.	63 yr. male, not listed 38 yr. female, H.S.+1.5 48 yr. female, H.S.+6 41 yr. male, H.S.+2 41 yr. male, H.S. 51 yr. female, H.S. 56 yr. female, 11th grade 59 yr. female, H.S. 24 yr. male, H.S.+2.5
Monsanto v. Mycogen and Novartis 6/15/98	96–133	panel 19 (H.S.+4=4) strikes (H.S.+4=2) jury 10 (H.S.+4=2)	66 yr. male, H.S.+4 37 yr. female, H.S.+3 55 yr. male, H.S.+2	65 yr. male, 8th grade 20 yr. male, H.S. 56 yr. female, H.S. 61 yr. male, H.S. 50 yr. male, H.S.+4 47 yr. male, H.S.+2	25 yr. female, H.S. 43 yr. female, H.S.+1 24 yr. male, H.S.+3 32 yr. female, H.S.+1+3 voc. 46 yr. female, H.S. 32 yr. male, H.S.+2 46 yr. female, H.S.+4

(*Continues*)

Table 25.5. *(Continued)*

Caption	C.A.No.	Striking	Jurors struck by Plaintiff	Jurors struck by Defendant	Jurors chosen
					64 yr. female, H.S.+1 voc. 46 yr. male, H.S.+3 70 yr. male, H.S.+4
Novartis v. Monsanto and DeKalb 10/26/98	97–39	panel 18 (H.S.+4=7) strikes (H.S.+4=2) jury 9 (H.S.+4=5)	45 yr. male, H.S.+2 51 yr. male, H.S.+1 59 yr. male, H.S.+4	Defendant Monsanto: 52 yr. male, 10th grade 27 yr. male, H.S. 49 female, H.S. Defendant DeKalb: 40 yr. female, H.S.+2.5 28 yr. female, H.S.+2 24 female, H.S.+4	50 yr. male, H.S.+4 30 yr. female, H.S.+2 58 yr. female, H.S.+2 49 yr. male, H.S.+3 59 yr. female, H.S.+6 49 yr. male, H.S.+6 34 yr. female, H.S.+4 57 yr. female, H.S.+4 29 yr. female, H.S.+2
LNP v. RTP 11/12/98	96–462	panel 16 (H.S.+4=4) strikes (H.S.+4=1) jury 10 (H.S.+4=3)	59 yr. female, not listed 34 yr. female, not listed 50 yr. female, H.S.+2+1	46 yr. male, H.S.+1 40 yr. male, not listed 26 yr. female, H.S.+6	47 yr. male, H.S. 50 yr. female, not listed 49 yr. female, H.S.+6 37 yr. female, H.S.+5 69 yr. female, H.S. 36 yr. male, H.S. 69 yr. male, H.S. 40 yr. female, H S.+4 50 yr. female, H.S. 63 yr. female, H.S.+3

KX Industries v. Culligan Water Technologies 3/8/ 99	98–88	panel 15 (H.S.+4=4) strikes (H.S.+4=1) jury 9 (H.S.+4=3)	52 yr. male, H.S.+7 60 yr. male, H.S.+3.5 43 yr. male, H.S.+1	59 yr. male, H.S. 50 yr. female, H.S. 44 yr. male, H.S.+2	24 yr. male, H.S.+4 61 yr. male, H.S. 32 yr. female, H.S.+1+4 43 yr. male, H.S.+5 40 yr. male, H.S.+3 33 yr. male, H.S. 65 yr. male, H.S. 48 yr. female, H.S. 47 yr. female, H.S.
Silicon Graphics v. Nvidia 7/12/99	98–188	panel 15 (H.S.+4=6) strikes (H.S.+4=3) jury 9 (H.S.+4=3)	18 yr. female, not listed 43 yr. male, H.S.+4 24 yr. male, H.S.+4	58 yr. male, H.S.+11 66 yr. male, H.S.+2 50 yr. male, 8th grade	37 yr. male, H.S. 40 yr. female, H.S.+2 58 yr. female, H.S. 57 yr. female, H.S. 42 yr. female, H.S.+4 33 yr. female, H.S.+1 27 yr. female, H.S.+5.5 38 yr. male, H.S.+6 61 yr. female, H.S.+2

Number of Cases	13
Total Number of Jurors in Panels	208
With College Degrees	59
% With College Degrees	28.4
Strikes	
Total Number Struck	90
With College Degrees	26
% with College Degrees Struck	28.9
In Jury	
Number of Jurors	118
With College Degrees	33
% of Jurors with College Degrees	27.9 or 3.1 college graduates per jury of 9.

trial, as I had given the jury an incorrect instruction on how the claims of the patent should be construed. Looking at this table, it appears the defendant wanted women on the jury.

Last, I want to share a few comments on the different roles scientists can take on during a patent infringement trial. Typically, we see and hear from scientists when they appear in our courtroom as expert witnesses, hired by one party to offer to the judge or jury opinions on technical or scientific matters at issue in the trial. We recognize that these experts are paid advocates, with an allegiance to one side or the other. Our rules of civil procedure also provide that a trial judge can retain a person to give technical advice and assistance to the judge. Not many judges use this rule, perhaps because it does not fit comfortably in our adversarial system. Lawyers are nervous about judges receiving and acting on information that is not shared with the lawyers. We prefer to have facts and opinions that a judge or jury may rely on that have been tested by cross examination. While some judges have developed procedures that address this problem when they retain an expert, it is not clear those procedures work well or that other judges and lawyers will adopt them. This same concern—having a transparent process where each side is aware of the information a judge receives and acts on—applies to why it may not be a good idea to put scientists, as opposed to judges, in a position of resolving these disputes. It may be a good idea to have scientist trained as judges, as they can bring their general knowledge and skills to the decision-making process. But it is probably not a good idea to have a scientist take on the role of judge, if what the scientist is bringing to the process is his or her knowledge of the science at issue. Looking to the scientist as decision-maker will mean there is a risk the scientist will resolve the dispute based on knowledge, perceptions, and opinions not tested in the adversarial process. At a minimum, incorporating the scientist into the process as a decision-maker would very much change not just who is deciding cases but also how we decide cases.

26

How Ordinary Judges and Juries Decide the Seemingly Complex Technological Questions of Patentability over the Prior Art

F. Scott Kieff
Associate Professor of Law
Washington University School of Law
St. Louis, Missouri 63130

John M. Olin Senior Research Fellow in Law, Economics, and Business
Harvard Law School
Cambridge, Massachusetts 02138

ABSTRACT

Determinations of patentability over the prior art are often thought to raise questions that are so technologically complex that they require special training and judgment to answer, especially in fast-moving fields like modern biotechnology. This essay explores the somewhat counterintuitive argument that under the U.S. system they do not and should not. According to this view, determinations of patentability over the prior art are based entirely on factual inquiries that are best made by lay judges and juries, just like the factual determinations these people regularly make in any ordinary nonpatent trial. This is good because judges and juries are adept at these determinations, and because appellate courts

0065-2660/03 $35.00

are adept at reviewing them. It leads to a system that is cheaper overall in allowing private litigants to better predict outcomes that also better approximate the correct answer than would be possible under other regimes premised upon the expert technological knowledge and judgment of a decisionmaker.

I. INTRODUCTION: A THEORY OF DESERT

Commentators have long tried to figure out what scope of protection a patent applicant deserves.[1] For example, Thomas Jefferson, who oversaw the administration of the U.S.'s first patent system, thought that for each patent application we should ask whether its subject matter was "worth to the public the embarrassment of an exclusive patent."[2] More recently, the contemporary law and economics literature has included a number of works that endeavor to better align the benefit a patent confers on the patentee on the one hand, with the benefit an invention confers on society on the other hand.[3] Inquiries like these are generally viewed as being premised on a theory of the patent system known as the "reward theory" because they try to determine what inventions deserve to be rewarded. They pose difficult theoretical problems because they require some theory of desert and some framework for measuring it. They also pose difficult implementation problems.[4] Allowing each party to present its own technological expert will degenerate into mere battles-of-the-experts. Allowing the judge to be or employ her own technological expertise runs afoul of the limited role that is played by the judge in the American civil justice system as a transparent and neutral arbiter, and the more powerful role played by the jury as a popular check on the governmental power of the judge.[5]

Based upon my other work in the field,[6] this essay explores the counterintuitive argument that our patent system does and should try to figure out

[1] For more on the normative and positive analysis of the law and economics of the patentability rules see, F. Scott Kieff, *The Case for Registering Patents and the Law & Economics of Present Patent-Obtaining Rules*, HARVARD LAW SCHOOL, THE HARVARD JOHN M. OLIN CENTER FOR LAW, ECONOMICS, AND BUSINESS, DISCUSSION PAPER NO. 415, available at http://ssrn.com/abstract=392202 (last visited, May 18, 2003). Much of the materials provided here have their origin in that working paper.

[2] Letter from Thomas Jefferson to Isaac McPherson (Aug. 13, 1813), *reprinted in* JEFFERSON WRITINGS 1291–92 (M. Peterson ed., 1894).

[3] *See, e.g.*, Steven Shavell & Tanguy Van Ypersele, *Rewards Versus Intellectual Property Rights*, NATIONAL BUREAU OF ECON. RESEARCH WORKING PAPER NO. 6956 (1999) (discussing ways to improve the match between social surplus of the invention and the amount an inventor will recoup); Michael Kremer, *Patent Buy-Outs: A Mechanism For Encouraging Innovation*, NATIONAL BUREAU OF ECON. RESEARCH WORKING PAPER NO. 6304 (1997) (same).

[4] For more on the roles of judge, jury, and technological expert in the American civil justice system, see Roderick R. McKelvie, "Some Empirical Evidence on How Technologically Complex Issues Are Decided In United States District Court" (Chapter 25).

something quite different from what scope of protection a patent applicant deserves. The system does and should try to figure out what scope of protection from patents those other than the patentee deserve. Under this view, patent law's rules regarding the prior art—the Section 102[7] and Section 103[8] requirements that a patentable invention be novel and nonobvious—protect investment-backed expectations of those other than the patentee in ways that involve remarkably few administrative costs by ensuring that a patent right to exclude will not extend to anything those in the art are doing or are about to do. This type of inquiry is premised on a theory of the patent system known as the "registration theory" because it would allow for patent applications to be merely registered by the Patent Office instead of being subjected to examination on the merits by the present technologically expert staff of this administrative agency.[9] Determinations about what people verifiably have done or are about to do implicate surprisingly simple questions of fact that do not require special technological training or judgment, and therefore are well-suited for adjudication by lay juries and judges. And these types of simple, factual questions turn out to be precisely the types of questions that are raised by the intricacies of the positive law rules for patentability discussed in the following pages.

II. NOVELTY

Under the law of Section 102, patentability is precluded if any single item that is determined to count as prior art under any single subsection of that statute is found to fully disclose the claimed invention. Under the Section 102 analysis, if a patent claim is directed to subject matter that is not new when compared to such a single piece of prior art, then that claim is said to be "anticipated by the prior art."[10] Also, under the Section 102 analysis, if a patent claim is directed to subject matter that, even if new at the time of invention, was disclosed in such a single piece of prior art by being exposed to the public more than a year before the application was filed, then that subject matter is said to be "statutorily barred."[11]

[5] *Id*, at last page. ("Looking to the scientist as decision-maker will mean there is a risk the scientist will resolve the dispute based on knowledge, perceptions, and opinions that are not tested in the adversarial process").

[6] *See* Kieff *supra* note 1 (discussing the law and economics of the patent-obtaining rules). *See also* F. Scott Kieff, *Property Rights and Property Rules for Commercializing Inventions*, 85 MINN. L. REV. 697 (2001). (discussing the law and economics of the patent-enforcing rules).

[7] 35 U.S.C. § 102 (novelty and statutory bars).

[8] 35 U.S.C. § 103 (nonobviousness).

[9] *See* Kieff *supra* note 1 (discussing the registration theory in comparison with the other major theories of the patent system such as the reward, prospect, and rent-dissipation theories).

[10] The maxim setting forth the so-called "classic infringement test for anticipation," which also applies to analysis under the statutory bar, is "[t]hat which will infringe if later, will

Table 26.1. Analysis Under § 102[14]

	PAR_1
E_1	✓
E_2	✓
$E_{\ldots}$	✓
E_n	✓
E_*	✓

These items of prior art may include things such as patents, printed publications, public uses, sales, and prior uses from which the claimed subject matter was derived.[12] But such a single piece of prior art will only prevent patentability under Section 102's novelty and bar provisions if it, standing alone, fully discloses the claimed subject matter. To make this determination, the substantive technological content of this piece of prior art must be compared to the substantive technological requirements of the patent claim using an analytical framework like that depicted in the schematic claim chart in Table 26.1.[13]

Table 26.1 compares the elements of a stylized claim against the prior art for a determination of potential unpatentability or invalidity under Section 102.[15] The substantive requirement for determining no valid patent claim under Section 102 is triggered only if a single prior art reference (PAR) discloses, either expressly

anticipate, if earlier." *See* Donald S. Chisum, Craig A. Nard, Herbert F. Schwartz, Pauline Newman, and F. Scott Kieff, PRINCIPLES OF PATENT LAW 414 (2nd ed. 2001) (citing *Knapp v. Morss*, 150 US 221 (1893)).

[11] This so-called "bar" may also be thought of as a 1-year grace period for filing after public disclosure. For a comparative analysis of grace periods under the different patent systems of the world, see Joseph Straus, GRACE PERIODS AND THE EUROPEAN AND INTERNATIONAL PATENT LAW: ANALYSIS OF KEY LEGAL AND SOCIO-ECONOMIC ASPECTS (2001) (study commissioned by the European Patent Organization to examine whether European patent law should provide a prefiling grace period) (collecting sources).

[12] Chisum *et al. supra* note 10 at 323–513 (reviewing statutory and case law relating to novelty and statutory bar).

[13] *See infra* notes 14–18 (discussing application of this test).

[14] E_1 through E_n represent the elements of the claim arbitrarily assigned numbers 1 through n. E_* represents enablement of the entire claim. PAR_1 represents any single prior art reference, such as a journal article, sample product, student thesis, etc.

[15] The term "invalidity" refers to the failure of a claim in an issued and successfully examined patent to satisfy one of the substantive patent-obtaining rules. The term "unpatentability" refers to the failure of a claim in a patent application to satisfy one of the substantive patent-obtaining rules. These terms are interchangeable if operating under a soft look system like the registration model, which does not involve any examination.

The representation of a claim as a listing of its several elements in claim charts like Table 1 has become so common in patent cases that the local rules of some courts that hear many or under principles of inherency, each and every element of the claim, plus enablement.[16] When mapped onto this table, this means that a proper holding of invalidity will only lie if a check mark can be found as a matter of fact for every row.[17]

To achieve a check mark there must be admissible evidence that, as a matter of fact the pertinent content was present at the pertinent time in the piece of prior art.[18] A lay judge or jury can evaluate this evidence because it involves simple determinations such as: Did the prior art journal article disclose the requisite details? To be sure, technological experts on both sides often are needed to teach the vocabulary of the art and point out the portions of such a piece of prior art that allegedly make, or do not make, the pertinent substantive disclosure. But, in the end, the fact-finder will have to decide whether the disclosure existed or not at a specific time in history, which is also something lay judges and juries do in any typical nonpatent litigation. For example, if the claim requires that a step in a process be performed at a certain temperature, the lay judge or jury can determine whether the prior art reference expressly or inherently discloses the same step at the same temperature, as well as whether that piece of prior art was available at the pertinent date.

---

patent cases, like the Northern District of California, have for some time required their use. Chisum *et al.*, *supra* note 10, at 848–849 (discussing local rules for claim charts). The identification of these elements turns largely on the interpretation, or construction, of a patent claim, which is treated as a matter of law for decision by the court, and which is the first step in any analysis of either validity or infringement because the claim must be construed the same for both purposes. *See generally*, *id.* at 829–73 (discussing the substantive and procedural law of claim interpretation after the Supreme Court decision in *Markman v. Westview Instruments, Inc.*, 517 U.S. 370 (1996)). The great degree of debate over the law of claim construction itself injects a degree of uncertainty into this otherwise relatively crisp analysis. Recent empirical work by Wagner suggests that this uncertainty may lessen over time as the court develops predictable trends in its case law. *See* www.claimconstruction.com (Web page discussing empirical work relating to trends in the court's law of claim construction) (last visited Mar. 15, 2003).

[16] *See Minnesota Mining and Mfg. v. Johnson & Johnson*, 976 F.2d 1559 (Fed. Cir. 1992) (Rich, J.) (invalidity under Section 102 is "a question of fact, and one who seeks such a finding must show that each element of the claim in issue is found, either expressly or under principles of inherency, in a single prior art reference"); *In re Paulson*, 20 F.3d 1475 (Fed. Cir. 1994) ("In addition, the reference must be enabling and describe the applicant's claimed invention sufficiently to have placed it in possession of a person of ordinary skill in the field of the invention"). *See also In re Robertson*, 169 F.3d 743, 745 (Fed.Cir.1999) ("To establish inherency, the extrinsic evidence 'must make clear that the missing descriptive matter is necessarily present in the thing described in the reference, and that it would be so recognized by persons of ordinary skill.' *Continental Can Co. v. Monsanto Co.*, 948 F.2d 1264, 1268 (Fed.Cir.1991). 'Inherency, however, may not be established by probabilities or possibilities. The mere fact that a certain thing may result from a given set of circumstances is not sufficient.' *Id.* at 1269.").

[17] This represents the presence of each element in the claim, plus enablement, which, as discussed in the case law *supra* note 16, is required for a finding of invalidity under section 102.

[18] As discussed in the case law *supra* note 16, invalidity under Section 102 requires the prior art disclosure to be in a single reference.

III. NONOBVIOUSNESS

Even the more involved analysis under the additional hurdle for determining patentability over the prior art, which is the requirement of nonobviousness, turns out to raise similarly factual inquiries that are entirely appropriate for decision by a lay judge or jury.[19] The nonobviousness requirement was written into the patent system through the 1952 Act to statutorily jettison the prior case law associated with the former vague and antipatent requirement called "the requirement for invention."[20] Although even the drafters of this new standard

[19] For history of the nonobviousness requirement in patent law, see John F. Witherspoon, NONOBVIOUSNESS—THE ULTIMATE CONDITION OF PATENTABILITY (1979); George M. Sirilla, *35 U.S.C. § 103: From Hotchkiss to Hand to Rich, the Obvious Patent Law Hall-of-Famers*, 32 J. MARSHALL L. REV. 437 (1999). During the first half of the 1900's when called the requirement for invention, before the 1952 Patent Act, it had become known as "the plaything of the judiciary." Giles S. Rich, *Why and How Section 103 Came to Be*, in WITHERSPOON, at 1:208. Even after Congress wrote the Section 103 nonobviousness into the statute in the 1952 Act, another 10 years passed before the Supreme Court applied the new standard of nonobviousness in *Graham* and its companion cases. *Graham v. John Deere Co.*, 383 U.S. 1 (1966) (consolidated with *Calmar, Inc. v. Cook Chem. Co.*, and *Colgate-Palmolive Co. v. Cook Chem. Co.*); and *United States v. Adams*, 383 U.S. 39 (1966). For an inside look at the *Graham* decision, see Tom Arnold, *Side Bar: the Way the Law of section 103 Was Made*, in Chisum *supra* note 10, at 549–554. Soon afterwards, The Court reinjected confusion by writing about synergism and combinations. *See Anderson's-Black Rock, Inc., v. Pavement Salvage Co.*, 396 U.S. 57, 61 (1969) (holding patent invalid because "[n]o such synergistic result is argued here"); *Sakraida v. Ag Pro, Inc.*, 425 U.S. 273, 282 (1976) (holding patent invalid because it was a mere combination of old elements and had no "synergistic effect"). These terms were not weeded back out of the law until the creation of the Federal Circuit in 1982. *See Sirilla*, at 543. As the Federal Circuit has reminded:

> A requirement for "synergism" or a "synergistic effect" is nowhere found in the statute, 35 U.S.C. When present, for example in a chemical case, synergism may point toward nonobviousness, but its absence has no place in evaluating the evidence on obviousness.
> The reference to a "combination patent" is equally without support in the statute. There is no warrant for judicial classification of patents, whether into "combination" patents and some other unnamed and undefined class or otherwise. Nor is there warrant for differing treatment or consideration of patents based on a judicially devised label. Reference to "combination" patents is, moreover, meaningless. Virtually all patents are "combination patents," if by that label one intends to describe patents having claims to inventions formed of a combination of elements. It is difficult to visualize, at least in the mechanical-structural arts, a "non-combination" invention, *i.e.*, an invention consisting of a single element. Such inventions, if they exist, are rare indeed.

Stratoflex, Inc. v. Aeroquip Corp., 713 F.2d 1530, 1540 (Fed.Cir.1983).

[20] *See* Giles S. Rich, *The Vague Concept of "Invention" as Replaced by Section 103 of the 1952 Patent Act*, in WITHERSPOON, *supra* note 19, at 1:401, reprinted from 46 J. PAT. OFF. SOC'Y. 855 (1964) (Judge Rich's speech upon receipt of the Kettering Award, in which he discusses the role

recognized that it did not, on its face, appear to be any more precise in application than the former requirement,[21] as shown next, it turns out to be remarkably well–adapted for decision by a lay judge or jury.

The analysis for a nonobviousness determination under Section 103 begins with the entire body of prior art determined to be available under Section 102.[22] But important areas of the prior art are then carved out and excluded from the nonobviousness analysis.[23]

First, only art that is analogous may be considered under the nonobviousness inquiry and nonanalogous art is excluded from the inquiry.[24] As

---

of nonobviousness in Section 103 as the replacement for the so-called requirement for invention); Giles S. Rich, *Laying the Ghost of the "Invention" Requirement*, in WITHERSPOON, *supra* note 19, at 1:501, reprinted from 1 AM. PAT. L. ASS'N. Q.J. 26, 26 (1972) (discussing the great lag between the arrival of the new standard in the statute and its adoption by the courts).

[21] *Compare* P. J. Federico, *Commentary on the New Patent Act*, TITLE 35, UNITED STATES CODE ANNOTATED (West 1954 ed.), reprinted in 75 J. PAT. & TRADEMARK OFF. SOC'Y 161, 183 (1993) (reviewing history and operation of our present patent system, which is largely based on the 1952 Patent Act, and noting that the requirement for invention "is an unmeasurable quantity having different meanings for different persons") with *id.*, at 184 ("The problem of what is obvious and hence unpatentable is still of necessity one of judgment.").

[22] *See* Federico *supra* note 21, at 180:

> In form this section is a limitation on section 102 and it should more logically have been made part of section 102, but it was made a separate section to prevent 102 from becoming too long and involved and because of its importance. The antecedent of the words "the prior art," which here appear in a statute for the first time, lies in the phrase "disclosed or described as set forth in section 102" and hence these words refer to the material specified in section 102 as the basic for comparison.

[23] Although all of the Section 102 art is initially available for analysis under Section 103, certain types of prior art are excluded. According to the registration theory, these carve outs exist to remove from consideration the prior art for which the inference of possible innocent third party reliance is not reasonable. *See infra* notes 24–27 and accompanying text (discussing carve outs).

[24] The statute provides that the analysis should look to a hypothetical "person having ordinary skill in the art to which [the claimed] subject matter pertains" and ask whether to that person "the invention as a whole would have been obvious" given the "differences between the subject matter sought to be patented and the prior art." 35 U.S.C. § 103. This, in turn, requires that several factual inquiries be made: "[T]he scope and content of the prior art are to be determined; differences between the prior art and the claims at issue are to be ascertained; and the level of ordinary skill in the pertinent art resolved." *Graham v. John Deehr Co.*, 383 U.S. 1, 17 (1966). A person having ordinary skill in the art according to this framework is sometimes called a PHOSITA, thanks to the coining of that term by Soans. Cyril A. Soans, *Some Absurd Presumptions in Patent Cases*, 10 IDEA 433, 436 (1966). The "pertinent art" is selected from among the entire set of prior art identified by Section 102 depending upon whether it is analogous or nonanalogous. According to the Federal Circuit:

> Two criteria have evolved for determining whether prior art is analogous: (1) whether the art is from the same field of endeavor, regardless of the problem addressed, and (2) if the

predicted by the registration theory, which looks to protect the reasonable investment-backed expectations of third parties, nonanalogous art is properly discarded because it is not likely to be the basis for any third-party reliance. This distinction between analogous and nonanalogous art is important, not as evidence of what the inventor himself or herself could have known about the art, but rather as evidence of what was provably knowable, and therefore potentially relied upon, to a hypothetical third party person having ordinary skill in the art (PHOSITA).[25]

Second, secret prior art that would count only under Sections 102(e), 102(f), and 102(g) has been statutorily excluded from the nonobviousness analysis if it is owned by the same entity whose patent claim is in issue.[26] The

---

> reference is not within the field of the inventor's endeavor, whether the reference still is reasonably pertinent to the particular problem with which the inventor is involved.

In re Clay, 966 F.2d 656, 658 (Fed.Cir.1992) (citations omitted). *See also*, *In re Paulson*, 30 F.3d 1475 (Fed. Cit. 1994) (affirming Patent Office rejection under Section 103 because references from the fields of cabinetry and desktop accessories are properly considered to be analogous art to a patent claim directed to a clamshell case for a laptop computer under the second of these two alternative criteria).

[25] *See* Soans *supra* note 24 (coining the term PHOSITA). Indeed, Judge Giles Rich, who coauthored Section 103, has portrayed this PHOSITA "as working in his shop with the prior art references—which he is presumed to know—hanging on the walls around him." *In re Winslow*, 365 F.2d 1017, 1020 (CCPA 1966) (Rich, J.) (this metaphor is referred to as the "*Winslow* Tableau"). *See also International Cellucotton Prod. Co. v. Sterilek Co.*, 94 F.2d 10, 13 (2d Cir.1938) (Hand, J.) ("[w]e must suppose the inventor to be endowed, as in fact no inventor is endowed; we are to impute to him knowledge of all that is not only in his immediate field, but in all fields nearly akin to that field."); *Custom Accessories, Inc. v. Jeffrey-Allan Industries, Inc.*, 807 F.2d 955, 962 (Fed.-Cir.1986) ("The person of ordinary skill is a hypothetical person who is presumed to be aware of all the pertinent prior art."). Judge Rich improved upon the *Winslow* Tableau in *In re Antle*, 444 F.2d at 1171–72:

> In Winslow we said that the principal secondary reference was "in the very same art" as appellant's invention and characterized all the references as "very pertinent art." The language relied on by the solicitor, quoted above, therefore, does not apply in cases where the very point in issue is whether one of ordinary skill in the art would have selected, without the advantage of hindsight and knowledge of the applicant's disclosure, the particular references which the examiner applied. As we also said in Winslow, "Section 103 requires us to presume full knowledge by the inventor of the prior art in the field of his endeavor," but it does not require us to presume full knowledge by the inventor of prior art outside the field of his endeavor, i.e., of "non analogous" art. In that respect, it only requires us to presume that the inventor would have that ability to select and utilize knowledge from other arts reasonably pertinent to his particular problem which would be expected of a man of ordinary skill in the art to which the subject matter pertains.

[26] *See* 35 U.S.C. § 103(c) (providing carve outs). The carve outs for 102(f) and 102(g) were added in 1984 to reverse the holding in *In re Bass*, 474 F.2d 1276 (Fed. Cir. 1973). 98 Stat.

exclusion of this art also makes sense under the registration theory because no third-party investments will have been made in art that is commonly owned and kept secret.[27]

The content of the remaining prior art as a whole must then be surveyed to determine whether it may have reasonably triggered investment-backed expectations in achieving the subject matter of the patent claim in issue. Such investments are most likely to have existed only when there can be found among these many remaining pieces of art each and every element of the claimed subject matter, along with sufficient teaching, motivation, or suggestion (TMS) for the pieces that contain those elements to be combined such that there would be a reasonable expectation of success (RES) in establishing the claimed subject matter when they are combined.[28] The practical operation of this analysis can be seen through the use of the schematic claim chart in Table 26.2. As with the analysis under Section 102 discussed previously, this analysis under Section 103 turns out to be based on factual inquiries that are perfectly accessible to a lay judge or jury.

Like Table 26.1, Table 26.2 compares the elements of a stylized claim against the prior art, but this time for a determination of nonobviousness under

---

3384. The carve outs for 102(e) was added in 1999 through Section 4807 of the American Inventors Protection Act of 1999. 113 Stat. 1501. For a discussion of the history of these carve outs, see Chisum *et al. supra* note 10, at 575–578.

[27] No carve out is needed for the novelty analysis because the co-owner can keep the information sufficiently secret before the later claim that the reference will not trigger any of the subsections of Section 102, except perhaps 102(f). For this subsection, derivation, the co-owner can seek a claim by naming the first inventor, whose activity is co-owned. If the earlier reference does not disclose enough to invalidate under a novelty analysis, then it would not have been possible for the subject matter to have been claimed at the time of the earlier reference, and the only opportunity to claim the subject matter is at the later time. The exclusion of the prior art from a nonobviousness analysis at that later time helps ensure leaves open the possibility of it being covered by a claim. Because the subject matter is co-owned with the prior art and not otherwise available under any of the other subsections of 102, it also is not the target of third-party investment.

[28] According to the Federal Circuit:

> The consistent criterion for determination of obviousness is whether the prior art would have suggested to one of ordinary skill in the art that this process should be carried out and would have a reasonable likelihood of success, viewed in the light of the prior art. Both the suggestion and the expectation of success must be founded in the prior art, not in the applicant's disclosure.

In re Dow Chem. Co., 837 F.2d 469, 473 (Fed. Cir. 1988) (citations omitted). *See also* Chisum *et al. supra* note 10, at 584–597 (discussing contours of this analysis in practice and collecting sources).

Table 26.2. Analysis Under § 103[29]

	PAR_1	PAR_2
E_1		✓
E_2	✓	
$E_{...}$		✓
E_n	✓	
E_*		✓
TMS	✓	
RES		✓

Section 103.[30] Invalidity under this rule of nonobviousness also requires the presence in the prior art reference either expressly or under principles of inherency of each and every element of the claim, plus enablement; but unlike the analysis under Section 102, the analysis under Section 103 allows the elements to be spread among two or more individual pieces of prior art as long as there is also present in those pieces of prior art some additional facts: teaching motivation or suggestion to combine those references to obtain the subject matter of the claim as a whole (TMS), plus a reasonable expectation of success

[29] As in Table 26.1, E_1 through E_n represent the elements of the claim arbitrarily assigned numbers 1 through n; and E_* represents enablement of the entire claim. In this table, PAR_1 and PAR_2 each represent any single prior art reference, such as a journal article, sample product, student thesis, etc. The key to the analysis under Section 103 is that it permits looking to more than one reference in the prior art to find all the elements of the claim plus enablement, but only if in those references there can also be found (1) a teaching, motivation, or suggestion (TMS in Table 26.2) for those references to be combined to form the claimed subject matter, as well as (2) a reasonable expectation of success (RES in the Table 26.2) that the claimed subject matter will result when the references are so combined.

The apparent crispness of this framework may be somewhat illusory for several reasons. First, as with Table 26.1, there is some uncertainty regarding claim construction. *See supra* note 29 (discussing uncertainty about the law of claim construction and its application in any given case). Second, as discussed *supra* note 24, the determination of obviousness is to be done from the perspective of a PHOSITA, and the case law leaves some substantial uncertainty as to how this hypothetical person is to be conceptualized. The Federal Circuit has provided a number of factors to consider when determining the characteristics of the PHOSITA:

> Factors that may be considered in determining level of ordinary skill in the art include: (1) the educational level of the inventor; (2) type of problems encountered in the art; (3) prior art solutions to those problems; (4) rapidity with which innovations are made; (5) sophistication of the technology; and (6) educational level of the workers in the field.

Environmental Designs, Ltd. v. Union Oil Co., 713 F.2d 693, 696 (Fed.Cir.1983). *See also* Chisum *et al. supra* note 10 at 597–600 (discussing the case law relating to the determination of the PHOSITA).

[30] *See supra* note 15 (discussing the validity and patentability analyses).

in achieving the claimed subject matter upon the combination (RES).[31] When mapped onto this table, this means that a proper holding of invalidity or unpatentability under Section 103 will only lie if a check mark can be found as a matter of fact for every row and at least some tie can be made across all columns using the TMS and RES that must be found in at least one of the rows.[32]

As with the analysis under Section 102, to achieve such a check mark, there must be admissible evidence that as a matter of fact the pertinent content was present at the pertinent time in the pieces of prior art. As with the analysis under Section 102, a lay judge or jury can evaluate this evidence because it involves simple determinations, such as: Did the prior art journal article disclose the requisite details at the requisite time?

Indeed, the courts have admonished that even the Patent Office, which is a technologically expert administrative agency, must provide express factual support in its substantive decisions on patentability over the prior art, such as those relating to nonobviousness:

> [D]eficiencies in the [disclosures in the specific pieces of prior art in the record] cannot be remedied by the [Patent Office's] general conclusions about what is "basic knowledge" or "common sense." The [Patent Office's] findings must extend to all material facts and must be documented on the record, lest the "haze of so-called expertise" acquire insulation from accountability.[33]

The Patent Office must set forth details in its written opinions, including citations to the record and elucidation of the logical connection between the determinations and the specifics of the record, because this allows the reviewing court to decide whether that agency has gotten the facts right.

Importantly, private parties will be able to predict the outcomes of decisions on patentability over the prior art—whether by the lay judges or juries of a trial court or by the technologically expert staff of the Patent Office—because they will, in the end, turn on factual inquiries. What is more, these predictable outcomes will be driven by facts that are extrinsic to the particular decisionmakers. In contrast, allowing the determinations to be made as matters of expert opinion would degenerate into mere battles-of-the-experts, just as

[31] For a discussion of the case law leading up to this composite test, see *supra* notes 22–28.

[32] The nonobviousness analysis is presently pertinent when determining patentability before the Patent Office and when determining validity in litigation, but under a soft-look system would only be relevant in litigation. *See supra* note 15.

[33] *In re Lee*, 277 F.3d 1338, 1344–45 (Fed. Cir. 2002).

allowing the judge to be or employ her own technological expertise would make the outcome more likely to be influenced by factors intrinsic to the judge.[34]

IV. CONCLUSION

Ordinary judges and juries can and should decide the patentability of inventions over the prior art using heuristics that are triggered by simple facts—such as whether a particular document set forth a particular content at a particular time in history. Indeed, this is what happens under the present positive law regime of the U.S. patent system. This makes sense because it inexpensively generates fair outcomes that are predictable by private parties.

Acknowledgments

I gratefully acknowledge the financial support of the Harvard Law School John M. Olin Center for Law, Economics, and Business and the Washington University School of Law. I also gratefully acknowledge contributions from participants in the "Conference on Intellectual Property and The Human Genome Project" held at Washington University in St. Louis on April 12–13, 2002, as well as more detailed comments provided by Michael Abramowicz, Chris Bracey, Richard Epstein, Doug Lichtman, Chuck McManis, Rob Merges, Troy Paredes, and John Witherspoon.

[34] *See supra* notes 4–5 and accompanying text. Interestingly, there may be other ways to reliably generate predictable results through various market-based adjudicative processes like those proposed by Abramowicz in Michael Abramowicz, *On Prizes and the Human Genome Project in Retrospect*, (Chapter 11). But it is not clear whether the administrative costs of the system Abramowicz proposes will turn out to be sufficiently low to justify adoption.

27

The Difficult Interface: Relations between the Sciences and the Law

Horace Freeland Judson
Director, Center for History of Recent Science
Research Professor of History
The George Washington University
Washington, DC 20052

I. INTRODUCTION

I am a journalist with pretensions to be an historian of science—of recent science, meaning, in effect, work done since the Second World War. Most of my writing about science has been in the history and practice of molecular biology. For the last several years, I have been looking into the problems of fraud in the sciences. The cases are many; I have examined scores that have arisen in the past quarter century. The details are astonishing, enthralling, and cumulatively horrific. Yet my aim is not to tell scandalous stories. Instead, I employ scientific misconduct as a window into the normal processes of science.

My interest in the relations between the sciences and the law thus goes considerably beyond the subject of patents, or, more generally, of intellectual property. I have come to question that relationship at a deep level. Pursuing fraud has led me to believe that the norms and needs of the sciences are ill-served by the American legal system and, in particular, by the courts or their

0065-2660/03 $35.00

alternatives, with their adversarial proceedings. To use a term that is fashionable, these two domains seem to me to be fundamentally incommensurate. So I would like to take this chapter to test out some ideas and to ask some questions. Though the matters I will raise may strike you as not immediately relevant, and my grasp of them as naïve, law is dependent on the sciences and in a way that is evolving fast. My questions have to do with the background of that evolution.

Consider first that defects can be windows into normal processes. In the sciences themselves, defects have often played a crucial role. When, early in the seventeenth century, the anatomist and physician William Harvey first attempted to analyze the motions of the heart and blood, he made crucial observations of animals and men that were in some relevant way defective. Such observations of the abnormal opened to Harvey's understanding the normal physiology of blood flow. They became essential components of his demonstration that the blood circulates.

The examples multiply. In the nineteenth century, physicians encountered cases of goiter, cretinism, and stunted growth, patients with galloping heartbeats and eyes so bulging that the lids would not close over them, along with instances of dwarfism, gigantism, and acromegaly. Physicians also took the first steps toward understanding the functions of thyroid and pituitary, and the study of such defects in the clinic engendered the science of endocrinology. Again, much of what neurobiologists know about the working of our brains comes from a century of attempts to use dysfunctions—the consequences of brain damage from accident, stroke, gunshot, or surgery in previously normal people—to analyze the neural basis of perception, thought, language, memory, and emotion.

Geneticists, from the start of the twentieth century to the present day, have constructed their science on the analysis of patterns of transmission of variant characters—mutations—down the generations. Most of these are genetic defects. Until the invention of the most recent techniques, the genetic map of fruitfly, mouse, or man was almost entirely the map not exactly of genes but of gene defects.

The conspicuous recent example of the place of defects in biological research is, of course, acquired immune deficiency syndrome (AIDS). The immune system is intricate, second only in complexity to the nervous system, and is recalcitrant to scientists' understanding. The human immunodeficiency virus (HIV) invades cells that command a central intersection of the pathways by which the entire immune system functions. The attack on AIDS requires fundamental, general research into the system of which it is a perturbation. Once again, the defect opens a window.

In sum: In the sciences, defects can give access to the processes they disrupt, processes otherwise inordinately large in scale, complex, quick, inaccessible; or where deliberate intervention is impossible or unethical. Now take a step back. All those characteristics are true of the enterprise of the sciences as they have developed in the latter half of the twentieth century. But fraud and related forms of misconduct are assuredly defects in the processes of science. The premise of my recent work is that scrutiny of the nature of fraud and other misconduct will reach to the heartbeat and pulse of what the sciences are and what scientists do.

II. THE NORMS OF SCIENCE

Fraud in science has greatly different import when considered in the domain of law and in the domain of science. Why should this be so? Perhaps we can best begin by raising the problem of the norms of science. The disjunction between the domains of science and of law is rooted in the norms. Most generally, for the sciences fraud is of course a violation of the norms. And taking instruction from the materials provided by Scott Kieff, I was interested to learn that jurists disputing the justifications for patenting scientific findings regularly appeal to such norms. That suggests that the dissonance between law and the sciences is acute in the patent regime, too.

The problem of norms has been central to the sociology of science from its first emergence as a distinct line of inquiry. One thinks of that emergence as beginning with Max Weber's celebrated lecture at the University of Munich, in 1918, "Science as a Vocation." Celebrated, but seldom read.

One quick definition of sociology is the study of the institutions—meaning the social structures—that we have erected in one sphere of human action or another, and that we take in and that shape our behavior within that sphere. For the sociologist, durable institutions are like Plato's axe. Plato had this axe, you see, the essence of axe, a fine bronze blade and a strong box-wood handle, and he bequeathed it to his intellectual epigones. At some later date, the handle splintered, and was replaced, but this time made of oak. Passed on down through the generations of philosophers, it made its way to Asia Minor, where, in the Dark Ages of the Christian West, the Arabs were the custodians of the philosophical and scientific wisdom of Greece. The blade was corroded by an attack of bronze disease, otherwise known as cuprous chloride, a simple chemical reaction that will eat its way through an old-fashioned English penny in 24 hours; so it was replaced with fine Damascus steel. Still later, Plato's axe made its way to the New World, where the handle, giving trouble again, was replaced with hickory. The axe, in its 2500 years, has gone through other repairs, and now resides in the basement of the Metropolitan Museum in New York, still sharp, still elegant, the essence of axe. (The curators, alas, do not realize what they have there.)

The question is, of course, in what sense is this eloquent object Plato's axe? Continuity and change: the parable of Plato's axe epitomizes the problem of the sociologist's study of institutions. Relating to the institutions of the law, in what sense is the jury system in American criminal and civil law today the same system prescribed in the Sixth and Seventh Amendments to the United States Constitution? Relating to the institutions of science, in what sense are the institutions of recent science, science today, the same as those of 1918 or of 1942?

For Weber, science as a vocation meant more than science as a career choice. He saw science as a calling, fully in the sense that one speaks of a religious vocation—a lifelong dedication, a total commitment. His concern was to contrast this vocation with others, the vocations of the politician, of the artist, of the religious believer. The first essential distinction is that scientific work is unique in being always subject to change, to being superseded. He said, "In science, each of us knows that what he has accomplished will be antiquated in 10, 20, 50 years. That is the fate to which science is subjected; it is the very *meaning* of scientific work. . . . Every scientific 'fulfillment' raises new 'questions'; it *asks* to be surpassed and outdated." And he said, "We cannot work without hoping that others will advance further than we have." Secondly, for two millennia and today at an ever-accelerating pace, it is science that has driven "rationalization and intellectualization and, above all, . . . the 'disenchantment of the world.' " Science has removed the supernatural from our view of the way things are.[1]

In sum, Weber's norms of science require total commitment, the acceptance of change and of being superseded, and above all the dedication to rationality and demystification. For him, the most important conclusion to draw was that science and religion are radically incompatible. "[T]he decisive characteristic of the positively religious man," Weber wrote, is what he termed "das intellectuelle Opfer," the "intellectual sacrifice"—the surrender of judgement to the prophet on to the church. "[T]he tension between the value spheres of 'science' and the sphere of 'the holy' is unbridgeable." Science for Weber was a high and ascetic calling.

This must seem remote from what we now mean by the norms of science. As I said, Weber distinguished science from other possible callings, commitments, and attitudes of mind, defining the difference that summed up all the others as the refusal to make the "intellectual sacrifice." In contrast, later sociologists have been concerned with distinctions within the scientific enterprise. To put it roughly, they are concerned with the attitudes of mind, enforced by the community, that make for good science rather than bad. The next

[1] Max Weber, *Science as a Vocation* (1919), *in* H. H. Gerth and C. Wright Mills, eds, FROM MAX WEBER: ESSAYS IN SOCIOLOGY (2nd Printing, 1949) Oxford University Press, Oxford, England, pp. 129–156.

statement of norms, the one we all know about, was established by Robert Merton, elegant progenitor of the modern sociology of science. Merton's sociological lineage traces back primarily not to Weber but to Talcott Parsons. In 1942, in an article "Science and Technology in a Democratic Order" (later reprinted several times as "The Normative Structure of Science") Merton wrote that he was responding to "incipient and actual attacks upon the integrity of science."[2] He continued, in part: "The ethos of science is that affectively toned complex of values and norms which is held to be binding on the man of science. The norms are expressed in the form of prescriptions, proscriptions, preferences, and permissions. They are legitimized in terms of institutional values."

Merton elaborated "four sets of institutional imperatives—universalism, communism, disinterestedness, organized skepticism."

- Universalism, briefly stated though his discussions were qualified and often subtle, asserts that science is impersonal and knows no boundaries of nation, ethnicity, or political ideology. Merton went on, "The acceptance or rejection of claims entering the lists of science is not to depend on the personal or social attributes of their protagonist. . . . Objectivity precludes particularism."
- Communism, in the sense of common ownership, Merton called "a second integral element of the scientific ethos." The results of science are owned in common: because "the substantive findings of science are a product of social collaboration," he said, "the equity of the individual producer is severely limited." Again, "Property rights in science are whittled down to a bare minimum by the rationale of the scientific ethic." The rewards proper to scientists, he mentioned here and pronounced upon elsewhere, are conferred by the scientific community, by one's peers. These rewards that drive scientists, as scientists, are two: to be recognized for originality, which necessarily puts high value on priority, and to be held in high esteem by other scientists.
- Disinterestedness, the third of Merton's norms, follows from the second, communism, but pushes it further. The scientist is to behave as though the possibility that he has a personal stake in his results is overriden by the power of the community. Merton's discussion of disinterestedness states most clearly how the norms of science are supposed to work. He held that it is "a distinctive pattern of institutional control of a wide range of motives which characterizes the behavior of scientists." He went on to state that "once the institution enjoins disinterested activity, it is to the interest of scientists to conform on pain of sanctions and, insofar as the norm has been internalized, on pain of psychological conflict."

[2] Robert K. Merton, *The Normative Structure of Science* (1942), *in* SOCIOLOGY OF SCIENCE (1973), University of Chicago Press, Chicago.

- Organized skepticism, the fourth norm, is the one that harks back most obviously to Weber and the refusal to make the intellectual sacrifice. Merton, too, recognizes the inherent fundamental conflict between the scientific ethos and "other attitudes toward these same data which have been crystallized and often ritualized by other institutions." But the central importance of organized skepticism is its service as the mechanism that enforces the other norms, notably disinterestedness. Science works, one can say, by the scrutiny of each by all.

To overstate the influence of Merton's essay would be difficult. Some great scientists have behaved as though acknowledging such norms. In 1969, Max Delbrück was awarded the Nobel Prize in physiology or medicine, along with Salvador Luria and Alfred Hershey. (That was the year the prize in literature went to Samuel Beckett, whose work Delbrück found to resonate with the deepest paradoxes of the scientist's endeavors. But Beckett did not come to Stockholm, which Delbrück always regretted.) Delbrück, in the lecture he gave upon receiving the prize, compared the scientist and the artist. He was speaking of the vocation of the scientist. Addressing his audience, with grave courtesy, in Swedish, he said, in part:

> The books of the great scientists are gathering dust on the shelves of learned libraries. And rightly so. The scientist addresses an infinitesimal audience of fellow composers. His message is not devoid of universality, but its universality is disembodied and anonymous. While the artist's communication is linked forever with its original form, that of the scientist is modified, amplified, fused with the ideas and results of others, and melts into the stream of knowledge and ideas which forms our culture. The scientist has in common with the artist only this: that he can find no better retreat from the world and also no stronger link with the world than his work.[3]

Alas, today the Mertonian norms seem sadly naïve, idealistic, old-fashioned. For some, indeed, that term "the Mertonian norms" has come to be used in derision. Can these four concepts genuinely be the norms of practicing scientists? Today it requires an active suspension of disbelief to think that they could have been thought so even during the second world war and before Vannevar Bush, the beginning of massive governmental funding for the sciences, and their exponential growth. In the last 20 years, every one of the Mertonian norms has come into question. Universality? Common ownership of the results? Disinterestedness? In language that saddens me, Merton wrote of the effects of disinterestedness as though describing the real world, declaring "[t]he virtual

[3] Nobel Lecture, 10 December 1969, *reprinted in* SCIENCE 168 (June 1970): 1312–1315.

absence of fraud in the annals of science, which appears exceptional when compared with the record of other spheres of activity." He went on to state:

> By implication, scientists are recruited from the ranks of those who exhibit an unusual degree of moral integrity. There is, in fact, no satisfactory evidence that such is the case; a more plausible explanation may be found in certain distinctive characteristics of science itself. Involving as it does the verifiability of results, scientific research is under the exacting scrutiny of fellow experts. Otherwise put . . . the activities of scientists are subject to rigorous policing, to a degree perhaps unparalleled in any other field of activity.

Science does not work that way. Outsiders speak of "the scientific method." An honest scientist will tell you, "You get there how you can." Sure, there are some rules, such as the design of control experiments, the need for rigor in statistical inference, and the techniques of double blinding in clinical trials. Yet many kinds of science must move from observation to hypothesis without the possibility of reproducibility. For example, you can hardly experiment with a pulsar or a supernova, and the kinds of things you can ethically do to a human nervous system are limited. But even in the broad middle ground, the laboratory where patentable discoveries are made and where Merton's "verifiability of results" would seem to reign, rarely are experiments replicated and the findings reported in a paper directly tested. The big exception is an addition to the cookbook—a new scientific technique. Significantly, a high proportion of Nobel prizes have been awarded for the invention of new techniques: from the ultracentrifuge by The Svedberg in the 1920s to the two Nobels to Fred Sanger, first for sequencing protein chains and then for sequencing DNA.

Of course, a high proportion of published papers die in infancy. Other scientists decide that the papers, on their face, are not good. They are not taken into the fabric of science. Otherwise, what happens is this: When you publish a significant paper, your closest competitor reads it, slaps his forehead and says, "Why didn't *I* think of that?" But then he says, "If that's true, then I can do such-and-such." He does not replicate your experiment, but rather he builds your result into his next step. Thus, that the problem of fraud is felt to be pervasive is evidenced by the practice, which I am told has arisen at some pharmaceutical companies, of duplicating every published report they might apply, before relying on it. One can stipulate three reasons significant laboratory results are not normally verified. Two of them are institutional: Repeating other scientists' work is not an enterprise that attracts funding, and journals rarely publish negative results. The third reason lies in the practical problems of laboratory research in that some experiments cannot be rerun.

You recall the David Baltimore affair. Baltimore, though not himself accused, was at the center of a notorious and protracted case involving charges of fraud in a paper of which he was a coauthor. The paper reported work in immunology, using mice, at his laboratory and another at the Massachusetts Institute of Technology. The experiment in question had been performed, if indeed it was performed, chiefly by one Thereza Imanishi-Kari. Margot O'Toole, a post-doctoral fellow in her laboratory, stumbled upon a notebook that demonstrated that the actual results had been misrepresented in the published paper. She asked that the paper be corrected or retracted. She was asked if she was charging fraud. She said no, but that the corrections must be made. Imanishi-Kari, Baltimore, and others in authority refused.

Over the next 3 years, the affair blossomed to the level of Congressional hearings. Meanwhile, Baltimore had become president of the Rockefeller University, and the faculty there used the scandal to force his resignation. I have given talks about the Baltimore affair to a variety of audiences. Several times, there has been a physicist in the audience who will rise with the obvious question: "Why not simply repeat the experiment?" The answer is that it could not be done. Biology is different; immunology is especially tricky. Imanishi-Kari's mice were a special inbred strain that could not reliably be recreated; her various reagents and procedures were too intricate and her use of them, indeed, haphazard. But a great deal of biological research can be like that—for a reason having nothing to do with chicanery. Significant work must often be out beyond the frontier, stretching the limits of what the available methods can reliably produce.

In the last decades of the twentieth century, the idea of the norms of science came under attack from another direction. The science wars have been exhaustively reported. Briefly, sociologists' attempts to analyze the workings of science were beset by bitter disputes over the extent to which scientific theories, or even what most people take as scientific facts, can claim a reliable grounding in a reality independent of presuppositions and prejudices of the scientists themselves. A camarilla of sociologists of science flourished in the 1980s and 1990s who declared it their program to show that in all sciences, the theories or hypotheses, methods and apparatus, and the facts themselves, are constructed by the agreement of scientists. This agreement was constructed, according to the strong version of the program, by nothing other than a consensus of scientists achieved by negotiation. And finally, no grounding of these in some putative objective reality can be proven. Scientists' agreement, their consensus about what is the case, by this doctrine is not about truth but about power—and in this sense, for some, all science must appear fraudulent. Social constructionism, the doctrine is called. Its practitioners congregate in programs or departments labeled "science and technology studies." Baldly put, the doctrine is compounded of elements of sociology, often of a Marxist tint, and aspects of a movement fashionable in academic literary studies called postmodernism or critical theory. Curiously, perhaps not coincidentally, social constructionism reached its apogee just in the period when cases of scientific fraud arose and multiplied. But the wave of social constructionism is receding.

Back to the question. How do norms work? Consider how they are passed on. Come with me into two laboratories. The first is that of Howard Temin, who in 1970 demonstrated the existence of what we now call reverse transcriptase, an enzyme that reads RNA back into DNA. The importance of the enzyme has steadily grown, as a multipurpose tool for genetic engineering but especially in the clinic, because it turns out that many viruses that infect animal cells are retroviruses, using the enzyme to incorporate their RNA genes into the DNA of the host. The headline example, of course, is HIV. The existence of the enzyme was demonstrated simultaneously and independently by Baltimore, and though the two men had not realized they were in competition until a fortnight before their papers reached NATURE, that June, the facts of the discoveries make an engrossing study in scientific styles. The second laboratory could be any of a number, but let us go to Robert Gallo's establishment when he was still at the National Cancer Institute. (You'll recall that Gallo and associates at his laboratory were accused of theft of intellectual property.)

Temin epitomized scientific probity. He embodied the norms. Temin rarely gave interviews, but when he knew he was dying of cancer we spent six hours in a span of two days going over his career—and by implication his style in science. His office was small, narrow, and dimly lit. The walls were full of shelves full of boxes full of papers and correspondence. The walls had no pictures. The adjoining laboratory space was small and lightly populated. To the side of his desk was a small wooden chair, armless and hard. There, each Friday, one by one, he asked his post-doctoral fellows and technicians to account for their week's work. There I sat during the interviews. At one point he said, "I have never started a biotech company. I have been offered positions as scientific adviser to pharmaceutical companies, even memberships of their boards. But I never took them. I do not believe that a scientist supported by public funds should exploit his work for private gain."

Gallo's office at the cancer institute was bright and large, with the walls adorned with framed blow-ups of awards and of newspaper and magazine articles about him. Prosperity, publicity, and no holds-barred competition were the themes. The laboratories were large and busy. Over a span of decades, Gallo had come up with valuable results in cancer virology—that is, the possibility that some cancers may be caused by viruses—and had repeatedly distorted and exaggerated the significance of those results. The pressures, the ethos, on the junior colleagues in his lab were reputedly harshly competitive. Even as Temin and I were talking, Gallo was under investigation for the misappropriation of intellectual property, related to the identification of the virus that causes AIDS, HIV.

Yet Temin was also a friend of Gallo, and defended him in public. Later that week, Temin also said, "You see my office. You have seen his. Bob Gallo and I are very different. Yet I believe that the unique, great strength of American science is its diversity. Different problems require different approaches, styles, personalities. There's a place for scientists like Gallo."

Norms are inculcated by example and are passed down through scientific lineages. Howard Temin as a student at the California Institute of Technology in the late 1950s was a junior member of the elite and rigorous circle around Delbrück. Delbrück, trained as a physicist in Germany before coming to the United States in 1938 and switching to biology, had been a member of the elite and rigorous circle around the quantum physicist Niels Bohr, in Copenhagen. In the early 1990s, as the polemics of the Baltimore affair reached their height, the chief focus of public criticism of Baltimore was a group of scientists at Harvard, plus James Watson: Many of these, including Watson, were part of that same Delbrück lineage. The Baltimore affair was at its bitter root a conflict over norms. It must be said that the classic norms of science did not prevail. The norms of Baltimore, or of Gallo, are very different from those described by Merton.

To understand why, we must turn to the overall history of the sciences since the Second World War.

III. RECENT HISTORY

Of all the scientists who have ever lived, eight out of 10, perhaps more, are alive today. Of the work in the sciences that has been done, published, and woven into the fabric of reliable knowledge, the bulk has been done by people who are alive today. The momentum of the sciences, their reach toward fundamental questions, their interest, has never been greater; the influence of sciences on everyday life is pervasive and increasing. The sciences are the dominant form of intellectual as well as practical activity of the latter half of this century. These are truisms.

Meanwhile, scientists are transforming the ways they work. We can state one aspect of the change in general, structural form: As the number of scientists and the complexity of their subjects increase, the networks of relationships among scientists in any particular field grow larger and more intricate. This is as true for, say, cell and developmental biology as for, say, high-energy theoretical physics.

Open communication is central to the norms of science. Most radically transformed by decades of exponential growth are the methods scientists use to communicate their work while it is in progress, both to each other and each to himself. Scientists used to write letters. They used to fill their daybooks up with speculations and inner dialogues. Newton and Darwin were unusual only in the enormous volume of their note-writing and (in Darwin's case) correspondence. But those practices have been superseded. Scientists today communicate at meetings, by telephone, and ever more by electronic mail, to the neglect of means that leave a permanent record. The communication that *is* committed to paper is more ephemeral. Files are thrown out or weeded. Rare is the scientist these days who retains his old correspondence. Even the traditional, rigorous standards for the keeping of notebooks in the laboratory, the observatory, and the field are relaxing. Paper records have been superseded by direct entry of data into computers. Where are the bound ledgers of times past? This is more than the particular lament of the historian: The changed standards here, too, redound upon the norms of science.

Communication is one side of the structural transformation. Competition is the other. In molecular biology, the field I know best, as late as the 1960s the most significant work was concentrated in a few places, chiefly the Medical Research Council Laboratory of Molecular Biology, in Cambridge, England; the California Institute of Technology, in Pasadena; and the Institut Pasteur in Paris, with outliers elsewhere. The number of scientists involved was in the low dozens. Typically, a particular line of research or set of problems was concentrated in one or two groups, and although they often published small steps, many of the central papers of those days were relatively longer—and more encompassing—than is common now. By the mid-1970s, the number of laboratories and workers had doubled and redoubled. Today, funding of the sciences continues to rise exponentially, the laboratories number in the hundreds and the workers in the many thousands. The characteristic paper presents what has derisively been called "the least publishable unit." In the laboratory from week to week the Mertonian competition for originality and the recognition of one's peers has been swamped by competition for funds. The governmental funding has steadily biased work in the biological sciences generally towards practical application, in effect, the conversion of science to engineering. This bias has been driven further by the commercialization of biotechnology—a few large firms and a number of new, small ones dangling the opportunity, which only a few scientists have realized, to get seriously rich. The transformation of the sciences we can sum up as their industrialization, or, in a phrase of neo-Marxist flavor, as their conversion to knowledge production.

The consequences cannot but be a change in the norms of science.

IV. FRAUD AND THE LAW

The problems of fraud bring into strong relief the incommensurability of the sciences and the law. Even to define scientific misconduct—fraud—has turned out to be difficult and bitterly contentious.

The ruling definitions of scientific fraud and other gross misconduct have been those of the National Science Foundation and of the National Institutes of Health—or, more properly, the Public Health Service, of which the institutes comprise the largest part. The two definitions are all but identical, and on casual reading seem to be simple enumerations. The language was set only in the 1980s, and was a response to the increasing number of publicized cases.

The National Science Foundation: "(1) Fabrication, falsification, plagiarism or other serious deviation from accepted practices in proposing, carrying out, or reporting results from activities funded by NSF; or (2) Retaliation of any kind against a person who reported or provided information about suspected or alleged misconduct and who has not acted in bad faith."[4]

The Public Health Service: " 'Misconduct' or 'Misconduct in Science' means fabrication, falsification, plagiarism, or other practices that seriously deviate from those that are commonly accepted within the scientific community for proposing, conducting, and reporting research. It does not include honest error or honest differences in interpretations or judgments of data."[5]

Fabrication is forging, or what biologists call "dry labbing"—making the data up. Falsification means altering the data or tendentiously selecting what to report. A miniature, emblematic instance was told to me a while ago by a senior biologist, a man who, after wide and international experience in immunology, has been director of research for a small startup company in Northern California exploring traditional Chinese medicines (many of which have been in use for centuries) to see whether any contain active ingredients that may prove genuinely efficacious. At a seminar of one of the laboratory groups, he had dropped in to sit at the back of the room. A junior investigator was presenting the data from a series of experiments, in preparation for writing a paper. She summarized her results as points on a graph, and offered an interpretation. The distribution of the data points looked good, much as expected, except for one result, which was far from the others and from the interpretation. She pointed this out. The lab chief, coincidentally also a woman, told her she should leave that result off the graph,

[4] National Science Foundation (1991). Misconduct in Science and Engineering: First Rule. FEDERAL REGISTER. 56 (May 14): 22286–90. *Cited in* U.S. Department of Health and Human Services, Public Health Service. INTEGRITY AND MISCONDUCT IN RESEARCH: REPORT OF THE COMMISSION ON RESEARCH INTEGRITY, pp. 1.

[5] 42 C.F.R. Part 50, Subpart A; August 8, 1989. *Cited in* COMM. ON RESEARCH INTEGRITY, pp. 1.

out of the paper. Whereupon the research director rose in wrath. "*No*." Even as he spoke of the episode several weeks later, his voice rose and his face twisted in distress and puzzlement. "I told them, 'You can never do that! You can reexamine the experimental conditions, or repeat the experiment, to see whether something went wrong. You can do other experiments to try to explain the result. If these don't solve it, you must confront the anomaly in the discussion section of the paper. But you cannot suppress a result.' " He shook his head and said, "They must know that." He is surely right—yet anomalous data of that type are called outliers, and suppression of outliers is by no means rare and is sometimes defended as a matter of good scientific judgment, part of the art of discovery.

We think of plagiarism as copying, what we were warned against as schoolchildren, but as we all know it comprises all ways of representing someone else's work as one's own: formally, theft of intellectual property. Fabrication, falsification, plagiarism—all three are fraud, and they're habitually grouped as FF&P. But plagiarism is different from the other two in purpose, in methods, in its incidence, and in its effects. After all, if you get away with faking the data or fudging it, you pollute the scientific record with something you know to be untrue; but you would never filch someone's work you thought false. Yet what plagiarism may lack in turpitude it more than makes up in ubiquity. At the level of anecdote, anyway, rare is the senior American scientist who has not suffered it. A widespread though unquantifiable form of plagiarism occurs in the process by which scientists review applications for grants to fund new research in their field. A senior scientist at Stanford University, in a conversation walking down the corridor: "It happens, but if you're productive it affects little of your work."

"Other serious deviation from accepted practices." The clause seems tacked onto the definitions. It provokes controversy. Many scientists and lawyers find it vague. Some find it threatening.

Attempts to improve the definition began early in 1989, as misconduct cases were becoming increasingly conspicuous, when the National Academy of Sciences gingerly took up the issues. Among senior scientists the instinct has always been strong to avoid alarming the community and frightening the Congress, and thus to downplay the seriousness and frequency of misconduct. The academy's loftiest organ, the Committee on Science, Engineering, and Public Policy, took months to come to agreement that a study was needed. Seven more months were used finding someone who would agree to be chairman. The next, and by no means minor, difficulty was to find willing panelists who did not have close ties to leading scientists involved in the current controversies—notably of course the Baltimore and the Gallo affairs. Only by the spring of 1990 was the academy panel assembled. Among its 22 members were 11 university scientists of various fields, 2 from industry, together with university administrators, ethicists, a lawyer, an historian, and the editor of a journal. From the start, according to reports, disagreement flourished. The panel took 2 years to produce its study; released in April of 1992, and it was anodyne. The panel balked at the question of incidence, saying rightly enough that no one had measured it. Their definition limited scientific misconduct to fabrication, falsification, and plagiarism, and they called on the National Science Foundation and the Public Health Service to eliminate the "other serious deviations" clause. Despite the safe, bland character of the report, two scientists on the panel dissented, saying that among other faults it exaggerated the importance of misconduct.

One of the dissenters was Howard Schachman. Schachman is a molecular biologist at the University of California, Berkeley, who has long, intelligently, and angrily defended science against what he considers to be the mortal danger of interference from government. He had put his argument before congressional committees as well as in the debates of the National Academy panel. He also campaigned publicly, privately, and pugnaciously with all who would listen. In July of 1993, in an article in SCIENCE,[6] reaching a wide audience, Schachman wrote:

> Many scientists, like others in our society, are ambitious, self-serving, opportunistic, selfish, competitive, contentious, aggressive, and arrogant; but that does not mean they are crooks. It is essential to distinguish between research fraud on the one hand and irritating or careless behavioral patterns of scientists, no matter how objectionable, on the other. We must distinguish between the crooks and the jerks.
> ... My concern is over vagueness of the term "misconduct in science" and how people with different orientations interpret various alleged abuses.
> In formulations of the term "misconduct in science" there is agreement on fabrication, falsification, and plagiarism. Scientists have emphasized that "misconduct in science" does not include factors intrinsic to the process of science, such as error, conflicts in data, or differences in interpretation or judgments of data or experimental design.

Shachman took particular exception to the inclusion of the phrase, from the Public Health Service, or NIH, version, "other practices that seriously deviate from those that are commonly accepted within the scientific community for proposing, conducting, or reporting research." As he read it:

> Not only is this language vague but it invites over-expansive interpretation. Also, its inclusion could discourage unorthodox, highly innovative approaches that lead to major advances in science. Brilliant, creative, pioneering research often deviates from that commonly accepted within the scientific community.

[6] Howard Schachman, *What is Misconduct in Science?* SCIENCE, July 9, 1993.

Schachman's conclusion was apocalyptic. Mindful of the Hitlerite rejection of the physics of relativity as "Jewish science" and the Stalinist consignment of Mendelian genetics and geneticists to the Gulag as not Marxist-Leninist science during the 15-year domination of Soviet biology by Trofim Lysenko, he wrote:

> History is full of examples of governmental promulgations of laws expressed in broad, open-ended terms that were elastic enough to be stretched to cover any individual action that irritated some officials. In this century alone it was a major offense in some countries to publish scientific papers that seriously deviated from accepted practice. The enforcement of such strictures virtually destroyed major areas of science in those countries. We should not expose science in this country to similar risks.

Other scientists have concurred. Among the more prominent is Bernadine Healey. President George H. W. Bush appointed Healey as director of the National Institutes of Health. Healey—conservative and politically ambitious, unpopular as an administrator—attacked the Public Health Service version of the "other serious deviation" clause because "tinkering, bold leaps, unthinkable experimental design, and even irritating challenges to accepted dogmatic standard" could be labeled misconduct. Healey's successor, Harold Varmus, a Nobel prize winner whom President Bill Clinton appointed in 1993, a smart, effective director who was liked by the scientific community as one of their own, called for dropping the clause because it is "too vague and could be used inappropriately."

The import and utility of the clause may be diametrically the reverse of the received view. What Schachman and others fail to understand is that the "other serious deviation from accepted practices" clause or something like it is essential to a definition of misconduct that can preserve the autonomy and self-governance of the sciences they think their objections are defending. Early in 1997, Karen Goldman and Montgomery Fisher published a tightly reasoned defense of the "other serious deviation" clause.[7] The two were lawyers in the Office of the Inspector General of the National Science Foundation (the office that handles misconduct charges serious enough to be brought to the foundation) at the time they posed the question in its toughest form: Is the clause constitutional?

Goldman and Fisher turned the usual reading of the National Science Foundation's definition end for end. Fabrication, falsification, and plagiarism, they wrote, are but three examples of misconduct, while "other serious deviation

[7] *See* Karen A. Goldman and Montgomery K. Fisher, "*The Constitutionality of the "Other Serious Deviation from Accepted Practices" Clause*," JURIMETRICS, Winter 1997.

from accepted practices," which they called the OSD clause, is the overriding and most significant part, "a general standard defining misconduct."

They wrote: "The OSD clause provides a legal basis for a finding of misconduct in all cases of serious breaches of scientific ethics, including cases that cannot be categorized as fabrication, falsification, or plagiarism. The clause relies on the practices of the scientific community to separate acceptable from unacceptable conduct."

This is the crucial point. The disputed clause throws the responsibility for determining standards back to the scientists. Although Goldman and Fisher did not say so, this is just what those who champion the independence of the sciences should want. They went on, "Only *serious* deviations from that standard constitute misconduct." Furthermore, the critics, including the Academy panel, had failed to address the fact that the agencies that give out money for research must "ensure that government funds are provided only to responsible individuals." For that reason, they wrote:

> [A] generalized, comprehensive provision such as the OSD clause is necessary because it is impossible to predict and list all the unethical actions that might warrant agency action. Conduct such as tampering with other scientists' experiments, serious exploitation of subordinates, misuse of confidential information obtained during the grant proposal review process, or misrepresentation of authorship, could, if sufficiently serious, be considered misconduct in science. Yet, such actions do not fit within the specified categories in the definition.

Nonetheless, the law recognizes vagueness as a reason a statute or regulation may be found void. Indeed, in constitutional law a void-for-vagueness doctrine has a recognized place since, as they point out, the due-process clause of the Fifth Amendment "requires adequate notice of what conduct is prohibited." Thus, a law or regulation must specify "with reasonable clarity the prohibited conduct," and this prevents—in the words of a decision Goldman and Fisher cited—"arbitrary and discriminatory enforcement." Now, lack of specificity and the likelihood of arbitrary enforcement are precisely the dangers alarming Schachman and all the other critics of the "other serious deviations" clause. Void for vagueness would be the one possible ground for attacking the clause.

However, open-ended, general language about misconduct, relying on accepted standards within a community, has repeatedly passed constitutional scrutiny. In a case in 1974, "the Supreme Court upheld broad provisions of the Uniform Code of Military Justice that prohibit 'conduct unbecoming an officer and a gentleman' and 'all disorders and neglects to the principles of good order and discipline in the armed forces.' " This case and others, Goldman and Fisher wrote, "make it clear that broad, undefined prohibitions are not void for vagueness when they are interpreted by reference to community standards of conduct."

Many professions are subject to such standards. Lawyers are ridden especially hard. The American Bar Association has a model code of professional responsibility, with a misconduct rule that winds up by saying that a lawyer must not engage "in any other conduct that adversely reflects on his fitness to practice law." The Federal Rules of Appellate Procedure have similar language: "conduct unbecoming of a member of the bar of the court" can subject a lawyer to suspension or disbarment. A Supreme Court rule penalizes "conduct unbecoming a member of the Bar of this Court." State courts typically have similar rules.

Public-school teachers and university professors, even when tenured, can be fired on such grounds. Goldman and Fisher quote the language of several cases, including one where "Rutgers University determined that the professor had misused grant funds and exploited, defrauded, threatened, and abused foreign students carrying out research in his laboratory." Rutgers had a regulation defining adequate cause for dismissal as "failure to maintain standards of sound scholarship and competent teaching, or gross neglect of established University obligations appropriate to the appointment, or incompetence. . . ." The professor sued. The United States Court of Appeals for the Third Circuit rejected the vagueness challenge in 1992, writing that "the academic community's shared professional standards"[8] permitted Rutgers to apply the regulation. The Supreme Court refused to entertain a further appeal. Physicians, nurses, pharmacists, architects, and engineers have been subject to similar rules, and across the country state courts have upheld them.

By far the most thoughtful and detailed inquiry into the definition of scientific misconduct was carried out in two years of meetings and consultations by a panel called the Commission on Research Integrity.[9] In the summer of 1993, Congress passed a National Institutes of Health Revitalization Act, which, among much else, directed that such a commission be set up, and the Secretary of Health and Human Services, Donna Shalala, chartered the commission on November 4, 1993. Kenneth J. Ryan, a physician and professor in the Department of Obstetrics, Gynecology, and Reproductive Biology at Harvard Medical School, and also chairman of the Ethics Committee at Brigham and Women's Hospital, served as chairman of the commission. Ryan is a tough-minded senior public figure. For the new commission, he recruited biomedical scientists, university administrators, others experienced in dealing with

[8] *San Filippo v. Bongiovanni*, 961 F.2d 1125 (3rd Cir. 1992), *cert. denied*, 506 U.S. 908 (1992), *cited* in Goldman and Fisher, *supra*, note 7.

[9] COMMISSION ON RESEARCH INTEGRITY: INTEGRITY AND MISCONDUCT IN RESEARCH (1995); Drummond Rennie and C. Kristina. Gunsalus, "*Scientific Misconduct: New Definition Procedures, and Office—Perhaps a New Leaf*," JAMA, 17 February 1993; Horace Freeland Judson, FRAUD AND THE GOVERNANCE OF SCIENCE, in preparation.

allegations of misconduct, a lawyer, and a member of that egregious new trade, an ethicist, for a grand total of 12. Their charge was to arrive at a new definition of misconduct, to recommend procedures for institutions' response to allegations, and to develop a regulation to protect whistleblowers.

From June 20, 1994 to October 25, 1995, they met 15 times. Meetings were held in characterless modern hotels in Washington satellite cities or at Dulles Airport and included expeditions to medical schools and research complexes of the University of California at San Francisco, DePaul University in Chicago, Harvard Medical School, and the University of Alabama at Birmingham. The meetings grew from single days to 2- and sometimes 3-day sessions. They were public, as the law requires of governmental commissions. They attracted few journalists, but a fringe of followers who were known victims of misconduct, mostly those whose work had been plagiarized or who had blown the whistle and suffered retaliation. At many of the earlier meetings the commission heard from new individuals—bitter tales of careers ruined and lives blasted. In Chicago, at a break after particularly pathetic testimony from one such witness, Ryan was seen in the men's room rinsing tears from his eyes. The panelists were earnest, hard-working, and increasingly harried, as they grew aware of the full extent of the problems and of the difficulty of educating, goading, and inciting the scientific community and the senior staffs of organizations where scientists work to anticipate or respond to the problems seriously and effectively.

The definition gave trouble. They soon moved away from fabrication, falsification, plagiarism and other serious deviations, seeking language at once more particular and more comprehensive. Over the months, too, they groped towards understanding that existing definitions of scientific misconduct have been rooted in the law, in the legal concept of fraud, and in the consequent obsession with due process, with adversarial proceedings, and with the guilt or innocence of individuals.

The most significant moment in the evolution of the definition came at the session where Karl J. Hittelman, then associate vice chancellor of academic affairs at the University of California, San Francisco, stated that the fundamental aim of the commission and of the definition was to protect the integrity of the scientific literature and of the scientific process. Here, in less than 20 words, was the clear assertion of the distinction between the attitudes of the law and the needs of the scientific community.

Later at that session, I offered from the floor a definition that attempted to capture the thinking of the panelists at that point in its evolution:

> Scientific misconduct is behavior that corrupts the scientific record or potentially compromises the objectivity or practices of science. All forms of scientific misconduct are inescapably and immediately the

> responsibility of research institutions and the scientific community. The most serious forms comprise research fraud: these are fabrication, falsification, and plagiarism, and if the research is funded by government, fraud in that research is also the concern of government. Furthermore, the testimony of whistleblowers is crucial to exposure of fabrication, falsification, or plagiarism: therefore, effective protection of whistleblowers is also the concern not only of the institution but of government. In the category of research fraud, fabrication and falsification can take many forms and occur at many stages, including (enumeration). Similarly, plagiarism has many forms, for example, (enumeration). In sum, plagiarism is theft or other misappropriation of intellectual property.
>
> But scientific misconduct includes other important forms of behavior that are not the concern of government yet cannot be ignored or played down. These behaviors group broadly under three headings: matters of authorship, of good practice, and of mentorship. This category of scientific misconduct includes, *inter alia*, misattribution of authorship or credit; failure to keep appropriate records; gross sloppiness in the conduct of research; unreasonable withholding of materials and results; and failure responsibly to train, supervise, and counsel research students and other junior colleagues.
>
> Finally, certain types of illegal or unethical behavior, which are not scientific misconduct as such, may occur in connection with research. These include, chiefly, failure to comply with rules that govern research with human subjects or with animals; sexual harassment; and fiscal misconduct. All of these—even such rare types as research sabotage—are already subject to regulations, laws, and punishments.

By this time, though, the interactions among panelists had taken on an intellectual dynamism that drove them towards a more radical reformulation. In a subtle—perhaps overly subtle—appeal to the norms of the community, their final report asserted that their "new definition introduces an ethical approach to behavior rather than serving as vehicle for containing or expanding the basis for blame or legal action." They offered their definition as a nested set:

> Research misconduct is significant misbehavior that improperly appropriates the intellectual property or contributions of others, that intentionally impedes the progress of research, or that risks corrupting the scientific record or compromising the integrity of scientific practices.

> Such behaviors are unethical and unacceptable in proposing, conducting, or reporting research, or in reviewing the proposals or research reports of others.[10]

To replace FF&P, they proposed three new categories, "misappropriation, interference, and misrepresentation, each a form of dishonesty or unfairness that, if sufficiently serious, violates the ethical principles on which the definition is based." The definition set these out:

> *Misappropriation:* An investigator or reviewer shall not intentionally or recklessly
> a. plagiarize, which shall be understood to mean the presentation of the documented words or ideas of another as his or her own, without attribution appropriate for the medium of presentation; or
> b. make use of any information in breach of any duty of confidentiality associated with the review of any manuscript or grant application.

The second category dealt with a kind of conduct that was turning up in a number of cases but that had not been characterized separately before this:

> *Interference:* An investigator or reviewer shall not intentionally and without authorization take or sequester or materially damage any research-related property of another, including without limitation the apparatus, reagents, biological materials, writings, data, hardware, software, or any other substance or device used or produced in the conduct of research.

Only with the third category did the commission come to the usual starting point of definitions of misconduct:

> *Misrepresentation:* An investigator or reviewer shall not with intent to deceive, or in reckless disregard for the truth,
> a. state or present a material or significant falsehood; or
> b. omit a fact so that what is stated or presented as a whole states or presents a material or significant falsehood.

In law, proof of fraud (as of any other first-degree crime) requires evidence of intent. In the commission's definition, the first and third of the categories offered an alternative to intent, namely, reckless behavior. The report explained, "[a]n intent to deceive is often difficult to prove." Therefore, "[o]ne commonly accepted principle, adopted by the Commission, is that an intent to

[10] COMM. ON RESEARCH INTEGRITY, *supra* note 4.

deceive may be inferred from a person's acting in reckless disregard for the truth."

The commission went on to define other forms of professional misconduct, comprising obstruction of investigations—meaning cover-ups, tampering with evidence, and intimidation of whistle-blowers and other witnesses—and noncompliance with regulations governing such things as research using biohazardous materials or human or animal subjects. These matters, though grave, are straightforward, even routine. The guts of the definition are in those three categories. In logic and rhetoric they are clear, concise, hierarchically inclusive, watertight. On thoughtful rereading, the definition emerges as masterly.

Yet having gone so far, the commission failed to go far enough. As one member pointed out, despite the appeal to an ethical base the language remains lawyer-bound. This is symptomatic. The panelists recognized, for example, the terrific problem of demonstrating intent, imposed by the legal system's definition of fraud. But they failed to address—and perhaps never fully perceived—how far the needs of the scientific system, in Hittelman's formulation to protect the integrity of the scientific literature and of the scientific process, diverge from the attitudes inherent in the Anglo-American legal system, and how profoundly important the divergence is.

Few scientists read the definition thoughtfully. Most dismissed it, some scathingly. The president of the National Academy of Sciences, Bruce Alberts, and six other members of the academy's governing council wrote attacking the Ryan Commission report. So did Ralph Bradshaw, president of the Federation of American Societies for Experimental Biology, an umbrella organization covering 50 professional groups, which claims to represent more than 285,000 scientists. The report was referred to a new panel, established in the Department of Health and Human Services to develop a common federal definition of research misconduct. There it was entombed.

Imanishi-Kari was charged with scientific misconduct by the Office of Scientific Integrity, in the National Institutes of Health. I have scrutinized the evidence and interviewed the participants in great detail and I have no doubt that she was guilty. She appealed the finding to the Departmental Appeals Board of the Department of Health and Human Services. The board is a quasi-judicial or administrative tribunal of the classic kind, and despite its name its three-person panel heard the case *de novo*, with witnesses and lawyers for both sides, and all the apparatus of a criminal trial save a jury. The Office of Scientific Integrity was not allowed to introduce the voluminous evidence they had built up over more than 3 years of investigation. Late in the afternoon of Friday, June 21, 1996, the panel made public their decision. The decision is 191 pages of thick-textured prose, but the first 2 sentences made its burden plain: "The Office of Research Integrity (ORI) did not prove its charges by a preponderance of the evidence" and no action was to be taken against Imanishi-Kari.[11]

The proceedings, which I attended at crucial moments, were, in my opinion, a travesty. The outcome, as a number of commentators said, could best be characterized as that dry Scottish verdict "not proven." Put that aside. Suppose the case had been heard properly in a proper court. Suppose the result had gone the other way. Was Imanishi-Kari's individual guilt or innocence the appropriate issue? The adversarial judicial system reduced it to that. But recall Hittelman: in considering fraud, the interest of the scientific community is in protection of the integrity of the scientific record and the scientific process. Recall the whistleblower: O'Toole was vilified and her career in science set back. But she was the one person who got it right. From the beginning, what she demanded was that the paper in question be corrected or retracted.

V. REMEDIES?

Possible remedies, even palliations, are not obvious. The use of testimony by scientists in complex cases is doubtless improving somewhat in the wake of *Daubert*[12] and ensuing decisions. The Federal Judicial Project has explored ways to drive such improvements further, notably in the experiment in June of 1997 in the court of Judge Sam C. Pointer, in Birmingham, Alabama, hearing a class-action suit about the effects on health of silicone breast-implants.[13] Briefly, the court appointed a panel of four scientists, whose specialties spanned the kinds of scientific evidence to be presented. Plaintiffs and defense put expert witnesses on the stand. Members of the panel questioned them, and elicited responses that were full and wide-ranging. Lawyers for the two sides were forbidden to object or otherwise intervene. Only when the panel was done with a witness were the attorneys allowed to cross-examine in the conventional manner. According to observers, it quickly became clear that the questioning by the panel members was far more productive and effective. But the experiment was costly and has not been repeated.

Other suggestions for dealing with complex scientific and technical cases have included the impanelling of expert juries, qualified in the fields concerned. The model is the merchant juries that were sometimes employed in the eighteenth century. Scott Kieff and Leslie Misrock have proposed that

[11] Department of Health and Human Services, Departmental Appeals Board, Research Integrity Adjudication Panel, Thereza Imanishi-Kari, PhD: DAB No. 1582 (1996), June 1996.

[12] *Daubert v. Merrell Dow Pharmaceuticals, Inc.*, 61 USLW 4805, 113 S. ct 2786 (1993).

[13] *In re Silicone Breast Implants Products Liability Litigation*, 996 F.Supp. 1110 (N. D. Ala. 1997).

the differences between the jury-trial language of the Sixth Amendment, for criminal cases, and the Seventh, for civil trials, provide room for such a change.[14]

In the United States and Britain, by legal practice, media accounts, crime fiction and decades of movies and television shows with dramatic trials, we are so deeply inculcated with the Anglo-American adversarial and jury system that we do not even have the vocabulary to express alternatives. However, civil proceedings in Germany, criminal trials in France, are conducted differently. In those, an examining magistrate—the actual titles vary—conducts the trial, chooses witnesses, and asks the questions, in effect directing an investigation. Lawyers, for the defendant in French criminal trials or for the two sides in a German civil case, may advise the magistrate and request that certain questions be put, but they do not question witnesses themselves. The system is called "inquisitorial" or "truth-seeking." It arouses the deepest skepticism among American and British jurists. None the less, could an adaptation of the inquisitorial system produce results in complex scientific cases? The experiment in Judge Pointer's courtroom perhaps takes a significant step toward such a method. Could moving further in that direction result in inquiries into cases of scientific fraud that would more effectively meet the needs of the scientific community?

One radical change in scientific practices is irreversibly here: the open publication of scientific papers on the Internet. This has come from none in 1990 to many tens of thousands of papers yearly today. They appear as the online versions of refereed journals, all the way to unrefereed preprints and drafts. Online publication offers remarkable advantages. Publication is orders of magnitude faster and cheaper. It establishes priority of discovery yet promotes prompt follow-on research by others. If these attributes of online publication can be accommodated in the patent law, much of the tension with the norms of science can be resolved. Online publication will reduce problems of fraud, as well. It permits far fuller presentation of data and refinement or qualification of conclusions than allowed by the space limitations of print journals. It invites full scrutiny of papers by any and all—in effect, open refereeing. It anticipates the temptation to plagiarize. Furthermore, it facilitates revisions, corrections, or retractions—for every time the anchor is pulled up, it brings all the seaweed with it. Indeed, Internet publication goes directly to the maintenance and reestablishment of the norms of science.

[14] F. Scott Kieff & S. Leslie Misrock, *Latent Cures For Patent Pathology: Do Our Civil Juries Promote Science and the Useful Arts?*, presented in *The Crisis of Science and the Law*, at *Science in Crisis at the Millennium (an International Symposium)*, The George Washington University Center for History of Recent Science, Sept. 19, 1996, and printed in Chisum *et al.*, PRINCIPLES OF PATENT LAW 1368 (1st ed. 1998), 1024 (2nd ed. 2001).

Conclusion

Members of the basic biological science research community and those who study that community are engaged in an intense debate about the impact of patenting on the exchanges of information and material among members of the community that are essential for allowing downstream research. Some—patent critics—argue that the patent right to exclude use will lead to too little use. Others—patent proponents—argue that this right to exclude use is essential for ensuring sufficient use. Within this broader debate, several discrete skirmishes have emerged.

As explored in Part 1, one set of skirmishes involves the conflict relating to all fields of technology between intellectual property policies on the one hand and various other legal policies on the other. At the interface between patent and antitrust policies there lies a basic tension between efforts to erect and eliminate barriers to entry. While this tension is most apparent when viewed from a static perspective, from a dynamic perspective over time, the tension fades because patents can serve as powerful pro-competitive devices. For example, a patent can be the essential slingshot in a David v. Goliath commercial battle between even an individual scientist and a large drug company. At the interface between patent and general property rights policies, there lies a basic tension between efforts to preserve or eliminate the control of private ownership. This tension about the basic strengths and weaknesses of capitalism must be kept in mind when considering biotechnology patents, precisely because they are a form of property designed to operate within a capitalism framework. Unfortunately, its resolution is beyond the scope of this volume.

As explored in Part 2, another set of skirmishes involves the conflict among those who take the capitalistic framework as a given and use economic analysis to study the operation of different legal institutions to protect the smooth operation of the free market. In a larger sense, this issue might be seen most easily by asking why, if the freedom of the market is to be protected, should the government intervene in the market to provide patent protection? A market proponent must therefore justify patents as a necessary some type of market failure. The central justification for patents is that patents are needed to provide incentives for some private market actors to undertake the risky and expensive costs of invention commercialization. Without such incentives, it is feared that while inventions might still be made, they might not be sufficiently publicized and developed into the commercial products and services that directly benefit society. Although patents might be able to achieve this benefit, the assertion of formal property rights such as patents in the products and processes of the Human Genome Project may cause a hyper-proliferation of tollbooths that unduly tax and retard flow down the cumulative and cooperative path of technological progress in this essential field of biomedical research. The social costs of such property rights therefore include both the potential decrease in use of the products and processes covered by a patent right to exclude use, plus the transaction costs of trying to obtain permission for use from the patentee. Given these costs, some offer other mechanisms for providing the needed incentives, such as tax credits and grants. In response, patent proponents argue that the social cost of administering such a reward system would be even greater than for the patent system. The theoretical economics debate therefore boils down to a balance between competing theories about different types and magnitudes of social costs, which may, in the end, not be determinative.

One response to this lack of clear outcome after a purely theoretical debate is the effort in Part 3 to draw meaningful analogies to real-world examples of industries outside of biotechnology. For example, the computer software industry evolved largely without meaningful patent protection. But this is partly because until the 1980 *Chakrabarty* case in the field of biotechnology and the 1981 *Diehr* case in the field of computer software, the software field was also considered to be ineligible for patent protection. Moreover, analogy to the software industry may not be without its detractions. Chief among these is the emergence in the software industry of the largest monopoly cases in recent history *U.S. v. Microsoft*. In that market, the absence of patents and the presence of a single large player may be causally related. As another example, the music and publishing industries have evolved complex institutions of fair use and compulsory licensing of their dominant form of intellectual property, copyrights. But these industries are characterized by large collective bargaining groups of artists and publishers that have been able to get substantial and frequent legislative reform of the copyright system as technologies and consumer preferences have shifted over time. It is not clear the biotechnology industry is presently organized into such large collective bargaining groups and it is equally unclear whether the legal institutions that work for an industry so organized will work as well for one that is not so organized. Furthermore, it is not clear that the present patent system, at least as it operates in the field of biotechnology, actually suffers from many of the practical problems critics attribute to it. Patents in this field may not be as terribly broad as patent critics suggest. As the ones best able to mitigate the transaction costs patent critics identify for overly broad patents, it is appropriate that patentees face substantial risk if they fail to do so.

Rather than look to other industries as possible sources of real-world input for the open theoretical debate, Part 4 looks instead to the actual transactions that take place in the biotechnology industry when patents are present and compare them to transactions that might occur when patents are not present. This raises the question of whether protective licensors are fundamentally engaging in behavior that is unethical and antisocial. It appears that lawyers can employ techniques for structuring these transactions to simultaneously maximize both private and public interests.

Moreover, the real-world data discussed in Part 5 about the difficulties in resolving disputes over exchanges among members of the basic science community under both regimes in which patents are available and in which they are not, suggest that the social costs may be lower under a regime like the present patent system. The costs of resolving such disputes using technologically trained members of the basic science community may not turn out to be lower than under the patent system. Similarly, the costs of resolving such disputes using the ordinary technique for resolving civil litigation in the U.S.—with a set of legal rules that require crisp resolution of discrete facts decided by a lay jury—may not turn out to be higher than under a regime of peer review.

In view of such conflicting views on both theoretical and practical levels, the answer to the question of whether patents facilitate or frustrate the promised developments and applications from the Human Genome Project is not yet determined. Although substantial disparity in normative views persists throughout this volume, a consensus has emerged that empirical work is essential for moving these normative and policy debates forward.

For example, survey data could be collected to determine the objective and subjective parameters that obtain when actual members of the basic biological science community participate in attempted exchanges of material or information, both when patents are at play among those individuals and when they are not. These data would provide the first empirical evidence about the frequency of exchange failure and consummation relating to the use of patents in the basic biological research community. A high rate of consummation may suggest one reason why the community has enjoyed such great success since 1980. A low rate of consummation may suggest one reason why such success might have been even greater, why the community might be expected to experience a lower rate of success in the future, and when research will be building upon the more foundational work from the period of initial development in this relatively new field. Perhaps more importantly, the data would also provide the first empirical evidence about the different types of market failures that might be occurring in the basic biological science community. This is important because different types of failure may suggest different policy responses. Some failures that might be occurring may suggest the need for the development of new institutional approaches. It may turn out that the individual hassle costs of dealing meaningfully with attempted exchange in each case are experienced by members of the community in much the same way we all experience the hassle costs of telemarketing for goods or services that actually are better deals than those we are presently using, and yet we rationally elect in each case simply to not reply. In such cases, one appropriate policy response would be the establishment of a new institution to facilitate exchanges using standardized terms like a clearing house analogous to the artist rights groups ASCAP and BMI in the music industry. Alternatively, some failures that might be occurring may suggest the need for legal reform. It may turn out that the fear of liability for patent infringement is keeping too many researchers from gaining access to needed materials and techniques. One response to such concerns that has been adopted in the regime of copyright law is to change the patent law so that it permits some fair-use exemption. A related concern may be that such perceptions about liability even under existing patent law may be misinformed because those who engage in activities without permission from the patent may not be suffering repercussions. That is, the patentees actually may be granting what the law treats as an implied license for such activities. In such a case, some failures that might be occurring may be due to a misunderstanding among ethical members of the community about how patenting actually operates. For example, some members of the community might be seeking and enforcing patents that would not be viewed as profitable to a rational, well-informed actor; and some members of the community might be seeking express permission that might not be worth obtaining to a rational, well-informed actor. In such cases, an appropriate policy response would be the creation of new educational programs about how individual members of the community could get the most out of the patent system in a way that also will provide net benefits for the community, such as the MIT patent acquisition and technology transfer practice, which tends to maintain the tightest private control over the inventions that can be most efficiently utilized from a societal point of view if subject to a single rights-holder.

This collection of diverse perspectives is designed to educate and stimulate the further work needed before the complex issues in this important field can be fully understood. It is hoped the efficient targeting of tomorrow's frontiers will benefit from this survey of the ground already covered through today.

Index

C

E

G

H

Q

R